AF851210

WAVES IN FLUIDS AND SOLIDS

Edited by **Rubén Picó Vila**

Waves in Fluids and Solids
http://dx.doi.org/10.5772/752
Edited by Ruben Pico Vila

Contributors

Leonardo Trujillo, Franklin Peniche, Xiaoping Jia, Jian-Chun Cheng, Bin Liang, Xin-ye Zou, Ying Yuan, Ovidiu Stoican, Hefeng Dong, Jens M Hovem, Alexander Feher, Eugen Syrkin, Igor Gospodarev, Sergey Feodosyev, Kirill Kravchenko, Alexander Kotlyar, Elena Manzhelii, Alexey Stovas, Yury Roganov, Dinghui Yang, Xiao Ma, Shan Chen, Meixia Wang, Nikolay Bunkin, Vladimir Gorelik, Irini Djeran-Maigre, Sergey Kuznetsov, Nikolaos L. Tsitsas, Valentine Cherednick, Michael Dvoesherstov

© The Editor(s) and the Author(s) 2011
The moral rights of the and the author(s) have been asserted.
All rights to the book as a whole are reserved by INTECH. The book as a whole (compilation) cannot be reproduced, distributed or used for commercial or non-commercial purposes without INTECH's written permission.
Enquiries concerning the use of the book should be directed to INTECH rights and permissions department (permissions@intechopen.com).
Violations are liable to prosecution under the governing Copyright Law.

(cc) BY

Individual chapters of this publication are distributed under the terms of the Creative Commons Attribution 3.0 Unported License which permits commercial use, distribution and reproduction of the individual chapters, provided the original author(s) and source publication are appropriately acknowledged. If so indicated, certain images may not be included under the Creative Commons license. In such cases users will need to obtain permission from the license holder to reproduce the material. More details and guidelines concerning content reuse and adaptation can be foundat http://www.intechopen.com/copyright-policy.html.

Notice
Statements and opinions expressed in the chapters are these of the individual contributors and not necessarily those of the editors or publisher. No responsibility is accepted for the accuracy of information contained in the published chapters. The publisher assumes no responsibility for any damage or injury to persons or property arising out of the use of any materials, instructions, methods or ideas contained in the book.

First published in Croatia, 2011 by INTECH d.o.o.
eBook (PDF) Published by IN TECH d.o.o.
Place and year of publication of eBook (PDF): Rijeka, 2019.
IntechOpen is the global imprint of IN TECH d.o.o.
Printed in Croatia

Legal deposit, Croatia: National and University Library in Zagreb

Additional hard and PDF copies can be obtained from orders@intechopen.com

Waves in Fluids and Solids
Edited by Ruben Pico Vila

p. cm.

ISBN 978-953-307-285-2

eBook (PDF) ISBN 978-953-51-6053-3

We are IntechOpen, the world's leading publisher of Open Access books

Built by scientists, for scientists

4,400+
Open access books available

117,000+
International authors and editors

130M+
Downloads

151
Countries delivered to

Our authors are among the
Top 1%
most cited scientists

12.2%
Contributors from top 500 universities

CLARIVATE ANALYTICS BOOK CITATION INDEX INDEXED

WEB OF SCIENCE™

Selection of our books indexed in the Book Citation Index in Web of Science™ Core Collection (BKCI)

Interested in publishing with us?
Contact book.department@intechopen.com

Numbers displayed above are based on latest data collected.
For more information visit www.intechopen.com

Meet the editor

Dr. Rubén Picó Vila is a researcher of the Applied Physics Department at the Polytechnic University of Valencia, UPV, (Spain) and member of the Research Institute for Integrated Management of Coastal (IGIC).
Doctor of physics at UPV and Doctor of Acoustics at the University of Maine, (Le Mans, France) in 2004. He has published in the field of acoustics and vibroacoustics in several journals indexed at JCR-SCI. He has participated in several national research projects concerning fundamental research of acoustic systems. He is the leader of a both regional fund projects based on vibroacoustography and periodic media.

Contents

Preface

Acoustics is a discipline that deals with many types of fields wave phenomena. Originally the field of Acoustics was consecrated to the sound, that is, the study of small pressure waves in air detected by the human ear. The scope of this field of physics has been extended to higher and lower frequencies and to higher intensity levels. Moreover, structural vibrations are also included in acoustics as a wave phenomena produced by elastic waves. This book is focused on acoustic waves in fluid media and elastic perturbations in heterogeneous media.

Acoustic wave propagation in layered media is very important topic for many practical applications including medicine, optics and applied geophysics. The key parameter controlling all effects in layered media is the scaling factor given by the ratio between the wavelength and the layer thickness. Existing theory mostly covers the solutions derived for the low-frequency and high-frequency limits. In practice, the wavelength could be comparable with the layer thickness, and application of both frequency limits is no longer valid. The frequency-dependent effects for acoustic waves propagating through the layered media are analyzed.

Solitons, or by the original terminology "waves of translation", are a special kind of hydrodynamic waves that can arise and propagate in narrow channels as solitary waves, resembling propagation of the wave front of shock waves. These waves can propagate without considerable attenuation, or change of form; or diminution of their speed. Motion of these waves can be described by a non-linear KdV differential equation. Soliton-like lamb waves are analyzed in the long-wave limits of Lamb waves propagating in elastic anisotropic plates.

The application of various layers on a piezoelectric substrate is a way of improving the parameters of propagating electroacoustic waves. For example, a metal film of certain thickness may provide the thermal stability of the wave for substrate cuts, corresponding to a high electromechanical coupling coefficient. The overlayer can vary the wave propagation velocity and, hence, the operating frequency of a device. The effect of the environment (gas or liquid) on the properties of the wave in the layered structure is used in sensors. The layer may protect the piezoelectric substrate against undesired external impacts. Multilayer compositions allow to reduce a velocity dispersion, which is observed in single-layer structures. In multilayer film bulk acoustic wave resonators (FBAR) many layers are necessary for proper work of such devices. Wave propagation characteristics in multilayer structures are analyzed by means of general methods of numerical calculations of the surface and bulk acoustic wave parameters in arbitrary multilayer structures.

Crystalline and disordered systems are analyzed as very peculiar systems. The most important elementary excitations appearing in them are acoustic phonons, which characterize vibration states in heterogeneous structures. In such systems, the crystalline regularity in the arrangement of atoms is either absent or its effect on the physical properties of the systems is weak, affecting substantially the local spectral functions of different atoms forming this structure.

Granular materials consist of a collection of discrete macroscopic solid particles interacting via repulsive contact forces. Classical examples are sand, powders, sugar, salt and gravel, which range from tens of micrometers to the macroscopic scale. Their physical behavior involves complex nonlinear phenomena, such as non equilibrium configurati. The elastic wave propagation in confined granular systems under external load is developed from both experimental and theoretical viewpoints.

Shear waves (S-wave) are essential in the field of seafloor geotechnical applications as they propagate in solids. More specifically interface waves and the use of the interface waves are important to estimate shear wave speed in the sediments as it provides a good indicator of sediment rigidity, as well as for sediment characterization, seismic exploration, and geohazard assessment. In addition, for environments with high seabed S-wave speeds, S-wave conversion from the compressional wave (also called P-wave) at the seafloor can represent an important ocean acoustic loss mechanism which must be accounted for in propagation modeling and sonar performance predictions.

Phononic Crystals are characterized by spatial periodic modulations of the sound velocity caused by the presence of the periodically settled elements of various materials (metals, polymers etc.) inside the sample. The properties of acoustic waves in Phononic Crystals are in many respects similar to the properties of electromagnetic waves in Photonic Crystals. Periodic media can be characterized by the dispersion dependences $\omega(k)$ for acoustic waves together with the dispersion dependences of their group velocities and effective mass of the corresponding acoustic phonons. The results of the theoretical analysis and the data of experimental studies of the optical and acoustic phenomena in PTC and PNC, including the studies of spectra of non-elastic scattering of light together with the experiments to observe the stimulated light scattering accompanying by the coherent oscillations of globules are reported.

The numerical solutions of the acoustic-wave equation via finite-differences, finite-elements, and other related numerical techniques are valuable tools for the simulation of wave propagation. These modeling techniques for the 1D and 2D cases are typically used as support for a sound interpretation when dealing with complex geology, or as a benchmark for testing processing algorithms, or used in more or less automatic inversion procedure by perturbation of the parameters characterizing the elastic medium until the synthetic records fit the observed real data. It is not advisable to apply these techniques in large-scale computation, especially for a large scale 3D simulation of seismic wave propagation because of an intensive use of Central Processing Unit (CPU) time and the requirement of a large amount of direct-access memory. A 3D numerical method is proposed to effectively suppress the numerical dispersion caused by the discretization of the acoustic- and elastic-wave equations through using both the local spatial difference-operator and the fourth-order Runge-Kutta (RK) method so that the numerical technique developed in this chapter has rapid computational speed and can save the memory storage.

The electrode systems generating a quadrupole electric field, having both static and variable components, under certain conditions, allow to maintain the charged particles in a well defined region of space without physical solid contact with the wall of a container. This process is sometime called levitation. Usually, these kinds of devices are known as quadrupole traps. The possibility to manipulate the stored microparticles by using an acoustic wave is investigated by experimental means. That means both controlling their position in space and performing a further selection of the stored microparticles.

There exists a special class of solid media called soft media (or weakly compressible media), for which the inequality $\lambda >> \mu$ is satisfied (λ and μ are the Lamé coefficients). Such media with very small shear stiffness are dynamically similar to liquids to a great extent and exhibit strongly a "water-like" characteristic. The class of soft media includes many common media in the fields of scientific research and practical applications (e.g. soft rubbers, tissues, or biomimetic materials), and air bubbles are often introduced due to artificial or non-artificial reasons. Propagation of acoustic waves in a bubbly soft medium is particularly different from that of a usual solid medium containing bubbles.

The interaction of a point-source spherical acoustic wave with a bounded obstacle possesses various attractive and useful properties in direct and inverse scattering theory. More precisely, concerning the direct scattering problem, the far-field interaction of a point source with an obstacle is, under certain conditions, stronger compared to that of a plane wave. In inverse scattering problems the distance of the point-source from the obstacle constitutes a crucial parameter, which is encoded in the far-field pattern and is utilized appropriately for the localization and reconstruction of the obstacle's physical and geometrical characteristics.

Rubén Picó Vila
Polytechnic University of Valencia
Spain

Part 1

Elastic Waves in Solids

1

Acoustic Waves in Layered Media - From Theory to Seismic Applications

Alexey Stovas[1] and Yury Roganov[2]
[1]*NTNU, Trondheim,*
[2]*USGPI, Kiev,*
[1]*Norway*
[2]*Ukraine*

1. Introduction

Acoustic wave propagation in layered media is very important topic for many practical applications including medicine, optics and applied geophysics. The key parameter controlling all effects in layered media is the scaling factor given by the ratio between the wavelength and the layer thickness. Existing theory mostly covers the solutions derived for the low-frequency and high-frequency limits. In the first limit, when the wavelength is much larger than the layer thickness, the layered medium is substituted by an effective medium with the properties given by special technique called the Backus averaging. In the second limit, when the wavelength is much smaller than the layer thickness, we can use the ray theory to compute both reflection and transmission responses.

In practice, the wavelength could be comparable with the layer thickness, and application of both frequency limits is no longer valid. In this chapter, we will mainly focus on the frequency-dependent effects for acoustic waves propagating through the layered media.

We show that there are distinct periodically repeated patterns consisted of the pass- and stop-bands of very complicated configuration defined in frequency-slowness or frequency-group angle domain that control the reflection and transmission responses. The edges between the pass- and stop-bands result in the caustics in the group domain. The quasi-shear waves in a homogeneous transversely isotropic medium could also results in the high-frequency caustics, but for the layered media, all wave modes can result in frequency-dependent caustics. The caustics computed for a specific frequency differ from those observed at the low- and high-frequency limits. From physics point of view, the pass-bands correspond to the effective medium, while the stop-bands correspond to the resonant medium. We distinguish between the effects of scattering and intrinsic attenuation in layered media. The propagation of acoustic waves in a layered medium results in the energy loss due to scattering effect. The intrinsic attenuation is an additional effect which plays very important role in seismic data inversion. We provide the theoretical and numerical study to compare both effects for a periodically layered medium. We also investigate the complex frequency roots of the reflection/transmission responses. We also derive the phase velocity approximations in a layered medium. As the trial model for layered medium, we widely use the periodically layered medium with the limited number of parameters. The propagation of acoustic waves through a periodic layered medium is analyzed by an eigenvalue decomposition of the propagator matrix. This reveals how the velocity and attenuation of the layered medium vary as function of the periodic structure, material parameters and frequency. We show that there is one more parameter controlling the wave propagation apart of the wavelength to layer thickness ratio that is the acoustic contrast between the layers. Multiple scattering in finely layered sediments is important in stratigraphic interpretation in seismic, matching of well log-data with seismic data and seismic modelling. Two methods have been used to treat this problem in seismic applications: the O'Doherty-Anstey approximation and Backus averaging. The O'Doherty-Anstey approximation describes the stratigraphic filtering effects, while the Backus averaging defines the elastic properties for an effective medium from the stack of the layers.

Using numerical examples, we show that there is a transition zone between the effective medium (low-frequency limit) and the time-average medium (high-frequency limit) and that the width of this zone depends on the strength of the reflection coefficient series. Assuming that a tubidite reservoir can be approximated by a stack of thin shale-sand layers we use standard AVO-attributes to estimate net-to-gross and oil saturation. Necessary input is Gassmann rock physics properties for sand and shale as well as the fluid properties for hydrocarbons. Required seismic input is AVO intercept and gradient. The method is based upon thin layer reflectivity modeling. It is shown that random variability in thickness and seismic properties of the thin sand and shale layers does not change the AVO attributes at top and base of the turbidite reservoir sequence significantly. The method is tested on seismic data from offshore Brazil, and the results show reasonable agreement between estimated and observed net-to-gross and oil saturation. The methodology can be further developed for estimating changes in pay thickness from time lapse seismic data. We propose the method of computation seismic AVO attributes (intercept and gradient) from ultra-thin geological model based on the SBED modelling software. The SBED software is based on manipulating sine-functions, creating surfaces representing incremental sedimentation. Displacement of the surfaces creates a three dimensional image mimicking bedform migration, and depositional environments as diverse as tidal channels and mass flows can be accurately recreated. The resulting modelled deposit volume may be populated with petrophysical information, creating intrinsic properties such as porosity and permeability (both vertical and horizontal). The Backus averaging technique is used for up-scaling within the centimetre scale (the intrinsic net-to-gross value controls the acoustic properties of the ultra-thin layers). It results in pseudo-log data including the intrinsic anisotropy parameters. The synthetic seismic modelling is given by the matrix propagator method allows us to take into account all pure mode multiples, and resulting AVO attributes become frequency dependent. Within this ultra-thin model we can test different fluid saturation scenarios and quantify the likelihood of possible composite analogues. This modelling can also be used for inversion of real seismic data into net-to-gross and fluid saturation for ultra-thin reservoirs.

There are many other issues related to wave propagation in layered media we do not discuss in this chapter. For further reading we suggest several books (Aki&Richards, 1980; Brekhovskih, 1960; Kennett, 1983; Tsvankin, 1995) that cover the problems we did not touch here.

2. System of differential equations

To describe the dynamic of the wave propagation in an elastic medium, it is common to use the Hook's law that defines the linear relation between the stress tensor τ_{ij} and the strains tensor e_{pq},

$$\tau_{ij} = c_{ijpq} e_{pq} , \tag{1}$$

where $e_{pq} = \frac{1}{2}\left(\frac{\partial u_p}{\partial x_q} + \frac{\partial u_q}{\partial x_p} \right)$ and c_{ijpq} is the stiffness tensor. If there are no volume forces within the medium, we can write the moment equation in the following way (Aki and Richards, 1980)

$$\rho \frac{\partial^2 u_i}{\partial t^2} = \frac{\partial \tau_{ij}}{\partial x_j} . \tag{2}$$

In equation (1) and (2), it is assumed that the summation is performed over the repeatable indices. Since the tensor c_{ijpq} is symmetrical with respect to index changing $(ij) \leftrightarrow (pq)$, $i \leftrightarrow j$, $p \leftrightarrow q$, the relation (1) can be written as

$$\tau_{ij} = c_{ijpq} \frac{\partial u_p}{\partial x_q} . \tag{3}$$

If we denote $v_i = \frac{\partial u_i}{\partial t}$ as the velocity of the particle movement, the equations (2) and (3) can be given as

$$\rho \frac{\partial v_i}{\partial t} = \frac{\partial \tau_{ij}}{\partial x_j} , \tag{4}$$

$$\frac{\partial \tau_{ij}}{\partial t} = c_{ijpq} \frac{\partial v_p}{\partial x_q} . \tag{5}$$

We can apply the Fourier transform for the variables ω, x_1, x_2 to equations (4) and (5) according to the following relations

$$\tilde{f}(p_1, p_2, \omega) = \frac{1}{8\pi^3} \iiint_{-\infty}^{\infty} f(x_1, x_2, t) e^{-i\omega(p_1 x_1 + p_2 x_2 - t)} dx_1 dx_2 dt ,$$

$$f(x_1, x_2, t) = \iiint_{-\infty}^{\infty} \omega^2 \tilde{f}(p_1, p_2, \omega) e^{i\omega(p_1 x_1 + p_2 x_2 - t)} dp_1 dp_2 d\omega .$$

After substituting the derivatives on x_1, x_2, t with the coefficients $i\omega p_1$, $i\omega p_2$, $-i\omega$, and excluding the variables τ_{11}, τ_{12}, τ_{22}, the system of equations (4)-(5) can be given in a vector-matrix form

$$\frac{d\mathbf{b}}{dz} = i\omega \mathbf{M}\mathbf{b} , \tag{6}$$

where vector $\mathbf{b} = (v_1, v_2, v_3, \tau_{13}, \tau_{23}, \tau_{33})^T$, $z \equiv x_3$, and matrix $\mathbf{M}$ is defined as

$$\mathbf{M} = -\begin{pmatrix} \mathbf{A} & \mathbf{C}_{33}^{-1} \\ \mathbf{B} & \mathbf{A}^T \end{pmatrix} , \tag{7}$$

with $\mathbf{C}_{mn}[p,q] = c_{mp,nq}$, $\mathbf{A} = \mathbf{C}_{33}^{-1}(p_1\mathbf{C}_{31} + p_2\mathbf{C}_{32})$ and $\mathbf{B} = \sum_{m,n=1,2} p_m p_n (\mathbf{C}_{m3}\mathbf{C}_{33}^{-1}\mathbf{C}_{3n} - \mathbf{C}_{mn}) + \rho\mathbf{I}$,

where p_1 and p_2 are the horizontal slowness in x_1 and x_2 direction, respectively, and ρ is the density. The matrices $\mathbf{C}_{33}^{-1}$ and $\mathbf{B}$ are symmetrical, and matrix $\mathbf{M}$ satisfies the equation

$$\mathbf{KMK} = \mathbf{M}^T, \tag{8}$$

where $\mathbf{M}^T$ is transposed matrix, $\mathbf{K} = \begin{pmatrix} \mathbf{0} & \mathbf{I} \\ \mathbf{I} & \mathbf{0} \end{pmatrix}$ and $\mathbf{I}$ is the unit 3x3 matrix. Equation (6) describes the wave propagation in the vertically heterogeneous medium with all parameters being dependent on z-coordinate only. Note, even if the parameters $\rho(z)$ and $c_{ijpq}(z)$ are discontinuous functions, the components of the vector $\mathbf{b}$ remain continuous functions of depth. If the medium is a homogeneous transversely isotropic with vertical symmetry axis (a VTI medium), the propagation of waves in qP-qSV system is independent from the qSH-waves propagation. In this case, there are two equations (6): with dimension 4x4 (qP-qSV system) and with dimension 2x2 (qSH wave). For vertical propagation, equation (6) is reduced into three independent equations (for qP-, qSV- and qSH-waves). The analytical form of matrix $\mathbf{M}$ for different types of media can be found in Aki and Richards (1980), Braga and Herrmann (1992) and Fryer and Frazer (1987).

3. Propagator matrix

The propagator matrix is the matrix $\mathbf{P}(z,z_0) \in \mathbf{GL}_6(\mathbf{C})$ that is the solution of initial-value problem

$$\frac{\partial \mathbf{P}(z,z_0)}{\partial z} = i\omega\mathbf{M}\mathbf{P}(z,z_0), \quad \mathbf{P}(z_0,z_0) = \mathbf{I}. \tag{9}$$

The propagator matrix satisfies the Volterra integral equation

$$\mathbf{P}(z,z_0) = \mathbf{I} + \int_{z_0}^{z} \mathbf{M}(\eta)\mathbf{P}(\eta,z_0)d\eta, \tag{10}$$

that can be solved by using the Picard iteration. This leads to the Peano series for $\mathbf{P}(z,z_0)$ (Peano, 1888; Pease, 1965)

$$\mathbf{P}(z,z_0) = \mathbf{I} + \int_{z_0}^{z} \mathbf{M}(\eta_1)d\eta_1 + \int_{z_0}^{z} \mathbf{M}(\eta_1)\int_{z_0}^{\eta_1} \mathbf{M}(\eta_2)d\eta_2 d\eta_1 + \dots. \tag{11}$$

The series (11) converges if $\|\mathbf{M}(\eta)\| \le c$ for all $\eta \in [z_0, z]$, where $\|...\|$ is some matrix norm. If $\mathbf{M}(\eta) = \mathbf{M}$, the matrix with constant elements, than equation (11) is reduced to the exponent

$$\mathbf{P}(z,z_0) = \exp\left[i\omega(z - z_0)\mathbf{M}\right]. \tag{12}$$

From equation (9), it follows that for any vector $\mathbf{b}(z_0)$, the vector $\mathbf{b}(z) = \mathbf{P}(z,z_0)\mathbf{b}(z_0)$ is the solution of equation (6). By substituting the basis vectors $\mathbf{e}_i$, $i = 1,2,3$, instead of the vector

$\mathbf{b}(z_0)$, we can prove that the columns of matrix $\mathbf{P}(z,z_0)$ are the linear independent solutions of equation (6), i.e. the propagator matrix describes the fundamental system of solutions of equation (6). Since,

$$\mathbf{b}(z_2) = \mathbf{P}(z_2,z_0)\mathbf{b}(z_0) = \mathbf{P}(z_2,z_1)\mathbf{P}(z_1,z_0)\mathbf{b}(z_0), \tag{13}$$

than,

$$\mathbf{P}(z_2,z_0) = \mathbf{P}(z_2,z_1)\mathbf{P}(z_1,z_0). \tag{14}$$

If the medium consists of N layers with matrices $\mathbf{M}_i$ and layer thickness h_i, the equations (12)-(14) gives the propagator matrix from the stack of the layers

$$\mathbf{P}(z_N,z_0) = \exp(i\omega h_N \mathbf{M}_N)...\exp(i\omega h_1 \mathbf{M}_1), \tag{15}$$

where $z_N = z_0 + h_1 + ... + h_N$.

Equation (15) was used by Haskell (1953) and Thomson (1950) in their works to describe the wave propagation in a layered isotropic medium.

4. Symmetry relations

Let us assume that matrices $\mathbf{M}_k$ can be diagonalized by using a similarity transformation, $\mathbf{M}_k = \mathbf{E}_k \operatorname{diag}\left(\theta_k^{(m)}\right)\mathbf{E}_k^{-1}$. In this case, $\exp\left(i\omega h_k \mathbf{M}_k\right) = \mathbf{E}_k \operatorname{diag}\left(\exp\left(i\omega h_k \theta_k^{(m)}\right)\right)\mathbf{E}_k^{-1}$. The columns of matrix $\mathbf{E}_k$ are the polarization vectors for all types of wave propagating in layer k with horizontal slownesses p_1 and p_2, while $\theta_k^{(m)}$ are corresponding vertical slownesses. Therefore, the vector $\mathbf{w}(z) = \mathbf{E}_k^{-1}\mathbf{b}(z)$ consists of up- and down-going wave amplitudes in layer k with $h_1 + ... + h_{k-1} < z < h_1 + ... + h_k$. Let us denote by $\mathbf{E}_0$ and $\mathbf{E}_{N+1}$ the matrices that consist of the eigen-vectors of matrices $\mathbf{M}_0$ and $\mathbf{M}_{N+1}$, respectively, which define the properties of the half-spaces surrounding the stack of layers. The matrix

$$\mathbf{Q} = \mathbf{E}_{N+1}^{-1}\mathbf{P}\mathbf{E}_0 \tag{16}$$

is called the amplitude propagator matrix, since the equation $\mathbf{w}_{N+1} = \mathbf{Q}\mathbf{w}_0$ provides the linear relation between the amplitudes over and under the stack of layers for all wave types. The amplitude propagator matrix can be blocked by 3x3 sub-matrices, $\mathbf{Q} = \left(\mathbf{Q}_{ij}\right)_{ij=1,2}$, and the amplitude vectors $\mathbf{w}_0$ and $\mathbf{w}_{N+1}$ can be blocked by three-component sub-vectors, corresponding to up- and down-going waves of different type,

$$\mathbf{w}_0 = \begin{pmatrix} \mathbf{u}_0 \\ \mathbf{d}_0 \end{pmatrix}, \quad \mathbf{w}_{N+1} = \begin{pmatrix} \mathbf{u}_{N+1} \\ \mathbf{d}_{N+1} \end{pmatrix}. \tag{17}$$

Therefore, we can write

$$\begin{pmatrix} \mathbf{u}_{N+1} \\ \mathbf{d}_{N+1} \end{pmatrix} = \begin{pmatrix} \mathbf{Q}_{11} & \mathbf{Q}_{21} \\ \mathbf{Q}_{21} & \mathbf{Q}_{22} \end{pmatrix} \begin{pmatrix} \mathbf{u}_0 \\ \mathbf{d}_0 \end{pmatrix}. \tag{18}$$

From equation (18), it follows that the amplitudes of waves going away from the stack of layers: $\mathbf{u}_0$ and $\mathbf{d}_{N+1}$ can be computed from the amplitudes of waves coming to the stack of layers: $\mathbf{u}_{N+1}$ and $\mathbf{d}_0$,

$$\begin{pmatrix} \mathbf{d}_{N+1} \\ \mathbf{u}_0 \end{pmatrix} = \begin{pmatrix} \mathbf{T}_D & \mathbf{R}_U \\ \mathbf{R}_D & \mathbf{T}_U \end{pmatrix} \begin{pmatrix} \mathbf{d}_0 \\ \mathbf{u}_{N+1} \end{pmatrix}, \tag{19}$$

where

$$\mathbf{T}_D = \mathbf{Q}_{22} - \mathbf{Q}_{21}\mathbf{Q}_{11}^{-1}\mathbf{Q}_{12}, \tag{20}$$

$$\mathbf{R}_U = \mathbf{Q}_{21}\mathbf{Q}_{11}^{-1}, \tag{21}$$

$$\mathbf{R}_D = -\mathbf{Q}_{11}^{-1}\mathbf{Q}_{12}, \tag{22}$$

$$\mathbf{T}_U = \mathbf{Q}_{11}^{-1}. \tag{23}$$

The matrices $\mathbf{R}_U$, $\mathbf{T}_U$ and, $\mathbf{R}_D$, $\mathbf{T}_D$ are called the reflection and transmission matrices for up- and down-going wave, respectively (Figure 1). From equations (20)-(23), the amplitude propagator matrix can be rewritten as

$$\mathbf{Q} = \begin{pmatrix} \mathbf{T}_U^{-1} & -\mathbf{T}_U^{-1}\mathbf{R}_D \\ \mathbf{R}_U\mathbf{T}_U^{-1} & \mathbf{T}_D - \mathbf{R}_U\mathbf{T}_U^{-1}\mathbf{R}_D \end{pmatrix}. \tag{24}$$

The matrix

$$\mathbf{S} = \begin{pmatrix} \mathbf{T}_D & \mathbf{R}_U \\ \mathbf{R}_D & \mathbf{T}_U \end{pmatrix} \tag{25}$$

is called the scattering matrix (Ursin, 1983) and contains all reflection and transmission coefficients for up- and down-going waves of all type as sub-matrices. If the layers posses the VTI (transverse isotropy with vertical symmetry axis) symmetry, the following relation is valid

$$\mathbf{Q}^T\mathbf{J}\mathbf{Q} = \mathbf{J}, \tag{26}$$

where $\mathbf{J} = \begin{pmatrix} \mathbf{0} & \mathbf{I} \\ -\mathbf{I} & \mathbf{0} \end{pmatrix}$. In order to prove the relation (26), we represent the amplitude propagator matrix $\mathbf{Q}$ as $\mathbf{Q} = \mathbf{F}_N\mathbf{\Lambda}_N \ldots \mathbf{\Lambda}_1\mathbf{F}_0$, where $\mathbf{F}_k = \mathbf{E}_{k+1}^{-1}\mathbf{E}_k$, $\mathbf{\Lambda}_k = \operatorname{diag}\left(\lambda_k^{(m)}\right)$ and $\lambda_k^{(m)} = \exp\left(i\omega h_k \theta_k^{(m)}\right)$. To validate the relation (26), it is sufficient to prove that $\mathbf{\Lambda}_k^T\mathbf{J}\mathbf{\Lambda}_k = \mathbf{J}$ and $\mathbf{F}_k^T\mathbf{J}\mathbf{F}_k = \mathbf{J}$. The first equality follows from the fact that the matrix $\mathbf{\Lambda}_k$ is diagonal, and the elements of matrix $\mathbf{\Lambda}_k$ are symmetric, $\lambda_k^{(m+3)} = 1/\lambda_k^{(m)}$. The second equality can be proven by substitution of equations (35) and (36) for matrices $\mathbf{E}_k$ and $\mathbf{E}_{k+1}^{-1}$ into this equation and using relations (37). The symmetry relations for reflection and transmission matrices for stack of layers sandwiched between two half-spaces,

$$\mathbf{R}_D = \mathbf{R}_D^T, \quad \mathbf{R}_U = \mathbf{R}_U^T, \quad \mathbf{T}_D = \mathbf{T}_U^T, \tag{27}$$

can be derived by substitution the expression for amplitude propagator matrix $\mathbf{Q}$ from equation (24) into equation (26). It is assumed that all layers and half-spaces posses the VTI symmetry, and the presence of attenuation is allowed. The symmetry relations (27) are also derived by Ursin (1983) for a stack of isotropic layers. If we consider the single interface between the half-spaces, than the diagonal sub-matrices in equation (24) are equal (see equations (35) and (36)), and, therefore, $\mathbf{R}_U\mathbf{T}_U^{-1} = -\mathbf{T}_U^{-1}\mathbf{R}_D$ and $\mathbf{T}_U^{-1} = \mathbf{T}_D - \mathbf{R}_U\mathbf{T}_U^{-1}\mathbf{R}_D$ (Ursin, 1983). These relations result in

$$\mathbf{T}_U\mathbf{R}_U + \mathbf{R}_D\mathbf{T}_U = 0, \quad \mathbf{T}_D\mathbf{T}_U + \mathbf{R}_U^2 = \mathbf{I}. \tag{28}$$

By using $\mathbf{R}_U = -\mathbf{T}_U^{-1}\mathbf{R}_D\mathbf{T}_U$ and equations (27) we can derive two additional symmetry relations

$$\mathbf{R}_U\mathbf{T}_D + \mathbf{T}_D\mathbf{R}_D = 0, \quad \mathbf{T}_U\mathbf{T}_D + \mathbf{R}_D^2 = \mathbf{I}. \tag{29}$$

The symmetry relations (28) and (29) are derived by Frasier (1970), Kennett et al. (1978) and Ursin (1983).

5. Reflection and transmission responses of layered transversely isotropic media

This chapter is mostly based on the paper Stovas and Ursin (2003). The major point in this chapter is wave field decomposion into up- and down-going waves (Kennett, 1983; Ursin, 1983) scaled such that, for an elastic medium, the vertical energy flux is constant. This results in important symmetries in the transformation matrix and also in the reflection and transmission (R/T) coefficients (Chapman, 1994; Ursin and Haugen, 1996). The equations of motion and Hook's law in a source free VTI medium (radially symmetric about the vertical, z axis) for qP- and qSV-waves in X_1X_3 plane after applying a Fourier transform can be expressed as an ordinary first-order matrix-vector differential equation as shown in equation (6), where ω denotes circular frequency, $\mathbf{b} = (v_3, \tau_{13}, \tau_{33}, v_1)^T$ is the displacement velocity - stress vector, the superscript "T" indicating the transpose, and the matrix $\mathbf{M}$ has the blocked structure,

$$\mathbf{M} = -\begin{pmatrix} 0 & \mathbf{A} \\ \mathbf{B} & 0 \end{pmatrix}, \tag{30}$$

composed of the 2x2 symmetric matrices $\mathbf{A}$ and $\mathbf{B}$ are given by

$$\mathbf{A} = \begin{pmatrix} c_{33}^{-1} & pc_{13}c_{33}^{-1} \\ pc_{13}c_{33}^{-1} & \rho - p^2\left(c_{11} - c_{13}^2c_{33}^{-1}\right) \end{pmatrix}, \quad \mathbf{B} = \begin{pmatrix} \rho & p \\ p & c_{55}^{-1} \end{pmatrix}, \tag{31}$$

which are dependent on the horizontal slowness p, the stiffness coefficients c_{ij} from the stiffness matrix and the density ρ. In order to decompose into up- and down-going waves (Ursin, 1983, Ursin and Stovas, 2002), we make the linear transformation

$$\mathbf{b} = \mathbf{E}\begin{pmatrix} \mathbf{u} \\ \mathbf{d} \end{pmatrix}. \tag{32}$$

Then equation (6) becomes

$$\frac{d}{dz}\begin{pmatrix} \mathbf{u} \\ \mathbf{d} \end{pmatrix} = \left[\begin{pmatrix} -i\omega\mathbf{\Delta} & \mathbf{0} \\ \mathbf{0} & i\omega\mathbf{\Delta} \end{pmatrix} + \begin{pmatrix} \mathbf{F} & \mathbf{G} \\ \mathbf{G} & \mathbf{F} \end{pmatrix}\right]\begin{pmatrix} \mathbf{u} \\ \mathbf{d} \end{pmatrix}, \tag{33}$$

where $\mathbf{u}$ and $\mathbf{d}$ are the up- and down-going wave amplitudes, and

$$\mathbf{\Delta} = diag\left(\theta_{qP}, \theta_{qSV}\right) \tag{34}$$

contains the vertical slownesses for qP- and qSV-waves, θ_{qP} and θ_{qSV}. The transformation matrix

$$\mathbf{E} = \frac{1}{\sqrt{2}}\begin{pmatrix} \mathbf{E}_1 & \mathbf{E}_1 \\ \mathbf{E}_2 & -\mathbf{E}_2 \end{pmatrix} \tag{35}$$

is normalized with respect to the vertical energy flux so that the inverse has the simple form

$$\mathbf{E}^{-1} = \frac{1}{\sqrt{2}}\begin{pmatrix} \mathbf{E}_1^{-1} & \mathbf{E}_2^{-1} \\ \mathbf{E}_1^{-1} & -\mathbf{E}_2^{-1} \end{pmatrix} = \frac{1}{\sqrt{2}}\begin{pmatrix} \mathbf{E}_2^{\mathrm{T}} & \mathbf{E}_1^{\mathrm{T}} \\ \mathbf{E}_2^{\mathrm{T}} & -\mathbf{E}_1^{\mathrm{T}} \end{pmatrix}. \tag{36}$$

Matrices $\mathbf{E}_1$ and $\mathbf{E}_2$ posses the symmetries

$$\mathbf{E}_2^T\mathbf{E}_1 = \mathbf{E}_2\mathbf{E}_1^T = \mathbf{I}. \tag{37}$$

Therefore, the layer propagator matrix can be given as

$$\mathbf{P}_k = \exp\left(i\omega h\mathbf{M}_k\right) = \mathbf{E}\exp\left(i\omega h\mathbf{\Lambda}\right)\mathbf{E}^{-1} = \begin{pmatrix} \mathbf{P}_{11}^{(k)} & i\mathbf{P}_{12}^{(k)} \\ i\mathbf{P}_{21}^{(k)} & \mathbf{P}_{11}^{(k)} \end{pmatrix}, \tag{38}$$

where $\mathbf{P}_{mm}^{(k)} = \mathbf{E}_m\cos\left(\omega h\mathbf{\Delta}\right)\mathbf{E}_m^{-1}$ and $\mathbf{P}_{mn}^{(k)} = \mathbf{E}_m\sin\left(\omega h\mathbf{\Delta}\right)\mathbf{E}_n^{-1}$, $m \neq n,\ m,n = 1,2$. From equation (37) and (38), it follows that

$$\mathbf{P}_k^{-1} = \exp\left(-i\omega h\mathbf{M}_k\right) = \mathbf{E}\exp\left(-i\omega h\mathbf{\Lambda}\right)\mathbf{E}^{-1} = \begin{pmatrix} \mathbf{P}_{11}^{(k)} & -i\mathbf{P}_{12}^{(k)} \\ -i\mathbf{P}_{21}^{(k)} & \mathbf{P}_{11}^{(k)} \end{pmatrix}, \tag{39}$$

and

$$\begin{aligned} \mathbf{P}_{11}^{(k)T} &= \left(\mathbf{E}_1\cos\left(\omega h\mathbf{\Delta}\right)\mathbf{E}_1^{-1}\right)^T = \mathbf{E}_2\cos\left(\omega h\mathbf{\Delta}\right)\mathbf{E}_2^{-1} = \mathbf{P}_{22}^{(k)} \\ \mathbf{P}_{21}^{(k)T} &= \left(\mathbf{E}_2\sin\left(\omega h\mathbf{\Delta}\right)\mathbf{E}_1^{-1}\right)^T = \mathbf{E}_2\sin\left(\omega h\mathbf{\Delta}\right)\mathbf{E}_1^{-1} = \mathbf{P}_{21}^{(k)} \\ \mathbf{P}_{12}^{(k)T} &= \left(\mathbf{E}_1\sin\left(\omega h\mathbf{\Delta}\right)\mathbf{E}_2^{-1}\right)^T = \mathbf{E}_1\sin\left(\omega h\mathbf{\Delta}\right)\mathbf{E}_2^{-1} = \mathbf{P}_{12}^{(k)} \end{aligned}. \tag{40}$$

The product of matrices of this type indicates that the propagator matrix $\mathbf{P}$ for the stack of layers can be blocked as follows

$$\mathbf{P} = \begin{pmatrix} \mathbf{P}_{11} & i\mathbf{P}_{12} \\ i\mathbf{P}_{21} & \mathbf{P}_{22} \end{pmatrix}, \tag{41}$$

where the elements of 2x2 matrices $\mathbf{P}_{kl}$ are real functions of horizontal slowness p and frequency ω. The elements of matrices $\mathbf{P}_{kl}(\omega)$ are even/odd functions of frequency if $k+l$ is an even/odd number, respectively. Matrix $\mathbf{M}_k$ possesses the following symmetries

$$\mathbf{K}\mathbf{M}_k\mathbf{K} = \mathbf{M}_k^T,\ \mathbf{T}\mathbf{M}_k\mathbf{T} = -\mathbf{M}_k\ , \tag{42}$$

where $\mathbf{T} = \begin{pmatrix} \mathbf{I} & \mathbf{0} \\ \mathbf{0} & -\mathbf{I} \end{pmatrix}$. Therefore,

$$\mathbf{K}\mathbf{P}_k\mathbf{K} = \exp\left(i\omega h\mathbf{M}_k^T\right) = \mathbf{P}_k^T,\ \mathbf{T}\mathbf{P}_k\mathbf{T} = \exp\left(-i\omega h\mathbf{M}_k\right) = \mathbf{P}_k^{-1}. \tag{43}$$

For an elastic medium, the elements of matrix $\mathbf{M}_k$ are real, and, therefore, $\mathbf{P}_k^* = \mathbf{P}_k^{-1}$, and equations (43) can be rewritten as

$$\mathbf{K}\mathbf{P}_k\mathbf{K} = \left(\mathbf{P}_k^*\right)^{-T},\ \mathbf{T}\mathbf{P}_k\mathbf{T} = \mathbf{P}_k^*. \tag{44}$$

Multiplications of equations (44) for $k = 1,...,N$, gives similar symmetry relations for propagator matrix for the stack of layers,

$$\mathbf{K}\mathbf{P}\mathbf{K} = \left(\mathbf{P}^*\right)^{-T},\ \mathbf{T}\mathbf{P}\mathbf{T} = \mathbf{P}^*. \tag{45}$$

The scattering matrices $\mathbf{F}$ and $\mathbf{G}$ are given by

$$\mathbf{F} = -\frac{1}{2}\left[\mathbf{E}_2^T\frac{d\mathbf{E}_1}{dz} + \mathbf{E}_1^T\frac{d\mathbf{E}_2}{dz}\right],\ \mathbf{G} = -\frac{1}{2}\left[\mathbf{E}_2^T\frac{d\mathbf{L}_1}{dz} - \mathbf{E}_1^T\frac{d\mathbf{E}_2}{dz}\right], \tag{46}$$

with symmetries

$$\mathbf{F} = -\mathbf{F}^T,\quad \mathbf{G} = \mathbf{G}^T, \tag{47}$$

so that $\mathbf{F}$ and $\mathbf{G}$ have the form

$$\mathbf{F} = \begin{pmatrix} 0 & f \\ -f & 0 \end{pmatrix},\ \mathbf{G} = \begin{pmatrix} g_{11} & g_{12} \\ g_{12} & g_{22} \end{pmatrix}. \tag{48}$$

To compute the scattering matrices $\mathbf{F}$ and $\mathbf{G}$, we express them as functions of perturbations in the elements of the matrices $\mathbf{A}$ and $\mathbf{B}$ in equation (31). The eigenvector matrices satisfy the following equations

$$\mathbf{A}\mathbf{E}_2 = \mathbf{E}_1\mathbf{\Delta},\ \mathbf{B}\mathbf{E}_1 = -\mathbf{E}_2\mathbf{\Delta}. \tag{49}$$

From equations (35) and (36), we obtain

$$\mathbf{E}_1' = -\mathbf{E}_1\left(\mathbf{G}+\mathbf{F}\right),\ \mathbf{E}_2' = \mathbf{E}_2\left(\mathbf{G}-\mathbf{F}\right)\cdot \tag{50}$$

Differentiating equations (49) and substituting equations (50), we obtain

$$\begin{aligned}
\boldsymbol{\Delta}\mathbf{G}+\mathbf{G}\boldsymbol{\Delta} &= \begin{pmatrix} 2g_{11}\theta_{qP} & g_{12}\left(\theta_{qP}+\theta_{qSV}\right) \\ g_{12}\left(\theta_{qP}+\theta_{qSV}\right) & 2g_{22}\theta_{qSV} \end{pmatrix} = -\frac{1}{2}\left(\mathbf{E}_2^T\mathbf{A}'\mathbf{E}_2+\mathbf{E}_1^T\mathbf{B}'\mathbf{E}_1\right) \\
\boldsymbol{\Delta}\mathbf{F}-\mathbf{F}\boldsymbol{\Delta} &= \begin{pmatrix} 0 & f\left(\theta_{qP}-\theta_{qSV}\right) \\ f\left(\theta_{qP}-\theta_{qSV}\right) & 0 \end{pmatrix} = \frac{1}{2}\left(\mathbf{E}_2^T\mathbf{A}'\mathbf{E}_2-\mathbf{E}_1^T\mathbf{B}'\mathbf{E}_1\right)-\boldsymbol{\Delta}'
\end{aligned}\ . \tag{51}$$

From equations (51) we can compute all elements of matrices $\mathbf{F}$ and $\mathbf{G}$.
Equation (6) can be solved not only by propagator matrix technique described in previous chapter, but also by a layer-recursive scheme described in Ursin and Stovas (2002). We let $\mathbf{R}_{Dj}$, $\mathbf{T}_{Dj}$, $\mathbf{R}_{Uj}$ and $\mathbf{T}_{Uj}$ denote the R/T matrices for the interface at $z = z_j$ (Figure 1). For the layer thickness $\Delta z_j = z_j - z_{j-1}$, the layer propagator matrix is given by

$$\mathbf{L}_j = \exp\left(i\omega\Delta z_j\boldsymbol{\Lambda}_j\right)\cdot \tag{52}$$

Assume that the generalized reflection matrix from the top of layer *j*+1 to the bottom of the stratified medium, $\mathbf{R}_D\left(z_j^+, z_N^+\right)$, is known; then the new reflection matrix up to the top of layer j is given by the recursive relation (Ursin, 1983)

$$\mathbf{R}_D\left(z_{j-1}^+, z_N^+\right) = \mathbf{L}_j\left\{\mathbf{R}_{Dj}+\mathbf{T}_{Uj}\mathbf{R}_D\left(z_j^+, z_N^+\right)\left[\mathbf{I}+\mathbf{R}_{Dj}\mathbf{R}_D\left(z_j^+, z_N^+\right)\right]^{-1}\mathbf{T}_{Dj}\right\}\mathbf{L}_j\cdot \tag{53}$$

It is started with the initial value $\mathbf{R}_D\left(z_j^+, z_N^+\right) = 0$, satisfying the radiation condition in the elastic half-space. By considering a plane interface between two homogeneous media, we can define the transmission and reflection coefficient matrices. Continuity of the wavefields gives the 2x2 reflection and transmission matrices (Ursin, 1983):

$$\mathbf{T}_D = \begin{pmatrix} t_{DPP} & t_{DPS} \\ t_{DSP} & t_{DSS} \end{pmatrix} = 2\left(\mathbf{C}+\mathbf{D}\right)^{-1},\ \mathbf{R}_D = \begin{pmatrix} r_{DPP} & r_{DPS} \\ r_{DSP} & r_{DSS} \end{pmatrix} = \left(\mathbf{C}-\mathbf{D}\right)\left(\mathbf{C}+\mathbf{D}\right)^{-1}, \tag{54}$$

with

$$\mathbf{C} = \left[\mathbf{E}_2^{(1)}\right]^T\mathbf{E}_1^{(2)},\ \mathbf{D} = \left[\mathbf{E}_1^{(1)}\right]^T\mathbf{E}_2^{(2)}, \tag{55}$$

where superscripts $^{(1)}$ and $^{(2)}$ denote the upper and lower medium, respectively.

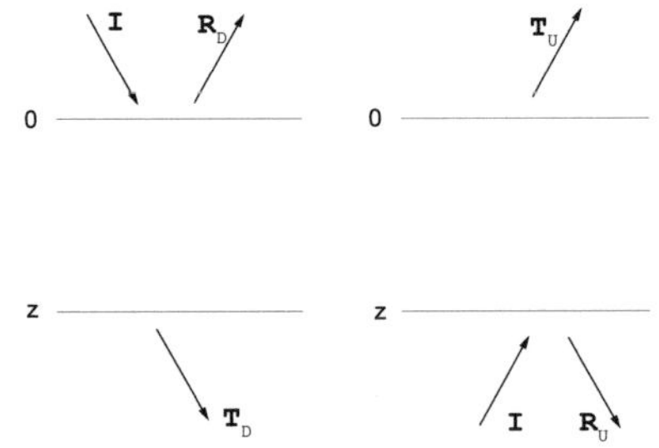

Fig. 1. Definition of the reflection and transmission response matrices (Ursin&Stovas, 2002).

The symmetry equations (37) imply that

$$\mathbf{C}\mathbf{D}^T = \mathbf{I}\,. \tag{56}$$

Using equation (55), the reflection and transmission matrices are given by

$$\begin{aligned}
\mathbf{T}_D &= \frac{2}{\det(\mathbf{C}+\mathbf{D})\det\mathbf{C}}\begin{pmatrix} c_{11}+c_{22}\det\mathbf{C} & c_{21}-c_{12}\det\mathbf{C} \\ c_{12}-c_{21}\det\mathbf{C} & c_{22}+c_{11}\det\mathbf{C}\end{pmatrix} \\
\mathbf{R}_D &= \frac{1}{\det(\mathbf{C}+\mathbf{D})\det\mathbf{C}}\begin{pmatrix} (\det\mathbf{C})^2-1+c_{11}^2+c_{12}^2-c_{21}^2-c_{22}^2 & 2(c_{11}c_{21}+c_{12}c_{22}) \\ 2(c_{11}c_{21}+c_{12}c_{22}) & (\det\mathbf{C})^2-1-c_{11}^2-c_{12}^2+c_{21}^2+c_{22}^2\end{pmatrix}
\end{aligned} \tag{57}$$

where c_{ij} are elements of matrix $\mathbf{C}$ and

$$\det(\mathbf{C}+\mathbf{D})\det\mathbf{C} = (\det\mathbf{C})^2+1+c_{11}^2+c_{12}^2+c_{21}^2+c_{22}^2\,. \tag{58}$$

The matrices $\mathbf{R}_U$ and $\mathbf{T}_U$ can be computed by interchanging the superscripts 1 and 2 or using the symmetry relations (37) and (56) (Ursin and Stovas, 2001), which gives

$$\mathbf{T}_U = \mathbf{T}_D^T, \quad \mathbf{R}_U = -\mathbf{T}_D\mathbf{R}_D\mathbf{T}_D^{-1}\,. \tag{59}$$

The weak-contrast approximations for reflection and transmission matrices an be derived by assuming weak contrast in elastic parameters above and below the interface. We consider a plane interface with a discontinuity in the parameters

$$\Delta m_k = m_k^{(2)} - m_k^{(1)} \tag{60}$$

and average parameters

$$m_k = \frac{m_k^{(2)}+m_k^{(1)}}{2}, \tag{61}$$

where $m_k^{(1)}$ and $m_k^{(2)}$ characterize the medium above and below the interface, respectively.

To approximate the reflection and transmission matrices in equation (54), we proceed as in Stovas and Ursin (2001) and expand the matrices $\mathbf{E}_k^{(j)}$, $k=1,2$, $j=1,2$, into second-order Taylor series with respect to the average medium. This gives the second-order approximations

$$\mathbf{T}_D^{(2)} \approx \begin{pmatrix} 1-\frac{1}{2}(g_{11}^2+g_{12}^2+f^2) & f-\frac{1}{2}g_{12}(g_{11}+g_{22}) \\ -f-\frac{1}{2}g_{12}(g_{11}+g_{22}) & 1-\frac{1}{2}(g_{12}^2+g_{22}^2+f^2)\end{pmatrix},\ \mathbf{R}_D^{(2)} \approx -\begin{pmatrix} g_{11}-g_{12}f & g_{12}+\frac{1}{2}f(g_{11}-g_{22}) \\ g_{12}+\frac{1}{2}f(g_{11}-g_{22}) & g_{22}+g_{12}f\end{pmatrix}. \tag{62}$$

From the symmetries of the matrices $\mathbf{E}_1$ and $\mathbf{E}_2$, the second-order derivatives of these matrices with respect to the medium parameters cancel; therefore they do not appear in the second-order approximations for the R/T coefficients. In the equations above, the elements f and g_{ij} are as defined in equation (46), but the derivatives of the medium parameters $\frac{dm_k}{dz}$ are replaced by Δm_k. Neglecting the second-order terms gives the linear approximations

$$\mathbf{T}_D^{(1)} \approx \mathbf{I} + \mathbf{F},\ \mathbf{R}_D^{(1)} \approx -\mathbf{G}\,. \tag{63}$$

These correspond to ones given in Ursin and Haugen (1996) for VTI media and in Aki and Richards (1980) for isotropic media, except that they are normalized with respect to the vertical energy flux and not with respect to amplitude.

6. Periodically layered media

Let us introduce the infinite periodically layered VTI medium with the period thickness $H = \sum_{j=1}^{N} h_j$, where h_j is the thickness of jth layer in the sequence of N layers comprising the period. The dispersion equation for this N layered medium is given by (Helbig, 1984)

$$\det\left(\mathbf{P} - \exp\left(i\omega H\theta\right)\mathbf{I}\right) = 0\,, \tag{64}$$

and the period propagator matrix $\mathbf{P}$ is specified by formula (15). The equation (64) is known as the Floquet (1883) equation.

The parameter $\theta = \theta\left(p,\omega\right)$ is effective and generally complex vertical component of the slowness vector. For plane waves with horizontal slowness p, the real part of θ which satisfies equation (64), $\operatorname{Re}\theta = q$, defines the vertical slowness of the envelope, while the imaginary part, $\operatorname{Im}\theta = \gamma$, characterizes the attenuation due to scattering. Note that for propagating waves, $\lim_{\omega\to 0} \gamma = 0$. This indicates that there is no scattering in the low frequency limit.

The low and high frequency limits

In the low-frequency asymptotic of the propagator matrix $\mathbf{P}$ has the following form $\mathbf{P} = \exp\left(i\omega H\tilde{\mathbf{M}}\left(\omega\right)\right)$ with

$$\tilde{\mathbf{M}}\left(\omega\right) = \frac{1}{H}\sum_{k=1}^{N} h_k\mathbf{M}_k + \frac{i\omega}{2H^2}\sum_{j>\ell} h_j h_\ell\left(\mathbf{M}_j\mathbf{M}_\ell - \mathbf{M}_\ell\mathbf{M}_j\right) + o\left(\omega\right) \tag{65}$$

Therefore, the dispersion equation (64) in the low-frequency limit has roots similar to those defined for a homogeneous VTI medium given by the averaged matrix

$$\tilde{\mathbf{M}}\left(0\right) = \frac{1}{H}\sum_{k=1}^{N} h_k\mathbf{M}_k\,. \tag{66}$$

One can see from observing the elements of the matrices in equation (31) that equation (66) is equivalent to the Backus averaging. The propagator matrix $\tilde{\mathbf{P}}$, which defines the propagation of mode m_k in the kth layer, $k = 1,\dots,N$, can be defined as

$$\tilde{\mathbf{P}} = \exp\left(i\omega h_N\mathbf{F}_N\right)\dots\exp\left(i\omega h_k\mathbf{F}_k\right)\dots\exp\left(i\omega h_1\mathbf{F}_1\right), \tag{67}$$

where $\mathbf{F}_k = \theta_k^{(m_k)}\mathbf{n}_k^{(m_k)}\mathbf{m}_k^{(m_k)T}$ is a 4x4 matrix of rank one, $\mathbf{m}_k^{(m_k)T}$ and $\mathbf{n}_k^{(m_k)}$ are the left- and right hand side eigenvectors of matrix $\mathbf{M}_k$ with eigenvalue $\theta_k^{(m)}$. Substituting $\mathbf{F}_k$ into equation (67) results in

$$\tilde{\mathbf{P}} = \exp\left[i\omega \sum_{k=1}^{N} h_k \theta_k^{(m_k)} \right] \left(\mathbf{n}_N^{(m_N)} \mathbf{m}_N^{(m_N)T} \ldots \mathbf{n}_1^{(m_1)} \mathbf{m}_1^{(m_1)T} \right) = \exp\left[i\omega \sum_{k=1}^{N} h_k \theta_k^{(m_k)} + \beta \right] \left(\mathbf{n}_N^{(m_N)} \mathbf{m}_1^{(m_1)T} \right), \quad (68)$$

where the number $\beta = \ln\left(\mathbf{m}_N^{(m_N)T} \ldots \mathbf{n}_1^{(m_1)} \right)$. In this case, the dispersion equation (64), which defines the vertical slowness for the period of the layered medium, has the root given by

$$\theta = \frac{1}{H} \sum_{k=1}^{N} h_k \theta_k^{(m_k)} - \frac{i\beta}{\omega H}, \quad (69)$$

where the term $i\beta / \omega H$ is responsible for the transmission losses for propagating waves which is frequency independent. This can be shown by considering the single mode plane wave,

$$\exp\left[i\omega \left(pr + \theta \quad z - t \right) \right] = \exp\left[i\omega \left(pr + q \quad z - t \right) \right] \exp\left[\beta \quad z / H \right], \quad (70)$$

where

$$q = \frac{1}{H} \sum_{k=1}^{N} h_k q_k^{(m_k)}. \quad (71)$$

This equation defines the vertical slowness for a single mode transmitted wave initiated by a wide-band δ - pulse, since it is frequency-independent. The caustics from multi-layered VTI medium in high-frequency limits are discussed in Roganov and Stovas (2010). Note, that propagator matrix in equation (68) describes the downward plane wave propagation of a given mode within each layer, i.e. the part of the full wave field. All multiple reflections and transmissions of other modes are ignored. Therefore, this notation is valid for the case of the frequency independent single mode propagation of a wide band δ pulse. In the low frequency limit, the wave field consists of the envelope with all wave modes. For an accurate description of this envelope and obtaining the Backus limit we have to use the formula (15) for complete propagator $\mathbf{P}$.

6.1 Dispersion equation analysis

From the relations (45), one can see that the matrices $\mathbf{P}$, $\mathbf{P}^*$ and $\left(\mathbf{P}^*\right)^{-T}$ are similar. These matrices have the same eigen-values. So, if x is eigen-value of matrix $\mathbf{P}$, than x^*, x^{-1} and $\left(x^{-1}\right)^*$ are also eigen-values. Additionally, taking into account the identity, $\det(\mathbf{P}) = 1$, it can be shown, that equation

$$\det(\mathbf{P} - x\mathbf{I}) = 0 \quad (72)$$

reduces to

$$\left(x + x^{-1}\right)^2 - a_1\left(x + x^{-1}\right) + a_2 - 2 = 0, \quad (73)$$

and the roots of equation (64) corresponding to qP- and qSV-waves, $\pm\theta_P$ and $\pm\theta_S$, satisfy the equations

$$\cos\left(\omega H\theta_{qP}\right)+\cos\left(\omega H\theta_{qSV}\right)=\frac{1}{2}a_1\left(p,\omega\right),\ \cos\left(\omega H\theta_{qP}\right)\cos\left(\omega H\theta_{qSV}\right)=\frac{1}{4}a_2\left(p,\omega\right)-\frac{1}{2}. \quad (74)$$

The real functions $a_1(p,\omega)$ and $a_2(p,\omega)$ can be computed using the trace and the sum of the principal second order minors of the matrix $\mathbf{P}$, respectively. Using equation (41) and taking into account that $\mathbf{P}_{11}(\omega)$ and $\mathbf{P}_{22}(\omega)$ are even functions of frequency, and $\mathbf{P}_{12}(\omega)$ and $\mathbf{P}_{21}(\omega)$ are odd functions of frequency, the functions $a_1(p,\omega)$ and $a_2(p,\omega)$ are even functions of frequency and horizontal slowness. The system of equations (74) defines the continuous branches of functions $\pm\mathrm{Re}\,\theta_{qP}=\pm q_{qP}(p,\omega)$ and $\pm\mathrm{Re}\,\theta_{qSV}=\pm q_{qSV}(p,\omega)$ which specify the vertical slowness of four envelopes with horizontal slowness p and frequency ω. Let us denote $b_1(p,\omega)=a_1(p,\omega)/4$, $b_2(p,\omega)=a_2(p,\omega)/4-1/2$ and $y=\dfrac{x+x^{-1}}{2}$. Note that the functions $b_1(p,\omega)$ and $b_2(p,\omega)$ are also even functions of frequency and horizontal slowness.

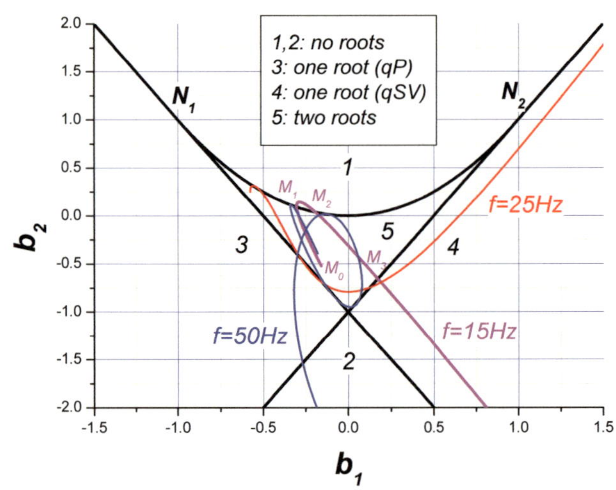

Fig. 2. Propagating and evanescent regions for $qP-$ and $qSV-$ waves in the (b_1,b_2) domain. The points $N_1(-1,1)$ and $N_2(1,1)$ denote the crossings between $b_2=-1\pm 2b_1$ and $b_2=b_1^2$. The paths corresponding to $f=const$ are given for frequencies of $f=15$, 25 and $50\,Hz$ are shown in magenta, red and blue, respectively. The starting point M_0 (that corresponds to zero horizontal slowness) and the points corresponding to crossings of the path and boundaries between the propagating regions, M_j, $j=1,2,3$, are shown for the frequency $f=15\,Hz$. Points M_4 and M_5 are outside of the plotting area (Roganov&Stovas, 2011).

All envelopes are propagating, if the roots of quadratic equation

$$y^2-2b_1y+b_2=0 \quad (75)$$

are such that $|y_1|\le 1$ and $|y_2|\le 1$. On the boundaries between propagating and evanescent envelopes, we have $y=\pm 1$ or discriminator of equation (75), $D(p,\omega)=b_1^2-b_2=0$. In the first case we have, $b_2=-1\pm 2b_1$, and in the second case, $b_2=b_1^2$ (Figure 2). If $|y|>1$, the equation $y=\cos(\omega H\theta)$ has the following solutions

$$\begin{aligned}\theta &= \pm\frac{1}{\omega H}\left[2\pi n + i\ln\left(y+\sqrt{y^2-1}\right)\right],\ n\in\mathbf{Z},\ y>1\\ \theta &= \pm\frac{1}{\omega H}\left[(2n+1)\pi + i\ln\left(-y+\sqrt{y^2-1}\right)\right],\ n\in\mathbf{Z},\ y<-1\end{aligned}, \qquad (76)$$

and $q=\mathrm{Re}\,\theta=const$ in this area. The straight lines $b_2=-1\pm 2b_1$ and the parabola $b_2=b_1^2$ defined between the tangent points $N_1(-1,1)$ and $N_2(1,1)$ split the coordinate plane (b_1,b_2) into five regions (Figure 2). If parameters b_1 and b_2 are such that the corresponding point (b_1,b_2) is located in region 1 or 2, the system of equations (74) has no real roots, and corresponding envelopes do not contain the propagating wave modes. The envelopes with one propagating wave of qP-or qSV- wave mode correspond to the points located in region 3 or region 4, respectively. The points from region 5 result in envelopes with both propagating qP- and qSV-wave modes. If a specific frequency is chosen, for instance, $\omega=30\pi Hz$ (or $f=15Hz$), and only the horizontal slowness is varied, the point with coordinates (b_1,b_2) will move along some curve passing through the different regions. Consequently, the number of propagating wave modes will be changed. In Figure 2, we show using the points M_i $(i=0,\dots 5)$ with the initial position M_0 defined by $p=0$ and the following positions crossing the boundaries for the regions occurred at $p_1=0.172\,s/km$, $p_2=0.217\,s/km$ and $p_3=0.246\,s/km$. This curve will also cross the line $b_2=-1+2b_1$ at $p_4=0.332\,s/km$ and $p_5=0.344\,s/km$. Between the last two points, the curve is located in the region 2 with no propagating waves for both modes. The frequency dependent positions of the stop bands for $p=const$ can be investigated using the curve, $b(\omega)=\left[b_1(\omega),b_2(\omega)\right]$. Since the propagator in the zero frequency limit is given by the identity matrix, $\lim\limits_{\omega\to 0}\mathbf{P}(\omega)=\mathbf{I}$, than $b_1(0)=b_2(0)=1$, and all curves $b(\omega)$ start at the point $N_2(1,1)$. For propagating waves, the functions $b_1(\omega)$ and $b_2(\omega)$ are given by linear combinations of trigonometric functions and therefore are defined only in a limited area in the (b_1,b_2) domain.

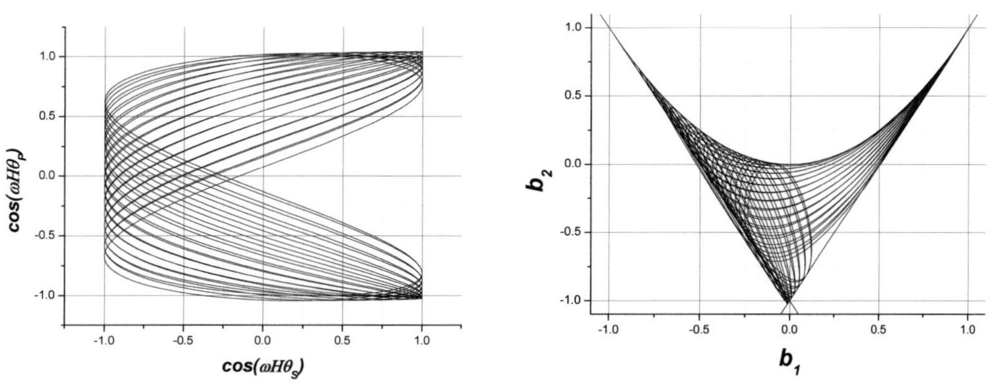

Fig. 3. The normal incidence case (p = 0). The dependence of $\cos\left(\omega H\theta_{qP}\right)$ on $\cos\left(\omega H\theta_{qSV}\right)$ (a Lissajous curve) is shown (left) and similar curve is plotted in the (b_1,b_2) domain (right). Both of these plots correspond to frequency range $0-50Hz$. Note, that the stop bands exist only for $qP-$ wave and can be seen for $\left|\cos\left(\omega H\theta_{qP}\right)\right|>1$ in the left plot and for $b_2<-1\pm 2b_1$ in the right one (Roganov&Stovas, 2011).

The simplest case occurs at the normal incidence where $p = 0$. At this point the quadratic equation $y^2 - 2b_1(\omega)y + b_2(\omega) = 0$ has two real roots $y_{qP}(\omega)$ and $y_{qSV}(\omega)$ for each value of ω. The functions $y_{qP}(\omega) = \cos(\omega H\theta_{qP})$ and $y_{qSV}(\omega) = \cos(\omega H\theta_{qSV})$ are the right side of the dispersion equation for qP- and qSV- wave, respectively. If these trigonometric functions have incommensurable periods, the parametric curve $(y_{qSV}(\omega), y_{qP}(\omega))$ densely fills the area that contains rectangle $[-1,1]\times[-1,1]$ and is defined as a Lissajous curve (Figure 3, left). The mapping $(y_{qSV}, y_{qP}) \to \left(\frac{y_{qSV} + y_{qP}}{2}, y_{qSV}y_{qP}\right)$ has the Jacobian $\left(\frac{y_{qSV} - y_{qP}}{2}\right)$ with a singularity at $y_{qSV} = y_{qP}$. This point is located at the discriminant curve, $b_2 = b_1^2$. We can prove that the curve $\left(\frac{y_{qSV}(\omega) + y_{qP}(\omega)}{2}, y_{qSV}(\omega)y_{qP}(\omega)\right)$ is tangent to parabola $b_2 = b_1^2$ at the singular point and is always located in the region $b_2 \le b_1^2$. In fact, if, $b_1(\omega) = \frac{y_{qSV}(\omega) + y_{qP}(\omega)}{2}$, $b_2(\omega) = y_{qSV}(\omega)y_{qP}(\omega)$ and $y_{qP}(\omega_0) = y_{qSV}(\omega_0)$ than $b_1^2(\omega) - b_2(\omega) = \frac{1}{4}(y_{qP}(\omega) - y_{qSV}(\omega))^2 = 0(\omega - \omega_0)$. In Figure 3 (left), it is shown the parametric curve $(y_{qSV}(\omega), y_{qP}(\omega))$ computed for our two layer model described in Table 1. Since both layers have the same vertical shear wave velocity and density, $y_{qSV}(\omega) = \cos\omega t_{qSV}$ with $t_{qSV} = (h_1 + h_2)/\beta_0$. In the qP- wave case, $y_{qP}(\omega) = \left(\cos\omega(t_{qP1} + t_{qP2}) - r\cos\omega(t_{qP1} - t_{qP2})\right)/(1 - r^2)$ where $t_{qP1} = h_1/\alpha_{01}$, $t_{qP2} = h_2/\alpha_{02}$ and $r = (\alpha_{02} - \alpha_{01})/(\alpha_{01} + \alpha_{02})$. The solutions of this equation and has been studied by Stovas and Ursin (2007) and Roganov and Roganov (2008). The plot of this curve in (b_1, b_2) domain is shown in Figure 3 (right). It can be seen that the stop bounds are characterized by the values $b_2 < -1 \pm 2b_1$. If $D(p,\omega) < 0$, equation (64) has the complex conjugate and dual roots. Let us denote one of them as $y_1 \in \mathbf{C}$. Then, equation (74) has four complex roots: θ, $-\theta$, θ^* and $-\theta^*$, where $\cos(\omega H\theta) = y_1$. In these cases, the energy envelope equals zero. The up going and down going wave envelopes have different signs for $\gamma = \operatorname{Im}\theta$ that correspond to exponentially damped and exponentially increasing terms.

6.2 Computational aspects

The computation of the slowness surface at different frequencies is performed by computing the propagator matrix (15) for the entire period and analysis of eigenvalues of this matrix. To define the direction for propagation of the envelope with eigenvectors $\mathbf{b} = (v_3, \tau_{13}, \tau_{33}, v_1)^T$ and non-zero energy is done in accordance with sign of the vertical energy flux (Ursin, 1983; Carcione, 2001)

$$E = -\frac{1}{2}\operatorname{Re}\left(v_1\tau_{13}^* + v_3\tau_{33}^*\right). \tag{77}$$

If $E=0$, the direction of the envelope propagation depends on the absolute value of $\exp(i\omega H\theta)$; $|\exp(i\omega H\theta)|>1$ (up going envelope) and $|\exp(i\omega H\theta)|<1$ (down going envelope). The mode of envelope can be defined by computing the amplitude propagators

$$\mathbf{Q}=\mathbf{E}_1\mathbf{P}\mathbf{E}_1^{-1}. \tag{78}$$

The absolute values of the elements of the matrix $\mathbf{Q}$ are the amplitudes of the different wave modes composing the envelope and defined in the first layer within the period. Therefore, the envelope of a given mode contains the plane wave of the same mode with the maximum amplitude (when compared with other envelopes).

6.3 Asymptotic analysis of caustics

Let us investigate the asymptotic properties for the vertical slowness of the envelope in the neighborhood of the boundary between propagating and evanescent waves when approaching this boundary from propagating region.

If $y(p_0)=1$ and $dy/dp_0=\alpha\neq 0$, than in the neighborhood of the point $p=p_0$ the following approximation of equation $y=\cos(\omega H\theta)$ is valid

$$1-\frac{\omega^2H^2d\theta^2}{2}\approx 1+\alpha dp, \tag{79}$$

where $dp=p-p_0$, $d\theta=\theta-\theta_0$ and $\theta_0=\theta(p_0)$. Therefore, $d\theta=O\left(\sqrt{dp}\right)$, $d\theta/dp=O\left(1/\sqrt{dp}\right)$, and the curve $\theta(p)$ at the $p=p_0$ has the vertical tangent line, $\lim_{p\to p_0}(d\theta/dp)=\pm\infty$. In the group space $(x,t(x))$, it leads to an infinite branch represented by caustic. In the area of propagating waves, we have $\lim_{p\to p_0}(d\theta/dp)=-\infty$. Therefore, $x(p)=-H(d\theta/dp)\to+\infty$ and $t(p)=H\theta(p)+px(p)\to+\infty$. Furthermore, for large values of x, $t(x)=p_0x+H\theta(p_0)$. This fact follows from existence of limit, $\lim_{p\to p_0}(\theta(p))=\theta(p_0)$. As a consequence, every continuous branch of the slowness surface limited by the attenuation zones (stop bands) results in the caustic in group space which looks like an open angle sharing the same vertex (Figure 4). When we move from one point of discontinuity to another in the increasing direction of p, the plane angle figure rotates clockwise since the slope of the traveltime curve $dt/dx=p_0$ is increasing. The case where $y(p_0)=-1$ can be discussed in the same manner. If $D(p_0)=0$, than $\cos(\omega h\theta)=b_1\pm\sqrt{D(p)}=b_1\pm O\left(\sqrt{p}\right)$. Therefore, the asymptotic behavior of $\theta(p)$ as $p\to p_0$ is the same as discussed above.

6.4 Low frequency caustics

In Figure 5 we show the propagating, evanescent and caustic regions in $p-f$ domain for qP- and qSV- waves ($f=\omega/2\pi$). Figure 5 displays contour plots of the vertical energy flux in the $p-f$ domain for qP- and qSV- waves.

From Figure 5 one can see that the caustic area has weak frequency dependence in the low frequency range (almost vertical structure for caustic region in (p,f) domain, Figures 5 and 6). This follows from more general fact that for VTI periodic medium, $\theta(\omega)$ is even function

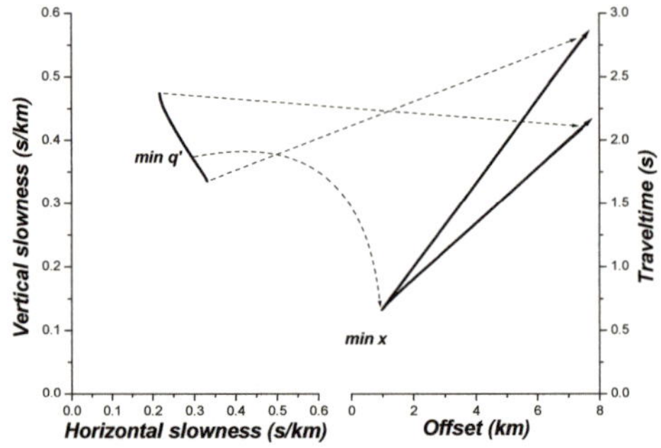

Fig. 4. Sketch for the stop band limited branch of the slowness surface and corresponding branch on the traveltime curve. The correspondence between characteristic points is shown by dotted line (Roganov&Stovas, 2011).

of frequency. Last statement is valid because $y(\omega)$ satisfies the equation (75) and functions $y(\omega)=\cos(\omega H\theta)$, $b_1(\omega)$ and $b_2(\omega)$ are even. Therefore,

$$\theta(\omega)=\theta(0)+o(\omega), \tag{80}$$

and the slowness surface at low frequencies is almost frequency independent.

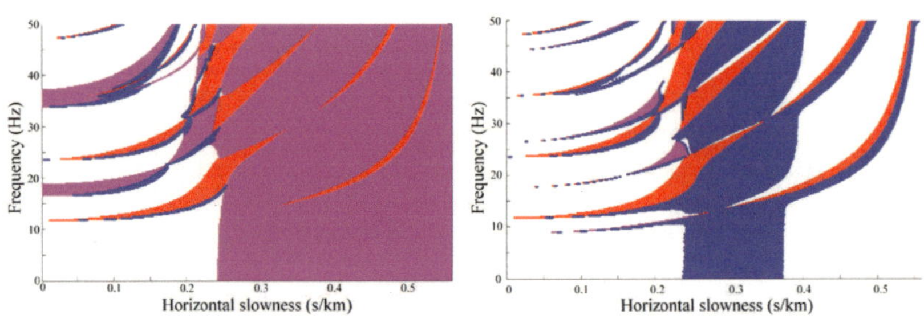

Fig. 5. The propagating, evanescent and caustic regions for the $qP-$ wave (left) and the $qSV-$ wave (right) are shown in the (p, f) domain. The regions are indicated by colors: red - no waves, white - both waves, magenta - $qSV-$ wave only and blue - caustic (Roganov&Stovas, 2011).

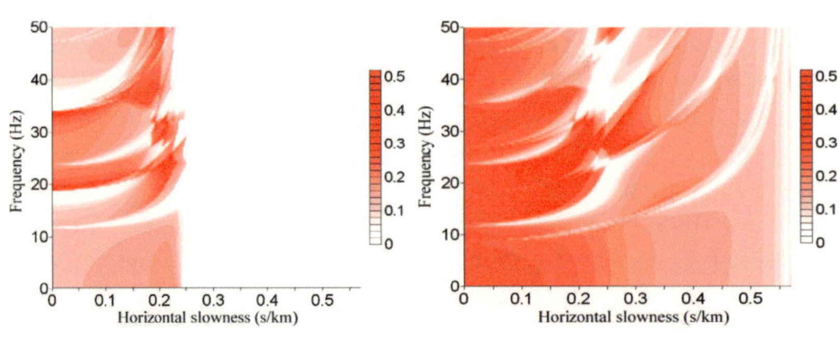

Fig. 6. The vertical energy flux for $qP-$ wave (left) and qSV - wave (right) shown in the (p, f) domain. The zero energy flux zones correspond to evanescent waves (Roganov&Stovas, 2011).

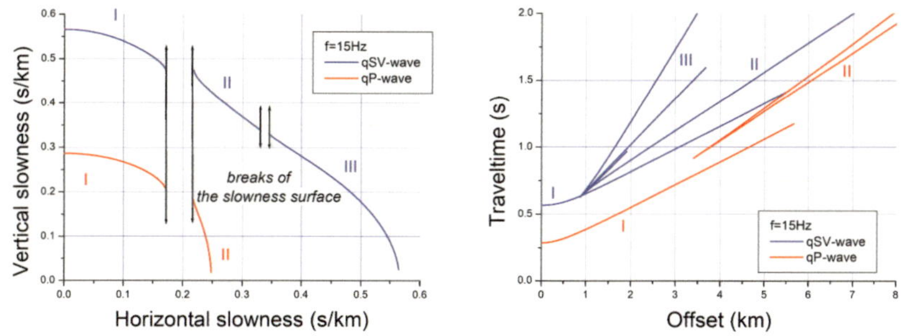

Fig. 7. The $qP-$ and $qSV-$ wave slowness surfaces (left) and the corresponding traveltime curves (right) corresponding frequency of $15Hz$. The branches on the slowness surfaces and on the traveltime curves are denoted by I, II and III (for the $qSV-$ wave) and I and II (for the $qP-$ wave) (Roganov&Stovas, 2011).

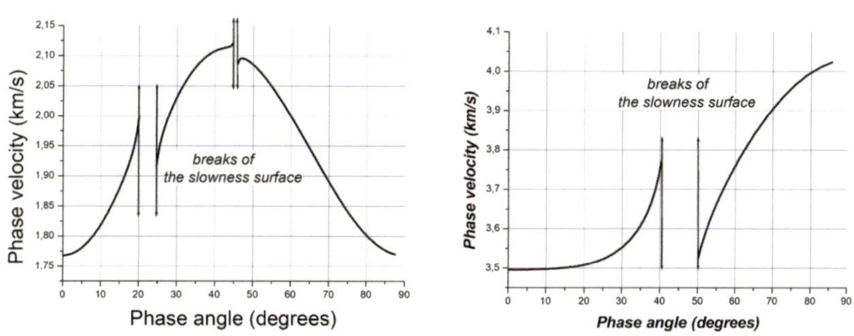

Fig. 8. The phase velocities for $qSV-$ wave (left) and $qP-$ wave (right) computed for a frequency of $15Hz$ (Roganov&Stovas, 2011).

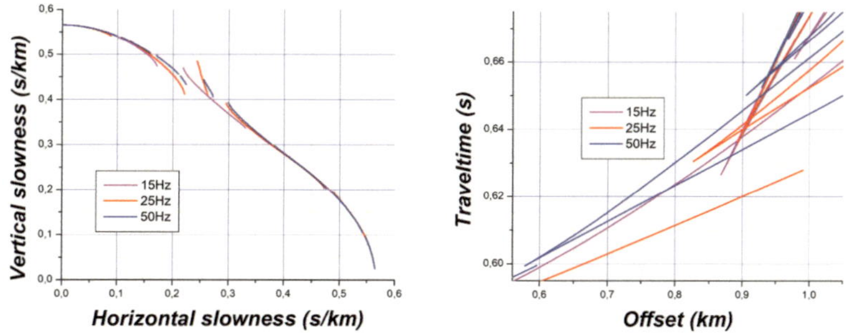

Fig. 9. Comparison of the $qSV-$ slowness surface and traveltime curves computed for frequencies of $f = 15$, 25 and $50Hz$ (shown in magenta, red and blue colors, respectively) (Roganov&Stovas, 2011).

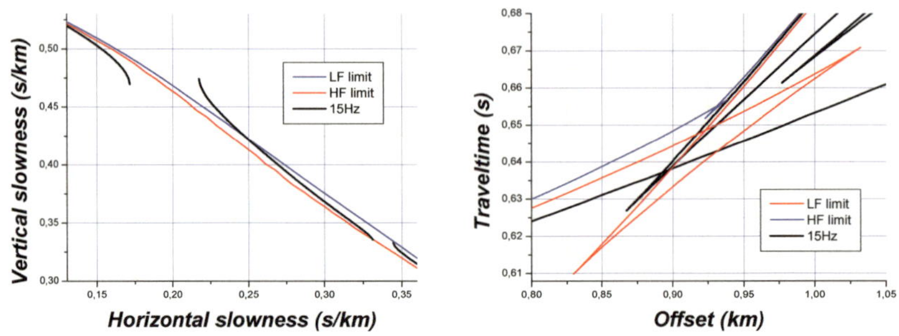

Fig. 10. Comparison of the qSV – slowness surfaces and traveltime curves computed for frequencies of 15, low and high frequency limits (shown in black, red and blue, respectively). Note, the both effective media in low and high frequency limits have triplications for traveltime curves (Roganov&Stovas, 2011).

To illustrate the method described above we choose two-layer transversely isotropic medium with vertical symmetry axis which we used in our previous paper (Roganov and Stovas, 2010). The medium parameters are given in Table 1. Each single VTI layer in the model has its own qSV- wave triplication. In Figure 7 (left), we show the slowness surfaces for the qP- and qSV- waves computed for a single frequency of 15 Hz. The discontinuities in both slowness surfaces correspond to the regions with evanescent waves or zero vertical energy flux, $E=0$ (equation (77)). The first discontinuity has the same location on the slowness axis for both qP- and qSV- wave slowness surfaces. In the group space (Figure 7, right), we can identify each traveltime branch with correspondent branch of the slowness surface. In Figure 8, we show the phase velocities for qP- and qSV- waves versus the phase angle ψ. The discontinuities in the phase velocity are clearly seen for both qP- and qSV- waves in different phase angle regions. Comparisons of the qSV- wave slowness surface and traveltime curves computed for different frequencies, $f=15,25$ and $50Hz$ are given in Figure 9. One can see that higher frequencies result in more discontinuities in the slowness surface. Only the branches near the vertical and horizontal axis remain almost the same. In Figure 10, we show the slowness surfaces and traveltime curves computed for frequency $f=15Hz$ and those computed in the low and high frequency limits. The vertical slowness and traveltime computed in low and high frequency limits are continuous functions of horizontal slowness and offset, respectively.

7. Reflection/transmission responses in periodicaly layered media

The problem of reflection and transmission responses in a periodically layered medium is closely related to stratigraphic filtering (O'Doherty and Anstey,1971; Schoenberger and Levin, 1974; Morlet et al., 1982a, b; Banik et al., 1985a, b; Ursin, 1987; Shapiro et al., 1996; Ursin and Stovas, 2002; Stovas and Ursin, 2003; Stovas and Arntsen, 2003). Physical experiments were performed by Marion and Coudin (1992) and analyzed by Marion et al (1994) and Hovem (1995). The key question is the transition between the applicability of low- and high-frequency regimes based on the ratio between wavelength (λ) and thickness (d) of one cycle in the layering. According to different literature sources, this transition

occurs at a critical λ/d value which Marion and Coudin (1992) found to be equal to 10. Carcione et al. (1991) found this critical value to be about 8 for epoxy and glass and to be 6 to 7 for sandstone and limestone. Helbig (1984) found a critical value of λ/d equal to 3. Hovem (1995) used an eigenvalue analysis of the propagator matrix to show that the critical value depends on the contrast in acoustic impedance between the two media. Stovas and Arntsen (2003) showed that there is a transition zone from effective medium to time-average medium which depends on the strength of the reflection coefficient in a finely layered medium.
To compute the reflection and transmission responses, we consider a 1D periodically layered medium. Griffiths and Steinke (2001) have given a general theory for wave propagation in periodic layered media. They expressed the transmission response in terms of Chebychev polynomials of the second degree which is a function of the elements of the propagator matrix for the basic two-layer medium. They also provided an extensive reference list.

7.1 Multi-layer transmission and reflection responses

We consider one cycle of a binary medium with velocities v_1 and v_2, densities ρ_1 and ρ_2 and the thicknesses h_1 and h_2 as shown in Figure 11. For a given frequency f the phase factors are: $\theta_k = 2\pi f h_k / v_k = 2\pi f \Delta t_k$, where Δt_k is the traveltime in medium *k* for one cycle. The normal incidence reflection coefficient at the interface between the layers is given by

$$r = \frac{\rho_2 v_2 - \rho_1 v_1}{\rho_2 v_2 + \rho_1 v_1}. \tag{81}$$

The amplitude propagator matrix for one cycle is computed for an input at the bottom of the layers (Hovem, 1995)

$$\mathbf{Q} = \frac{1}{1-r^2}\begin{pmatrix} e^{i\theta_1} & 0 \\ 0 & e^{-i\theta_1} \end{pmatrix}\begin{pmatrix} 1 & r \\ r & 1 \end{pmatrix}\begin{pmatrix} e^{i\theta_2} & 0 \\ 0 & e^{-i\theta_2} \end{pmatrix}\begin{pmatrix} 1 & -r \\ -r & 1 \end{pmatrix} = \begin{pmatrix} a & b \\ b^* & a^* \end{pmatrix}, \tag{82}$$

and

$$a = \frac{e^{i(\theta_1+\theta_2)}\left(1 - r^2 e^{-2i\theta_2}\right)}{1-r^2}, \quad b = -\frac{re^{i(\theta_1+\theta_2)}\left(1 - e^{-2i\theta_2}\right)}{1-r^2} = \frac{2ir\sin\theta_2}{1-r^2}e^{i\theta_1}. \tag{83}$$

We also compute the real and imaginary part of a (Brekhovskikh, 1960) and absolute value of b, resulting in

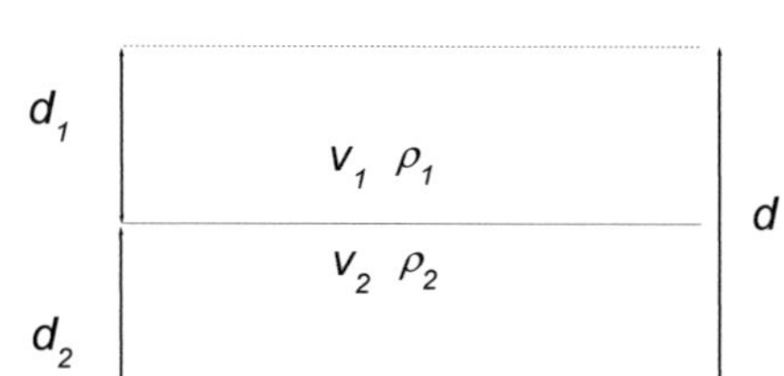

Fig. 11. Single cycle of the periodic medium (Stovas&Ursin, 2007).

$$\operatorname{Re} a = \cos(\theta_1 + \theta_2) - \frac{2r^2}{1-r^2}\sin\theta_1 \sin\theta_2 = \cos\theta_1\cos\theta_2 - \frac{1+r^2}{1-r^2}\sin\theta_1\sin\theta_2$$

$$\operatorname{Im} a = \sin(\theta_1 + \theta_2) + \frac{2r^2}{1-r^2}\cos\theta_1 \sin\theta_2 = \sin\theta_1\cos\theta_2 + \frac{1+r^2}{1-r^2}\cos\theta_1\sin\theta_2 \qquad (84)$$

$$|b| = \frac{2r\sin\theta_2}{1-r^2}$$

We note, that $\det \mathbf{Q} = |a|^2 - |b|^2 = 1$ as shown also by Griffiths and Steinke (2001). The amplitude propagator matrix can be represented by the eigenvalue decomposition (Hovem, 1995)

$$\mathbf{Q} = \mathbf{E}\,\boldsymbol{\Sigma}\,\mathbf{E}^{-1}, \qquad (85)$$

where $\boldsymbol{\Sigma} = diag[\sigma_1, \sigma_2]$ with

$$\sigma_{1,2} = \begin{cases} \operatorname{Re} a \pm i\sqrt{1-(\operatorname{Re} a)^2} & for\,|\operatorname{Re} a| < 1 \\ \operatorname{Re} a \mp \sqrt{(\operatorname{Re} a)^2 - 1} & for\,|\operatorname{Re} a| \geq 1 \end{cases} \qquad (86)$$

and the matrix

$$\mathbf{E} = \begin{pmatrix} 1 & 1 \\ (\sigma_1 - a)/b & (\sigma_2 - a)/b \end{pmatrix} \qquad (87)$$

A stack of M cycles of total thickness $D = Mh = M(h_1 + h_2)$ has the propagator matrix

$$\mathbf{Q}(M) = \mathbf{Q}^M = \mathbf{E}\,\boldsymbol{\Sigma}^M\,\mathbf{E}^{-1} = \frac{1}{u_{22} - u_{21}} \begin{pmatrix} \sigma_1^M u_{22} - \sigma_2^M u_{21} & \sigma_2^M - \sigma_1^M \\ -u_{21}u_{22}(\sigma_2^M - \sigma_1^M) & \sigma_2^M u_{22} - \sigma_1^M u_{21} \end{pmatrix} \qquad (88)$$

with $u_{21} = (\sigma_1 - a)/b$ and $u_{22} = (\sigma_2 - a)/b$. Another way to compute the propagator or transfer matrix is to exploit the Cailey-Hamilton theorem to establish relation between $\mathbf{Q}^2$ and $\mathbf{Q}$ (Wu et al., 1993) which results in the recursive relation for Chebychev polynomials. The transmission and reflection responses for a down-going wave at the top of the layers are (Ursin, 1983)

$$t_D = p_{22}^{-1} = \frac{\sigma_2 - \sigma_1}{(\sigma_2 - a)\sigma_2^M - (\sigma_1 - a)\sigma_1^M}, \quad r_D = p_{12}p_{22}^{-1} = \frac{(\sigma_2^M - \sigma_1^M)b}{(\sigma_2 - a)\sigma_2^M - (\sigma_1 - a)\sigma_1^M} \qquad (89)$$

with p_{ij}, $i, j = 1, 2$ being the elements of propagator matrix $\mathbf{Q}(M)$ given in (88). After algebraic manipulations equation (89) can be written as

$$t_D = \frac{\sin\varphi}{\sin\varphi\cos M\varphi - i\,\mathrm{Im}\,a\sin M\varphi} = \frac{\cos M\varphi + i\sin M\varphi\dfrac{\mathrm{Im}\,a}{\sin\varphi}}{1+\sin^2 M\varphi\left[\left(\dfrac{\mathrm{Im}\,a}{\sin\varphi}\right)^2 - 1\right]} = \frac{e^{i\alpha}}{\sqrt{1+C^2}}, \quad (90)$$

$$r_D = \frac{b\sin M\varphi}{\sin\varphi\cos M\varphi - i\,\mathrm{Im}\,a\sin M\varphi} = t_D b\frac{\sin M\varphi}{\sin\varphi} = \frac{Ce^{i\left(\alpha+\theta_1+\frac{\pi}{2}\right)}}{\sqrt{1+C^2}}$$

where α and C are the phase and amplitude factors, respectively, and φ is the phase of the eigen-value. The equation for transmission response in periodic structure was apparently first obtained in the quantum mechanics (Cvetich and Picman, 1981) and has been rediscovered several times. For extensive discussion see reference 13 in Griffiths and Steinke (2001). The reflection and transmission response satisfy

$$|t_D|^2 + |r_D|^2 = 1, \quad (91)$$

which is conservation of energy. When $|\mathrm{Re}\,a| < 1$, the eigen-values give a complex phase-shift, representing a propagating regime. Then equation (86) gives

$$\sigma_{1,2} = e^{\pm i\varphi} \quad (92)$$

with $\cos\varphi = \mathrm{Re}\,a$, which may be obtained from Floquet solution for periodic media, but for first time appeared in Brekhovskikh (1960), equation (7.25). Then we use

$$\cos\alpha = \frac{\cos M\varphi}{\sqrt{1+C^2}}, \; C = |b|\frac{\sin M\varphi}{\sin\varphi}, \quad (93)$$

in equation (90). Equation (93) for the amplitude factor is given in a form of Chebychev polynomials of the second kind written in terms of sinusoidal functions. When $|\mathrm{Re}\,a| > 1$, the eigen-values are a damped or increasing exponential function, representing an attenuating regime. Then equation (86) gives

$$\sigma_{1,2} = e^{\mp\varphi} \quad (94)$$

with $\cosh\varphi = \mathrm{Re}\,a$. Then the reflection and transmission responses are still given by equation (90) but with phase and amplitude factors now given by

$$\cos\alpha = \frac{\cosh M\varphi}{\sqrt{1+C^2}}, \; C = |b|\frac{\sinh M\varphi}{\sinh\varphi}, \quad (95)$$

For the limiting cases with $|\mathrm{Re}\,a| = 1$, there is a double root

$$\sigma_{1,2} = \mathrm{Re}\,a \quad (96)$$

and then we must use

$$\cos\alpha = \frac{1}{\sqrt{1+|b|^2 M^2}}, \; C = |b|\,M \tag{97}$$

Note, that in this case $|b|^2 = (\operatorname{Im} a)^2$. To compute expressions (93) and (95) for even number M we use (Gradshtein and Ryzhik, 1995, equation 1.382)

$$\cos\alpha = \frac{1}{\sqrt{1+C^2}} \prod_{k=1}^{\frac{M}{2}} \left(1 - \frac{1-(\operatorname{Re} a)^2}{\sin^2 \frac{(2k-1)\pi}{2M}} \right), \; C = |b|\,M(\operatorname{Re} a) \prod_{k=1}^{\frac{M-2}{2}} \left(1 - \frac{1-(\operatorname{Re} a)^2}{\sin^2 \frac{k\pi}{M}} \right) \tag{98}$$

The transmission response from equation (84) can be expressed via the complex phase factor $\beta(\omega)$:

$$t_D(\omega) = e^{i\beta(\omega)} \tag{99}$$

with

$$\beta(\omega) = \alpha(\omega) + \frac{1}{2} i \ln\left(1 + C^2\right). \tag{100}$$

The angular wavenumber is denoted k. With $kD = \operatorname{Re}\beta(\omega)$ or $\cos kD = \cos\alpha(\omega)$, the phase velocity is given by

$$v(\omega) = \omega/k = D\omega/\alpha(\omega). \tag{101}$$

Using notations from Carcione (2001), the dispersion equation can be written as $F(k,\omega) = \cos kD - \cos\alpha(\omega) = 0$, and the expression for group velocity is

$$V(\omega) = -\frac{\partial F/\partial k}{\partial F/\partial\omega} = \frac{D}{\partial\alpha(\omega)/\partial\omega}. \tag{102}$$

7.2 Equivalent time-average and effective medium

The behaviour of the reflection and transmission responses is determined by $\operatorname{Re} a$ which is one for $f = 0$. The boundaries between a propagating and attenuating regime are at $\operatorname{Re} a = -1$ (see equation (84)) given by the equation

$$\tan\frac{\theta_1}{2} \tan\frac{\theta_2}{2} = \frac{1 \mp |r|}{1 \pm |r|}. \tag{103}$$

For low frequencies the stack of the layers behaves as an effective medium with a velocity defined by (Backus, 1962; Hovem, 1995) and can be defined as the zero-frequency limit ($v_{EF} = v(\omega \to 0)$ from equation (101))

$$\frac{1}{v_{EF}^2} = \frac{1}{v_{TA}^2} + \frac{4h_1h_2}{h^2}\frac{1}{v_1v_2}\frac{r^2}{1-r^2}. \tag{104}$$

This occurs for frequencies below the first root of the equation $\operatorname{Re} a = -1$. For higher frequencies the stack of the layers is characterized by the time-average velocity defined by the infinite frequency limit ($v_{TA} = v(\omega \to \infty)$ from equation (101))

$$\frac{1}{v_{TA}} = \frac{1}{h_1 + h_2}\left(\frac{h_1}{v_1} + \frac{h_2}{v_2}\right). \tag{105}$$

This occurs for frequencies above the second root of the equation $\operatorname{Re} a = -1$. There is a transition zone between these two roots in which the stack of layers partly blocks the transmitted wave.

The behaviour of the medium is characterized by the ratio between wavelength and layer thickness. This is given by

$$\gamma = \frac{\lambda}{h} = \frac{v_{TA}}{f(h_1 + h_2)} = \frac{1}{f\Delta t}, \tag{106}$$

where Δt is the traveltime through the two single layers. To estimate the critical ratio of wavelength to layer thickness we assume $\Delta t_1 = \Delta t_2 = \Delta t/2$. The effective medium limit the occurs at

$$\gamma_1 = \pi\left(a\tan\sqrt{\frac{1-|r|}{1+|r|}}\right)^{-1}, \tag{107}$$

and the time-average limit the occurs at

$$\gamma_2 = \pi\left(\frac{\pi}{2} - a\tan\sqrt{\frac{1-|r|}{1+|r|}}\right)^{-1}, \tag{108}$$

For small values of $|r| \ll 1$ we obtain

$$\gamma_{1,2} = \pi\left(\frac{\pi}{4} \mp a\tan\frac{|r|}{2}\right)^{-1} \approx 4\left(1 \pm \frac{2|r|}{\pi}\right). \tag{109}$$

The transition between an effective medium to time average medium is schematically illustrated in Figure 12. Since the boundaries for the transmission zone in equation (103) are periodic functions of frequency, the low wavelength zone (high frequencies) is more complicated than shown in this figure.

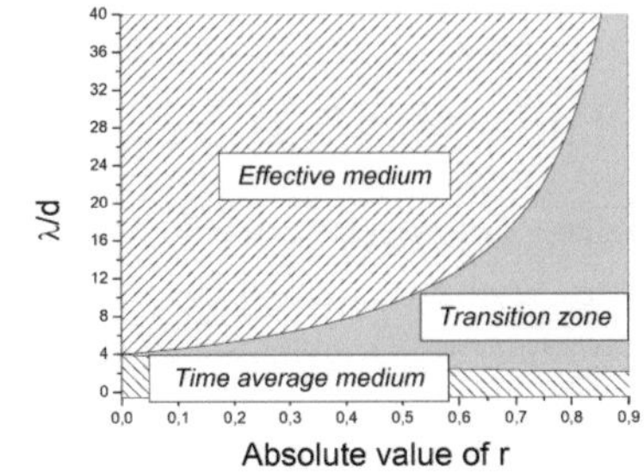

Fig. 12. Schematic representation of the critical $\gamma = \lambda / d$ ratio as function of reflection coefficient ($\theta_1 = \theta_2$) (Stovas&Ursin, 2007).

7.3 Reflection and transmission responses versus layering and layer contrast

We use a similar model as in Marion and Coudin (1992) with three different reflection coefficients: the original $r = 0.87$ and $r = 0.48$ and $r = 0.16$. We use m and Hz instead of mm and kHz. The total thickness of the layered medium is $D = M_k h_k = 51\ m$ is constant. M_k, $k = 1, 2, 4, ..., 64$ is the number of cycles in the layered medium, so that the individual layer thickness is decreasing as k is increasing. The ratio $\theta_1 / \theta_2 = \Delta t_1 / \Delta t_2 = (h_1 v_2)/(h_2 v_1) = 0.91$. The other model parameters are given in Stovas and Ursin (2007).

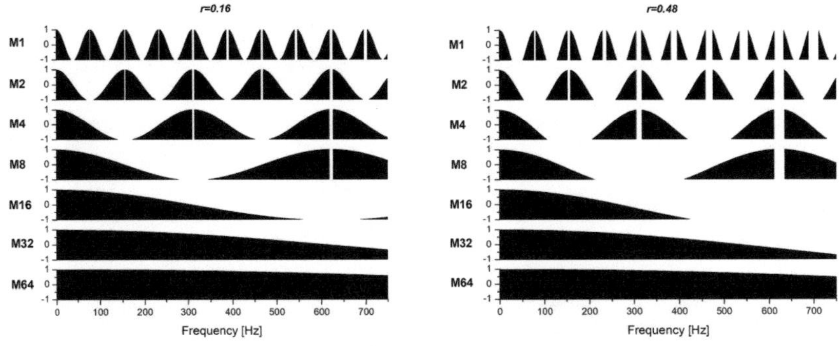

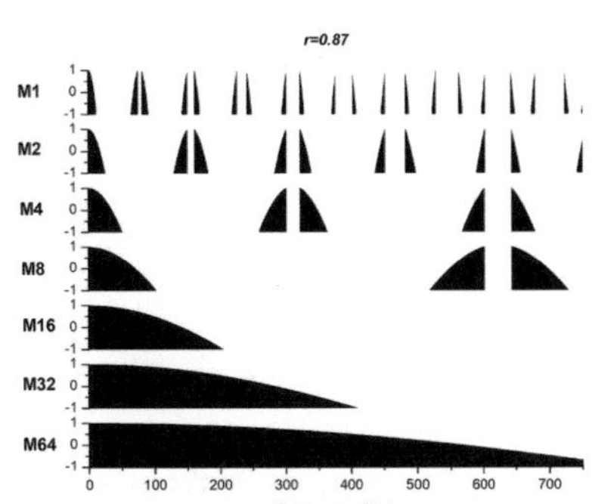

Fig. 13. $\mathrm{Re}\,a$ as function of frequency. $|\mathrm{Re}\,a| < 1$ is only plotted with area filled under the curve (Stovas&Ursin, 2007).

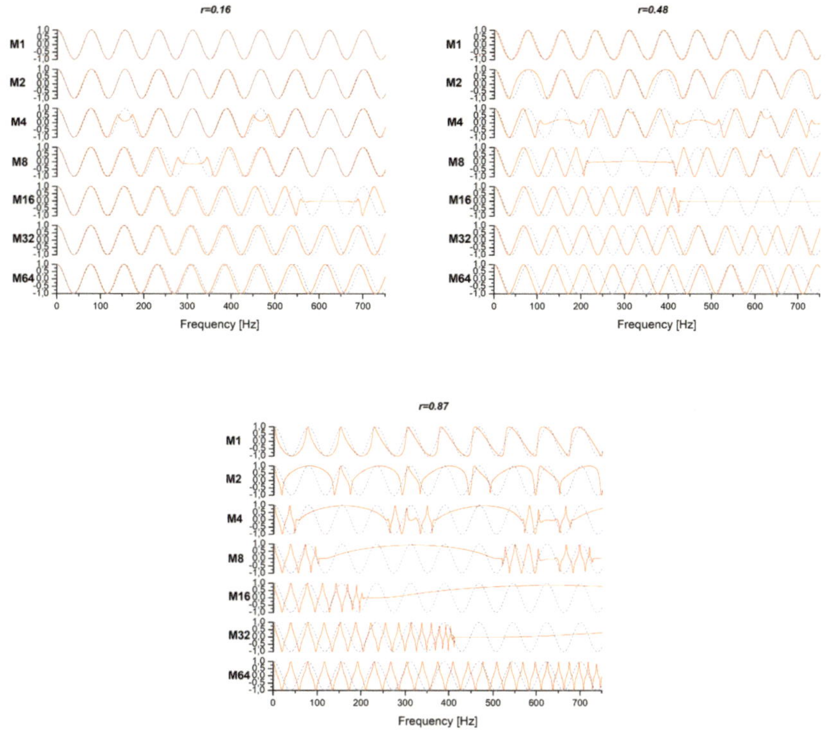

Fig. 14. The phase function $\cos\alpha$ as a function of frequency (from equation (98)) shown by solid line and $\cos\alpha$ corresponding to the single layer with time-average velocity shown by dashed line (Stovas&Ursin, 2007).

The very important parameter that controls the regime is $\mathrm{Re}\,a$ (equation (84)). The plots of $\mathrm{Re}\,a$ versus frequency are given in Figure 13 for different models. One can see that the propagating and attenuating regimes are periodically repeated in frequency. The higher reflectivity the more narrow frequency bands are related to propagating regime ($\mathrm{Re}\,a > -1$). One can also follow that the first effective medium zone is widening as the index of model increases and reflection coefficient decreases, and that the wavelength to layer thickness ratio γ is the parameter which controls the regime. The gaps between the propagating regime bands become larger with increase of reflection coefficient. These gaps correspond to the blocking or attenuating regime. The graphs for the phase factor $\cos\alpha$ and amplitude factor C (equation (98)) are shown in Figure 14 and 15, respectively. The dotted lines in Figure 14 correspond to the time-average phase behaviour. One can see when the computed phase becomes detached from the time-average phase. Note also the anomalous phase behaviour in transition zones. The amplitude factor C (Figure 15) has periodic structure, and periodicity increase with increase of reflection coefficient. In transition zones the amplitude factor reaches extremely large values which correspond to strong dampening. The transmission and reflection amplitudes are shown in Figure 16. The larger reflection coefficient the more frequently amplitudes change with frequency. The transition zones can be seen by attenuated values for transmission amplitudes. With increase of reflection

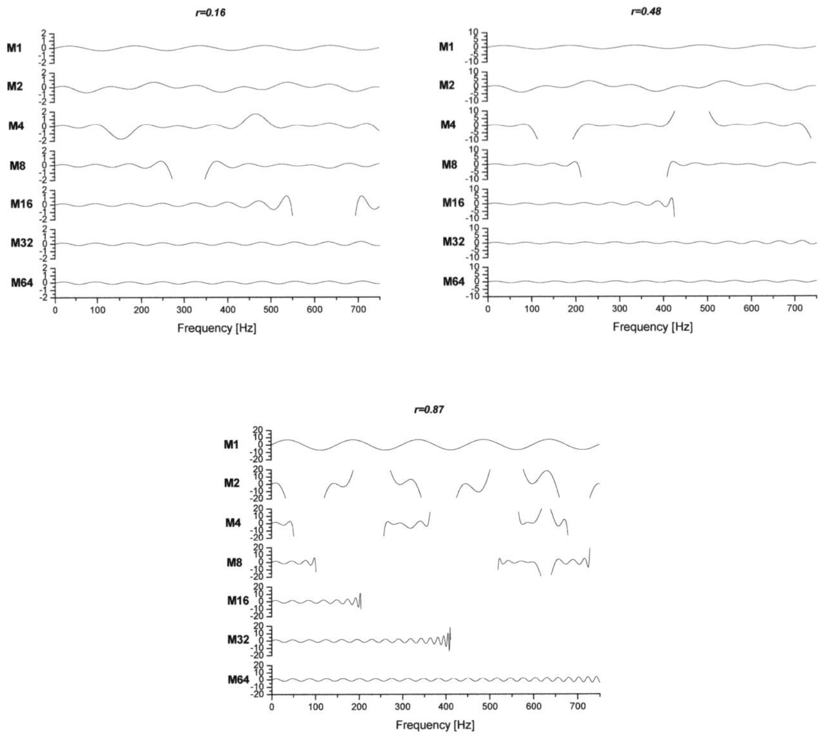

Fig. 15. The amplitude function C (from equation (98)) as a function of frequency (very large values of C are not shown) (Stovas&Ursin, 2007).

coefficient the dampening in transmission amplitudes becomes more dramatic. The exact transmission and reflection responses are computed using a layer recursive algorithm (Ursin and Stovas, 2002). We use a Ricker wavelet with a central frequency of 500 Hz. The transmission and reflection responses are shown in Figure 17 and 18, respectively. No amplitude scaling was used. One can see that these plots are strongly related to the behaviour of $\operatorname{Re} a$ (Figure 13). The upper seismogram in Figure 17 is similar to the Marion and Coudin (1992) experiment and the Hovem (1995) simulations. The effects related to the effective medium (difference between the first arrival traveltime for model M_1 and M_{64} and the transition between effective and time average medium, models $M_4 - M_{16}$) are the more pronounced for the high reflectivity model. For this model the first two traces (models M_1 and M_2) are composed of separate events, and then the events become more and more interferential as the thickness of the layers decrease. Model M_8 gives trains of nearly sinusoidal waves (tuning effect). The transmission response for model M_{16} is strongly attenuated, and models M_{32} and M_{64} behave as the effective medium. From Figures 13-16 and the transmission and reflection responses (Figures 17 and 18) one can distinguish between time average, effective medium and transition behaviour. This behaviour can be seen for any reflectivity, but a decrease in the reflection coefficient results in the convergence of the traveltimes for time-average and effective medium. This makes the effective medium arrival very close to the time average one. Note also that for the very much-pronounced

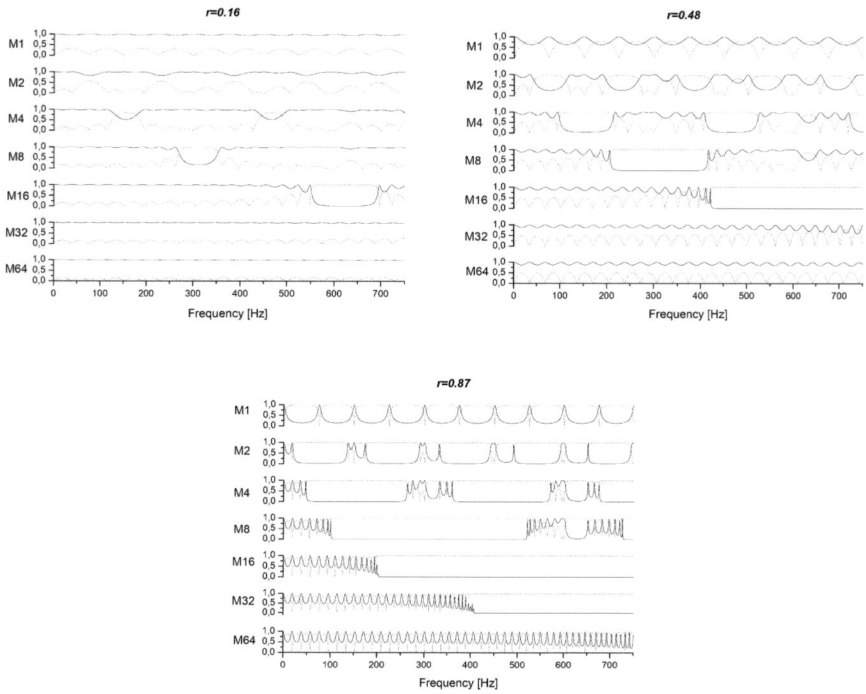

Fig. 16. The transmission amplitude (solid line) and reflection amplitude (dotted line) as function of frequency (Stovas&Ursin, 2007).

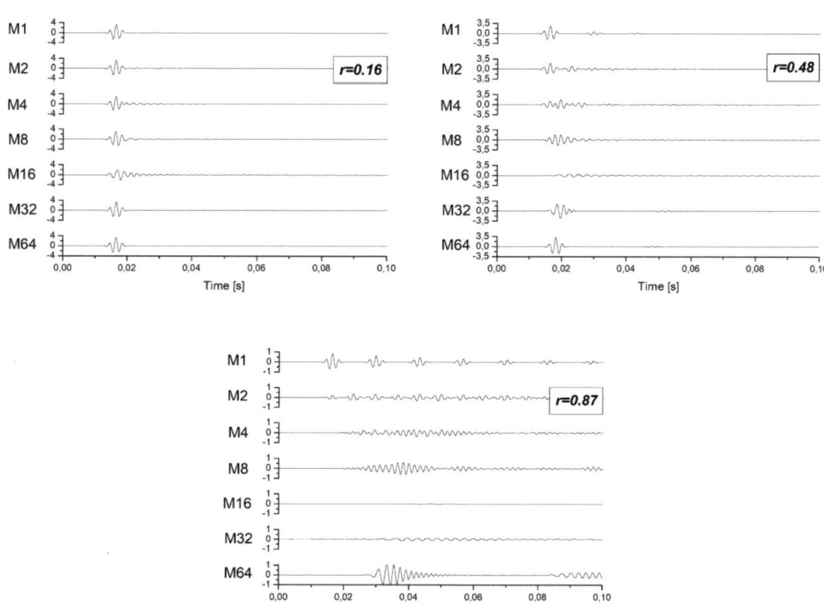

Fig. 17. Numerical simulations of the transmission response (Stovas&Ursin, 2007).

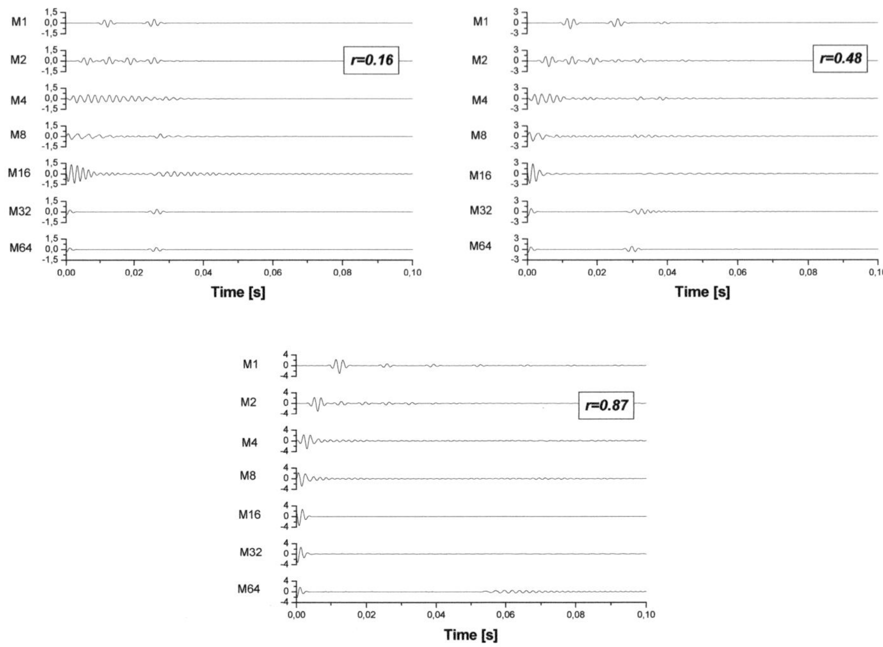

Fig. 18. Numerical simulations of the reflection response (Stovas&Ursin, 2007).

effective medium (model M_{64}, $r = 0.87$) one can see the effective medium multiple on the time about 0.085s. The reflection responses (Figure 18) demonstrate the same features as the transmission responses (for example strongly attenuated transmission response is related to weak attenuated reflection response). Effective medium is represented by the reflections from the bottom of the total stack of the layers. The phase velocities are computed from the phase factor (equation (100)) and shown in Figure 19 as a function of frequency. The phase velocity curve starts from the effective medium velocity and at the critical frequency it jumps up to the time-average velocity. One can see that the width of the transition zone is larger for larger values of reflection coefficient. The difference between the effective medium velocity and time-average velocity limits also increases with reflection coefficient increase.

In Figure 20, one can see time-average velocity, effective medium velocity, phase velocity (equation (100)) and group velocity (equation (102)) computed for reflection coefficients 0.16, 0.48 and 0,87 and model M_{64}. In this case we are in the effective medium zone. The larger reflection coefficient is, the lower effective medium velocity, the larger difference between the phase and group velocity and the velocity dispersion becomes more pronounced. The effective medium velocity limit also depends on the volume fraction ($\phi = h_2/(h_1 + h_2)$) for one cycle (see equation (104)). This dependence is illustrated in Figure 21 for different values of reflection coefficient. The maximum difference between the time-average and effective medium velocity reaches 2.191 km/s at $\phi = 0.18$ ($r = 0.87$), 0.513 km/s at $\phi = 0.24$ ($r = 0.48$) and 0.053 km/s at $\phi = 0.26$ ($r = 0.16$). One can see that for the small values of reflection coefficient the difference between the time-average velocity (high-frequency limit) and effective medium velocity (low frequency limit) becomes very small. For large values of reflection coefficient and certain range of volume fraction the effective medium velocity is

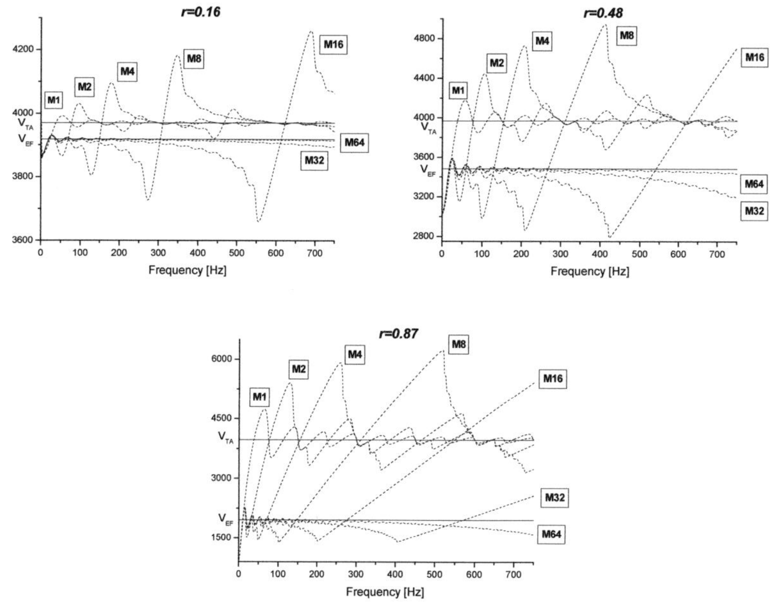

Fig. 19. The phase velocity (equation (101)) in m/s as function of frequency (Stovas&Ursin, 2007).

smaller than the minimum velocity of single layer constitutes. The effective medium is related to the propagating regime. The effective medium velocity depends on the reflection coefficient. The lesser contrast the higher effective medium velocity (the more close to the time-average velocity). One can also distinguish between effective medium, transition and time average frequency bands (Figure 22). These bands are separated by the frequencies given by conditions from the first two roots of equation $\mathrm{Re}\, a = -1$. The first root (equation (107)) gives the limiting frequency for effective medium. The second one (equation (108)) gives the limiting frequency for time average medium. From Figure 22, one can see that the transition zone converges to the limit $\gamma = 4$ with decreasing reflection coefficient. The first transition zone results in the most significant changes in the phase velocity and the amplitude factor C.

8. High-frequency caustics in periodically layered VTI media

The triplications (caustics) in a VTI medium can also be observed for high-frequencies. They are physically possible for qSV-wave propagation only. The qSV-wave triplications in a homogeneous transversely isotropic medium with vertical symmetry axis (VTI medium) have been discussed by many authors (Dellinger, 1991; Schoenberg and Daley, 2003; Thomsen and Dellinger, 2003; Vavrycuk, 2003; Tygel and et. 2007; Roganov, 2008). The condition for incipient triplication is given in Dellinger (1991) and Thomsen and Dellinger (2003). According to Musgrave (1970), we consider axial (on- axis vertical), basal (on-axis horizontal) and oblique (off-axis) triplications. He also provided the conditions for generation of all the types of triplications. The approximate condition for off-axis triplication is derived in Schoenberg and Daley (2003) and Vavrycuk (2003). The condition for on-axis triplication in multi-layered VTI medium is shown in Tygel et al. (2007).

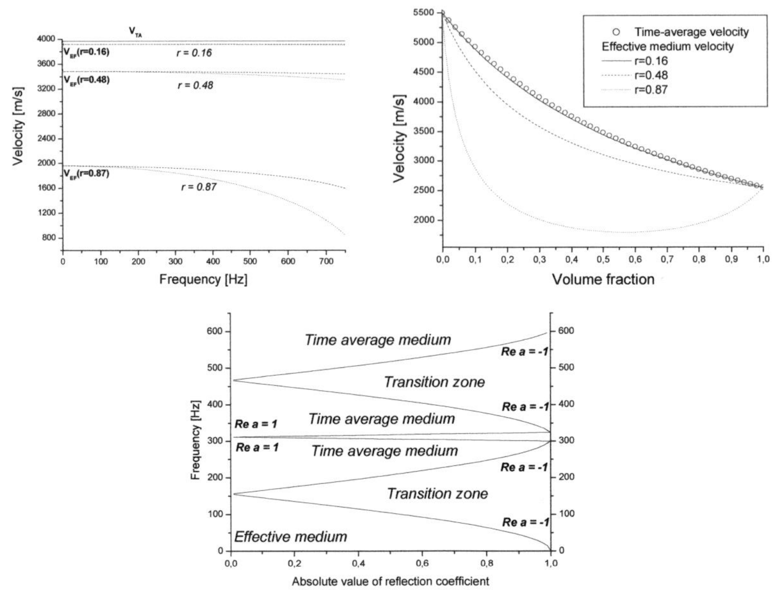

Fig. 20. The time average velocity (equation (105), solid line), the effective medium velocity (equation (104)), the phase velocity (equation (101), dashed line) and the group velocity (equation (102), dotted line) as function of frequency computed for model M_{64} and different reflection coefficients (Stovas&Ursin, 2007) shown to the left. The time-average velocity (equation (105)) and effective medium velocity (equation 104) as a function of volume fraction shown in the middle. Behaviour of the model M_4 as function of frequency and reflection coefficient shown to the right.

8.1 qSV- wave in a homogeneous VTI medium

In a homogeneous anisotropic medium, the plane wave with the slowness surface defined by $q = q(p)$ and the normal $(p,\ q)$ is given by

$$px + q\left(p\right)h = t\,. \tag{110}$$

The envelope of a family of plane waves given in (110) can be found by differentiation of equation (110) over the horizontal slowness p

$$x + q'\left(p\right)h = 0\,. \tag{111}$$

Equations (110) and (111) define the parametric offset-traveltime equations in a this medium at the depth h and can be written as follows

$$x\left(p\right) = -h\,q',\ t\left(p\right) = -h\left[pq' - q\right], \tag{112}$$

where $q' = dq/dp$ is the derivative of vertical slowness. The condition for triplication (caustic in the group space or concavity region on the slowness curve) is given by setting the curvature of the vertical slowness to zero, $q'' = 0$, or by setting the first derivative of offset to zero, $x' = 0$. These points at slowness surface have the curvature equal to zero. If the triplication region is degenerated to a single point, it is called incipient triplication. At that point we have $q''=0$ and $q'''=0$. For the incipient horizontal on-axis triplication, it is not true, because at that point q' equal to infinity, and we have to take the corresponding limit. The triplications can be also related to the parabolic points on the slowness surface with one of the principal curvatures being zero. The exception is the incipient vertical on-axis triplication with both principal curvatures equal to zero. The phase velocity can be written (Schoenberg and Daley, 2003) as a function of phase angle θ

$$v^2(\theta)=\frac{v_{0S}^2}{2g}f(\theta), \tag{113}$$

where v_{0S} is the vertical S-wave velocity, and function $f(\theta)$ is defined as

$$f(\theta)=f_{P(S)}(\theta)=1-eu+g\pm\sqrt{(1-eu-g)^2-E(1-u^2)}, \tag{114}$$

where + and - correspond to qP- and qSV-wave, respectively, $u=\cos 2\theta$, and parameters $e=\dfrac{c_{11}-c_{33}}{c_{11}+c_{33}}$, $g=\dfrac{2c_{55}}{c_{11}+c_{33}}$ and $E=\dfrac{4(c_{11}-c_{55})(c_{33}-c_{55})-4(c_{13}+c_{55})^2}{(c_{11}+c_{33})^2}$, which can be written in terms of Thomsen (1986) anisotropy parameters

$$e=\varepsilon/(1+\varepsilon),\ g=\gamma_0^2/(1+\varepsilon),\ E=2(\varepsilon-\delta)(1-e)(1-e-g) \tag{115}$$

and $\gamma_0=v_{0S}/v_{0P}$ is the S- to P-wave vertical velocity ratio. Note, that $0<g<1$ and $|e|<1-g$. Parameter E is also known as anelliptic parameter (Schoenberg and Daley, 2003) since it is proportional to the parameter σ, $\sigma=(\varepsilon-\delta)/\gamma_0^2=E/(2g(1-e-g))$, which is responsible for the anellipticity of the slowness surface and for the non-hyperbolicity of the traveltime equation. Vavrycuk (2003) used parameter σ to estimate the critical strength of anisotropy for the off-axis triplications in a homogeneous VTI medium. For the waves propagating for entire range of the phase angles, it is required that the function $f_s(u,E)$ and the expression under the square-root in equation (114)

$$s(u,E)=(1-eu-g)^2-E(1-u^2) \tag{116}$$

should be non-negative for all $|u|\le 1$. Solving equations $f_s(u,E)=0$ and $s(u,E)=0$ for the parameter E, results in the following explicit functions $E_f(u)=-\dfrac{4g(1-eu)}{1-u^2}$ and $E_s(u)=\dfrac{(1-eu-g)^2}{1-u^2}$, where the sub-indices f and s indicate the solutions for the equations $f=0$ and $s=0$, respectively. Function $E_f(u)$ defines the minimum plausible values for the parameter E in order to satisfy the condition $v^2(\theta)\ge 0$, while the function $E_s(u)$ defines the maximum plausible values for the parameter E in order to satisfy the condition $\operatorname{Im} v^2(\theta)=0$. By setting derivatives $dE_f/du=0$ and $dE_s/du=0$, we obtain that, for the range $|u|<1$, implicit functions $f_S(u,E)=0$ and $s(u,E)=0$ have the maximum and the minimum at the points A_1 (from $f=0$ and $dE_f/du=0$) and A_2 (from $s=0$ and $dE_s/du=0$), respectively (Figure 21). The coordinates for these points are

$$u_{A1}=(1-\sqrt{1-e^2})/e,\ E_{A1}=-2g(1+\sqrt{1-e^2}),\ u_{A2}=e/(1-g),\ E_{A2}=(1-g)^2-e^2. \quad (117)$$

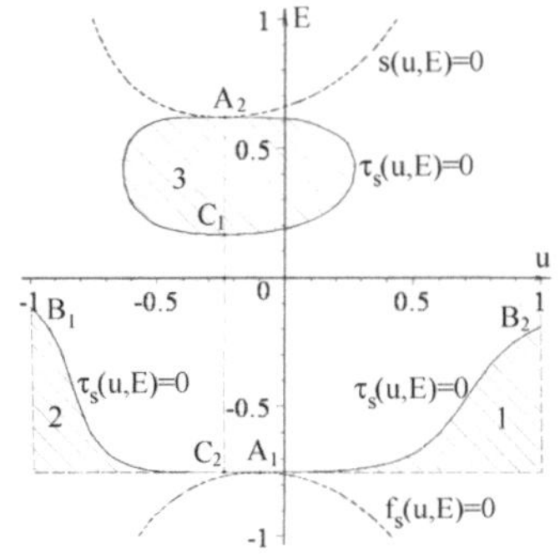

Fig. 21. Schematic plot of the triplication conditions on (u,E) space. The graphs for $f_S(u,E)=0$ and $s(u,E)=0$ are shown by dash line, and the graph for $\tau_S(u,E)=0$ is shown by solid line. Points A_1 and A_2 correspond to the extrema of the functions $f_S(u,E)=0$ and $s(u,E)=0$, respectively. Points B_1 and B_2 are the limiting points on the lower branch of $\tau_S(u,E)=0$ from the left and from right, respectively. Points A_2, C_1 and C_2 correspond to the extrema of the function $\tau_S(u,E)=0$ which has two branches limiting the triplication areas: 1 – vertical on-axis, 2 – horizontal on-axis and 3 – off-axis. The condition $E_{A1}\le E\le E_{A2}$ is a necessary and sufficient condition for existence of qP- and qSV-waves for entire range of the phase angles (Roganov&Stovas, 2010).

Therefore, the parameter E is limited as follows $E_{A1}\le E\le E_{A2}$. Note, that $u_{A1}=0$ if $e=0$. The equation for u_{A2} was shown in Schoenberg and Daley (2003). The lower limit yields the condition $v^2(\theta)\ge 0$, while the upper limit is related to the Thomsen's (1986) definition of parameter δ, i.e. $1+2\delta-\gamma_0^2\ge 0$. The range for parameter E yields a necessary and sufficient condition for the Christoffel matrix being positive definite for the entire range of the phase angles. Therefore, this condition is valid for all physically plausible medium parameters. Using equation (113) and

$$p=\sin\ \theta/V(\theta),\ q=\cos\ \theta/V(\theta),\ u=\cos 2\theta,$$

we obtain that for both types of waves we have the following equalities p

$$p=\frac{1}{v_{0S}}\sqrt{\frac{(1-u)g}{f}},\ q=\frac{1}{v_{0S}}\sqrt{\frac{(1+u)g}{f}}, \quad (118)$$

where $f=f_{P(S)}$ depending on the wave-mode (see equation (114)). The first and second derivatives of the vertical slowness are given by

$$\frac{dq}{dp} = \frac{p(-f + f_u^{/} \cdot (1+u))}{q(f + f_u^{/} \cdot (1-u))}, \tag{119}$$

$$\frac{d^2q}{dp^2} = \frac{2\tau g}{v_{0S}^2 q^3 \chi^3}, \tag{120}$$

where $f_u^{/} = df/du$ and

$$\chi = \chi_{P(S)} = (1-e-g)(1-eu-g) - E(1-u) \pm (1-e+g)\sqrt{s}, \quad \tau = \tau_{P(S)} = m \mp n\sqrt{s} \tag{121}$$

with s given in equation (116) and

$$\begin{aligned} m &= 2E\left[-e\left(e^2 - 2g + 2g^2\right)u^3 + 3e^2(1-g)u^2 - 3e\left(1-g^2\right)u + (1+g)(1-g)^2 + 2e^2 g\right] - 2\left(1-e^2-g^2\right)(1-eu-g)^3. \\ n &= -\left(1-u^2\right)E^2 - \left[\left(1+e^2+g^2-6g\right)u^2 - 4e(1-g)u + e^2 + (1+g)^2\right]E + 2\left(1+g^2-e^2\right)(1-eu-g)^2 \end{aligned} \tag{122}$$

For qSV and qP waves propagating in a homogeneous VTI medium, the triplication condition is given by (Roganov, 2008)

$$\tau_S(u, E) = 0, \tag{123}$$

$$\tau_P(u, E) = 0. \tag{124}$$

Equations (121)-(122) are too complicated to define the influence of parameters e and g on the form of the curves given by equations (123) and (124). Nevertheless, these equations can be used for numerical estimation of the position for triplications with any given values for e and g, as well as for the following theoretical analysis. It is well known that qP waves never have triplications in a homogeneous VTI medium (Musgrave, 1970; Dellinger, 1991; Vavrycuk, 2003), and, therefore, the equation (124) has no roots for propagating qP waves. By taking all the physically possible values for u, e, g and E, one can prove that if $s(u,E) \ge 0$, $|u| \le 1$, $0 < g < 1$ and $|e| < 1 - g$, the following inequality always takes place, $\tau_P(u,E) < 0$.
The product of $\tau_P(u,E)$ and $\tau_S(u,E)$

$$\tau(u,E) = \tau_P(u,E)\tau_S(u,E) = m^2 - sn^2 \tag{125}$$

given by polynomial with the sixth order in u and fifth order in E, can also be used to define the triplications for qSV-waves. For elliptical anisotropy, $E = 0$, we have the equalities $f_S(u,E) \equiv 2g$ and $V^2(\theta) \equiv c_{55}$. In this case, both the slowness and the group velocity surfaces are circles, and qSV-waves have no triplications. The straight line given by $E = 0$, separate the plane (u, E) into sub-planes with non-crossing branches of the curve $\tau_S(u,E) = 0$. One of these branches is located in the range of values $E_{A1} < E < 0$ and $|u| \le 1$ and is limited from the left and the right by the points B_1 and B_2, respectively. The coordinates of these points are (Figure 23)

$$u_{B1} = -1,\ E_{B1} = -g\left(1+e-g\right),\ u_{B2} = 1,\quad E_{B2} = -g\left(1-e-g\right), \tag{126}$$

while the second branch is closed and located in the range of values $0 < E < E_{A2}$. The limited values $E_{B1} < 0$ and $E_{B2} < 0$, that follows from inequalities $0 < g < 1$ and $|e| < 1 - g$. Under the lower branch of the curve $\tau_S(u,E) = 0$ on the plane (u,E), there are two parameter areas resulting in the on-axis triplications (both for vertical and horizontal axis), while the upper branch of the curve $\tau_S(u,E) = 0$ defines the parameter area for the off-axis triplications. In Figure 23, these areas are denoted by numbers 1, 2 and 3, respectively. Therefore, the on-axis triplications can simultaneously happen for both axis, if $E < 0$, while the off-axis triplications can exist only alone, if $E > 0$. Let us define the critical values for parameter E in (u,E) space with horizontal tangent line (E_{A1}, E_{A2}, E_{C1} and E_{C2} in Figure 23) that define the incipient triplications. In order to do so we need to solve the system of equations

$$\begin{cases} \tau_S(u,E) = 0 \\ \dfrac{\partial \tau_S(u,E)}{\partial u} = 0 \end{cases}. \tag{127}$$

For the range of parameters $E_{A1} \le E \le E_{A2}$ and $|u| \le 1$, system of equations (127) has four solutions, where the first two solutions $\{u_{A1},\ E_{A1}\}$ and $\{u_{A2},\ E_{A2}\}$ are defined above. For the second two solutions $\{u_{C1},\ E_{C1}\}$ and $\{u_{C2},\ E_{C2}\}$ we have $u_{C1} = u_{C2} = u_{A2}$, while E_{C1} and E_{C2} are the largest (always positive) and intermediate (always negative) roots of the cubic equation

$$t\left(E\right) = E^3 + \left(3g^2 - 3e^2 + 2g + 3\right)E^2 + 8g\left(1+g^2-e^2\right)E - 16g^2\left(1-e^2\right)\ \left(\left(1-g\right)^2 - e^2\right) = 0. \tag{128}$$

Equations similar to our equation (128) are derived in (Peyton, 1983; Schoenberg and Helbig, 1997; Thomsen and Dellinger, 2003; Vavrycuk, 2003; Roganov, 2008). To prove that equation $t(E) = 0$ gives the positions for the critical points of the curve $\tau(u,E) = 0$ one can substitute $u = u_{A2}$. Consequently, we have

$$\tau(u_{A2},E) = \frac{\left(\left(1-g\right)^2 - e^2\right)}{\left(1-g\right)^2} t(E) p^2(u_{A2},\ E) \tag{129}$$

and

$$\frac{\partial \tau(u_{A2},E)}{\partial u} = \frac{6e(1-g)}{(1-g)^2 - e^2}\tau(u_{A2},E). \tag{130}$$

Therefore, if $t(E) = 0$ and $u = u_{A2}$, then system of equations (127) is obeyed. The least root of equation (128) is located in the area defined by $E < -1$ and holds also equation $\tau_P(u,E) = 0$. It is located outside the region $E_{A1} \le E \le E_{A2}$ and does not define any qSV wave triplication. The intermediate root defines the critical point C_2 on the lower branch of the curve (123). The minimum point A_1 (dividing the triplication domain 1 and 2) is also located on the same branch. The largest root corresponds to the minimum point C_1 on the upper branch of the curve (123). If $E_{C1} < E < E_{A1}$, we have off-axis triplications. The coordinates E_{C1} and E_{C2} are given by equations

$$E_{C1} = 2Q\sin\left(2\psi/3 + \pi/6\right) - d, \quad E_{C2} = 2Q\sin\left(2\psi/3 + 5\pi/6\right) - d, \tag{131}$$

where

$$d = 1 + 2g/3 + g^2 - e^2, \quad Q = \left|1 - e^2 - 2g/3 + g^2\right|, \quad \psi = \arccos\frac{\sqrt{[(1-g)^2 - e^2](1 - e^2 - g^2)}}{Q^{3/2}}. \tag{132}$$

The case for the qSV-wave vertical on-axis incipient triplication can be obtained by setting $u = 1$ ($\theta = 0$) with condition (120) being simplified to $E = E_{B2}$ or $\sigma = -0.5$ or $v_{nmo}^2 = 0$ (Tygel et al., 2007). The case for the qSV-wave horizontal on-axis incipient triplication can be obtained by setting $u = -1$ ($\theta = \pi/2$) with condition (118) being simplified to $E = E_{B1}$ or $\sigma = -\left(1 + 2\varepsilon - \gamma_0^2\right)/2\left(1 - \gamma_0^2\right)$. If $E_{A1} < E < \min(E_{B1}, E_{B2})$, we have both on-axis triplications. If $e = 0$ (or $\varepsilon = 0$), then we have the following equality $E_{B1} = E_{B2}$, and, therefore, both on-axis triplications are incipient.

8.2 Extension of qSV-wave triplications for multilayered case

From the ray theory it follows that for any vertically heterogeneous medium including horizontally layered medium, kinematically effective vertical slowness is always the average of the vertical slownesses from the individual layers. We have to stress that our approach is based on the high-frequency limit of the wave propagation, not on the low-frequency one which results in effective medium averaging. Since the wave propagates through the layered medium with the same horizontal slowness p, the effective vertical slowness has very simple form

$$\tilde{q} = \langle q \rangle, \tag{133}$$

where $\langle\ \rangle$ denotes the arithmetic thickness averaging, $\langle m \rangle = \sum m_i h_i / \sum h_i$, with $h_i, i = 1, N$ being the thickness of layer i in the stack of N layers. With notation (133), equations (112) are valid for the multilayered case. Similar approach is used in Stovas (2009) for a vertically heterogeneous isotropic medium. If a layered VTI medium results in more then one caustic, there is no any kinematically effective VTI medium given in equation (133), which can reproduce the same number of caustics. This statement follows from the fact that a homogeneous VTI medium might have only one off-axis triplication. Therefore, the second derivative of the effective vertical slowness is given by

$$\frac{d^2\tilde{q}}{dp^2} = \left\langle \frac{2\tau\ g}{v_{0S}^2 q^3 \chi^3} \right\rangle. \tag{134}$$

With equation (134) the condition for off-axis, vertical on-axis and horizontal on-axis triplications in multi-layered VTI medium takes the form (Roganov and Stovas, 2010)

$$\left\langle \frac{\tau_S g}{v_{0S}^2 q_S^3 \chi_S^3} \right\rangle = 0 . \tag{135}$$

To obtain the condition for incipient vertical triplications, we have to substitute $u = 1$ и $p = 0$ into equation (135). After some algebraic manipulations, we obtain

$$\left. \frac{d^2 \tilde{q}}{dp^2} \right|_{u=1, p=0} = -\left\langle \frac{v_{0S}(E_{B2} - E)}{E_{B2}} \right\rangle = 0 . \tag{136}$$

Similar equation can be derived by using the traveltime parameters. Tygel et al. (2007) shown that the vertical on-axis triplications in the multilayered VTI medium are defined by the normal moveout velocity (representing the curvature of the traveltime curve $t(x)$ taken at zero offset): $\tilde{v}_{nmo}^2 = 0$, where $\tilde{v}_{nmo}^2 = \langle v_{nmo}^2 v_{0S}^{-1} \rangle / \langle v_{0S}^{-1} \rangle$ is the overall normal moveout velocity squared. In order to use equation (135), the function $u(p)$ has to be defined in terms of horizontal slowness for each layer

$$u(p) = u_s(p) = \frac{a + p^2 v_{0S}^2 \sqrt{b}}{c} , \tag{137}$$

where

$$\begin{aligned} a &= g^2 - g(1+e+g) p^2 v_{0S}^2 + 2egp^4 v_{0S}^4 \\ b &= g^2 (1-e-g)^2 + 2g\left[2eg(1-e-g) - (1-e+g)E\right] p^2 v_{0S}^2 + \left(E^2 + 4Eg + 4e^2 g^2\right) p^4 v_{0S}^4 . \\ c &= g^2 - 2egp^2 v_{0S}^2 - Ep^4 v_{0S}^4 \end{aligned} \tag{138}$$

Function $b(p) > 0$ if $E > 0$. We are going to prove that the function $b = b(p)$ from equation (138) is positive for all physically plausible parameters e and g, if anelliptic parameter $E > 0$. Solving bi-quadratic equation $b(p) = 0$ yields

$$p_{1,2,3,4} = \pm \frac{1}{v_{0,S}} \sqrt{\frac{Eg(1-e+g) - 2eg^2(1-e-g) \pm 2\sqrt{E(1-e)\left(e^2 + E - (1-g)^2\right)}}{E^2 + 4Eg + 4e^2 g^2}} \tag{139}$$

The expression under the inner square root in equation (139) can be written as

$$e^2 + E - (1-g)^2 = -4\left(1-\gamma_0^2\right)\left(1+2\delta-\gamma_0^2\right) \tag{140}$$

Note, that $1+2\delta-\gamma_0^2 \geq 0$ (it follows from Thomsen's (1986) definition of parameter δ). Taking into account that $e < 1$, and $b(p=0) = g^2(1-e-g)^2 > 0$ and $b(p = 1/v_{0S}) = (1-e)^2 (1+e-g)^2 > 0$, one can see that if $E > 0$, the expression under the square root in equation (139) is negative, and the equation $b(p) = 0$ has no roots. Function $c = c(p)$ can take zero value at

$$p = \tilde{p} = \pm \frac{1}{v_{0S}} \sqrt{\frac{g}{E}\left(-e + \sqrt{e^2 + E}\right)} \tag{141}$$

To compute $u = u(\tilde{p})$ from equation (137) we need to take the limit given by

$$\lim_{p \to \tilde{p}} u(p) = u(\tilde{p}) = \frac{4e^2 g + E\left(e + e^2 + 2g + eg\right) + E^2 - \sqrt{e^2 + E}\left(4eg + E\left(1 + e + g\right)\right)}{4e^3 g + E\left(e + e^2 + 3eg\right) + E^2 - \sqrt{e^2 + E}\left(4e^2 g + E\left(1 + e + g\right)\right)}. \tag{141a}$$

If $c(p) = 0$, that happens at

$$p = \tilde{p} = \frac{1}{v_{0S}} \sqrt{\frac{g}{E}\left(-e + \sqrt{e^2 + E}\right)}, \tag{142}$$

function $u(p)$ takes the value

$$u(\tilde{p}) = \frac{4e^2 g + E\left(e + e^2 + 2g + eg\right) + E^2 - \sqrt{e^2 + E}\left(4eg + E\left(1 + e + g\right)\right)}{4e^3 g + E\left(e + e^2 + 3eg\right) + E^2 - \sqrt{e^2 + E}\left(4e^2 g + E\left(1 + e + g\right)\right)}. \tag{143}$$

Note that in the presence of on-axis triplication (for the horizontal axis), function $u(p)$ has two branches when $|p| > 1/v_{S0}$, and the second branch is defined by $u(p) = u_p(p) = \left(a - p^2 v_{0S}^2 \sqrt{b}\right)/c$. The incipient off-axis triplication condition in a multi-layered medium is given by equation (Roganov and Stovas, 2010)

$$\left\langle \frac{g}{v_{0S}} \frac{d\left(\dfrac{\tau_S}{q_S^3 \chi_S^3}\right)}{dp} \right\rangle = 0. \tag{144}$$

Functions q_S and τ_S, χ_S defined in equations (118) and (121), respectively, are given in terms of u. To compute the derivatives in equation above one need to exploit equation (117) for $u = u(p)$ and apply the chain rule, i.e. $dq_S/dp = (dq_S/du)(du/dp)$. For a given model this equation can be resolved for horizontal slowness and used to estimate the limits for the vertical slowness approximation or traveltime approximation. For multilayered case, the parametric offset-traveltime equations (112) take the following form

$$\tilde{x}(p) = -H\,\tilde{q}', \quad \tilde{t}(p) = -H\left[p\tilde{q}' - \tilde{q}\right], \tag{145}$$

where $H = \sum h_i$ is the total thickness of the stack of layers.

8.3 Converted wave case

In the special case of converted qP-qSV waves (C-waves) in a homogeneous VTI medium, the condition (113) reduces to

$$\frac{\tau_S}{q_S^3 \chi_S^3} + \frac{\tau_P}{q_P^3 \chi_P^3} = 0 . \qquad (146)$$

To compute functions τ_P, q_P and χ_P we need to define $u_P(p)$ which can be computed similar to equation (117)

$$u_P(p) = \frac{a - p^2 v_{0S}^2 \sqrt{b}}{c}, \qquad (147)$$

where functions a, b and c can be computed from equation (138). One can show that for the range of horizontal slowness corresponding to propagating qP-wave, the sum

$$\frac{\tau_S}{q_S^3 \chi_S^3} + \frac{\tau_P}{q_P^3 \chi_P^3} < 0 , \qquad (148)$$

which means that the converted qP-qSV waves in a homogeneous VTI medium have no triplications. In Figure 22 one can see the functions $2\tau_S / q_S^3 \chi_S^3$ (controlling the triplications for qSV-wave), $2\tau_P / q_P^3 \chi_P^3$ (controlling the triplications for qP-wave) and $\tau_S / q_S^3 \chi_S^3 + \tau_P / q_P^3 \chi_P^3$ (controlling the triplications for converted waves). The model parameters are taken from the case 1 model 1. One can see that the only function crossing the u – axis is the qSV-wave related one.

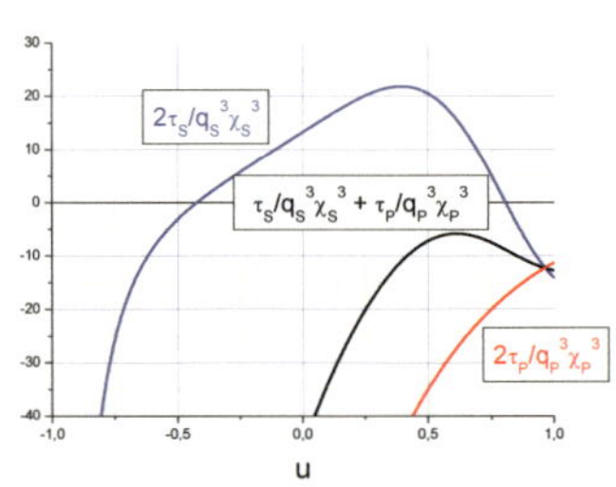

Fig. 22. The functions controlled the qP- (red line), qSV- (blue line) and qPqSV-wave (black line) triplications. The data are taken from the case 1 model 1 (Roganov&Stovas, 2010).

8.4 Single-layer caustics versus multi-layer caustics

For our numerical tests we consider the off-axis triplications only, because the vertical on-axis triplications were discussed in details in Tygel et. al (2007), while the horizontal on-axis triplications have only theoretical implications.

First we illustrate the transition from the vertical on-axis triplication to the off-axis triplication by changing the values for parameter E only, $E = -0.3, -0.2, \ldots, 0.5$. Since the other parameters remain constant, this change corresponds to the changing in Thomsen's

(1986) parameter δ. The slowness surfaces, the curvature of the slowness surfaces and the traveltime curves are shown in Figure 23. One can see how the anomaly in the curvature moves from zero slowness to non-zero one.

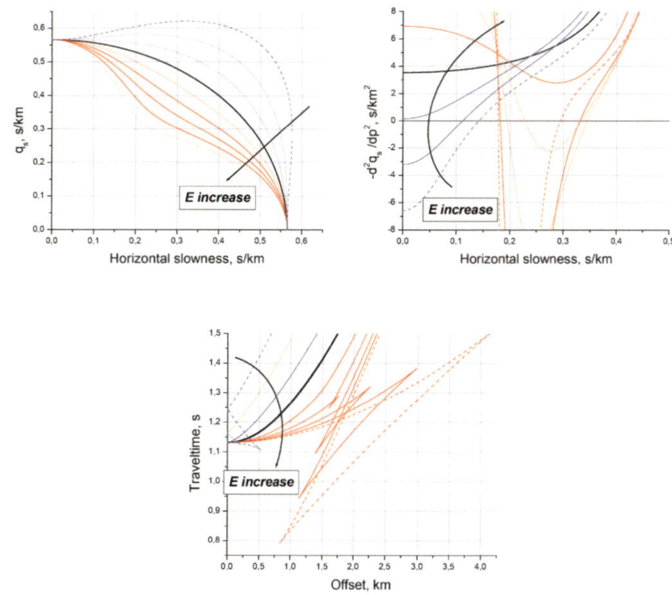

Fig. 23. The slowness surface (to the left), the curvature of slowness surface (in the middle) and the traveltime versus offset (to the right) from the homogeneous VTI media with change in parameter E only. The model parameters are taken from the model 1 in Table 1. Parameter E takes the values -0.3, -0.2,..., 0.5. The curves with positive and negative values for E are shown with red lines and blue lines, respectively. The elliptically isotropic case, $E = 0$, is shown by black line (Roganov&Stovas, 2010).

Next we test the qSV-wave slowness-surface approximations from Stovas and Roganov (2009). The slowness-surface approximations for qSV waves (similar to acoustic approximation for qP waves) are used for processing (in particular, phase-shift migration) and modeling purpose with reduced number of medium parameters. With that respect, it is important to know how the slowness-surface approximations reproduce the triplications. We notice that if the triplication is located for short offset, it can partly be shown up by approximation 1 (short spread approximation). The wide-angle approximations 2 and 3 can not treat the triplications.

In the numerical examples provided in Roganov and Stovas (2010), we considered four cases with two layer models when each layer has parameters resulting in triplication for qSV-wave. With changing the fraction ratio from 0 to 1 with the step of 0.1, we can see the transition between two different triplications for cases 1-4. For given numbers of the fraction ratio we can observe the different cases for two-layer triplications. For the overall propagation we can have no triplication (case 1), one triplication (case 2), two triplications (case 3) and one "pentaplication" or two overlapped triplications (case 4). Intuitively, we can say that the most complicated caustic from N VTI layers can be composed from N

overlapped triplications or one "(2*N*+1)-plication". The examples shown in Roganov and Stovas (2010) provide the complete set of situations for off-axis triplications in two-layer VTI media and give a clue what we can expect to see from multilayered VTI media.

9. Phase velocity approximation in finely layered sediments

The effect of multiple scattering in finely layered sediments is of importance for stratigraphic interpretation, matching of well log-data with seismic data and seismic modelling. This problem was first studied in the now classical paper by O'Doherty and Anstey (1971) and further investigated by Shapiro and Treitel (1997). In this paper I derive a new approximation for the phase velocity in an effective medium which depends on three parameters only and show how it depends on the strength of the reflection coefficients (Stovas, 2007). Approximation is tested on the real well log data example and shows very good performance.

9.1 Vertical propagation through the stack of the layers

The transmission and reflection responses of normal-incident plane wave from the stack of N layers are given by the following expressions (Stovas and Arntsen, 2006)

$$t_D^{(N)} = \frac{e^{i\vartheta_N}\prod_{k=1}^{N}(1-r_k)}{[1+\Phi]}, \quad r_D^{(N)} = \frac{e^{2i\vartheta_N}\left(\sum_{j=1}^{N} r_j e^{-2i(\vartheta_N-\vartheta_j)} + \ldots\right)}{[1+\Phi]}, \tag{149}$$

where r_k are reflection coefficients, the cumulated phases $\vartheta_i = \omega\sum_{i=1}^{N}\tau_i = \omega\sum_{i=1}^{N} h_i/V_i$, with h_j and V_j are thickness and velocity in the layer j, respectively, and the reflection coefficient correlation function

$$\Phi = \sum_{k=1}^{N-1}\sum_{j=k+1}^{N} r_k r_j e^{2i(\vartheta_j-\vartheta_k)} + \ldots \tag{150}$$

The exponential factors in denominators for transmission and reflection response are the phase delays for direct wave, the product function in transmission response gives the direct transmission loss and the sum function in reflection response corresponds to contributions from the primary reflections (first order term) and interbedded multiples (higher order terms). The phase velocity is given by

$$\frac{1}{V(\omega)} = \frac{1}{V_{TA}} - \frac{1}{\omega D} a\tan\frac{\operatorname{Im}\Phi}{1+\operatorname{Re}\Phi} = \frac{1}{V_{TA}} - \frac{1}{\omega D} a\tan\frac{\sum_{k=1}^{N-1}\sum_{j=k+1}^{N} r_k r_j \sin 2(\vartheta_j-\vartheta_k)+\ldots}{1+\sum_{k=1}^{N-1}\sum_{j=k+1}^{N} r_k r_j \cos 2(\vartheta_j-\vartheta_k)+\ldots}, \tag{151}$$

where D is the total thickness of the stack and $V_{TA} = \omega D/\vartheta_N$ is time-average velocity. The velocity in zero-frequency limit is given by (Stovas and Arntsen, 2006)

$$\frac{1}{V_0} = \lim_{\omega \to 0} \frac{1}{V(\omega)} = \frac{1}{V_{TA}} - \frac{2}{D} \frac{\sum_{k=1}^{N-1} \sum_{j=k+1}^{N} r_k r_j (\tau_j - \tau_k) + \dots}{1 + \sum_{k=1}^{N-1} \sum_{j=k+1}^{N} r_k r_j + \dots} . \quad (152)$$

9.2 Weak-contrast approximation

The weak-contrast approximation means that we neglect the higher order terms in the scattering function Φ (equation 150),

$$\Phi = \sum_{k=1}^{N-1} \sum_{j=k+1}^{N} r_k r_j e^{2i(\vartheta_j - \vartheta_k)} . \quad (153)$$

This function can be expanded into Taylor series

$$\Phi = \sum_{n=0}^{\infty} \frac{(2i\omega)^n}{n!} u_n \quad (154)$$

with coefficients

$$u_n = \sum_{k=1}^{N-1} \sum_{j=k+1}^{N} r_k r_j (\tau_j - \tau_k)^n , \quad (155)$$

which can be considered as correlation moments for reflection coefficients series. To approximate equation (155) we use

$$u_n = u_0 \tau_N^n e^{-\alpha n}, \quad n = 0, 1, 2, \dots , \quad (156)$$

where $\tau_N = \vartheta_N / \omega$ is total one-way propagation time and parameter α will be explained later. The form of approximation (156) has been chosen due to the exponential nature of the reflection coefficient correlation moments (O'Doherty and Anstey, 1971), and the term τ_N^n is introduced simply to preserve the dimension for u_n. Substituting (156) into (154) results in

$$\Phi = u_0 \sum_{n=0}^{\infty} \frac{(2i\omega\tau_N)^n}{n!} e^{-\alpha n} . \quad (157)$$

Equation (151) in weak contrast approximation is reduced to (Stovas, 2007)

$$\frac{1}{V(\omega)} = \frac{1}{V_{TA}} - \frac{1}{\omega D} \operatorname{Im} \Phi = \frac{1}{V_{TA}} \left[1 - \beta \left(1 + S(\omega) \right) \right] \quad (158)$$

with $\beta = 2u_0 e^{-\alpha} = 2u_1/\tau_N$ and $S(\omega) = \sum_{n=1}^{\infty} \frac{(-1)^n (2\omega\tau_N)^{2n}}{(2n+1)!} e^{-2n\alpha}$, where u_0 being considered as the zero-order auto-correlation moment for reflection coefficients series $u_0 = \sum_{k=1}^{N-1} \sum_{j=k+1}^{N} r_k r_j$ and α is the parameter in correlation moments approximation. For practical use we need the limited number of terms M in equation (160). The zero-frequency limit from equation (152) is given by $V_0^{-1} = V_{TA}^{-1} - 2u_1/D = (1-\beta)/V_{TA}$.
Substituting this limit into equation (158) we obtain

$$\frac{1}{V(\omega)} = \frac{1}{V_0}\left[1 - \frac{\beta}{1-\beta} S(\omega)\right]. \tag{163}$$

Parameter $\beta \le 0$, therefore, describes the relation between two limits $\beta = 1 - V_{TA}/V_0$ and function $S(\omega)$ can be interpreted as the normalized relative change in the phase slowness $-S(\omega) = \left(V_0^{-1} - V^{-1}\right)/\left(V_0^{-1} - V_{TA}^{-1}\right)$.
The phase velocity approximation is described by three parameters only: one-way propagation time τ_N; 2) parameter β which is ratio of low and high frequency velocity limits; 3) parameter α which describes the structure of the stack.

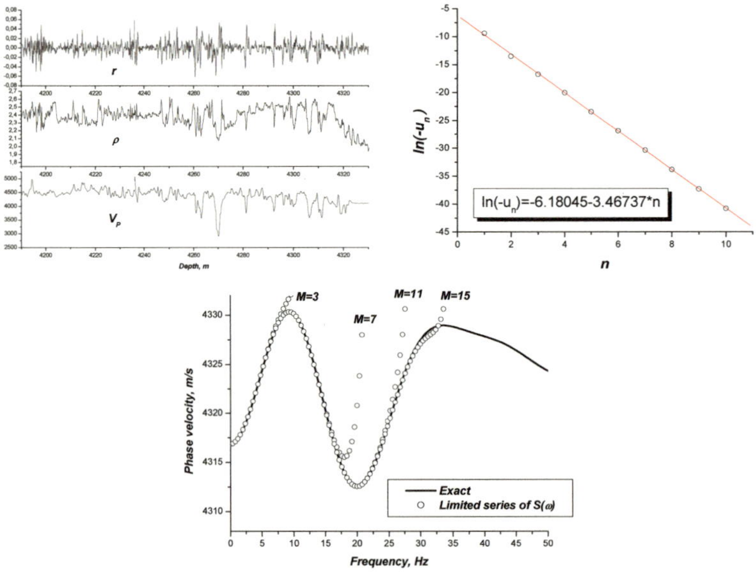

Fig. 24. Elastic parameters and reflection coefficients for Tilje formation (to the left), the correlation moments approximation (in the middle) and the phase velocity and its approximations computed from limited series of $S(\omega)$. (Stovas, 2007).

For numerical application we use 140m of the real well-log data sampled in 0.125m (Figure 24). This interval related to the Tilje formation from the North Sea. In Figure 24, we also show how to compute parameters for approximation (156). The one way traveltime is

$\tau_N = 0.0323\,s$, $\beta = -0.04$ and $\alpha = 0.03468$. In particular it means that the time-average velocity is only 4% higher than the zero frequency limit. The results of using this approximation with the limited number of terms (M = 3, 7, 11 and 15) in equation (157) are shown in Figure 24. The exact phase velocity function is obtained from the transmission response computed by the matrix propagator method (Stovas and Arntsen, 2006). One can see that with increase of M the quality of approximation increases with frequency.

10. Estimation of fuid saturation in finely layered reservoir

The theory of reflection and transmission response from a stack of periodically layered sediments can be used for inversion of seismic data in turbidite reservoirs. In this case, the model consists of sand and shale layers with quasi-periodical structure. The key parameters we invert for are the net-to-gross ratio (the fractural amount of sand) and the fluid saturation in sand. The seismic data are decomposed into the AVO (amplitude versus offset) or AVA (amplitude versus incident angle) attributes. The following notations are used: AVO intercept is the normal reflectivity and AVO gradient is the first order term in Taylor series expansion of reflectivity with respect to sine squared of incident angle.

For simultaneous estimation of net-to-gross and fluid saturation we can use the PP AVO parameters (Stovas, Landro and Avseth, 2006). To model the effect of water saturation we use the Gassmann model (Gassmann, 1951). Another way of doing that is to apply the poroelastic Backus averaging based on the Biot model (Gelinsky and Shapiro, 1997). Both net-to-gross and water saturation can be estimated from the cross-plot of AVO parameters. This method is applied on the seismic data set from offshore Brazil. To build the AVO cross-plot for the interface between the overlaying shale and the turbidite channel we used the rock physics data. These data were estimated from well logs. The AVO cross-plot contains the contour lines for intercept and gradient plotted versus net-to-gross and water saturation. The discrimination between the AVO attributes depends on the discrimination angle (angle between the contour lines, see Stovas and Landrø, 2004).One can see that the best discrimination is observed for high values of net-to-gross and water saturation, while the worst discrimination is for low net-to-gross and water saturation (where the contour lines are almost parallel each other). Note, that the inversion is performed in the diagonal band of AVO attributes. Zones outside from this band relate to the values which are outside the chosen sand/shale model. Our idea is that the top reservoir reflection should give relatively high values for net-to-gross regardless to water saturation values. The arbitrary reflection should give either low values for net-to-gross with large uncertainties in water saturation or both net-to-gross and saturation values outside the range for the chosen model. The data outside the diagonal band are considered as a noise. To calibrate them we use well-log data from the well. The P-wave velocity, density and gamma ray logs are taken from the well-log. One can say that the variations in the sand properties are higher than we tested in the randomization model. Nevertheless, the range of variations affects more on the applicability of the Backus averaging (which is weak contrast approximation) than the value for the Backus statistics. The AVO attributes were picked from the AVO sections (intercept and gradient), calibrated to the well logs and then placed on the cross-plot. One might therefore argue that the AVO-attributes themselves can be used as a hydrocarbon indicator, and this is of course being used by the industry. However, the attractiveness of the proposed method is that we convert the two AVO-attributes directly into net-to-gross and saturation attributes, in a fully deterministic way. Furthermore the results are quantitative, given the limitations and simplifications in the model being used.

11. Seismic attributes from ultra-thin reservoir

Here we propose the method of computation seismic AVO attributes (intercept and gradient) from ultra-thin geological model based on the SBED modelling software (Stovas, Landro and Janbo, 2007). The SBED software is based on manipulating sine-functions, creating surfaces representing incremental sedimentation. Displacement of the surfaces creates a three dimensional image mimicking bedform migration, and depositional environments as diverse as tidal channels and mass flows can be accurately recreated. The resulting modelled deposit volume may be populated with petrophysical information, creating intrinsic properties such as porosity and permeability (both vertical and horizontal). The Backus averaging technique is used for up-scaling within the centimetre scale (the intrinsic net-to-gross value controls the acoustic properties of the ultra-thin layers). It results in pseudo-log data including the intrinsic anisotropy parameters. The synthetic seismic modelling is given by the matrix propagator method allows us to take into account all pure mode multiples, and resulting AVO attributes become frequency dependent. Within this ultra-thin model we can test different fluid saturation scenarios and quantify the likelihood of possible composite analogues. This modelling can also be used for inversion of real seismic data into net-to-gross and fluid saturation for ultra-thin reservoirs.

11.1 SBED model

The SBED software is based on manipulating sine-functions, creating surfaces representing incremental sedimentation (Wen, 2004; Nordahl, 2005). Displacement of the surfaces creates a three-dimensional image mimicking bedform migration, and depositional environments as diverse as tidal channels and mass flows can be accurately recreated. Due to the high-resolution output, common practice is to generate models that are volumetrically slightly larger than real core data (30 x 30 cm in x and y directions). The resulting modelled deposit volume may be populated with petrophysical information, creating intrinsic properties such as porosity and permeability (both vertical and horizontal). These petrophysical properties are based on empirical Gaussian distributions that can be further customized to fit observed data. In addition, a detailed net-to-gross ratio is produced for each modelled case.

11.2 AVO attributes

To test our method we use the porosity and net-to-gross synthetic logs computed in SBED model with sedimentation conditions based on the turbidite system from the Glitne Field. In Figure 25, we show these plots for 80 m thickness of reservoir. First, we consider the homogeneous fluid saturation in reservoir. The anisotropy parameters logs are computed by using available rock physics data. The water saturation results in increase in both anisotropy parameters, but parameter δ remains negative. Water saturation results in amplitude increase in the mid-reservoir section for both central frequencies. The oil-water contact (OWC) scenario (20% water saturation above and 90 percent water saturation below the OWC) results in elastic properties can easily be seen on the upscaled log data. The position for OWC is quite pronounced in elastic properties. The synthetic near- and far-offset traces results in more smooth reflection in the mid-reservoir section.The advantages of proposed technology are following: 1) the sedimentology scenario, 2) the fluid saturation scenario, 3) the AVO attributes from ultra-thin layered reservoirs taking into account the interbedded multiples.

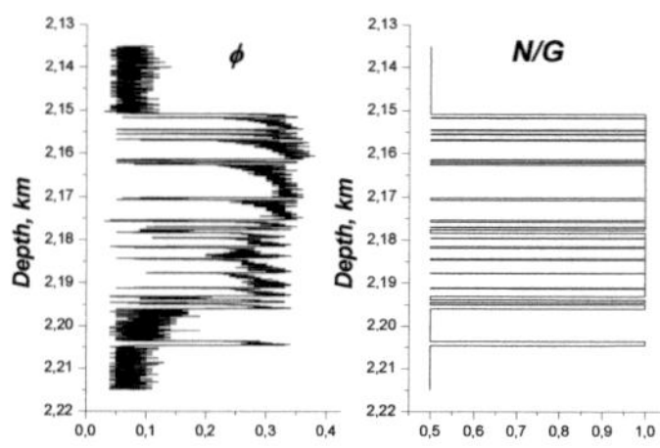

Fig. 25. The porosity (to the left) and net-to-gross (to the right) vertical profiles generated by SBED for the reservoir zone (Stovas et al., 2007).

12. Conclusions

In this chapter we discuss different issues related to wave propagation in layered media with major focus on finely (thin) layered media. We widely use the matrix propagator technique and discuss very important symmetries of propagator and reflection/ transmission matrices. The weak-contrast reflection and transmission coefficients are derived in first- and second-order approximations. The periodically layered medium is a fundamental example to illustrate the effect of periodicity on the wavefield, and we use this example to derive reflection and transmission responses. We analyze the caustics of the shear waves in a single layer and in multilayered media. Few seismic applications mostly related to seismic upscaling problem are discussed at the end of this chapter.

13. Acknowledgments

Alexey Stovas acknowledges the ROSE project at NTNU for financial support. Yury Roganov acknowledges Tesseral Technologies Inc. for financial support.

14. References

Aki, K. & Richards, P. (1980). *Quantitative seismology. Theory and methods*, W.H. Freeman and Company, San Francisco, ISBN 9781891389634.

Backus, G.E. (1962). Long-wave elastic anisotropy produced by horizontal layering. *J.Geophys.Res.*, Vol.67, pp. 4427-4440, ISSN 0148-0227.

Bamberger, A.; Chavent, G.; Hemon, Ch. & Lailly, P. (1982). Inversion of normal incidence seismograms. *Geophysics*, Vol.47, pp. 757-770, ISSN 0016-8033.

Banik, N.C.; Lerche, I. & Shuey, R.T. (1985a). Stratigrafic filtering, Part I: Derivation of the O'Doherty-Anstey formula. *Geophysics*, Vol.50, pp. 2768-2774, ISSN 0016-8033.

Banik, N.C.; Lerche, I.; Resnik, J.R. & Shuey, R.T. (1985b). Stratigrafic filtering, Part II: Model spectra. *Geophysics*, Vol.50, pp. 2775-2783, ISSN 0016-8033.

Braga, A.M. & Herrmann G. (1992). Floquet waves in anisotropic periodically layered composites. *J. Acoust. Soc. Am.*, Vol.91, pp. 1211-1227, ISSN 0001-4966.

Brekhovskikh, L.M. (1960). *Waves in layered media*, Academic Press, Applied Mathematics and Mechanics, Vol.6, New York.

Carcione, J.M.; Kosloff, D. & Behle, A. (1991). Long-wave anisotropy in stratified media: A numerical test. *Geophysics*, Vol.56, pp. 245-254, ISSN 0016-8033.

Carcione, J.M. (2007). *Wavefields in real media: Wave propagation in anisotropic, anelastic and porous media* (second edition), Elsevier, ISBN 0950-1401, Amsterdam-London-Paris.

Chapman, C.H. (1994). Reflection/transmission coefficients reciprocities in anisotropic media. *Geophysical Journal International*, Vol.116, pp. 498-501, ISSN *0956-540X.*

Cvetich, M. & Picman, L. (1981). Scattering states for a finite chain in one dimension. *J.Phys.A.*, Vol.14, pp. 379.382, ISSN 0305-4470.

Dellinger, J.A. (1991). Anisotropic seismic wave propagation. PhD Thesis, Stanford, CA, USA (http://sepwww.stanford.edu/theses/sep69/).

Fryer G. J. & Frazer L. N. (1987). Seismic waves in stratified anisotropic media-II. Elastodynamic eigensolutions for some anisotropic systems. *Geophy. J. R. astr. Soc.*, Vol.91, pp. 73-101, ISSN 0305-8719.

Gassmann, F. (1951). Uber die Elastitat poroser Medien, *Vier. Der Natur*. Gesellschaft in Zurich, Vol.96, pp. 1-23.

Gelinsky, S. & Shapiro, S.A. (1997). Poroelastic Backus averaging for anisotropic layered fluid- and gas-saturated Sediments. Geophysics, Vol.62, pp. 1867-1878, ISSN 0016-8033.

Gradshtein, I.S. & Ryzhik, I.M. (1995). *Tables of integrals, series and products* (seventh edition), ISBN 0-12-059876-0 ,Academic Press.

Griffiths, D.J. & Steinke, C.A. (2001). Waves in locally periodic media. *Am.J.Phys.*, Vol.69, pp. 137-154, ISSN 0096-0322.

Haskell, N.A. (1953). The dispersion of surface waves in multilayered media. *Bulletin of the Seismological Society of America*, Vol.43, pp. 17-34, ISSN 0037-1106.

Helbig, K. (1984). Anisotropy and dispersion in periodically layered media. *Geophysics*, Vol.49, pp. 364-373, ISSN 0016-8033.

Hovem, J.M. (1995). Acoustic waves in finely layered media. *Geophysics*, Vol.60, pp. 1217-1221, ISSN 0016-8033.

Kennett, B.L.N. (1983). Seismic wave propagation in stratified media. Cambridge University Press.

Marion, D. & Coudin, P. (1992) From ray to effective medium theories in stratified media: An experimental study, *Expanded Abstract of 62nd Ann.Internat.Mtg., Soc.Expl.Geophys.*, pp. 1341-1343, New Orleans, USA.

Marion, D.; Mukerji, T. & Mavko, G. (1994). Scale effects on velocity dispersion: From ray to effective theories in stratified media. *Geophysics*, Vol.59, pp. 1613-1619, ISSN 0016-8033.

Morlet, J.; Atens, G.; Fourgeau, E., & Giard, D. (1982a). Wave propagation and sampling theory - Part I: Complex signal andscattering in multilayered media. *Geophysics*, Vol.47, pp. 203-221, ISSN 0016-8033.

Musgrave M.J.P. (1970). Crystal Acoustics. San Francisco: Holden — Day.

Nordahl, K.; Ringrose, P.S. & Wen, R. (2005). Petrophysical characterization of a heterolithic tidal reservoir interval using a process-based modeling tool. *Petroleum Geoscience* Vol.11, pp. 17-28.

O'Doherty, R.F. & Anstey, N.A. (1971). Reflections on amplitudes. *Geophys. Prospect.*, Vol.19, pp. 430-458, ISSN 0016-8025.

Peano, G. (1888). Int´egration par s´eries des ´equations diff´erentielles lin´eaires. *Math. Ann.* Vol.32, pp. 450–456.

Pease, M.C. (1965). *Methods of Matrix Algebra*, Academic Press, New York.

Peyton, R.G. (1983). *Elastic wave propagation in transversely isotropic media*, Martinus Nijhoff Publishers, The Netherlands.

Roganov, Y. & Roganov, V. (2008). Dispersion of low-frequency waves in periodically layered media. *Geophysical Journal* Vol.30, pp. 27-33, ISSN 0203-3100 (in Russian).

Roganov, Yu. (2008). Caustics on the wave fronts in transversely isotropic media. *Geoinformatika* N3, 29-34, ISSN 1684-2189 (in Russian).

Roganov, Yu. & Stovas, A. (2010). On shear wave triplications in a multi-layered VTI medium. *Geophysical Prospecting*, Vol.58, N4, pp. 549-559, ISSN 0016-8025.

Roganov, Yu. & Stovas, A. (2011). Caustics in a periodically layered transversely isotropic medium with vertical symmetry Axis. *Geophysical Prospecting*, (accepted, early view), ISSN 0016-8025.

Schoenberger, M. & Levin, F.K. (1974). Apparent attenuation due to intrabed multiples. *Geophysics*, Vol.39, pp. 278-291, ISSN 0016-8033.

Schoenberg, M. & Daley T.M. (2003). qSV wavefront triplication in transversely isotropic material. *Expanded Abstracts of 73th Annual International Meeting, SEG*, pp. 137-140, Dallas, USA, October 26-31.

Schoenberg, M. & Helbig K. (1997). Orthorhombic media: Modeling elastic wave behavior in vertically fractured earth. *Geophysics*, Vol.62, pp. 1954—1974, ISSN 0016-8033.

Shapiro, S.A.; Hubral, P. & Ursin, B. (1996). Reflectivity/transmissivity for one dimensional inhomogeneous random elastic media: dynamic-equivalent medium approach. *Geoph. J. Int.*, Vol.126, pp. 184-196, ISSN 0956-540X.

Shapiro, S.A. & Treitel, S. (1997). Multiple scattering of seismic waves in multilayered structures. Physics of the Earth and Planetary Interiors, Vol.104, pp. 147-159.

Stovas, A. & Ursin, B. (2001). Second-order approximations of the reflection and transmission coefficients between two visco- elastic isotropic media. *Journal of Seismic Exploration*, Vol.9, pp. 223-233, ISSN: 0963-0651.

Stovas, A. & Ursin, B. (2003). Reflection and transmission responses of a layered transversely isotropic viscoelastic medium. Geophysical Prospecting, Vol.51, No.5, pp. 447-477, ISSN 0016-8025.

Stovas, A. & Landrø, M. (2004). Optimal use of PP and PS time-lapse stacks for fluid-pressure discrimination. *Geophysical Prospecting. Vol.*52, pp. 1-12, ISSN 0016-8025.

Stovas, A. & Arntsen, B. (2006). Vertical propagation of low frequency waves in finely layered medium. *Geophysics*, Vol.71, N3, pp. T87-T94, ISSN 0016-8033.

Stovas, A.; Landrø, M. & Avseth, P. (2006). AVO attribute inversion for finely layered reservoirs. *Geophysics*, Vol.71, N3, pp. C25-C36, ISSN 0016-8033.

Stovas, A. (2007). Phase velocity approximation in a finely layered sediments. *Geophysics*, Vol.72, N5, pp. T57-T59, ISSN 0016-8033.

Stovas, A. & Ursin, B. (2007). Equivalent time-average and effective medium for periodic layers, *Geophysical Prospecting, Vol.*55, N6, pp. 871-882, ISSN 0016-8025.

Stovas, A.; Landro, M. & Janbu, M. (2007). Seismic attributes from ultra-thin layered reservoirs. *Oil and Gas*, Vol.62, N2, pp. 147-154.

Stovas, A. (2009). Power-gradient velocity model. *Geophysics*, Vol.74, No.5, pp. U13-U33, ISSN 0016-8033.

Stovas, A. & Roganov, Yu. (2009). Slowness surface approximations for qSV-waves in transversely isotropic media. *Geophysical Prospecting, Vol.* 57, pp. 1-11, ISSN 0016-8025.

Thomson, W.T. (1950). Transmission of elastic waves through a stratified solid. *J. Appl. Phys.* Vol. 21, pp. 89-93, ISSN 0021-8979.

Thomsen, L. (1986). Weak elastic anisotropy. *Geophysics,* Vol.51, pp. 1954-1966, ISSN 0016-8033.

Thomsen, L. & Dellinger, J. (2003). On shear-wave triplication in transversely isotropic media. *Journal of Applied Geophysics*, Vol.54, pp. 289—296, ISSN: 0926-9851.

Tsvankin, I. (1995). Seismic wavefields in layered isotropic media, Samizdat Press, Golden, USA.

Tygel, M.; Ursin, B. & Stovas, A. (2007). Convergence of the traveltime power series for a layered transversely isotropicmedium. *Geophysics,* Vol. 72, No.2, pp. D21-D28, ISSN 0016-8033.

Ursin, B. (1983). Review of elastic and electromagnetic wave propagation in horizontally layered media. *Geophysics*, Vol.48,No.8, pp.1063-1081, ISSN 0016-8033.

Ursin, B. (1987). The plane-wave reflection and transmission response of a vertically inhomogeneous acoustic medium. In Bernabini, M.; Carrion, P.; Jacovitti, G.; Rocca, F.; Treitel, S. & Worthington, M., eds., Deconvolution and inversion, pp. 189-207 , Blackwell Sc. Publ., Oxford, ISBN: 9780632019014.

Ursin, B. & Haugen, G.U. (1996). Weak-contrast approximation of the elastic scattering matrix in anisotropic media. *Pure and Applied Geophysics,* Vol.148, pp. 685- 714, ISSN 0033-4553.

Ursin, B. & Stovas, A. (2002). Reflection and transmission responses of a layered isotropic visco-elastic medium. *Geophysics* Vol.67, pp. 307-323, ISSN 0016-8033.

Vavrycuk, V. (2003). Generation of triplication on transversely isotropic media. *Physical Review B.* Vol.68, 054107, ISSN 1098-0121.

Wen, R. (2004). 3D modeling of stratigraphic heterogeneity in channelized reservoirs: methods and applications in seismic attribute facies classification. *CSEG Recorder,* Vol.29, pp. 38-45.

Wu, H.; Sprung, D.W.L. & Martorell, J. (1993). Periodic quantum wires and their quasi-one-dimensional nature. *J.Phys.D,* Vol.26, pp. 798-803, ISSN 0022-3727.

2

Soliton-Like Lamb Waves in Layered Media

I. Djeran-Maigre[1] and S. V. Kuznetsov[2]
[1]*University of Lyon, INSA Lyon, LGCIE,*
[2]*Institute for Problems in Mechanics, Moscow,*
[1]*France*
[2]*Russia*

1. Introduction

Solitons, or by the original terminology *waves of translation*, were for the first time observed and described by Scott Russel (1845) as a special kind of the hydrodynamic waves that can arise and propagate in narrow channels. Solitons are: (i) solitary waves, resembling propagation of the wave front of shock waves; (ii) these waves can propagate without considerable attenuation, or (iii) change of form; or (iv) diminution of their speed; see, Craik (2004). It was shown later on, that motion of these waves can be described by a non-linear KdV differential equation; see the work by the originators of the KdV-equation Korteweg and de Vries (1885) and the subsequent works by Lax (1968), Miles (1981), and Zwillinger (1997), where some of the analytical solutions are presented and the main properties of the KdV equation are analyzed.

Herein, we analyze the long-wave limits of Lamb waves propagating in multilayered elastic anisotropic plates at vanishing frequency $\omega \to 0$, or in terms of the wave number r, at $r \to 0$. These vanishing frequency Lamb waves satisfy conditions (i) - (iv), and thus, resemble the solitons. But, in contrast to the genuine solitons in hydrodynamics or their nonlinear analogues propagating in elastic solids; see, Eckl et al. (2004), Kawahara (1972), Kliakhandler, Porubov, and Verlande (2000), Planat and Hoummady (1989), Porubov et al. (1998), Samsonov (2001), our soliton-like waves are described by *linear* vectorial differential equations, known as the Christoffel equations for Lamb waves.

Studies of Lamb waves, as solutions of linear equations of motion for the infinite plates, and the corresponding soliton-like *linear* waves traveling with the finite phase speed at vanishing frequency have quite a long history. Presumably, the first asymptotic analysis of the waves propagating at vanishing frequency in an *isotropic* plate with the traction-free outer planes was performed by Gogoladze (1947). He obtained an analytical expression for the phase speed of such a wave by asymptotic analysis of the approximate equation of motion related to the theory of plates based on the Bernoulli - Euler hypotheses. Later on, the similar approach and a more elaborate one allowing to consider plates with different boundary conditions at outer planes, but still based on the approximated theories of plates, were exploited by Mindlin (1951a, b, 1958, 1960), Mindlin and Medick (1959), Mindlin and Onoe (1957), Onoe (1955), and Tolstoy and Usdin (1953). The latter authors reported highly intricate behavior of the disperse curves in the vicinity of the zero frequency. See also a more recent work by Pagneux and Maurel (2001), where the dispersion relations in the complex space were analyzed, and a paper by Kaplunov and Nolde (2002), where an asymptotic method was developed for analyzing the limiting case of the flexural mode. The behavior of the lowest branches of the disperse curves at $\omega \to 0$ for the traction-free *isotropic* plate was studied numerically by Lyon (1955), who used the classical theory of plates.

Along with the approximate theories of plates, a more general approach based on the Papkovich - Neuber elastic potentials for solving equations of motion was used. It allowed obtaining dispersion relations for different wave modes, not necessary flexural; see, Holden (1951). This approach was especially useful for analyzing behavior of the dispersion curves at $\omega \to 0$; see also works by Auld (1990) Ewing, Jardetzki, and Press (1957), Graff (1975). The Papkovich - Neuber potentials written in cylindrical coordinates allowed obtaining dispersion relations for elastic rods in the frame of Pochhammer - Chree theory for waves in *isotropic* rods; see, Pochhammer (1876), Chree (1889), Davies (1948), Meeker and Meitzler (1964), Miklowitz (1978), Mindlin and McNiven (1960), Onoe, McNiven, and Mindlin (1962). It should be noted that for rods a similar intricate behavior at $\omega \to 0$ of the lowest branches of the disperse curves for longitudinal and torsional waves was observed. The generalizations of the Papkovich - Neuber potentials to cover media with elastic anisotropy were also worked out; see Barber (2006), however the generalized potentials became so complicated that no analytical solutions obtained with them are known.

Analysis of Lamb waves propagating in *anisotropic* plates and their soliton-like counterparts relies on reducing the second-order vectorial equations of motion to the first-order systems via different variants of the six-dimensional formalisms. Such a reduction can be referred to as the first step of the generalized Hamiltonian formalism; see Arnold (1989). From these formalisms the Stroh (1958, 1962) formalism is the most widely used, but there are also some other variants, among which we mention Lekhintskij (1963) formalism; see, also works by Barnett and Kirchner (1997) and Ting (1996, 1999, 2000) discussing equivalence of Stroh and Lekhnitskii formalisms. There are also different variants of the genuine six-dimensional Hamiltonian formalism applied to analysis of the surface acoustic wave; see, works by Tarn (2002a, b), Yan-ze Pen (2003), a recent paper by Fu (2007), and works by Kuznetsov (2002, 2003, and 2006). In the framework of the generalized Hamiltonian formalisms, several asymptotic approaches have been developed to study the limiting SH waves, propagating at $\omega \to 0$; see, Kuznetsov and Djeran-Maigre (2008); the lower modes of Lamb waves; see, Li and Romanowicz (1995); and the flexural modes of Lamb waves; see, Poncelet et al. (2006).

Another interesting variant of the asymptotic analysis is developed by Simonetti (2003), who studied behavior of propagation modes of Lamb and SH waves in a single-layered (infinite) plate with different types of boundary conditions by considering a two-layered plate and taking limits in material properties of one of the contacting layers.

Remarks 1.1. a) Analytical and numerical data; see Graff (1975), reveal that in the vicinity of the limiting phase speed c_s the corresponding dispersion curve $c(\omega)$ satisfies a condition

$$\left| c(\omega) - c_s \right| = O(\omega^n), \quad \omega \to +0, \tag{1.1}$$

where $n > 0$ is a positive number. However, by numerical analyses it is not possible to determine the exponent n. Below, a condition for obtaining the limiting speed c_s will be developed.

b) Low or vanishing frequencies of Lamb waves traveling with the phase speed satisfying condition (1.1), need in a small amount of energy needed for excitation. Indeed, the specific kinetic energy is determined by the following expression:

$$E_{kin} \equiv \frac{1}{2}\rho|\dot{\mathbf{u}}|^2 = \frac{1}{2}\rho|\mathbf{m}|^2\omega^2, \tag{1.2}$$

where $\mathbf{m}$ is the wave amplitude (possibly varying along depth of a layer). The right-hand side of (1.2) ensures that at finite values of the amplitudes and at $\omega \to 0$, the specific kinetic energy vanishes. It can be shown that the specific potential energy is also proportional to square of amplitude and frequency, thus, vanishing at $\omega \to 0$, as well.

c) Importance of the limiting waves is underlined by the fact that they resemble propagation of the wave front (WF) in a layer; see Treves (1982, Ch.V, §1) for definition of the WF and Achenbach (1973, Ch.IV, §4.5) for the corresponding notion used in acoustical applications.

Following Lamb (1917), the displacement field of the wave traveling in an *isotropic* layer can be represented by the following

$$\mathbf{u}(\mathbf{x},t) = \left(\sum_{p=1}^{4} \mathbf{m}_p C_p e^{ir\gamma_p x'}\right) e^{ir(\mathbf{n}\cdot\mathbf{x}-ct)}, \tag{1.3}$$

where $\mathbf{u}$ is the displacement field, and $\mathbf{m}_p \in \mathbb{R}^3$ are the unit amplitudes (polarizations). It is assumed that each vector $\mathbf{m}_p$ belongs to the sagittal plane. This plane is determined by the unit normal $\mathbf{w} = \mathbf{n} \times \mathbf{v}$, where $\mathbf{n}$ is the unit normal to the wave front and $\mathbf{v}$ is the unit normal to the median plane of the plate. In (1.3) $x' \equiv \mathbf{v} \cdot \mathbf{x}$ is a coordinate along vector $\mathbf{v}$; r is the wave number; c is the phase speed; t is time. The Christoffel parameters γ_p will be introduced later on. In representation (1.3)

$$\mathbf{u}^p(\mathbf{x},t) = \mathbf{m}_p e^{ir\gamma_p x'} e^{ir(\mathbf{n}\cdot\mathbf{x}-ct)} \tag{1.4}$$

are the partial waves. The unknown coefficients C_p in (1.3) are determined up to a multiplier by the traction-free boundary conditions:

$$x' = \pm h: \quad \mathbf{t}_\mathbf{v} \equiv \mathbf{v} \cdot \mathbf{C} \cdot\cdot \nabla_\mathbf{x} \mathbf{u} = 0, \tag{1.5}$$

where $\mathbf{C}$ is the fourth-order elasticity tensor (for isotropic medium tensor $\mathbf{C}$ is determined by two independent constants); and $2h$ is the depth of a plate. Exponential multiplier $e^{ir(\mathbf{n}\cdot\mathbf{x}-ct)}$ in (1.3) and (1.4) stands for propagation of the plane wave front $\mathbf{n} \cdot \mathbf{x} = const$.

Remark 1.2. Representation (1.3) is also valid in a case of *anisotropic* plate, provided: (A) the elasticity tensor has an axis of elastic symmetry, and (B) the wave travels in the direction of such an axis. Condition (A) is equivalent to monoclinic symmetry of the elasticity tensor, meaning that the elasticity tensor contains 13 independent decomposable components. At violating conditions (A) or (B), the amplitudes of partial waves may not belong to the sagittal plane. If that is the case, the six partial waves compose Lamb wave, instead of four partial waves used in (1.3); see, Kuznetsov (2002).

If a multilayered plate is concidered, the solution is usualy constructed by one of the following methods: (i) the transfer matrix (TM) method, known also as Thomson – Haskell method due to its originators; see, Thomson (1950), Haskell (1953) and more recent papers by Ryden et al. (2006) and Lowe (2008); and, (ii) the global matrix (GM) method; see,

Knopoff (1964) and Mal and Knopoff (1968). The TM method is based on a sequential solution of the boundary-value problems on the interfaces and constructing the transfer matrices. The TM method will be discussed in a more detail in the subsequent sections. The GM method is based on solving a system of the governing differential equations with the piecewise constant coefficients, resulting in construction of the special "global matrix".
Herein, a variant of the modified TM (MTM) method will be developed. That is associated with construction of the fundamental exponential matrices and satisfying interface conditions in terms of these matrices. The MTM method allows us to analyze both phase speed and polarization of Lamb waves propagating at vanishing frequencies in anisotropic multilayered plates.

2. Basic notations

All the layers of a multilayered plate are assumed homogeneous and hyperelastic. Equations of motion for a homogeneous elastic anisotropic medium can be written in the following form

$$\mathbf{A}(\partial_x, \partial_t)\mathbf{u} \equiv \mathrm{div}_x \mathbf{C} \cdot\cdot \nabla_x \mathbf{u} - \rho \ddot{\mathbf{u}} = 0\,, \tag{2.1}$$

where the elasticity tensor $\mathbf{C}$ is assumed to be positively definite:

$$(\mathbf{A} \cdot\cdot \mathbf{C} \cdot\cdot \mathbf{A}) \equiv \sum_{i,j,\,m,n} A_{ij} C^{ijmn} A_{mn} > 0, \qquad \underset{\mathbf{A} \in \mathrm{sym}(R^3 \otimes R^3),\ \mathbf{A} \neq 0}{\forall \mathbf{A}}\,, \tag{2.2}$$

In expression (2.2) $\mathrm{sym}\,\mathbf{A} = \frac{1}{2}(\mathbf{A} + \mathbf{A}^t)$.

Remark 2.1. For isotropic medium the positive definiteness of the elasticity tensor yields:

$$\mu > 0, \quad \lambda > -\frac{2}{3}\mu\,, \tag{2.2´}$$

where λ and μ are Lamé constants.
Following Kuznetsov (2002, 2003) we consider a more general than (1.3) representation for Lamb waves, that is suitable for layers with arbitrary elastic anisotropy:

$$\mathbf{u}(\mathbf{x}, t) \equiv \mathbf{f}(x'') e^{ir(\mathbf{n}\cdot\mathbf{x} - ct)}\,, \tag{2.3}$$

where $x'' = irx'$ is a dimensionless coordinate; and $\mathbf{f}$ is the unknown vectorial function defining variation of the amplitude at the wave front. Substituting representation (2.3) into Eq. (2.1), yields the ordinary differential equation with respect to $\mathbf{f}$. This is known as the Christoffel equation for Lamb waves:

$$-r^2\left(\mathbf{A}_1 \partial_{x''}^2 + \mathbf{A}_2 \partial_{x''} + \mathbf{A}_3\right) \cdot \mathbf{f} = 0\,, \tag{2.4}$$

where

$$\mathbf{A}_1 = \mathbf{v} \cdot \mathbf{C} \cdot \mathbf{v}, \quad \mathbf{A}_2 = \mathbf{v} \cdot \mathbf{C} \cdot \mathbf{n} + \mathbf{n} \cdot \mathbf{C} \cdot \mathbf{v}, \quad \mathbf{A}_3 = \mathbf{n} \cdot \mathbf{C} \cdot \mathbf{n} - \rho c^2 \mathbf{I}\,. \tag{2.5}$$

By introducing an auxiliary function $\mathbf{w} = \partial_{x''}\mathbf{f}$, Eq. (2.4) can be reduced to the matrix ODE of the first order:

$$\partial_{x''}\begin{pmatrix} \mathbf{f} \\ \mathbf{w} \end{pmatrix} = \mathbf{G}\cdot\begin{pmatrix} \mathbf{f} \\ \mathbf{w} \end{pmatrix}, \tag{2.6}$$

where $\mathbf{G}$ is the matrix of the sixth rank for arbitrary elastic anisotropy, and of the fourth rank for the case described by conditions A and B in Remark 1.2:

$$\mathbf{G} = \begin{pmatrix} \mathbf{0} & \mathbf{I} \\ -\mathbf{A}_1^{-1}\cdot\mathbf{A}_3 & -\mathbf{A}_1^{-1}\cdot\mathbf{A}_2 \end{pmatrix}. \tag{2.7}$$

It can easily be deduced from (2.7)

$$\det(\mathbf{G}) = \det(\mathbf{A}_3)\cdot\det^{-1}(\mathbf{A}_1). \tag{2.8}$$

In the right-hand side of (2.7) $\mathbf{0}$ and $\mathbf{I}$ are the corresponding 3×3 matrices. By means of (2.7), the general solution of Eq. (2.6) can be represented in the form

$$\begin{pmatrix} \mathbf{f} \\ \mathbf{w} \end{pmatrix}_0 = e^{ir\mathbf{G}x'}\cdot\vec{C}, \tag{2.9}$$

where $\vec{C}$ is the six-dimensional complex vector, defined up to a scalar multiplier by boundary conditions (1.5). Taking into account (2.9), representation (2.3) takes the form

$$\begin{pmatrix} \mathbf{u}(\mathbf{x},t) \\ \mathbf{v}(\mathbf{x},t) \end{pmatrix} = \left(e^{ir\mathbf{G}x'}\cdot\vec{C}\right)e^{ir(\mathbf{n}\cdot\mathbf{x}-ct)}, \tag{2.10}$$

where $\mathbf{v}(\mathbf{x},t) = \mathbf{w}(x'')e^{ir(\mathbf{n}\cdot\mathbf{x}-ct)}$.

Remarks 2.2. *a*) Representation (2.10) remains valid if matrix $\mathbf{G}$ is a *non-semisimple* matrix, i.e. when matrix $\mathbf{G}$ has Jordan blocks in its Jordan normal form.

b) Computing exponential matrix $e^{ir\mathbf{G}x'}$ can be done by different numerical methods; see, Moler and Van Loan (1978, 2003) Higham (2001) and Zanna and Munthe-Kaas (2002), where different numerical schemes are discussed. For analytical purposes the exponential matrix can be constructed by applying two alternative methods: (1) the Taylor series expansion, or (2) reducing matrix $\mathbf{G}$ to the Jordan canonical form and taking exponent of the diagonal matrix (assuming that $\mathbf{G}$ is a semisimple matrix)

$$e^{ir\mathbf{G}x'} = \mathbf{W}^{-1}\cdot e^{ir\mathbf{D}x'}\cdot\mathbf{W}. \tag{2.11}$$

where $\mathbf{D}$ is diagonal matrix, and $\mathbf{W}$ is a non-degenerate matrix needed to reduce $\mathbf{G}$ to the Jordan canonical form; see, Meyer (2002). If matrix $\mathbf{G}$ is not semisimple, representation (2.11) changes; see, Meyer (2002, §7.3).

3. Vanishing frequency Lamb wave in a homogeneous anisotropic plate

Substituting solution (2.10) into boundary conditions (1.5) yields

$$\mathbf{M}\cdot\vec{C} = 0, \tag{3.1}$$

where

$$\mathbf{M}=\begin{pmatrix} (\mathbf{A}_4,\mathbf{A}_1)\cdot e^{+irGh} \\ -(\mathbf{A}_4,\mathbf{A}_1)\cdot e^{-irGh} \end{pmatrix}. \tag{3.2}$$

In (3.2)

$$\mathbf{A}_4=\mathbf{\nu}\cdot\mathbf{C}\cdot\mathbf{n}\,. \tag{3.3}$$

Existence of a non-trivial solution for Eq. (3.2) is equivalent to the following condition

$$\det(\mathbf{M})=0 \tag{3.4}$$

Equation (3.4) is known as the dispersion equation for Lamb wave, since it implicitly defines speed of propagation as a function of frequency or wave number.
Proposition 3.1. At $r=0$ and at arbitrary anisotropy, Eq. (2.14) is trivially satisfied.
Proof flows out Eq. (3.2), which ensures at $r=0$:

$$\mathbf{M}=\begin{pmatrix} \mathbf{A}_4 & \mathbf{A}_1 \\ -\mathbf{A}_4 & -\mathbf{A}_1 \end{pmatrix}. \tag{3.5}$$

It is clear that for matrix (3.5) condition (3.4) is satisfied.
However, the obtained at $r=0$ solution is meaningless; firstly, it does not satisfy Eq. (3.4) at small $r\to 0$; and secondly, it does not define speed of the wave at $r=0$. To construct the solution valid at $r=0$, the condition (1.1) will be used. Taking into account (3.4) and Proposition 3.1, condition (1.1) can be rewritten as a sequence of the following conditions imposed on the phase speed $c(r)$, that is implicitely defined by Eq. (3.4)

$$\frac{d^k}{dr^k}c(r)\equiv-\left(\partial_r^k\det(\mathbf{M})\right)/\left(\partial_c\det(\mathbf{M})\right)\Big|_{r=0}=0,\quad k=1,\ldots,n\,. \tag{3.6}$$

Conditions (3.6) are equivalent to

$$\partial_r^k\det(\mathbf{M})\Big|_{r=0}=0,\qquad k=1,\ldots,n\,. \tag{3.7}$$

Taking Taylor's expansion (with respect to r) of the exponential mappings in (3.2), yields

$$\begin{aligned}\mathbf{M}=&\begin{pmatrix} \mathbf{A}_4 & \mathbf{A}_1 \\ -\mathbf{A}_4 & -\mathbf{A}_1 \end{pmatrix}+\frac{irh}{1!}\begin{pmatrix} -\mathbf{A}_3 & \mathbf{A}_4-\mathbf{A}_2 \\ -\mathbf{A}_3 & \mathbf{A}_4-\mathbf{A}_2 \end{pmatrix}+\\ &+\frac{(irh)^2}{2!}\begin{pmatrix} -\mathbf{A}_4\mathbf{A}_1^{-1}\mathbf{A}_3+\mathbf{A}_2\mathbf{A}_1^{-1}\mathbf{A}_3 & -\mathbf{A}_4\mathbf{A}_1^{-1}\mathbf{A}_2-\mathbf{A}_3+\mathbf{A}_2\mathbf{A}_1^{-1}\mathbf{A}_2 \\ \mathbf{A}_4\mathbf{A}_1^{-1}\mathbf{A}_3-\mathbf{A}_2\mathbf{A}_1^{-1}\mathbf{A}_3 & \mathbf{A}_4\mathbf{A}_1^{-1}\mathbf{A}_2+\mathbf{A}_3-\mathbf{A}_2\mathbf{A}_1^{-1}\mathbf{A}_2 \end{pmatrix}+\\ &+\frac{(irh)^3}{3!}\begin{pmatrix} \mathbf{A}_4\mathbf{A}_1^{-1}\mathbf{A}_2\mathbf{A}_1^{-1}\mathbf{A}_3+\mathbf{A}_3\mathbf{A}_1^{-1}\mathbf{A}_3- & -\mathbf{A}_4\mathbf{A}_1^{-1}\mathbf{A}_3+\mathbf{A}_4\left(\mathbf{A}_1^{-1}\mathbf{A}_2\right)^2+\mathbf{A}_3\mathbf{A}_1^{-1}\mathbf{A}_2+ \\ -\left(\mathbf{A}_2\mathbf{A}_1^{-1}\right)^2\mathbf{A}_3 & +\mathbf{A}_2\mathbf{A}_1^{-1}\mathbf{A}_3-\left(\mathbf{A}_2\mathbf{A}_1^{-1}\right)^2\mathbf{A}_2 \\ \mathbf{A}_4\mathbf{A}_1^{-1}\mathbf{A}_2\mathbf{A}_1^{-1}\mathbf{A}_3+\mathbf{A}_3\mathbf{A}_1^{-1}\mathbf{A}_3- & -\mathbf{A}_4\mathbf{A}_1^{-1}\mathbf{A}_3+\mathbf{A}_4\left(\mathbf{A}_1^{-1}\mathbf{A}_2\right)^2+\mathbf{A}_3\mathbf{A}_1^{-1}\mathbf{A}_2+ \\ -\left(\mathbf{A}_2\mathbf{A}_1^{-1}\right)^2\mathbf{A}_3 & +\mathbf{A}_2\mathbf{A}_1^{-1}\mathbf{A}_3-\left(\mathbf{A}_2\mathbf{A}_1^{-1}\right)^2\mathbf{A}_2 \end{pmatrix}+O(r^4)\end{aligned}\,. \tag{3.8}$$

Substituting the first four terms of Taylor's series (3.8) into (3.7) and applying Schur's formulas; see, Meyer (2002), yields conditions (3.7) in the form

$$\partial_r^k \det(\mathbf{M})\Big|_{r=0} \equiv \partial_r^k\left(\det(\mathbf{W})\det\left(\mathbf{Z}-\mathbf{X}\mathbf{W}^{-1}\mathbf{Y}\right)\right)\Big|_{r=0} = 0, \quad k=1,...,n, \tag{3.9}$$

where

$$\begin{aligned}
\mathbf{W} &= \mathbf{A}_4 - (irh)\mathbf{A}_3 + \frac{(irh)^2}{2}\left(-\mathbf{A}_4\mathbf{A}_1^{-1}\mathbf{A}_3 + \mathbf{A}_2\mathbf{A}_1^{-1}\mathbf{A}_3\right) + \\
&+ \frac{(irh)^3}{3!}\left(\mathbf{A}_4\mathbf{A}_1^{-1}\mathbf{A}_2\mathbf{A}_1^{-1}\mathbf{A}_3 + \mathbf{A}_3\mathbf{A}_1^{-1}\mathbf{A}_3 - \left(\mathbf{A}_2\mathbf{A}_1^{-1}\right)^2\mathbf{A}_3\right) \\
\mathbf{X} &= -\mathbf{A}_4 - (irh)\mathbf{A}_3 + \frac{(irh)^2}{2}\left(\mathbf{A}_4\mathbf{A}_1^{-1}\mathbf{A}_3 - \mathbf{A}_2\mathbf{A}_1^{-1}\mathbf{A}_3\right) + \\
&+ \frac{(irh)^3}{3!}\left(\mathbf{A}_4\mathbf{A}_1^{-1}\mathbf{A}_2\mathbf{A}_1^{-1}\mathbf{A}_3 + \mathbf{A}_3\mathbf{A}_1^{-1}\mathbf{A}_3 - \left(\mathbf{A}_2\mathbf{A}_1^{-1}\right)^2\mathbf{A}_3\right) \\
\mathbf{Y} &= \mathbf{A}_1 + (irh)\left(\mathbf{A}_4 - \mathbf{A}_2\right) + \frac{(irh)^2}{2}\left(-\mathbf{A}_4\mathbf{A}_1^{-1}\mathbf{A}_2 - \mathbf{A}_3 + \mathbf{A}_2\mathbf{A}_1^{-1}\mathbf{A}_2\right) + \\
&+ \frac{(irh)^3}{3!}\left(-\mathbf{A}_4\mathbf{A}_1^{-1}\mathbf{A}_3 + \mathbf{A}_4\left(\mathbf{A}_1^{-1}\mathbf{A}_2\right)^2 + \mathbf{A}_3\mathbf{A}_1^{-1}\mathbf{A}_2 + \mathbf{A}_2\mathbf{A}_1^{-1}\mathbf{A}_3 - \left(\mathbf{A}_2\mathbf{A}_1^{-1}\right)^2\mathbf{A}_2\right) \\
\mathbf{Z} &= -\mathbf{A}_1 + (irh)\left(\mathbf{A}_4 - \mathbf{A}_2\right) + \frac{(irh)^2}{2}\left(\mathbf{A}_4\mathbf{A}_1^{-1}\mathbf{A}_2 + \mathbf{A}_3 - \mathbf{A}_2\mathbf{A}_1^{-1}\mathbf{A}_2\right) + \\
&+ \frac{(irh)^3}{3!}\left(-\mathbf{A}_4\mathbf{A}_1^{-1}\mathbf{A}_3 + \mathbf{A}_4\left(\mathbf{A}_1^{-1}\mathbf{A}_2\right)^2 + \mathbf{A}_3\mathbf{A}_1^{-1}\mathbf{A}_2 + \mathbf{A}_2\mathbf{A}_1^{-1}\mathbf{A}_3 - \left(\mathbf{A}_2\mathbf{A}_1^{-1}\right)^2\mathbf{A}_2\right)
\end{aligned} \tag{3.10}$$

Matrices in (3.9) and (3.10) are correctly defined, if the phase speed c does not coincide with any of the bulk wave speeds propagating in the direction of the wave normal $\mathbf{n}$. Henceforth, this is assumed to hold. Equations (3.9) are the necessary and sufficient conditions for existing a vanishing frequency Lamb wave that satisfies (1.1).

Remark 3.1. Parameter $n \geq 1$ in conditions (3.6) and (3.7) is dependent on anisotropy, and it characterizes attenuation of the phase speed $c(r)$ at $r \to 0$. Necessity of conditions (3.6) can be explained by analyzing Taylor's expansion of $\det(\mathbf{M})$ at small r, yielding

$$\det(\mathbf{M}) = r^n V_n + o(r^n), \quad r \to 0, \tag{3.11}$$

where V_n is an independent on r constant. Taking into account (3.11), it becomes clear that conditions (3.6) and (3.7) define the phase speed, at which vanishes the lowest non-trivial coefficient V_n of expansion (3.11).

4. Vanishing frequency Lamb wave in a homogeneous isotropic plate

For an *isotropic* elastic plate

$$\begin{aligned}
&\mathbf{A}_1 = (\lambda + 2\mu)\mathbf{v}\otimes\mathbf{v} + \mu(\mathbf{n}\otimes\mathbf{n} + \mathbf{w}\otimes\mathbf{w}), \quad \mathbf{A}_2 = (\lambda+\mu)(\mathbf{v}\otimes\mathbf{n} + \mathbf{n}\otimes\mathbf{v}) \\
&\mathbf{A}_3 = (\lambda + 2\mu - \rho c^2)\mathbf{n}\otimes\mathbf{n} + (\mu - \rho c^2)(\mathbf{v}\otimes\mathbf{v} + \mathbf{w}\otimes\mathbf{w}) \\
&\mathbf{A}_4 = \lambda\mathbf{v}\otimes\mathbf{n} + +\mu\mathbf{n}\otimes\mathbf{v}
\end{aligned} \quad , \tag{4.1}$$

where $\mathbf{w} = \mathbf{v} \times \mathbf{n}$.
Substituting matrices (4.1) into (2.7) gives matrix $\mathbf{G}$ in a form

$$\mathbf{G} = \begin{pmatrix} 0 & 0 & 0 & 1 & 0 & 0 \\ 0 & 0 & 0 & 0 & 1 & 0 \\ 0 & 0 & 0 & 0 & 0 & 1 \\ \dfrac{\rho c^2 - \mu}{\lambda + 2\mu} & 0 & 0 & 0 & -\dfrac{\lambda + \mu}{\lambda + 2\mu} & 0 \\ 0 & \dfrac{\rho c^2 - (\lambda + 2\mu)}{\mu} & 0 & -\dfrac{\lambda + \mu}{\mu} & 0 & 0 \\ 0 & 0 & \dfrac{\rho c^2 - \mu}{\mu} & 0 & 0 & 0 \end{pmatrix}. \tag{4.2}$$

For the isotropic plate the fundamental matrix $e^{ir\mathbf{G}x'}$ can also be constructed explicitly by reducing matrix $\mathbf{G}$ to the Jordan normal form

$$\mathbf{G} = \mathbf{W} \cdot \mathbf{D} \cdot \mathbf{W}^{-1}, \tag{4.3}$$

where $\mathbf{W}$ is a matrix containing (right) eigenvectors of matrix $\mathbf{G}$ stored columnwise

$$\mathbf{W} = \begin{pmatrix} 1 & 1 & 1 & 1 & 0 & 0 \\ -a & a & b^{-1} & -b^{-1} & 0 & 0 \\ 0 & 0 & 0 & 0 & a^{-1} & -a^{-1} \\ a & -a & b & -b & 0 & 0 \\ -a^2 & -a^2 & 1 & 1 & 0 & 0 \\ 0 & 0 & 0 & 0 & 1 & 1 \end{pmatrix}, \tag{4.4}$$

and $\mathbf{D}$ is a diagonal matrix

$$\mathbf{D} = \mathrm{diag}(a, -a, b, -b, a, -a). \tag{4.5}$$

In (4.4), (4.5) parameters a and b take the following values

$$a = \sqrt{c^2 / c_S^2 - 1}, \qquad b = \sqrt{c^2 / c_P^2 - 1}, \tag{4.6}$$

where

$$c_P = \sqrt{\frac{\lambda + 2\mu}{\rho}}, \qquad c_S = \sqrt{\frac{\mu}{\rho}} \tag{4.7}$$

are speeds of bulk primary (c_P) and secondary (c_S) waves.
It can be proved that at any admissible values of λ and μ satisfying condition (2.2′), matrix (4.2) is a semisimple matrix. Taking into account Eqs. (4.3) – (4.6) the fundamental matrix takes the form given by (2.11). Now, combining Eqs. (3.2) and (4.1 - (4.7), it is possible to represent matrix $\mathbf{M}$ in a complicated, but closed form.

Considering Eqs. (4.2), (4.5), and (4.7), the dispersion equation (3.7) gives the following values for the phase speed of the vanishing frequency waves propagating in a homogeneous isotropic plate:

$$c_{s_1} = 2\sqrt{\frac{\mu(\lambda+\mu)}{\rho(\lambda+2\mu)}}, \qquad c_{s_2} = \sqrt{\frac{\mu}{\rho}}. \tag{4.8}$$

Remarks 4.1. *a*) Ewing, Jardetsky, and Press (1957) determined speed c_{s_1} by applying asymptotic analysis based on Papkovich - Neuber potentials (and thus, confined to the isotropic plate only).

b) It can be shown from analyzing Eqs. (4.1) - (4.7) that a wave propagating with speed c_{s_2} is polarized in direction normal to the sagittal plane (SH wave). Soliton-like SH-waves were studied in (Kuznetsov and Djeran-Maigre, 2008).

c) The phase speed c_{s_1} does not depend upon depth of the layer. Analysis of (4.8) shows, that at any admissible values of Lamé's constants λ and μ, the speed c_{s_1} lies in the interval $c_T^{bulk} < c_{s_1} \le c_L^{bulk}$, where c_T^{bulk}, c_L^{bulk} are speeds of the transverse and longitudinal bulk waves respectively. The phase speed c_{s_1} coincides with c_L^{bulk} only at $\lambda = 0$.

d) At c_{s_1} parameters a and b in (4.6) take the following values:

$$a = \sqrt{\frac{3\lambda+2\mu}{\lambda+2\mu}}, \qquad b = i\frac{|\lambda|}{\lambda+2\mu}. \tag{4.9}$$

The inequality (2.2′) ensures parameter a in (4.9) to be real.

The eigenvectors (4.4) enable to obtain polarization of the vanishing frequency Lamb wave. Substituting the wave number $r = 0$ and the phase speed $c = c_s$ into matrix $\mathbf{M}$, yields (up to a scalar constant) two eigenvectors $\vec{C}$ corresponding to the (multiple) zero-eigenvalue of matrix $\mathbf{M}$:

$$\vec{C}_1 = (1, 0, 0, -1, 0, 0); \qquad \vec{C}_2 = \left(0, 1, -\frac{\lambda}{\lambda+2\mu}, 0, 0, 0\right). \tag{4.10}$$

The first eigenvector $\vec{C}_1$ ensures existence at $r = 0$ the Lamb wave, linearly polarized in the $\mathbf{n}$-direction. Such a wave resembles the longitudinal bulk wave with respect to polarization, but naturally differs in the phase speed. According to (4.9) the second eigenvector in (4.10) also leads to a linearly polarized wave with the following complex (not normalized) amplitude:

$$\mathbf{m} = \left(\frac{2\mu}{\lambda+2\mu}\right)\mathbf{v} + \left(\sqrt{\frac{3\lambda+2\mu}{\lambda+2\mu}} + i\,\mathrm{sign}(\lambda)\right)\mathbf{n}. \tag{4.11}$$

The real part of (4.11) leads to the slanted wave with respect to vectors $\mathbf{v}$ and $\mathbf{n}$, while the imaginary part corresponds to a wave defined by the first eigenvector $\vec{C}_1$.

Since both $\vec{C}_1$ and $\vec{C}_2$ correspond to the zero eigenvalue, we can make a liner combination of them. This allows us to construct a vanishing frequency wave arbitrary (indefinitely) polarized in the sagittal plane. Summarizing, we arrive at

Proposition 4.1. For the arbitrary isotropic traction-free plate and at $\omega \to 0$ there exists a nontrivial wave propagating with the phase speed c_{s_1} independent of the thickness of a plate and *indefinetely* polarized in the sagittal plane.

5. Vanishing frequency Lamb wave in a multilayered anisotropic plate

At first a two-layered plate will be considered, and aftrewards the generalization to a plate with arbitrary number of anisotropic layeres will be given.
Let the two-layered plate consists of two homogeneous anisotropic layers with the ideal mechanical conact at the interface:

$$\begin{cases} \mathbf{u}(-h_1) = \mathbf{u}(+h_2) \\ \mathbf{t}_{-\nu}(-h_1) = -\mathbf{t}_\nu(+h_2) \end{cases}, \tag{5.1}$$

where $2h_k$, $k = 1,2$ are the depths of the corresponding layers.
The outer surfaces of the plate are assumed to be traction-free:

$$\begin{cases} \mathbf{t}_\nu(+h_1) = 0 \\ \mathbf{t}_{-\nu}(-h_2) = 0 \end{cases}. \tag{5.2}$$

By analogy with (2.10), the six-dimensional field in each of the layers can be represented in terms of the fundamental matrices $e^{ir\mathbf{G}_k x'}$:

$$\begin{pmatrix} \mathbf{u}_k(\mathbf{x},t) \\ \mathbf{v}_k(\mathbf{x},t) \end{pmatrix} = \left(e^{ir\mathbf{G}_k x'} \cdot \vec{C}_k\right) e^{ir(\mathbf{n}\cdot\mathbf{x} - ct)}. \tag{5.3}$$

Substituting representation (5.3) into interface conditions (5.1) yields:

$$\begin{pmatrix} \mathbf{I} & \mathbf{0} \\ (\mathbf{A}_4)_1 & (\mathbf{A}_1)_1 \end{pmatrix} \cdot \left(e^{-ir\mathbf{G}_1 h_1} \cdot \vec{C}_1\right) = \begin{pmatrix} \mathbf{I} & \mathbf{0} \\ (\mathbf{A}_4)_2 & (\mathbf{A}_1)_2 \end{pmatrix} \cdot \left(e^{+ir\mathbf{G}_2 h_2} \cdot \vec{C}_2\right) \tag{5.4}$$

It is easy to see that under condition of positive definiteness (2.2) for tensors $\mathbf{C}_k$, $k = 1,2$, all 6×6 matrices appearing in (5.4), are non-degenerate. That allows us to represented the six-dimensional vector $\vec{C}_2$ in terms of $\vec{C}_1$:

$$\vec{C}_2 = \left(e^{-ir\mathbf{G}_2 h_2}\right) \cdot \begin{pmatrix} \mathbf{I} & \mathbf{0} \\ (\mathbf{A}_4)_2 & (\mathbf{A}_1)_2 \end{pmatrix}^{-1} \cdot \begin{pmatrix} \mathbf{I} & \mathbf{0} \\ (\mathbf{A}_4)_1 & (\mathbf{A}_1)_1 \end{pmatrix} \cdot \left(e^{-ir\mathbf{G}_1 h_1}\right) \cdot \vec{C}_1 \tag{5.5}$$

Remark 5.1. Expression (5.5) constitutes the basis of the Modified Transfer Matrix method, while the matrices appearing in the right-hand side of (5.5) are known as the transfer matrices.
Taking into account (5.5), the boundary conditions (5.2) can be expressed in the following form:

$$\mathbf{M}\cdot\vec{C}_1 = 0, \tag{5.6}$$

where 6×6 matrix $\mathbf{M}$ is

$$\mathbf{M} = \begin{pmatrix} \left((\mathbf{A}_4)_1, (\mathbf{A}_1)_1\right)\cdot\left(e^{+ir\mathbf{G}_1 h_1}\right) \\ -\left((\mathbf{A}_4)_2, (\mathbf{A}_1)_2\right)\cdot\left(e^{-2ir\mathbf{G}_2 h_2}\right)\cdot\begin{pmatrix} \mathbf{I} & \mathbf{0} \\ (\mathbf{A}_4)_2 & (\mathbf{A}_1)_2 \end{pmatrix}^{-1}\cdot\begin{pmatrix} \mathbf{I} & \mathbf{0} \\ (\mathbf{A}_4)_1 & (\mathbf{A}_1)_1 \end{pmatrix}\cdot\left(e^{-ir\mathbf{G}_1 h_1}\right) \end{pmatrix}. \tag{5.7}$$

In (5.7) $\left((\mathbf{A}_4)_k, (\mathbf{A}_1)_k\right)$, $k = 1,2$ are 3×6 matrices. Existing (at $r > 0$) the nontrivial solutions for Eq. (5.6) is equivalent to satisfying condition (3.4). However, for the vanishing frequency wave propagating at $r = 0$, condition (3.4) becomes meaningless, as it was for a single homogeneous layer, for such a wave the additional conditions (3.6) should be applied to matrix (5.7).

For a plate consisting of $n > 2$ homogeneous monoclinic layers in a contact, the secular matrix $\mathbf{M}$ becomes:

$$\mathbf{M} = \begin{pmatrix} \left((\mathbf{A}_4)_1, (\mathbf{A}_1)_1\right)\cdot\left(e^{+ir\mathbf{G}_1 h_1}\right) \\ -\left((\mathbf{A}_4)_n, (\mathbf{A}_1)_n\right)\cdot\prod_{k=2}^{n}\left[\left(e^{-2ir\mathbf{G}_k h_k}\right)\cdot\begin{pmatrix} \mathbf{I} & \mathbf{0} \\ (\mathbf{A}_4)_k & (\mathbf{A}_1)_k \end{pmatrix}^{-1}\cdot\begin{pmatrix} \mathbf{I} & \mathbf{0} \\ (\mathbf{A}_4)_{k-1} & (\mathbf{A}_1)_{k-1} \end{pmatrix}\right]\cdot\left(e^{-ir\mathbf{G}_1 h_1}\right) \end{pmatrix} \tag{5.8}$$

6. Vanishing frequency Lamb wave in a multilayered isotropic plate

Adopting the general method developed in the previous section and applying Eqs. (4.1) - (4.7) to construct the fundamental matrices, we arrive at the following two values for the limiting phase speed:

$$c_{s_1} = 2\sqrt{\left(\sum_{k=1}^{n}\mu_k h_k \frac{\lambda_k+\mu_k}{\lambda_k+2\mu_k}\right)\Big/\left(\sum_{k=1}^{n}\rho_k h_k\right)}, \qquad c_{s_2} = \sqrt{\left(\sum_{k=1}^{n}\mu_k h_k\right)\Big/\left(\sum_{k=1}^{n}\rho_k h_k\right)}. \tag{6.1}$$

Analysing polarization of the corresponding waves reveals that a wave propagating with speed c_{s_1} is polarized in the sagittal plane, whereas wave propagating with speed c_{s_2} is a SH wave.

Confining ourselves to the genuine Lamb wave propagating with speed c_{s_1}, we can formulate:

Proposition 6.1. *a*) Let $c_{s_1}^{\max}$ and $c_{s_1}^{\min}$ be maximal and minimal limiting wave speeds in the distinct layers (according to Proposition 4.1 these speeds are independent of thickness of the layers), then

$$c_{s_1}^{\min} \le c_{s_1} \le c_{s_1}^{\max}. \tag{6.2}$$

b) Supposing that depth of the n-th layer tends to infinity (halfspace) we arrive at the following value for the limiting wave speed

$$c_{s_1} = 2\sqrt{\frac{\mu_n(\lambda_n+\mu_n)}{\rho_n(\lambda_n+2\mu_n)}}. \tag{6.3}$$

Proof a) flows out from considering the right-hand-side of (6.1), it ensures that all the terms

$$\underset{k=1,\dots,n}{\forall k} \qquad \frac{\mu_k(\lambda_k+\mu_k)}{\rho_k(\lambda_k+2\mu_k)}>0 \tag{6.4}$$

are positive at the assumption of positive definiteness of the elasticity tensor. Proof *b*) also follows from the right-hand-side of (6.1) by passing to a limit at $h_n \to \infty$.

Remarks 6.1. *a*) Expression $(6.1)_1$ for the limiting speed c_{S_1} was apparently obtained for the first time; expression for the limiting speed c_{S_2} was obtained by Kuznetsov (2006) and Kuznetsov and Djeran-Maigre (2008) with a different asymptotic scheme.

b) It follows from the right-hand side of (6.3) that the corresponding limiting speed is independent of physical and geometrical properties of other layers. It can be said that the limiting wave is insensitive to the layers of finite thickness in a contact with a halfspace.

c) Assuming in Eq. $(6.1)_1$ that the plate is single-layered with $n=1$ and taking $\mu_1=1$, $\rho_1=1$, and $h_1=1$ we arrive at the following one-parametric expression for the speed c_{S_1}:

$$c_{S_1}=\frac{1}{\sqrt{2}}\sqrt{\frac{1}{1-\nu}}, \tag{6.5}$$

where ν is Poisson's ratio. The plot on Fig.1 shows variation of the longitudinal bulk wave speed and the limiting speed c_{S_1} versus Poisson's ratio. The plot reveals that in the whole admissible range of $\nu\in(-1;\tfrac{1}{2})$, the speed c_{S_1} remains substantially lower than the longitudinal bulk wave speed. The speed c_{S_1} approaches speed of the shear bulk wave only at $\nu\to 1/2$, where actually $c_{S_1}=c_S$.

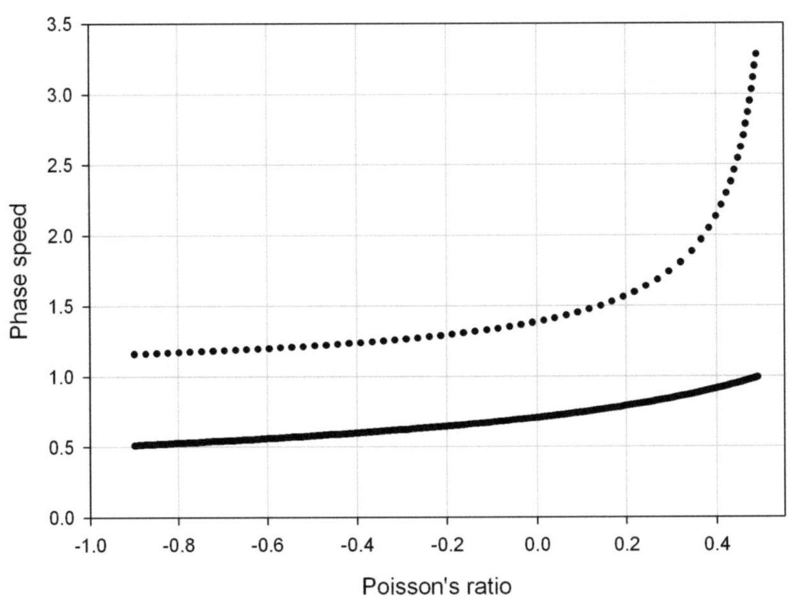

Fig. 1. Single layered isotropic plate: dependencies of the limiting speed c_{S_1} (bold curve) and the longitudinal bulk wave speed (dotted curve) on Poisson's ratio.

d) For a triple-layered plate with the outer layers of the same physical and geometrical properties (such a case often occurs in practice) the limiting speed c_{s_1} is

$$c_{s_1} = 2\sqrt{\left(2\mu_1 h_1 \frac{\lambda_1 + \mu_1}{\lambda_1 + 2\mu_1} + \mu_2 h_2 \frac{\lambda_2 + \mu_2}{\lambda_2 + 2\mu_2}\right) / \left(2\rho_1 h_1 + \rho_2 h_2\right)} \tag{6.6}$$

and c_{s_2} is

$$c_{s_2} = \sqrt{\frac{2\mu_1 h_1 + \mu_2 h_2}{2\rho_1 h_1 + \rho_2 h_2}}, \tag{6.7}$$

where index 1 is referred to the outer layers, and 2 corresponds to the inner layer. Assuming in Eq. (6.7) that $h_1 \ll h_2$, while other physical properties of the layers have comparable values, yields coincidence of c_{s_2} with the shear bulk wave speed of the inner layer.
Remarks 6.2. *a*) Expression $(6.1)_1$ for the limiting speed c_{s_1} was apparently obtained for the first time; expression for the limiting speed c_{s_2} was obtained by Kuznetsov (2006) and Kuznetsov and Djeran-Maigre (2008) with a different asymptotic scheme.

7. Acknowledgements

Authors thank INSA de Lyon (France) and the Russian Foundation for Basic Research (Grants 08-08-00855 and 09-01-12063) for partial financial support.

8. References

Achenbach J.D. Wave Propagation in Elastic Solids, North-Holland Publ., Amsterdam - London, 1973.

Arnol'd V.I. Mathematical Methods of Classical Mechanics, Springer-Verlag, N.Y., 1989.

Auld B.A. Acoustic Fields and waves in Solids, Vol.2, 2nd edition, Krieger Pub Co, Malabar FL, 1990.

Barber J.R. Three-dimensional elasticity solutions for isotropic and generally anisotropic bodies, Applied Mechanics and Materials, 5-6 (2006) 541-550.

Barber J.R. and Ting T.C.T. Three-dimensional solutions for general anisotropy, J. Mech. Phys. Solids, 55 (2007) 1993 - 2006

Chree C. The equations of an isotropic elastic solid in polar and cylindrical coordinates, their solutions and applications, Trans. Camb. phil. Soc. Math. Phys. Sci., 14 (1889) 250.

Craik A.D.D. The origins of water wave theory, Annual Review of Fluid Mechanics 36 (2004) 1 - 28.

Davies R. M. A critical study of the Hopkinson pressure bar, Phil. Trans. R. Soc. A240 (1948) 375 - 457.

Eckl C., Kovalev A.S., Mayer A.P., Lomonosov A.M., and *Hess P.* Solitory surface acoustic waves, Physical Rev., E70 (2004) 1 - 15.

Ewing W.M., Jardetzky W.S., and *Press F.* Elastic Waves in Layered Media, McGraw-Hill Inc., N.Y., 1957.

Fu Y.B. Hamiltonian interpretation of the Stroh formalism in anisotropic elasticity, Proc. Roy. Soc. A. 463 (2007) 3073 - 3087.

Gogoladze V.G. Dispersion of Rayleigh Waves in a Layer (in Russian), Publ. Inst. Seism. Acad. Sci. U.R.S.S. 119 (1947) 27 - 38.

Graff K. F. Wave Motion in Elastic Solids, Dover Inc., N.Y., 1975.

Hartman P. Ordinary differential equations, John Wiley & Sons, Inc., N.Y., 1964.

Haskell N.A. Dispersion of surface waves on multilayered media, Bull. Seismol. Soc. America. 43 (1953) 17 - 34.

Higham N. J. Evaluating Padé approximants of the matrix logarithm, SIAM J. Matrix Anal. Appl., 22 (2001) 1126 - 1135.

Holden A.N. Longitudinal modes of elastic waves in isotropic cylinders and slabs, Bell Sys. Tech. J. 30 (1951) 956 - 969.

Kaplunov J.D. and Nolde, E.V. Long-Wave Vibrations of a Nearly Incompressible Isotropic Plate with Fixed Faces, Quart. J. Mech. Appl. Math. 55 (2002): 345 - 356.

Kawahara T. Oscillatory solitary waves in dispersive media, J. Phys. Soc. Japan 33 (1972) 260 - 268.

Kliakhandler I.L., Porubov A.V., and *Velarde M.G.* Localized finite-amplitude disturbances and selection of solitary waves, Physical Review E 62 (2000) 4959-4962.

Korteweg D.J. and *de Vries F.* On the Change of Form of Long Waves Advancing in a Rectangular Canal, and on a New Type of Long Stationary Waves, Philosophical Magazine 39 (1895) 422 - 443.

Knopoff L. A matrix method for elastic wave problems, Bull. Seismol. Soc. Am. 54 (1964) 431 - 438.

Kuznetsov S.V. Subsonic Lamb waves in anisotropic plates, Quart. Appl. Math. 60 (2002) 577 - 587.

Kuznetsov S.V. Surface waves of non-Rayleigh type, Quart. Appl. Math. 61 (2003) 575 - 582.

Kuznetsov S.V. SH-waves in laminated plates, Quart. Appl. Math., 64 (2006) 153 - 165.

Kuznetsov S.V. and *Djeran-Maigre I.*, Solitary SH waves in two layered traction free plates, Comptes Rendus Acad Sci., Paris, Ser. Mecanique, №336(2008) pp. 102 - 107.

Lamb H. On waves in an elastic plate, Proc. Roy. Soc. A93 (1917) 114 - 128.

Lange J.N. Mode conversion in the long-wavelength limit, J. Acoust. Soc. Am., 41 (1967) 1449-1452.

Lax P. Integrals of Nonlinear Evolution Equations and Solitary Waves, Comm. Pure Appl. Math. 21 (1968) 467 - 490.

Lekhnitskii S.G. Theory of Elasticity of an Anisotropic Elastic Body. Holden-Day, San Francisco 1963.

Li, X.-D., and *Romanowicz B.* Comparison of global waveform inversions with and without considering cross-branch modal coupling, Geophys. J. Int., 121 (1995) 695 - 709.

Lowe M.J.S. Matrix techniques for modeling ultrasonic waves in multilayered media, IEEE Transactions on Ultrasonics, Ferroelectrics, and Frequency Control 42 (1995) 525 - 542.

Lowe M.J.S. A computer model to study the properties of guided waves., J. Acoust. Soc. Am., 123 (2008) 3654 - 3654.

Lyon R.H. Response of an elastic plate to localized driving forces, J. Acoust. Soc. Am., 27 (1955) 259 - 265.

Mal A.K., Knopoff L. A differential equation for surface waves in layers with varying thickness, J. Math. Anal. Appl. 21 (1968) 431 - 441.

Meeker T.R. and *Meitzler A.H.* Guided wave propagation in elongated cylinders and plates, in: Physical Acoustics, (Ed. W. P. Mason) Vol. 1, Part A, Chap. 2. Academic Press, New York 1964.

Meyer C.D. Matrix Analysis and Applied Linear Algebra, SIAM, N.Y., 2002.

Miles J.W. The Korteweg-de Vries equation, a historical essay, J. Fluid Mech. 106 (1981) 131 - 147.

Miklowitz J. Elastic waves and waveguides, North-Holland, Amsterdam, 1978.

Mindlin R.D. The thickness shear and flexural vibrations of crystal plates. J. Appl. Phys. 22 (1951) 316.

Mindlin R.D. Influence of rotatory inertia and shear on flexural motions of isotropic, elastic plates, J. Appl. Mech., 18 (1951) 316.

Mindlin R.D. Vibrations of an infinite elastic plate and its cut-off frequencies, Proc Third US Nat. Congr. Appl. Mech., (1958) 225.

Mindlin R.D. Waves and vibrations in isotropic, elastic plates. In Structural Mechanics (Editors J.N. Goodier and N. Hoff). (1960) 199 - 232.

Mindlin R.D. and *McNiven H.D.* Axially symmetric waves in elastic rods, J. Appl. Mech., 27 (1960) 145 - 151.

Mindlin R.D. and *Medick M.A.* Extensional vibrations of elastic plates, J. Appl. Mech., 26 (1959) 561 - 569.

Mindlin R.D. and *Onoe M.* Mathematical theory of vibrations of elastic plates, In: Proceedings of the XI Annual Symposium on Frequency Control U. S. Army Signal Corps Engineering Laboratories, Fort Monmouth, New Jersey (1957) 17 - 40.

Moler C. and *Van Loan Ch.* Nineteen Dubious Ways to Compute the Exponential of a Matrix, SIAM Review, 20 (1978) 801 - 836.

Moler C. and Van Loan Ch. Nineteen Dubious Ways to Compute the Exponential of a Matrix, Twenty-Five Years Later SIAM Review, 45 (2003) 3 - 48.

Onoe M. A study of the branches of the velocity-dispersion equations of elastic plates and rods, Report: Joint Commitee on Ultrasonics of the Institute of Electrical Communication Engineers and the Acoustical Society of Japan, (1955) 1 - 21.

Onoe M., McNiven H.D., and *Mindlin R.D.* Dispersion of axially symmetric waves in elastic rods, J. Appl. Mech., 29 (1962) 729 - 734.

Pagneux V., Maurel A. Determination of Lamb mode eigenvalues, J. Acoust. Soc. Am., 110(3) (2001) Sep.

Planat M., Hoummady M. Observation of soliton-like envelope modulation generated in an anisotropic quartz plate by metallic in interdigital transducers, Appl. Phys. Lett, 55 (1989) 103 - 114.

Pochhammer L. Uber die Fortpflanzungsgeschwindigkeiten kleiner Schwingungen in einem unbegrenzten istropen Kreiszylinder, J. reine angew. Math. 81 (1876) 324 - 336.

Poncelet O., Shuvalov A.L., and Kaplunov J.D. Approximation of the flexural velocity branch in plates, Int. J. Solid Struct., 43 (2006) 6329 - 6346.

Porubov I.V., Samsonov A.M., Velarde M.G., and *Bukhanovsky A.V.* Strain solitary waves in an elastic rod embedded in another elastic external medium with sliding, Phys.Rev. Ser. E, 58 (1998) 3854 - 3864.

Russell Scott J. Report on waves, In: Fourteenth meeting of the British Association for the Advancement of Science, York, 1844 (London 1845), 311 - 390.

Ryden N., Lowe M.J.S., Cawley P., and *Park C.B.* Evaluation of multilayered pavement structures from measurements of surface waves, Review of Progress in Quantitative Nondestructive Evaluation, 820 (2006) 1616 - 1623.

Simonetti F. Sound Propagation in Lossless Waveguides Coated with Attenuative Materials, PhD thesis, Imperial Colledge, 2003.

Soerensen M.P., Christiansen P.L., and *Lomdahl P.S.* Solitary waves on nonlinear elastic rods. I. J. Acoust. Soc. Amer., 76 (1984) 871 - 879.

Stroh A.N. Dislocations and cracks in anisotropic elasticity. Phil. Mag. 3 (1958) 625-646.

Stroh A.N. Steady state problems in anisotropic elasticity, J. Math. Phys. 41 (1962) 77 - 103.

Tanuma K. Stroh Formalism and Rayleigh Waves (Reprinted from J. Elasticity, 89, 2007), Springer, N.Y., 2007.

Tarn J.-Q. A state space formalism for anisotropic elasticity. Part. I. Rectilinear anisotropy. Int. J. Solids Structures, 39 (2002) 5157 - 5172.

Tarn J.-Q. A state space formalism for anisotropic elasticity. Part. II. Cylindrical anisotropy. Int. J. Solids Structures, 39 (2002) 5143 - 5155.

Thomson W.T. Transmission of elastic waves through a stratified solid medium, J. Appl. Phys. 21 (1950) 89 - 93.

Ting T.C.T. Anisotropic elasticity: theory and applications. Oxford University Press, New York, 1996.

Ting T.C.T. A modified Lekhnitskii formalism á la Stroh for anisotropic elasticity and classifications of the 6X6 matrix N, Proc. Roy. Soc. London, A455 (1999) 69 - 89.

Ting T.C.T. Recent developments in anisotropic elasticity. Int. J. Solids Structures, 37 (2000) 401 - 409.

Ting T.C.T. and *Barnett D.M.* Classifications of surface waves in anisotropic elastic materials. Wave Motion 26 (1997) 207 - 218.

Tolstoy I. and *Usdin E.* Dispersive properties of stratified elastic and liquid media: a ray theory. Geophysics 18 (1953) 844 - 870.

Treves F. Introduction to Pseudodifferential and Fourier Integral Operators, Vol.1, Plenum Press, N.Y. and London, 1982.

Yan-ze Peng Exact periodic wave solutions to a new Hamiltonian amplitude equation, J. Phys. Soc. Japan 72 (2003) 1356 - 1359.

Zanna A. and *Munthe-Kaas H.Z.* Generalized polar decompositions for the approximation of the matrix exponential, SIAM J. Matrix Anal. Appl., 23 (2002) 840 - 862.

Zwillinger D. Handbook of Differential Equations, Academic Press, Third Edition, 1998.

3

Surface and Bulk Acoustic Waves in Multilayer Structures

V. I. Cherednick and M. Y. Dvoesherstov
Nizhny Novgorod State University
Russia

1. Introduction

The application of various layers on a piezoelectric substrate is a way of improving the parameters of propagating electroacoustic waves. For example, a metal film of certain thickness may provide the thermal stability of the wave for substrate cuts, corresponding to a high electromechanical coupling coefficient. The overlayer can vary the wave propagation velocity and, hence, the operating frequency of a device. The effect of the environment (gas or liquid) on the properties of the wave in the layered structure is used in sensors. The layer may protect the piezoelectric substrate against undesired external impacts. Multilayer compositions allow to reduce a velocity dispersion, which is observed in single-layer structures. In multilayer film bulk acoustic wave resonators (FBAR) many layers are necessary for proper work of such devices. Therefore, analysis and optimization of the wave propagation characteristics in multilayer structures seems to be topical. General methods of numerical calculations of the surface and bulk acoustic wave parameters in arbitrary multilayer structures are described in this chapter.

2. Surface acoustic waves in multilayer structures

In the linear theory of piezoelectricity and in the quasistatic electric approximation the system of differential equations, describing the mechanical displacements u_i along the three spatial coordinates x_i (i = 1, 2, 3) and the electric potential φ in the solid piezoelectric medium, may be written in such view (Campbell and Jones, 1968):

$$c_{ijkl}\frac{\partial^2 u_k}{\partial x_i \partial x_l}+e_{kij}\frac{\partial^2 \varphi}{\partial x_k \partial x_i}=\rho\frac{\partial^2 u_j}{\partial t^2} \quad (1)$$

$$e_{ikl}\frac{\partial^2 u_k}{\partial x_i \partial x_l}-\varepsilon_{ik}\frac{\partial^2 \varphi}{\partial x_i \partial x_k}=0 \qquad i, j, k, l = 1, 2, 3 \quad (2)$$

In these equations c_{ijkl} is the forth rank tensor of the elastic stiffness constants, e_{ijk} is the third rank tensor of the piezoelectric constants, ε_{ij} is the second rank tensor of the dielectric constants, ρ - the mass density, t – time, and the summation convention for repeated indices is used. The expression (1) contains three equations and (2) gives one more equation, totally four equations. These equations must be solved for each medium of all the multilayer system, which is shown in Fig. 1.

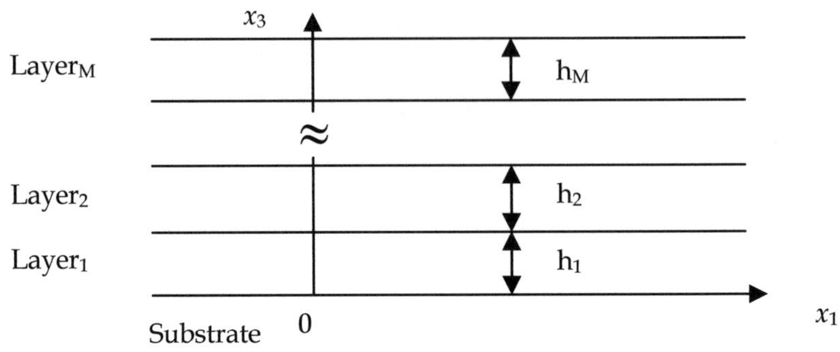

Fig. 1. Multilayer structure - substrate and M layers.

The coordinate axis x_1 direction coincides with the wave phase velocity v, the coordinate axis x_3 is normal to the substrate surface and the axis origin is set on this surface, as shown in Fig 1. A solution of equations (1) and (2) we will seek in the following form:

$$u_j = \alpha_j \exp[ik(b_i x_i - vt)] \qquad i, j = 1, 2, 3 \qquad (3)$$
$$\varphi = \alpha_4 \exp[ik(b_i x_i - vt)]$$

Here α_j - amplitudes of the mechanical displacements, α_4 - the amplitude of the electric potential, b_i - directional cosines of the wave velocity vector along the corresponding axises, $k = \omega/v = 2\pi/\lambda$ - the wave number, ω - a circular frequency, λ - a wavelength. Substitution of (3) into (1) and (2) gives the system of four linear algebraic equations for wave amplitudes:

$$c_{ijkl} b_i b_l \alpha_k + e_{kij} b_k b_i \alpha_4 = \rho v^2 \alpha_j \qquad (4)$$

$$e_{ikl} b_i b_l \alpha_k - \varepsilon_{ik} b_i b_k \alpha_4 = 0 \qquad (5)$$

The detailed form of these equations is following:

$$\begin{array}{ccccccccc} (\Gamma_{11} - \rho v^2)\alpha_1 & + & \Gamma_{12}\alpha_2 & + & \Gamma_{13}\alpha_3 & + & \Gamma_{14}\alpha_4 & = & 0 \\ \Gamma_{21}\alpha_1 & + & (\Gamma_{22} - \rho v^2)\alpha_2 & + & \Gamma_{23}\alpha_3 & + & \Gamma_{24}\alpha_4 & = & 0 \\ \Gamma_{31}\alpha_1 & + & \Gamma_{32}\alpha_2 & + & (\Gamma_{33} - \rho v^2)\alpha_3 & + & \Gamma_{34}\alpha_4 & = & 0 \\ \Gamma_{41}\alpha_1 & + & \Gamma_{42}\alpha_2 & + & \Gamma_{43}\alpha_3 & + & \Gamma_{44}\alpha_4 & = & 0 \end{array} \qquad (6)$$

Here:

$$\Gamma_{jk} = \Gamma_{kj} = c_{ijkl} b_i b_l, \quad \Gamma_{j4} = \Gamma_{4j} = e_{ikj} b_i b_k, \quad \Gamma_{44} = -\varepsilon_{ik} b_i b_k \qquad i,j,k,l = 1,2,3 \qquad (7)$$

For the existence of a nontrivial solution of the system (6) a determinant of this system must be equal to zero:

$$\begin{vmatrix} \Gamma_{11}-\rho v^2 & \Gamma_{12} & \Gamma_{13} & \Gamma_{14} \\ \Gamma_{21} & \Gamma_{22}-\rho v^2 & \Gamma_{23} & \Gamma_{24} \\ \Gamma_{31} & \Gamma_{32} & \Gamma_{33}-\rho v^2 & \Gamma_{34} \\ \Gamma_{41} & \Gamma_{42} & \Gamma_{43} & \Gamma_{44} \end{vmatrix} = 0 \quad (8)$$

This equation allows to determine the unknown directional cosine b_3, if the values v, b_1, and b_2 are set. For flat pseudo-surface acoustic wave the values of the directional cosines are following:

$$b_1 = 1 + i\delta,\ \ b_2 = 0,\ \ b_3 = b, \quad (9)$$

where δ is the wave attenuation coefficient along the propagation direction. For surface acoustic wave the attenuation is absent and $\delta = 0$. The equation (8) with taking into account (9) gives the following eighth power polynomial equation with respect to the b value:

$$a_8 b^8 + a_7 b^7 + a_6 b^6 + a_5 b^5 + a_4 b^4 + a_3 b^3 + a_2 b^2 + a_1 b + a_0 = 0 \quad (10)$$

Coefficients a_i of this equation are represented by very complicated expressions, depending on material constants of the medium, a phase velocity v, and the attenuation coefficient δ. For pseudo-surface acoustic waves $\delta \neq 0$ and therefore coefficients a_i are complex values. For surface acoustic waves $\delta = 0$ and coefficients a_i are pure real values. In this case roots of the equation (10) are either real or complex conjugated pairs. If $\delta \neq 0$, roots of the equation (10) are complex but not conjugated. So, solving (numerically certainly) the equation (10), we get eight roots $b^{(n)}$ (n = 1, 2, …, 8), which are complex values in general case. These values are the eigenvalues of the problem. Substituting each of these values into (7) and then into equation system (6), we can define all four complex amplitudes $\alpha_j^{(n)}$ for each root $b^{(n)}$. Values $\alpha_j^{(n)}$ represent the eigenvectors of the problem. This procedure must be performed for the substrate and for each layer. Found solutions are the partial solutions of the problem or partial modes.

The general solution for each medium is formed as a linear combination of partial solutions (partial modes). Quantity partial modes in the general solution for each medium must be equal to quantity of boundary conditions on its surfaces. Four boundary conditions on each surface are used, namely three mechanical and one electrical one. The substrate is semi-infinite, i.e. it has only one surface. Hence only four partial solutions are required for forming the general solution for the substrate. It means that some procedure of roots selection is required for substrate. For surface acoustic wave four roots with negative imaginary parts are selected from four complex conjugated pairs. This condition of roots selection corresponds to decreasing of the wave amplitude along the $-x_3$ direction (into the depth of the substrate), i.e. to condition of the localization of the wave near the surface. Practically the procedure of roots sorting with increasing imaginary parts order is performed and then four first roots are used for forming of the general solution.

For pseudo-surface wave roots are not complex conjugated, but they also contain four roots with negative imaginary part and also these four roots are first in the sorted roots sequence. In this case the roots selection rule is some different. Three first roots in the sorted sequence are selected, but the fourth root of this sequence is replaced with the fifth one (with the positive imaginary part of minimal value). This condition corresponds to increasing of the wave amplitude into the depth of the substrate and provide the energy conservation law satisfaction (wave attenuates along the propagation direction x_1 due to nonzero value of δ in the direction cosine b_1, see (9)). For high velocity pseudo-surface wave (the second order pseudo-surface wave or quasi-longitudinal pseudo-surface wave) only two first roots of the sorted sequence are selected, the third and the fourth roots are replaced with the fifth and the sixth ones.

All these rules of roots selection are applied for substrate only. For each layer of the structure shown in Fig. 1 there is no problem of roots selection, because each layer has two surfaces and all eight roots (all eight partial modes) are used for forming of the general solution for each layer.

One must to note, that in some special cases the quantity of partial modes may be less, than four for substrate and less, than eight for layers. This must be taken into account at forming of the general solution for corresponding case.

So, the general solution for each medium is formed as a linear combination of corresponding partial modes:

$$(u_j)_m = \sum_{n=N_{m-1}+1}^{N_m} C_n(\alpha_j^{(n-N_{m-1})})_m \exp\left\{ik\left[(b_i^{(n-N_{m-1})}x_i)_m - vt\right]\right\} \tag{11}$$

$$(\varphi)_m = \sum_{n=N_{m-1}+1}^{N_m} C_n(\alpha_4^{(n-N_{m-1})})_m \exp\left\{ik\left[(b_i^{(n-N_{m-1})}x_i)_m - vt\right]\right\} \tag{12}$$

Here m is the medium number, $N_m = n_0 + n_1 + \ldots + n_m$, n_m – the quantity of partial modes in the medium number m ($m = 0$ corresponds to a substrate, $m = 1$ corresponds to the 1st layer etc., $N_{0\text{-}1} = n_{0\text{-}1} = 0$), C_n – unknown coefficients and a continuous numeration is used for them (strange upper indices support this continuous numeration here and further).

The substrate is assumed the piezoelectric medium in all the cases and $n_0 = 4$ in general case (or less in some special cases). There are eight partial modes for each layer in the general case if it is piezoelectric or six modes in the general case, if the layer is anisotropic nonpiezoelectric or isotropic medium (dielectric or metal). For isotropic medium the second component of the mechanical displacement u_2 is decoupled with u_1 and u_3 and may be arbitrary, for example one can set $u_2 = 1$.

Unknown coefficient C_n in (11) and (12) can be determined using the boundary conditions on all the internal boundaries and on the external surface of the upper layer. Unfortunately it is impossible to formulate boundary conditions in the universal form, applicable to all the combinations of the substrate and layers materials. Therefore we must investigate different variants of material combinations separately.

For piezoelectric layers conditions of continuity of the mechanical displacements, electric potential, normal components of the stress tensor and the electric displacement must be satisfied for all the internal boundaries. On the external surface of the top layer normal components of the stress tensor must be equal to zero. If this surface is open (free), the continuity of the normal component of the electric displacement must be satisfied, if this surface is short circuited, then electric potential must be equal to zero. The stress tensor and electric displacement in piezoelectric medium can be calculated by means of following expressions:

$$T_{ij} = c_{ijkl}\frac{\partial u_k}{\partial x_l} + e_{kij}\frac{\partial \varphi}{\partial x_k}, \qquad i, j, k, l = 1, 2, 3 \tag{13}$$

$$D_i = -\varepsilon_{ij}\frac{\partial \varphi}{\partial x_j} + e_{ijk}\frac{\partial u_j}{\partial x_k}, \qquad i, j, k = 1, 2, 3 \tag{14}$$

Substituting (11) and (12) into (13) and (14) we can get following boundary conditions equations:

$$\sum_{n=N_{m-1}+1}^{N_m} C_n\left(\alpha_j^{(n-N_{m-1})}\right)_m \exp[ik(b_3^{(n-N_{m-1})})_m x_3^{(m)}] = \sum_{n=N_m+1}^{N_{m+1}} C_n\left(\alpha_j^{(n-N_m)}\right)_{m+1} \exp[ik(b_3^{(n-N_m)})_{m+1} x_3^{(m)}] \tag{15a}$$

$$\begin{aligned}&\sum_{n=N_{m-1}+1}^{N_m} C_n\left(c_{3jkl}\alpha_k^{(n-N_{m-1})}b_l^{(n-N_{m-1})} + e_{k3j}\alpha_4^{(n-N_{m-1})}b_k^{(n-N_{m-1})}\right)_m \exp[ik(b_3^{(n-N_{m-1})})_m x_3^{(m)}] = \\ &= \sum_{n=N_m+1}^{N_{m+1}} C_n\left(c_{3jkl}\alpha_k^{(n-N_m)}b_l^{(n-N_m)} + e_{k3j}\alpha_4^{(n-N_m)}b_k^{(n-N_m)}\right)_{m+1} \exp[ik(b_3^{(n-N_m)})_{m+1} x_3^{(m)}]\end{aligned} \tag{15b}$$

$$\sum_{n=N_{m-1}+1}^{N_m} C_n(\alpha_4^{(n-N_{m-1})})_m \exp[ik(b_3^{(n-N_{m-1})})_m x_3^{(m)}] = \sum_{n=N_m+1}^{N_{m+1}} C_n(\alpha_4^{(n-N_m)})_{m+1} \exp[ik(b_3^{(n-N_m)})_{m+1} x_3^{(m)}] \tag{15c}$$

$$\begin{aligned}&\sum_{n=N_{m-1}+1}^{N_m} C_n\left(e_{3jk}\alpha_j^{(n-N_{m-1})}b_k^{(n-N_{m-1})} - \varepsilon_{3j}\alpha_4^{(n-N_{m-1})}b_j^{(n-N_{m-1})}\right)_m \exp[ik(b_3^{(n-N_{m-1})})_m x_3^{(m)}] = \\ &= \sum_{n=N_m+1}^{N_{m+1}} C_n\left(e_{3jk}\alpha_j^{(n-N_m)}b_k^{(n-N_m)} - \varepsilon_{3j}\alpha_4^{(n-N_m)}b_j^{(n-N_m)}\right)_{m+1} \exp[ik(b_3^{(n-N_m)})_{m+1} x_3^{(m)}]\end{aligned} \tag{15d}$$

In these equations $j, k, l = 1, 2, 3$, $m = 0, 1, 2, \ldots$ M-1 (not up to M!), where M is the quantity of layers, $x_3^{(m)} = h_1 + h_2 + \ldots + h_m$, $x_3^{(0)} = 0$. Equations (15a) represent the continuity of mechanical displacements, (15b) - the continuity of the stress normal components, (15c) - the continuity of the electrical potential, (15d) - the continuity of the electric displacement normal component. If surface $x_3 = x_3^{(m)}$ is short circuited by metal layer of zero thickness, equations (15c) and (15d) must be changed. The right part of the (15c) must be replaced with zero, the left part of (15d) also must be replaced with zero and the right part of (15d) must be replaced with the right part of (15c).

The boundary conditions equations for stress on the external surface of the top layer (m = M) can be obtained from equations (15b) by replacing the right part of this equation with zero. Analogously by replacing the right part with zero the equation (15c) gives electric boundary condition for the short circuited external surface. In order to formulate the boundary condition on the free external surface, the potential in the free space must be written in the following form:

$$\varphi^{(f)} = \varphi^{(M)} e^{-kb_1(x_3 - x_3^{(M)})}, \qquad x_3 \geq x_3^{(M)} \tag{16}$$

Here $\varphi^{(M)}$ is the potential of the external surface ($x_3 = x_3^{(M)}$). The potential (16) satisfies Laplace equation (that can be checked by direct substitution of (16) into this equation) and vanishes at $x_3 \to \infty$.

The normal component of the electric displacement in the free space:

$$D_3^{(f)} = -\varepsilon_0 \frac{\partial \varphi^{(f)}}{\partial x_3} = kb_1\varepsilon_0\varphi^{(M)} e^{-kb_1(x_3 - x_3^{(M)})} \tag{17}$$

Here ε_0 is the dielectric permittivity of the free space. Using the expression (17) we can get the condition of the continuity of the normal component of the electric displacement on the free (open) external surface:

$$\begin{aligned} i \sum_{n=N_{M-1}+1}^{N_M} C_n \left(e_{3jk}\alpha_j^{(n-N_{M-1})} b_k^{(n-N_{M-1})} - \varepsilon_{3j}\alpha_4^{(n-N_{M-1})} b_j^{(n-N_{M-1})} \right)_M \exp[ik(b_3^{(n-N_{M-1})})_M x_3^{(M)}] = \\ = b_1\varepsilon_0 \sum_{n=N_{M-1}+1}^{N_M} C_n (\alpha_4^{(n-N_{M-1})})_M \exp[ik(b_3^{(n-N_{M-1})})_M x_3^{(M)}] \end{aligned} \tag{18}$$

The system of the boundary conditions equations contains $n_0 + n_1 + n_2 + \ldots + n_M$ equations with the same number of unknown coefficients C_n. In general case $n_0 = 4$, $n_1 = n_2 = \ldots = n_M = 8$. For metal layers mechanical boundary conditions are the same as for the previous case (only one must take into account, that piezoelectric constants of layers are zero) and the electric boundary condition is formulated only for the substrate surface:

$$\sum_{n=1}^{n_0} C_n (\alpha_4^{(n)})_0 = 0 \tag{19}$$

This variant of boundary conditions is also valid, if the first layer is metal and all other layers are non-piezoelectric dielectrics and metals in an arbitrary combination. For this variant in the general case $n_0 = 4$, $n_1 = n_2 = \ldots = n_M = 6$.

For isotropic dielectric layers the mechanical boundary conditions are the same as for the previous case. Electric boundary conditions became complicated and multi-variant because any boundary may be either free or short circuited. Only the single variant is simple - the first boundary is short circuited. For this variant the electric boundary condition is presented by the single equation (19), such as for previous case.

In general case the dependence of the potential in the free space is defined by equation (16) and inside the *m*-th dielectric isotropic layer it must be written as:

$$\varphi^{(m)}(x_3) = A_m e^{-kb_1(x_3 - x_3^{(m-1)})} + B_m e^{kb_1(x_3 - x_3^{(m-1)})}, \qquad x_3^{(m-1)} \le x_3 \le x_3^{(m)} \tag{20}$$

Coefficients A_m and B_m can be expressed by potentials on the layer boundaries, which depend on the electric conditions on this boundaries (free or short). Using conditions of the continuity of the potential and the normal component of the electric displacement one can exclude all the boundary potentials and express the potential $\varphi^{(1)}$ in the first layer as function of x_3. This function will content only $\varphi^{(0)}(x_3 = 0)$ - potential on the substrate surface. From the potential $\varphi^{(1)}$ one can express the normal component of the electric displacement on the substrate surface and use the condition of the continuity of this value for formulation of the electric boundary condition equation. This is the single equation, but its view significantly depends on the electric conditions on other boundaries.

If all the boundaries are electrically free and there is only the single layer, the equation, which describes the electric boundary conditions, can be written so:

$$i\sum_{n=1}^{n_0} C_n \left(e_{3jk}\alpha_j^{(n)}b_k^{(n)} - \varepsilon_{3j}\alpha_4^{(n)}b_j^{(n)}\right)_0 = \frac{b_1\varepsilon_1\varepsilon_0}{sh(kb_1h_1)} S_1 \sum_{n=1}^{n_0} C_n(\alpha_4^{(n)})_0 \tag{21a}$$

where

$$S_1 = ch(kb_1h_1) - \frac{\varepsilon_1}{\varepsilon_1 ch(kb_1h_1) + R_2 sh(kb_1h_1)} \tag{21b}$$

Here and hereinafter ε_m (m = 1, 2, ... M) is the relative permittivity of the m-th layer. R_2 in (21b) is the recurrent coefficient, which allows to obtain the equation for two layers from equations (21) for one layer. For the single layer $R_2 = 1$, and for two layers:

$$R_2 = \frac{\varepsilon_2}{sh(kb_1h_2)} S_2 \tag{22}$$

I.e. for two layers the electric boundary condition has the following view:

$$i\sum_{n=1}^{n_0} C_n \left(e_{3jk}\alpha_j^{(n)}b_k^{(n)} - \varepsilon_{3j}\alpha_4^{(n)}b_j^{(n)}\right)_0 = \frac{b_1\varepsilon_1\varepsilon_0}{sh(kb_1h_1)} \left[ch(kb_1h_1) - \frac{\varepsilon_1}{\varepsilon_1 ch(kb_1h_1) + \frac{\varepsilon_2 sh(kb_1h_1)}{sh(kb_1h_2)} S_2} \right] \sum_{n=1}^{n_0} C_n(\alpha_4^{(n)})_0 \tag{23a}$$

where:

$$S_2 = ch(kb_1h_2) - \frac{\varepsilon_2}{\varepsilon_2 ch(kb_1h_2) + R_3 sh(kb_1h_2)} \tag{23b}$$

The recurrent coefficient R_3 gives possibility to obtain the equation for three layers from equation for two layers:

$$R_3 = \frac{\varepsilon_3}{sh(kb_1h_3)} S_3 \tag{24}$$

For three layers:

$$S_3 = ch(kb_1h_3) - \frac{\varepsilon_3}{\varepsilon_3 ch(kb_1h_3) + R_4 sh(kb_1h_3)} \tag{25}$$

For three layers $R_4 = 1$, and for more than three:

$$R_4 = \frac{\varepsilon_4}{sh(kb_1h_4)} S_4 \tag{26}$$

And so on, i.e. the equation of electric boundary conditions for $m + 1$ layers may be obtained from the equation for m layers by using the recurrent coefficient R_{m+1} ($R_{M+1} = 1$, if M is the total number of layers). To obtain the equation for M layers one must write equation for one layer, then for two layers and so on until the equation for M layers will be obtained.

If one of the boundary surfaces $x_3 = x_3^{(m)}$ is short circuited (metalized), then electric conditions of all the further boundaries are unimportant, because the electric field outside the short circuited surface ($x_3 > x_3^{(m)}$) is equal to zero. The same result will be, if the layer $m + 1$ is metal and all the further layers are metals and dielectrics in arbitrary combination. To obtain the electric boundary condition equation in this case one has to get the equation for m layers with electrically free boundaries as described above. Then one must remain in the expression for S_m (for the last layer before the short circuited surface) only the first term $ch(kb_1h_m)$ and the second term, which contains R_{m+1}, replace with zero. The equation, obtained so, corresponds to the zero potential on the surface $x_3 = x_3^{(m)}$. For example, for case then the second boundary is short circuited, i.e. $\varphi^{(2)} = 0$, the boundary condition equation coincides with (23a), but $S_2 = ch(kb_1h_2)$ must be set in this equation instead of (23b).

So, the single electric boundary condition equation for multi-layer structure must be formulated by one of way, described above, and then full system of the boundary conditions equations must be solved. This equations system can be written in such form:

$$\begin{aligned} &a_{11}C_1 + a_{12}C_2 + ... + a_{1N}C_N = 0 \\ &a_{21}C_1 + a_{22}C_2 + ... + a_{2N}C_N = 0 \\ &\dots\dots\dots\dots\dots\dots\dots\dots\dots \\ &a_{N1}C_1 + a_{N2}C_2 + ... + a_{NN}C_N = 0 \end{aligned} \tag{27}$$

The order N of this system is equal to total quantity of partial modes of all the structure: $N = n_0 + n_1 + \dots n_M$.

For nontrivial solution of this system its determinant must be equal to zero:

$$\begin{vmatrix} a_{11} & a_{12} & \cdots & a_{1N} \\ a_{21} & a_{22} & \cdots & a_{2N} \\ \cdots & \cdots & \cdots & \cdots \\ a_{N1} & a_{N2} & \cdots & a_{NN} \end{vmatrix} = 0 \tag{28}$$

The simplest example is one metal layer on the piezoelectric substrate (or one arbitrary nonpiezoelectric layer with shorted (metalized) bottom surface). The order of boundary conditions determinant is 10 for this case and its coefficients a_{qn} have the such view:

$$\left.\begin{array}{ll} a_{qn} = (\alpha_j^{(n)})_0 & n=1,...,4 \\ a_{qn} = -(\alpha_j^{(n-4)})_1 & n=5,...,10 \end{array}\right\} \begin{array}{c} q=1,2,3 \\ j=q \end{array} \quad \left.\begin{array}{ll} a_{qn} = \left(c_{3jkl}\alpha_k^{(n)}b_l^{(n)} + e_{k3j}\alpha_4^{(n)}b_k^{(n)}\right)_0 & n=1,...,4 \\ a_{qn} = -\left(c_{3jkl}\alpha_k^{(n-4)}b_l^{(n-4)}\right)_1 & n=5,...,10 \end{array}\right\} \begin{array}{l} q=4,5,6 \\ j=q-3 \end{array}$$

$$\left.\begin{array}{ll} a_{qn} = 0 & n=1,...,4 \\ a_{qn} = \left(c_{3jkl}\alpha_k^{(n-4)}b_l^{(n-4)}\right)_1 \exp[ik(b_3^{(n-4)})_1 h_1] & n=5,...,10 \end{array}\right\} \begin{array}{c} q=7,8,9 \\ j=q-6 \end{array} \quad \left.\begin{array}{ll} a_{qn} = (\alpha_4^{(n)})_0 & n=1,...,4 \\ a_{qn} = 0 & n=5,...,10 \end{array}\right\} q=10 \tag{29}$$

Here the first three strings ($q = 1, 2, 3$) represent the continuity of the three components ($j = 1, 2, 3$) of mechanical displacements on the substrate surface ($x_3^{(0)} = 0$), the second three strings ($q = 4, 5, 6$) are the continuity of the three normal components ($j = 1, 2, 3$) of the mechanical stress on the substrate surface ($x_3^{(0)} = 0$), the third three strings ($q = 7, 8, 9$) are three ($j = 1, 2, 3$) zero normal components of the mechanical stress on the top surface of the layer ($x_3^{(1)} = h_1$), and the last string ($q = 10$) expresses the zero electric potential on the substrate surface ($x_3^{(0)} = 0$).

For two metal layers (or the first layer is metal and the second layer is an arbitrary nonpiezoelectric material, or two arbitrary nonpiezoelectric layers with shorted bottom surface of the first layer):

$$\left.\begin{array}{ll} a_{qn}=(\alpha_j^{(n)})_0 & n=1,...,4 \\ a_{qn}=-(\alpha_j^{(n-4)})_1 & n=5,...,10 \\ a_{qn}=0 & n=11,...,16 \end{array}\right\} \begin{array}{c} q=1,2,3 \\ j=q \end{array} \qquad \left.\begin{array}{ll} a_{qn}=\left(c_{3jkl}\alpha_k^{(n)}b_l^{(n)}+e_{k3j}\alpha_4^{(n)}b_k^{(n)}\right)_0 & n=1,...,4 \\ a_{qn}=-\left(c_{3jkl}\alpha_k^{(n-4)}b_l^{(n-4)}\right)_1 & n=5,...,10 \\ a_{qn}=0 & n=11,...,16 \end{array}\right\} \begin{array}{c} q=4,5,6 \\ j=q-3 \end{array}$$

$$\left.\begin{array}{ll} a_{qn}=0 & n=1,...,4 \\ a_{qn}=(\alpha_j^{(n-4)})_1 \exp[ik(b_3^{(n-4)})_1 h_1] & n=5,...,10 \\ a_{qn}=-(\alpha_j^{(n-10)})_2 \exp[ik(b_3^{(n-10)})_2 h_1] & n=11,...,16 \end{array}\right\} \begin{array}{c} q=7,8,9 \\ j=q-6 \end{array}$$

$$\left.\begin{array}{ll} a_{qn}=0 & n=1,...,4 \\ a_{qn}=\left(c_{3jkl}\alpha_k^{(n-4)}b_l^{(n-4)}\right)_1 \exp[ik(b_3^{(n-4)})_1 h_1] & n=5,...,10 \\ a_{qn}=-\left(c_{3jkl}\alpha_k^{(n-10)}b_l^{(n-10)}\right)_2 \exp[ik(b_3^{(n-10)})_2 h_1] & n=11,...,16 \end{array}\right\} \begin{array}{c} q=10,11,12 \\ j=q-9 \end{array} \qquad (30)$$

$$\left.\begin{array}{ll} a_{qn}=0 \quad n=1,...,10 & \\ a_{qn}=\left(c_{3jkl}\alpha_k^{(n-10)}b_l^{(n-10)}\right)_2 \exp[ik(b_3^{(n-10)})_2(h_1+h_2)] & n=11,...,16 \end{array}\right\} \begin{array}{c} q=13,14,15 \\ j=q-12 \end{array}$$

$$\left.\begin{array}{ll} a_{qn}=(\alpha_4^{(n)})_0 & n=1,...,4 \\ a_{qn}=0 & n=5,...,16 \end{array}\right\} q=16$$

The first six strings represent continuity of the displacements (q = 1, 2, 3) and the stresses (q = 4, 5, 6) on the bottom surface of the first layer, the second six strings - continuity of the displacements (q = 7, 8, 9) and stresses (q = 10, 11, 12) on the bottom surface of the second layer, the strings up 13 to 15 – zero stress on the top surface of the second (top) layer, and the last string (q = 16) – zero potential on the bottom surface of the first layer.

The next examples are the isotropic dielectric layers on the piezoelectric substrate.

For one isotropic dielectric layer with both open surfaces the first 9 strings of the boundary conditions determinant are the same as in (29) and the last string is:

$$\left.\begin{array}{ll} a_{qn}=i\left(e_{3jk}\alpha_j^{(n)}b_k^{(n)}-\varepsilon_{3j}\alpha_4^{(n)}b_j^{(n)}\right)_0-\dfrac{b_1\varepsilon_1\varepsilon_0}{sh(kb_1h_1)}S_1(\alpha_4^{(n)})_0 & n=1,...,4 \\ a_{qn}=0 & n=5,...,10 \end{array}\right\} q=10 \qquad (31)$$

where S_1 is represented by (21b) at $R_2 = 1$.

For one isotropic dielectric layer with the open bottom surface and the shorted top surface the expression (31) is valid, but $S_1 = ch(kb_1h_1)$.

For one isotropic dielectric layer with bottom shorted surface the boundary conditions determinant coincides with (29) completely.

For two isotropic dielectric layers with all open surfaces the first 15 strings of the boundary conditions determinant are the same as in (30) and the last string is:

$$\left.\begin{array}{ll} a_{qn}=i\left(e_{3jk}\alpha_j^{(n)}b_k^{(n)}-\varepsilon_{3j}\alpha_4^{(n)}b_j^{(n)}\right)_0-\dfrac{b_1\varepsilon_1\varepsilon_0}{sh(kb_1h_1)}S_1(\alpha_4^{(n)})_0 & n=1,\ldots,4 \\ a_{qn}=0 & n=5,\ldots,16 \end{array}\right\} q=16 \quad (32)$$

where one must use (21b) for S_1, (22) for R_2, and (23b) for S_2 ($R_3 = 1$ must be set in (23b)).
For two isotropic dielectric layers with the top shorted surface of the top layer (all other boundaries are open) the expression (32) is valid, but $S_2 = ch(kb_1h_2)$ instead of (23b).
For two isotropic dielectric layers with the bottom shorted surface of the top layer the expression (32) is valid, but $S_1 = ch(kb_1h_1)$ instead of (21b), and (22), (23b) are not needed.
If the bottom surface of the first layer is shorted, the boundary conditions determinant coincides with (30) completely.
And now we will consider some examples with piezoelectric layers.
For one piezoelectric layer with open surfaces the boundary conditions determinant contains 12 strings and 12 columns and elements of this determinant are:

$$\left.\begin{array}{ll} a_{qn}=(\alpha_j^{(n)})_0 & n=1,\ldots,4 \\ a_{qn}=-(\alpha_j^{(n-4)})_1 & n=5,\ldots,12 \end{array}\right\} \begin{array}{l} q=1,2,3 \\ j=q \end{array} \quad \left.\begin{array}{ll} a_{qn}=\left(c_{3jkl}\alpha_k^{(n)}b_l^{(n)}+e_{k3j}\alpha_4^{(n)}b_k^{(n)}\right)_0 & n=1,\ldots,4 \\ a_{qn}=-\left(c_{3jkl}\alpha_k^{(n-4)}b_l^{(n-4)}+e_{k3j}\alpha_4^{(n-4)}b_k^{(n-4)}\right)_1 & n=5,\ldots,12 \end{array}\right\} \begin{array}{l} q=4,5,6 \\ j=q-3 \end{array}$$

$$\left.\begin{array}{ll} a_{qn}=(\alpha_4^{(n)})_0 & n=1,\ldots,4 \\ a_{qn}=-(\alpha_4^{(n-4)})_1 & n=5,\ldots,12 \end{array}\right\} q=7 \quad \left.\begin{array}{ll} a_{qn}=\left(e_{3jk}\alpha_j^{(n)}b_k^{(n)}-\varepsilon_{3j}\alpha_4^{(n)}b_j^{(n)}\right)_0 & n=1,\ldots,4 \\ a_{qn}=-\left(e_{3jk}\alpha_j^{(n-4_m)}b_k^{(n-4)}-\varepsilon_{3j}\alpha_4^{(n-4)}b_j^{(n-4)}\right)_1 & n=5,\ldots,12 \end{array}\right\} q=8$$

$$\left.\begin{array}{ll} a_{qn}=0 & n=1,\ldots,4 \\ a_{qn}=\left(c_{3jkl}\alpha_k^{(n-4)}b_l^{(n-4)}+e_{k3j}\alpha_4^{(n-4)}b_k^{(n-4)}\right)_1\exp[ik(b_3^{(n-4)})_1h_1] & n=5,\ldots,12 \end{array}\right\} \begin{array}{l} q=9,10,11 \\ j=q-8 \end{array} \quad (33)$$

$$\left.\begin{array}{ll} a_{qn}=0 & n=1,\ldots,4 \\ a_{qn}=\left\{i\left(e_{3jk}\alpha_j^{(n-4)}b_k^{(n-4)}-\varepsilon_{3j}\alpha_4^{(n-4)}b_j^{(n-4)}\right)_1-b_1\varepsilon_0(\alpha_4^{(n-4)})_1\right\}\exp[ik(b_3^{(n-4)})_1h_1] & n=5,\ldots,12 \end{array}\right\} q=12$$

Here the first three strings (q = 1, 2, 3) represent the continuity of the three components (j = 1, 2, 3) of mechanical displacements on the substrate surface ($x_3^{(0)} = 0$), the next three strings (q = 4, 5, 6) are the continuity of the three normal components (j = 1, 2, 3) of the mechanical stress on the substrate surface ($x_3^{(0)} = 0$), the next string (q = 7) - continuity of the electric potential on the same surface, then (q = 8) – continuity of the normal component of the electric displacement on the substrate surface ($x_3^{(0)} = 0$), the next three strings (q = 9, 10, 11) are three (j = 1, 2, 3) zero normal components of the mechanical stress on the top surface of the layer ($x_3^{(1)} = h_1$), and the last string (q = 12) expresses the continuity of the normal component of the electric displacement on the open top surface of the layer ($x_3^{(1)} = h_1$).
For one piezoelectric layer with shorted bottom surface and open top one the expressions (33) are valid, excepting the strings 7 and 8 (q = 7 and 8), which must be replaced with:

$$\left.\begin{array}{ll} a_{qn}=(\alpha_4^{(n)})_0 & n=1,...,4 \\ a_{qn}=0 & n=5,...,12 \end{array}\right\} q=7 \qquad \left.\begin{array}{ll} a_{qn}=0 & n=1,...,4 \\ a_{qn}=(\alpha_4^{(n-4)})_1 & n=5,...,12 \end{array}\right\} q=8 \qquad (34a)$$

These expressions represent the zero electric potential of the bottom surface of the layer (the substrate surface).

For one piezoelectric layer with shorted top surface and open bottom one the expressions (33) are valid, excepting the last string (q = 12), which must be replaced with:

$$\left.\begin{array}{ll} a_{qn}=0 & n=1,...,4 \\ a_{qn}=(\alpha_4^{(n-4)})_1\exp[ik(b_3^{(n-4)})_1h_1] & n=5,...,12 \end{array}\right\} q=12 \qquad (34b)$$

which corresponds to the zero electric potential of the top surface of the layer.

For one piezoelectric layer with both shorted surface one can use expressions (33), in which strings 7 and 8 must be replaced with (34a) and string 12 - with (34b).

For two piezoelectric layers on the piezoelectric substrate with all open surfaces the boundary conditions determinant contains the following 20 strings:

$$\left.\begin{array}{ll} a_{qn}=(\alpha_j^{(n)})_0 & n=1,...,4 \\ a_{qn}=-(\alpha_j^{(n-4)})_1 & n=5,...,12 \\ a_{qn}=0 & n=13,...,20 \end{array}\right\} \begin{array}{c} q=1,2,3 \\ j=q \end{array} \qquad \left.\begin{array}{ll} a_{qn}=\left(c_{3jkl}\alpha_k^{(n)}b_l^{(n)}+e_{k3j}\alpha_4^{(n)}b_k^{(n)}\right)_0 & n=1,...,4 \\ a_{qn}=-\left(c_{3jkl}\alpha_k^{(n-4)}b_l^{(n-4)}+e_{k3j}\alpha_4^{(n-4)}b_k^{(n-4)}\right)_1 & n=5,...,12 \\ a_{qn}=0 & n=13,...,20 \end{array}\right\} \begin{array}{c} q=4,5,6 \\ j=q-3 \end{array}$$

$$\left.\begin{array}{ll} a_{qn}=(\alpha_4^{(n)})_0 & n=1,...,4 \\ a_{qn}=-(\alpha_4^{(n-4)})_1 & n=5,...,12 \\ a_{qn}=0 & n=13,...,20 \end{array}\right\} q=7 \qquad \left.\begin{array}{ll} a_{qn}=\left(e_{3jk}\alpha_j^{(n)}b_k^{(n)}-\varepsilon_{3j}\alpha_4^{(n)}b_j^{(n)}\right)_0 & n=1,...,4 \\ a_{qn}=-\left(e_{3jk}\alpha_j^{(n-4_m)}b_k^{(n-4)}-\varepsilon_{3j}\alpha_4^{(n-4)}b_j^{(n-4)}\right)_1 & n=5,...,12 \\ a_{qn}=0 & n=13,...,20 \end{array}\right\} q=8$$

$$\left.\begin{array}{ll} a_{qn}=0 & n=1,...,4 \\ a_{qn}=(\alpha_j^{(n-4)})_1\exp[ik(b_3^{(n-4)})_1h_1] & n=5,...,12 \\ a_{qn}=-(\alpha_j^{(n-12)})_2\exp[ik(b_3^{(n-12)})_2h_1] & n=13,...,20 \end{array}\right\} q=9,10,11 \;\; j=q-8$$

$$\left.\begin{array}{ll} a_{qn}=0 & n=1,...,4 \\ a_{qn}=\left(c_{3jkl}\alpha_k^{(n-4)}b_l^{(n-4)}+e_{k3j}\alpha_4^{(n-4)}b_k^{(n-4)}\right)_1\exp[ik(b_3^{(n-4)})_1h_1] & n=5,...,12 \\ a_{qn}=-\left(c_{3jkl}\alpha_k^{(n-12)}b_l^{(n-12)}+e_{k3j}\alpha_4^{(n-12)}b_k^{(n-12)}\right)_2\exp[ik(b_3^{(n-12)})_2h_1] & n=13,...,20 \end{array}\right\} q=12,13,14 \;\; j=q-11$$

$$\left.\begin{array}{ll} a_{qn}=0 & n=1,...,4 \\ a_{qn}=(\alpha_4^{(n-4)})_1\exp[ik(b_3^{(n-4)})_1h_1] & n=5,...,12 \\ a_{qn}=-(\alpha_4^{(n-12)})_2\exp[ik(b_3^{(n-12)})_2h_1] & n=13,...20 \end{array}\right\} q=15 \qquad (35)$$

$$\left.\begin{array}{ll} a_{qn}=0 & n=1,...,4 \\ a_{qn}=\left(e_{3jk}\alpha_j^{(n-4)}b_k^{(n-4)}-\varepsilon_{3j}\alpha_4^{(n-4)}b_j^{(n-4)}\right)_1 \exp[ik(b_3^{(n-4)})_1 h_1] & n=5,...,12 \\ a_{qn}=-\left(e_{3jk}\alpha_j^{(n-12)}b_k^{(n-12)}-\varepsilon_{3j}\alpha_4^{(n-12)}b_j^{(n-12)}\right)_2 \exp[ik(b_3^{(n-12)})_2 h_1] & n=13,...,20 \end{array}\right\} q=16$$

$$\left.\begin{array}{ll} a_{qn}=0 & n=1,...,12 \\ a_{qn}=\left(c_{3jkl}\alpha_k^{(n-12)}b_l^{(n-12)}+e_{k3j}\alpha_4^{(n-12)}b_k^{(n-12)}\right)_2 \exp[ik(b_3^{(n-12)})_2(h_1+h_2)] & n=13,...,20 \end{array}\right\} \begin{array}{l} q=17,18,19 \\ j=q-16 \end{array}$$

$$\left.\begin{array}{l} a_{qn}=0 \qquad\qquad n=1,...,12 \\ a_{qn}=\left\{i\left(e_{3jk}\alpha_j^{(n-12)}b_k^{(n-12)}-\varepsilon_{3j}\alpha_4^{(n-12)}b_j^{(n-12)}\right)_2-b_1\varepsilon_0(\alpha_4^{(n-12)})_2\right\}\times \\ \qquad\times\exp[ik(b_3^{(n-12)})_2(h_1+h_2)] \quad n=13,...,20 \end{array}\right\} q=20$$

Here the first three strings (q = 1, 2, 3) represent the continuity of the three components (j = 1, 2, 3) of mechanical displacements on the substrate surface ($x_3^{(0)} = 0$), the next three strings (q = 4, 5, 6) are the continuity of the three normal components (j = 1, 2, 3) of the mechanical stress on the substrate surface ($x_3^{(0)} = 0$), the next string (q = 7) - continuity of the electric potential on the same surface, then (q = 8) – continuity of the normal component of the electric displacement on the substrate surface ($x_3^{(0)} = 0$), strings 9, 10, 11 - the continuity of the three components (j = 1, 2, 3) of mechanical displacements on the surface between the first and the second layers ($x_3^{(1)} = h_1$), strings 12, 13, 14 - the continuity of the three components (j = 1, 2, 3) of mechanical stress on the surface between the first and the second layers ($x_3^{(1)} = h_1$), the next string (q = 15) - continuity of the electric potential on the same surface, then (q = 16) – continuity of the normal component of the electric displacement on the same surface, the next three strings (q = 17, 18, 19) are three (j = 1, 2, 3) zero normal components of the mechanical stress on the top surface of the top layer ($x_3^{(2)} = h_1 + h_2$), and the last string (q = 20) expresses the continuity of the normal component of the electric displacement on the open top surface of the top layer ($x_3^{(2)} = h_1 + h_2$).
For two piezoelectric layers on the piezoelectric substrate with shorted bottom surface of the first layer and with the open other surfaces strings number 7 and 8 in expressions (35) must be replaced with:

$$\left.\begin{array}{ll} a_{qn}=(\alpha_4^{(n)})_0 & n=1,...,4 \\ a_{qn}=0 & n=5,...,20 \end{array}\right\} q=7 \qquad \left.\begin{array}{ll} a_{qn}=0 & n=1,...,4 \\ a_{qn}=(\alpha_4^{(n-4)})_1 & n=5,...,12 \\ a_{qn}=0 & n=13,...,20 \end{array}\right\} q=8 \qquad (36a)$$

For two piezoelectric layers on the piezoelectric substrate with shorted bottom surface of the second layer and with open other surfaces strings number 15 and 16 in expressions (35) must be replaced with:

$$\left.\begin{array}{ll} a_{qn}=0 & n=1,...,4 \\ a_{qn}=(\alpha_4^{(n-4)})_1\exp[ik(b_3^{(n-4)})_1h_1] & n=5,...,12 \\ a_{qn}=0 & n=13,...20 \end{array}\right\} q=15$$

$$\left.\begin{array}{ll} a_{qn}=0 & n=1,...,12 \\ a_{qn}=(\alpha_4^{(n-12)})_2\exp[ik(b_3^{(n-12)})_2h_1] & n=13,...20 \end{array}\right\} q=16 \qquad (36b)$$

For two piezoelectric layers on the piezoelectric substrate with shorted top surface of the second layer and with open other surfaces the string number 20 in expressions (35) must be replaced with:

$$\left.\begin{array}{ll} a_{qn}=0 & n=1,...,12 \\ a_{qn}=(\alpha_4^{(n-12)})_2\exp[ik(b_3^{(n-12)})_2(h_1+h_2)] & n=13,...20 \end{array}\right\} q=20 \qquad (36c)$$

If two surfaces of three are shorted, then two corresponding expressions of (36a) - (36c) must be used for replacing the corresponding expressions of (35), taking into account, that (36a) "short-circuits" the first surface (the substrate surface), (36b) - the second surface, and (36c) - the third one (the top surface of the top layer).

If all three surfaces are shorted, all expressions (36a) - (36c) must be used for replacing the corresponding expressions in (35).

All the examples, considered above, allow to understand how to form the boundary conditions determinant and for more complicated structures with three, four, five etc. layers, if necessary.

Thus, the determinant of the boundary conditions is formed. Now we have to solve the equation (28). This means we need to find a value of wave velocity (or velocity and attenuation coefficient for the pseudo-surface wave), for which the boundary conditions determinant vanishes. The solution of equation (28) can be found by any available iterative procedure. In our case, we apply our own algorithm to search the global extremum of function of several variables (Dvoesherstov et. al., 1999). Solution corresponds to the global minimum of the function, which is the square of the absolute value of the boundary conditions determinant.

Another widely used method of finding solution is to calculate the effective dielectric permittivity (Adler, 1994):

$$\varepsilon_{eff}=\frac{D_3^{(m)}}{kb_1\varphi^{(m)}} \qquad (37)$$

Here $\varphi^{(m)}$ and $D_3^{(m)}$ - the potential and electric displacement on the top surface of the layer m. Corresponding string of the boundary conditions determinant is used for expression (37). For example, for top surface of the top layer under condition that this layer is piezoelectric, the effective permittivity technique gives the follow equation, which expresses continuity of the dielectric permittivity:

$$\frac{i\sum\limits_{n=N_{M-1}+1}^{N_M} C_n\left(e_{3jk}\alpha_j^{(n)}b_k^{(n)}-\varepsilon_{3j}\alpha_4^{(n)}b_j^{(n)}\right)\exp[ikb_3^{(n)}x_3^{(M)}]}{b_1\varepsilon_0\sum\limits_{n=N_{M-1}+1}^{N_M} C_n\alpha_4^{(n)}\exp[ikb_3^{(n)}x_3^{(M)}]}=\begin{cases}\varepsilon_0\\ \infty\end{cases} \qquad (38)$$

The top value in the right part of this equation corresponds to the open surface, the bottom value (∞) – to short-circuited one. One can see that coefficients C_n are needed for using of this technique. These coefficients are obtained by solving the equations system (27), from those the equation, corresponding to the surface number *m*, is excluded. For example, in our case one must exclude the last equation of this system (corresponding to the last string of the boundary conditions determinant). The system (27) is uniform and its solution is defined with an accuracy up to an arbitrary coefficient. Therefore after excluding one of the equation from this system we can set any C_n of any value, for example $C_N = 1$ and then solve the *N-1* power nonuniform system and to obtain all the coefficients C_n for using the equation (38). This procedure is repeating for different values of the wave velocity (or the velocity and the attenuation coefficients) until the equation (38) is satisfied. We used the global search procedure for equation (38) solving (Dvoesherstov et. al., 1999). Calculations by using the boundary conditions determinant (solving the system (27) in this case is not required) and by using the effective dielectric permittivity are mathematically equivalent each other and give the same result. But in some cases one technique gives result with better reliability than another, and in other cases – contrary. Our soft contains both techniques and one can easily switch from one to another by the single mouse click. When the wave velocity (or the velocity and the attenuation coefficient) is obtained, one can calculate all the coefficients C_n by solving the equation system (27) and then the wave amplitudes for any x_3 coordinate in any medium by substitution C_n into (11) and (12).

After the calculation of the wave phase velocity one can obtain all the wave propagation characteristics: an electromechanical coupling coefficient, a temperature coefficient of delay, a power flow angle, a diffraction parameter. Dependences of the layers thickness and theirs mass density on a temperature, which are needed for temperature coefficient of delay calculations, one can find, for example in (Shimizu et. al., 1976).

All the propagation characteristics can be modified by proper choice of the layer parameters. For example, Fig. 2 shows dependences of the temperature coefficient of delay (*TCD*) on quartz with single Al and Au layer on the second Euler angle and on the relative layer thickness. Material constants for quartz are taken from (Shimizu and Yamamoto, 1980), for

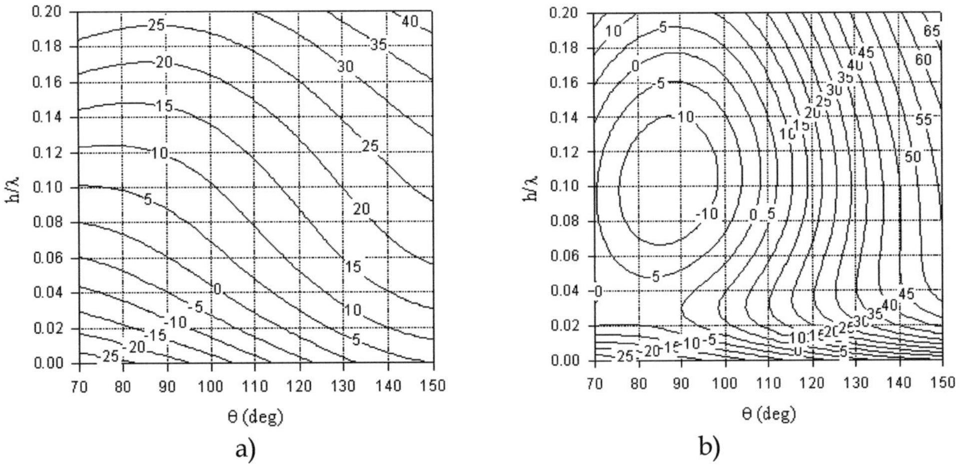

a) b)

Fig. 2. Dependence of *TCD* (ppm/°C) on the 2nd Euler angle θ and on the relative layer thickness h/λ for Al (a) and Au (b). The first and third Euler angles are equal to zero.

Al and Au - from (Ballandras et. al., 1997). One can see in Fig. 2, that negative values of *TCD* can be compensated by metallic layer. For example, orientation YX-quartz (0°,90°,0°) becomes thermostable if $h/\lambda = 0.061$ for Al layer and YX-quartz keeps the temperature stability in range $0.027 \le h/\lambda \le 0.032$ for Au layer.
So, multilayer structures can be used both for protection against external undesired influence and for improvement of the wave propagation characteristics, i.e. the SAW device properties. All these possibilities can be evaluated by means of calculation technique, described here.

3. Bulk acoustic waves in multilayer structures

Bulk acoustic waves are used in film bulk acoustic resonators. The simplest such resonator contains at least three layers, namely an active piezoelectric layer, in which transformation of the electric signal into the acoustic wave takes place, and two metallic (usually aluminum) electrodes, connected to the source of the electric signal. The structure of such resonator (named membrane type resonator) is schematically shown in Fig. 3a.

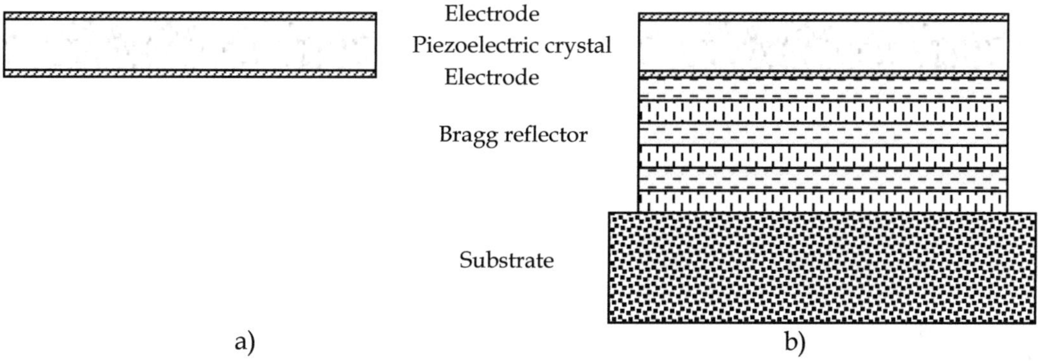

Fig. 3. Schematic view of the membrane type film bulk acoustic wave resonator (a) and of the SMR resonator (b).

FBAR resonators are used in the ultra high frequency range (several GHz and higher), therefore a thickness of the active layer is very small (microns and less). There are some problems with mounting of such small structures on the solid massive and relatively thick substrate. It is impossible to place membrane type FBAR on the substrate directly, because in this case the useful signal will be deformed by multiple spurious oscillation modes due to an acoustic interaction of the resonator and the substrate. To prevent this interaction more complicated constructions are required. In particular, an air gap between the bottom electrode and the substrate must be provided or cavity in substrate under a bottom electrode must be etched. These variants require rather complex technological processes application. Another possibility is mounting the multilayer Bragg reflector directly on the substrate and then mounting of the resonator directly on this reflector. Such construction is named a solid mounted resonator (SMR). The structure of such resonator schematically is shown in Fig. 3b.
The Bragg reflector contains several (3 - 5) pairs of two materials with different acoustic properties. The thickness of each layer in the reflector must be equal to a quarter of the

wavelength in its material. Such construction provides attenuation of the wave and prevents an acoustic interaction of the active zone of the resonator and the substrate.
Transversal sizes of the resonator are usually much larger than its total thickness, therefore an analysis of all the main properties may be performed in the one-dimensional approach. The most rigorous one-dimensional theory of such multilayer structures is presented in (Nowotny and Benes, 1987). The following description is based on this theory, some modified for expansion of its possibilities.
The wave equations, describing processes in the solid piezoelectric medium, are the same as for surface acoustic waves - see (1) and (2). Assuming that all the values depend only on the single spatial coordinate x_1 (mechanical displacements u_i along all the coordinates x_i take place in this case nevertheless), we can write simpler form of these equations:

$$c_{1jk1}\frac{\partial^2 u_k}{\partial x_1^2}+e_{11j}\frac{\partial^2 \varphi}{\partial x_1^2}=\rho\frac{\partial^2 u_j}{\partial t^2} \qquad j,k=1,2,3 \tag{39}$$

$$e_{11k}\frac{\partial^2 u_k}{\partial x_1^2}-\varepsilon_{11}\frac{\partial^2 \varphi}{\partial x_1^2}=0 \tag{40}$$

Complex material constants (with real and imaginary parts) can be used for modeling of electro-acoustic losses in the medium.
The solution for the electric potential φ can be obtained from (40) in such form:

$$\varphi=\frac{e_{11k}}{\varepsilon_{11}}u_k+\phi_1 x_1+\phi_0 \tag{41}$$

Here ϕ_0 and ϕ_1 are arbitrary unknown constants.
Substitution (41) into (39) gives:

$$\bar{c}_{1jk1}\frac{\partial^2 u_k}{\partial x_1^2}=\rho\frac{\partial^2 u_j}{\partial t^2} \tag{42}$$

Here $\bar{c}_{1jk1}$ are the stiffened elastic constants:

$$\bar{c}_{1jk1}=c_{1jk1}+\frac{e_{11j}e_{11k}}{\varepsilon_{11}} \tag{43}$$

We will seek the solution of these equations (j = 1, 2, 3) as a sinusoidal wave, propagating along the x_1 axis with the velocity v:

$$u_k(x_1,t)=\beta_k e^{i\omega\left(\frac{x_1}{v}-t\right)}=\beta_k e^{i\omega(\alpha x_1-t)}, \tag{44}$$

where $\alpha=1/v$ is a slowness.
Substitution of (44) into the equations (42) transforms them into the linear algebraic equations system:

$$\bar{c}_{1jk1}\,\beta_k=\bar{c}\,\beta_j, \tag{45}$$

where

$$\bar{c} = \frac{\rho}{\alpha^2} = \rho v^2 \quad (46)$$

In more detailed form the system (45) has the following view:

$$\begin{aligned} &(\bar{c}_{1111} - \bar{c})\beta_1 + \bar{c}_{1121}\beta_2 + \bar{c}_{1131}\beta_3 = 0 \\ &\bar{c}_{1211}\beta_1 + (\bar{c}_{1221} - \bar{c})\beta_2 + \bar{c}_{1231}\beta_3 = 0 \\ &\bar{c}_{1311}\beta_1 + \bar{c}_{1321}\beta_2 + (\bar{c}_{1331} - \bar{c})\beta_3 = 0 \end{aligned} \quad (47)$$

This is a system of linear equations for the three amplitudes β_1, β_2, β_3. This system can have a nontrivial solution only if the determinant of its coefficients is equal to zero:

$$\begin{vmatrix} \bar{c}_{1111} - \bar{c} & \bar{c}_{1121} & \bar{c}_{1131} \\ \bar{c}_{1211} & \bar{c}_{1221} - \bar{c} & \bar{c}_{1231} \\ \bar{c}_{1311} & \bar{c}_{1321} & \bar{c}_{1331} - \bar{c} \end{vmatrix} = 0 \quad (48)$$

It gives the third power polynomial equation for $\bar{c}$, i.e. for ρv^2. Three roots of this equation will represent the three eigenvalues $\bar{c}^{(n)}$ (n = 1, 2, 3), giving three values of the bulk wave velocity $v^{(n)}$ or three values of the slowness $\alpha^{(n)}$.

Three values $\beta_k^{(n)}$ (k = 1, 2, 3) correspond to each value $\bar{c}^{(n)}$. These values $\beta_k^{(n)}$ are obtained by solving the system (47) for each value $\bar{c}^{(n)}$ and represent the eigenvector. System (47) is homogeneous, so its solution is determined up to an arbitrary factor. Consequently, we can normalize each eigenvector by its modulus, and work further with the normalized dimensionless vector. The three normalized eigenvectors are complete and orthogonal:

$$\beta_k^{(n)}\beta_k^{(m)} = \delta_{nm}, \quad \sum_n \beta_k^{(n)}\beta_l^{(n)} = \delta_{kl} \quad (\delta_{kl} \text{ is the Kronecker symbol}) \quad (49)$$

The general solution of the equations system (42) we will seek in such view:

$$u_k(x_1, t) = u_k(x_1)e^{-i\omega t}, \quad (50)$$

where $u_k(x_1)$ is the linear combination of three bulk waves, obtained from equations (47) and (48):

$$u_k(x_1) = \sum_{n=1}^{3} \beta_k^{(n)}[A^{(n)}\cos(\alpha^{(n)}\omega x_1) + B^{(n)}\sin(\alpha^{(n)}\omega x_1)] \quad (51)$$

Here $A^{(n)}$ and $B^{(n)}$ are six unknown coefficients of the linear combination. Together with ϕ_0 and ϕ_1 we have the eight unknown coefficients to be defined further.
We need the eight boundary conditions for obtaining the eight unknown coefficients. We will use three normal components of the stress tensor, three components of the mechanical displacement, the normal component of the electric displacement and the electric potential for some concrete coordinate x_1, for example for $x_1 = 0$, as boundary values for unknown coefficients determination.
The mechanical displacements and the electric potential are determined by expressions (51) and (41) respectively, and for the stress tensor and for electric displacement the following expressions are valid:

$$T_{1j}=c_{1jk1}\frac{\partial u_k}{\partial x_1}+e_{11j}\frac{\partial \varphi}{\partial x_1}=\sum_n \beta_j^{(n)}\bar{c}^{(n)}\alpha^{(n)}\omega[-A^{(n)}\sin(\alpha^{(n)}\omega x_1)+B^{(n)}\cos(\alpha^{(n)}\omega x_1)]+e_{11j}\phi_1 \tag{52}$$

$$D_1=e_{11k}\frac{\partial u_k}{\partial x_1}-\varepsilon_{11}\frac{\partial \varphi}{\partial x_1}=-\varepsilon_{11}\phi_1 \tag{53}$$

Substituting $x_1 = 0$ into (41) and (51) - (53), we get the following eight equations for determination of $A^{(n)}$, $B^{(n)}$, ϕ_0, and ϕ_1:

$$u_j(0)=\sum_n \beta_j^{(n)}A^{(n)} \qquad T_{1j}(0)=\sum_n \beta_j^{(n)}\bar{c}^{(n)}\alpha^{(n)}\omega B^{(n)}+e_{11j}\varphi_1 \tag{54}$$

$$D_1(0)=-\varepsilon_{11}\phi_1 \qquad \varphi(0)=\frac{e_{11k}}{\varepsilon_{11}}u_k(0)+\phi_0 \tag{55}$$

Solving this system (taking into account the completeness and the orthogonality conditions (49)), we can get all the unknown coefficients:

$$A^{(n)}=\beta_k^{(n)}u_k(0) \quad B^{(n)}=\frac{1}{\bar{c}^{(n)}\alpha^{(n)}\omega}\beta_j^{(n)}\left[T_{1j}(0)+\frac{e_{11j}}{\varepsilon_{11}}D_1(0)\right] \quad \phi_0=\varphi(0)-\frac{e_{11k}}{\varepsilon_{11}}u_k(0) \quad \phi_1=-\frac{1}{\varepsilon_{11}}D_1(0) \tag{56}$$

These coefficients (with using (41), (51) - (53)) give the possibility to obtain all the values u_j, T_{1j}, D_1, and φ for any coordinate x_1, if these values are known for $x_1 = 0$ coordinate.
Let us consider in particular the single layer of thickness l, infinite in lateral directions - see Fig. 4.

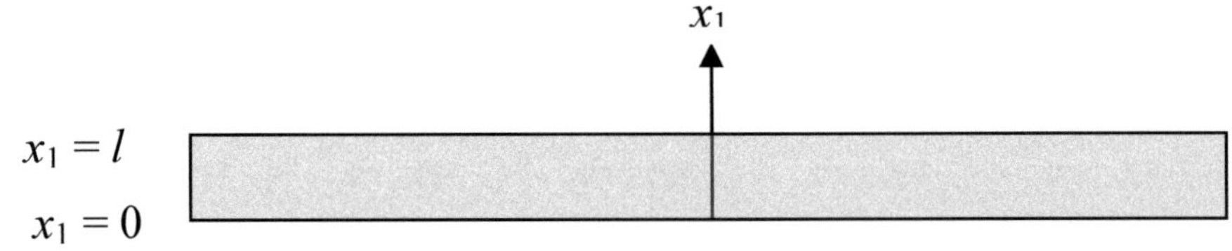

Fig. 4. The single layer of thickness l.

All the values u_j, T_{1j}, D_1, and φ for coordinate $x_1 = l$ can be expressed as a linear combination of these values for coordinate $x_1 = 0$ in the following matrix form:

$$\begin{pmatrix} u_1 \\ u_2 \\ u_3 \\ T_{11} \\ T_{12} \\ T_{13} \\ \varphi \\ D_1 \end{pmatrix}_{x_1=l} = \begin{pmatrix} M_{11}^{uu} & M_{12}^{uu} & M_{13}^{uu} & M_{11}^{uT} & M_{12}^{uT} & M_{13}^{uT} & 0 & M_1^{uD} \\ M_{21}^{uu} & M_{22}^{uu} & M_{23}^{uu} & M_{21}^{uT} & M_{22}^{uT} & M_{23}^{uT} & 0 & M_2^{uD} \\ M_{31}^{uu} & M_{32}^{uu} & M_{33}^{uu} & M_{31}^{uT} & M_{32}^{uT} & M_{33}^{uT} & 0 & M_3^{uD} \\ M_{11}^{Tu} & M_{12}^{Tu} & M_{13}^{Tu} & M_{11}^{TT} & M_{12}^{TT} & M_{13}^{TT} & 0 & M_1^{TD} \\ M_{21}^{Tu} & M_{22}^{Tu} & M_{23}^{Tu} & M_{21}^{TT} & M_{22}^{TT} & M_{23}^{TT} & 0 & M_2^{TD} \\ M_{31}^{Tu} & M_{32}^{Tu} & M_{33}^{Tu} & M_{31}^{TT} & M_{32}^{TT} & M_{33}^{TT} & 0 & M_3^{TD} \\ M_1^{\varphi u} & M_2^{\varphi u} & M_3^{\varphi u} & M_1^{\varphi T} & M_2^{\varphi T} & M_3^{\varphi T} & 1 & M^{\varphi D} \\ 0 & 0 & 0 & 0 & 0 & 0 & 0 & 1 \end{pmatrix} \cdot \begin{pmatrix} u_1 \\ u_2 \\ u_3 \\ T_{11} \\ T_{12} \\ T_{13} \\ \varphi \\ D_1 \end{pmatrix}_{x_1=0} = \mathbf{M} \cdot \begin{pmatrix} u_1 \\ u_2 \\ u_3 \\ T_{11} \\ T_{12} \\ T_{13} \\ \varphi \\ D_1 \end{pmatrix}_{x_1=0} \quad (57)$$

Here 8x8 matrix **M** is the transfer matrix of the single layer. This matrix allows to calculate the values u_j, T_{1j}, D_1, and φ on one surface of the layer via these values on another surface. The elements of the transfer matrix are defined by wave equations solutions (i.e. by material properties of the layer) by such a manner:

$$M_{ij}^{uu} = M_{ij}^{TT} = \sum_n \beta_i^{(n)} \beta_j^{(n)} \cos(\alpha^{(n)} \omega l) \quad M_{ij}^{uT} = \frac{1}{\rho\omega} \sum_n \beta_i^{(n)} \beta_j^{(n)} \alpha^{(n)} \sin(\alpha^{(n)} \omega l) \quad (58)$$

$$M_{ij}^{Tu} = -\rho\omega \sum_n \beta_i^{(n)} \beta_j^{(n)} v^{(n)} \sin(\alpha^{(n)} \omega l) \quad M_i^{uD} = M_i^{\varphi T} = \frac{1}{\omega\sqrt{\rho\varepsilon_{11}}} \sum_n \beta_i^{(n)} k^{(n)} \sin(\alpha^{(n)} \omega l) \quad (59)$$

$$M_i^{TD} = M_i^{\varphi u} = -2\sqrt{\frac{\rho}{\varepsilon_{11}}} \sum_n \beta_i^{(n)} k^{(n)} v^{(n)} \sin^2 \frac{\alpha^{(n)} \omega l}{2} \quad M^{\varphi D} = -\frac{l}{\varepsilon_{11}} \left[1 - \frac{1}{\omega l} \sum_n k^{(n)2} v^{(n)} \sin(\alpha^{(n)} \omega l) \right] \quad (60)$$

Here

$$k^{(n)} = \frac{\beta_j^{(n)} e_{11j}}{\sqrt{\bar{c}^{(n)} \varepsilon_{11}}} \quad (61)$$

In expressions (58) - (61) $i, j = 1, 2, 3$ (a number of the coordinate axis), $n = 1, 2, 3$ (a number of the partial solution of the wave equations). The values $k^{(n)}$, given by (61), are the dimensionless scalar coupling coefficients ($k^{(n)}$ are nonzero only for piezoelectric medium).
One can see from the previous equations that the transfer matrix approaches to the unit matrix if the layer thickness $l \to 0$.
If a layer is nonpiezoelectric dielectric, all the elements of its transfer matrix, containing the value $k^{(n)}$, are zero, excepting $M^{\varphi D}$, and the transfer matrix of the nonpiezoelectric dielectric layer has a simpler form:

$$\mathbf{M}=\begin{pmatrix} M_{11}^{uu} & M_{12}^{uu} & M_{13}^{uu} & M_{11}^{uT} & M_{12}^{uT} & M_{13}^{uT} & 0 & 0 \\ M_{21}^{uu} & M_{22}^{uu} & M_{23}^{uu} & M_{21}^{uT} & M_{22}^{uT} & M_{23}^{uT} & 0 & 0 \\ M_{31}^{uu} & M_{32}^{uu} & M_{33}^{uu} & M_{31}^{uT} & M_{32}^{uT} & M_{33}^{uT} & 0 & 0 \\ M_{11}^{Tu} & M_{12}^{Tu} & M_{13}^{Tu} & M_{11}^{TT} & M_{12}^{TT} & M_{13}^{TT} & 0 & 0 \\ M_{21}^{Tu} & M_{22}^{Tu} & M_{23}^{Tu} & M_{21}^{TT} & M_{22}^{TT} & M_{23}^{TT} & 0 & 0 \\ M_{31}^{Tu} & M_{32}^{Tu} & M_{33}^{Tu} & M_{31}^{TT} & M_{32}^{TT} & M_{33}^{TT} & 0 & 0 \\ 0 & 0 & 0 & 0 & 0 & 0 & 1 & M^{\varphi D} \\ 0 & 0 & 0 & 0 & 0 & 0 & 0 & 1 \end{pmatrix} \quad (62)$$

For a metal layer in an electrostatic approximation the electric potential is always the same on both its surfaces, therefore $M^{\varphi D} = 0$ for metal layer and the transfer matrix of the metal layer has the simplest form:

$$\mathbf{M}=\mathbf{M}_{Em}=\begin{pmatrix} M_{11}^{uu} & M_{12}^{uu} & M_{13}^{uu} & M_{11}^{uT} & M_{12}^{uT} & M_{13}^{uT} & 0 & 0 \\ M_{21}^{uu} & M_{22}^{uu} & M_{23}^{uu} & M_{21}^{uT} & M_{22}^{uT} & M_{23}^{uT} & 0 & 0 \\ M_{31}^{uu} & M_{32}^{uu} & M_{33}^{uu} & M_{31}^{uT} & M_{32}^{uT} & M_{33}^{uT} & 0 & 0 \\ M_{11}^{Tu} & M_{12}^{Tu} & M_{13}^{Tu} & M_{11}^{TT} & M_{12}^{TT} & M_{13}^{TT} & 0 & 0 \\ M_{21}^{Tu} & M_{22}^{Tu} & M_{23}^{Tu} & M_{21}^{TT} & M_{22}^{TT} & M_{23}^{TT} & 0 & 0 \\ M_{31}^{Tu} & M_{32}^{Tu} & M_{33}^{Tu} & M_{31}^{TT} & M_{32}^{TT} & M_{33}^{TT} & 0 & 0 \\ 0 & 0 & 0 & 0 & 0 & 0 & 1 & 0 \\ 0 & 0 & 0 & 0 & 0 & 0 & 0 & 1 \end{pmatrix} \quad (63)$$

The designation "$\mathbf{M}_{Em}$" will be explained further.
Now we can consider a multilayer system. Fig. 5 shows a multilayer structure with arbitrary quantity N of arbitrary layers.

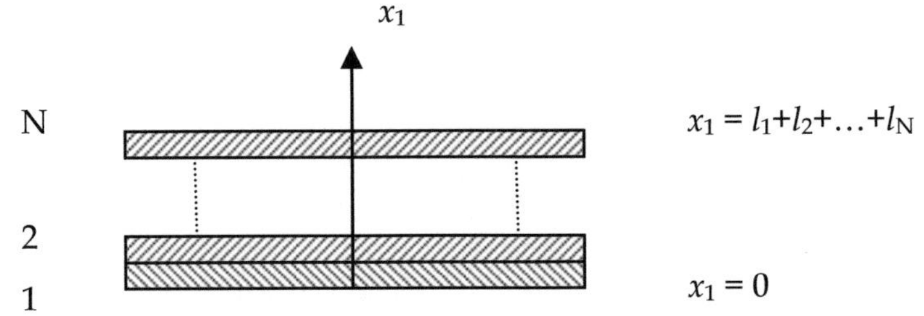

Fig. 5. Multilayer structure.

For multilayer structure the "output" values u_j, T_{1j}, D_1 and φ of the first layer are the "input" values for the second layer and so on. Therefore the transfer matrix of the multilayer structure is a multiplication of the transfer matrices of each layer:

$$\mathbf{M} = \mathbf{M}_N \cdot \ldots \cdot \mathbf{M}_2 \cdot \mathbf{M}_1 \quad (64)$$

The factors sequence must be namely such, as in (64), any transposition is impossible in general case, because A·B ≠ B·A for a matrices multiplication in general case. The matrix **M** in (64) transfers the values u_j, T_{1j}, D_1 and φ from the surface $x_1 = 0$ (bottom) to the surface $x_1 = l_1+l_2+\ldots+l_N$ (top).

All the layers may be arbitrary (piezoelectric, dielectric, metal), but if the layer is used as an electrode, its transfer matrix differs from matrices, described above. It is obviously, that only the metal layer can be used as an electrode. Therefore all the mechanical values and the electric potential of the electrode are transferred by the matrix (63). If the metal layer is not connected to the electric source and is electrically neutral, the matrix (63) transfer the normal component of the electric displacement correctly too, i.e. $(D_1)_{x1=l} = (D_1)_{x1=0}$ (but not inside the metal layer, where $D_1 = 0$). But if the metal layer is connected to the electric source and is used as an electrode, a discontinuity of the value D_1 takes place which is not represented in the matrix (63).

Therefore the special consideration is needed for electrodes. Fig. 6 shows two electrodes, connected to an external harmonic voltage source with an amplitude V and a frequency ω.

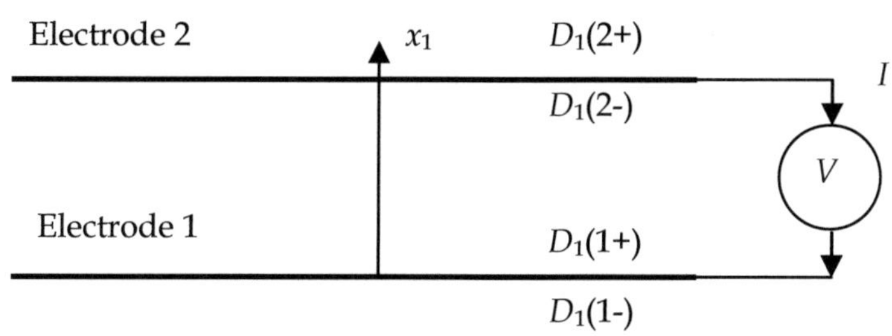

Fig. 6. Two electrodes, connected to an external harmonic voltage source with amplitude V and frequency ω.

First we will consider electrodes of zero thickness. Therefore all the mechanical values are transferred without changes (electric potential is transferred without changes always by metal layer of any thickness).

Values $D_1(1-)$ and $D_1(1+)$ on both sides of the first electrode are different, for the second electrode analogously. The difference $D_1(1+) - D_1(1-)$ is equal to the electric charge per unit area of the electrode (in the SI system). A time derivative of this value is the current density. Its multiplication on the electrode area A gives the total electrode current. For a harmonic signal the time derivative equivalent to a multiplication on $i\omega$. As a result the following expression takes place for a current I_1 of the electrode 1:

$$I_1 = i\omega A[D_1(1+) - D_1(1-)] \tag{65}$$

For electrode 2 analogously. If there are only two electrodes connected to one electric source, then $I = I_1 = - I_2$ and:

$$I = VY, \tag{66}$$

where $V = \varphi_1 - \varphi_2$ (φ_1 and φ_2 are electrode potentials) and Y is an admittance of two electrodes for the external electric source.

We are free in determining the zero point of the electric potential φ and we can choose it so:

$$\varphi_1 + \varphi_2 = 0, \quad \text{i.e. } V = 2\varphi_1 = -2\varphi_2 \tag{67}$$

As a result, we can obtain from (65) and (66):

$$D_1(1+) = \frac{2Y}{i\omega A}\varphi_1 + D_1(1-) \tag{68}$$

which expresses the value of D_1 at the upper side of the electrode as a linear function of the values of φ and D_1 at the lower side (φ has the same value on both sides of an electrode). It means that the transfer matrix of the electrode of zero thickness (an ideal electrode) has a following form:

$$\mathbf{M}_{\mathrm{Ee}} = \begin{pmatrix} 1 & 0 & 0 & 0 & 0 & 0 & 0 & 0 \\ 0 & 1 & 0 & 0 & 0 & 0 & 0 & 0 \\ 0 & 0 & 1 & 0 & 0 & 0 & 0 & 0 \\ 0 & 0 & 0 & 1 & 0 & 0 & 0 & 0 \\ 0 & 0 & 0 & 0 & 1 & 0 & 0 & 0 \\ 0 & 0 & 0 & 0 & 0 & 1 & 0 & 0 \\ 0 & 0 & 0 & 0 & 0 & 0 & 1 & 0 \\ 0 & 0 & 0 & 0 & 0 & 0 & \frac{2Y}{i\omega A} & 1 \end{pmatrix} \tag{69}$$

The metal electrode of a finite thickness (a real electrode) can be presented as a combination of two layers, one of which is the metal electrode of a zero thickness (an ideal electrode), transferring only electric values, and another one is a layer of a finite thickness, transferring only the mechanical values (mechanical layer) - see Fig. 7.

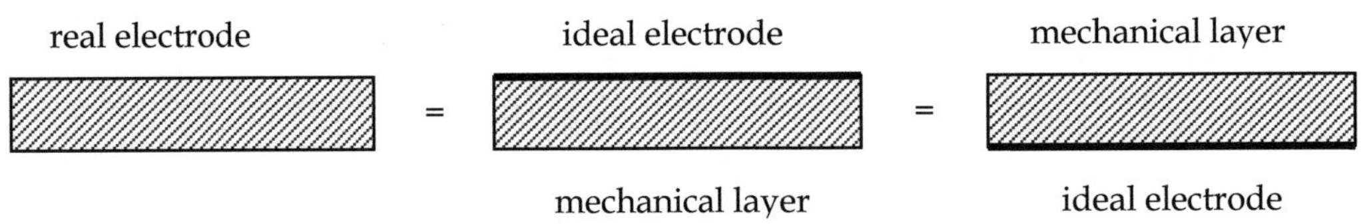

Fig. 7. Representation of a real electrode as a combination of an ideal electrode and a mechanical layer.

Therefore we can obtain the whole transfer matrix of the real electrode as a multiplication of a matrix of the ideal electrode (69) and a matrix, transferring only mechanical values and presented by expression (63):

$$\mathbf{M}_{\mathrm{E}} = \mathbf{M}_{\mathrm{Ee}} \cdot \mathbf{M}_{\mathrm{Em}} = \mathbf{M}_{\mathrm{Em}} \cdot \mathbf{M}_{\mathrm{Ee}} \tag{70}$$

As it was mentioned above, the matrices don't obey the commutative law in general case, but in this concrete case one can transpose these two matrices, what can be checked by direct multiplication. This means, in particular, that an ideal electrode can be placed on any side of the read electrode, as shown in Fig. 7. Physically more correctly to place the ideal electrode on the side which is a face of contact with the interelectrode space.

As a result, the multilayer bulk acoustic wave resonator, containing arbitrary quantity of arbitrary layers, but only with two electrodes, has a view, presented in Fig. 8.

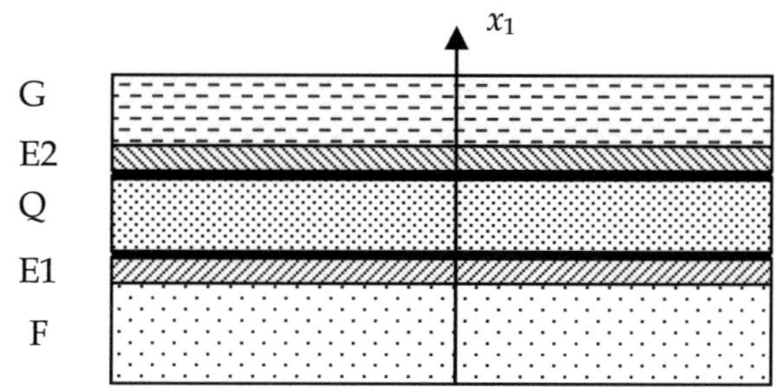

Fig. 8. Multilayer bulk acoustic wave resonator with two electrodes.

Here F is a combination of arbitrary quantity of arbitrary layers under electrodes, G is the same above electrodes, Q is the same between electrodes (at least one of layers in Q must be piezoelectric), E1 and E2 are the two electrodes of a finite thickness.
All the eight values u_j, T_{1j}, D_1 and φ are transferring from a lower surface of the whole construction to its upper surface by the whole transfer matrix, which is the multiplication of transfer matrices of each elements:

$$\mathbf{M}_{FE1QE2G} = \mathbf{M}_G \cdot \mathbf{M}_{E2} \cdot \mathbf{M}_Q \cdot \mathbf{M}_{E1} \cdot \mathbf{M}_F \tag{71}$$

Transfer matrices $\mathbf{M}_F$, $\mathbf{M}_Q$, $\mathbf{M}_G$ are calculated by (64) and matrices $\mathbf{M}_{E1}$ and $\mathbf{M}_{E2}$ – by (70).
Because of electrodes presence the total transfer matrix of the whole resonator $\mathbf{M}_{FE1QE2G}$ does not have generally the special form with 0 and 1 in the 7th column and the 8th row (as in (57)), but it is of the most general form:

$$\boldsymbol{M}_{FE1QE2G} = \begin{pmatrix} M_{11}^{uu} & M_{12}^{uu} & M_{13}^{uu} & M_{11}^{uT} & M_{12}^{uT} & M_{13}^{uT} & M_1^{u\varphi} & M_1^{uD} \\ M_{21}^{uu} & M_{22}^{uu} & M_{23}^{uu} & M_{21}^{uT} & M_{22}^{uT} & M_{23}^{uT} & M_2^{u\varphi} & M_2^{uD} \\ M_{31}^{uu} & M_{32}^{uu} & M_{33}^{uu} & M_{31}^{uT} & M_{32}^{uT} & M_{33}^{uT} & M_3^{u\varphi} & M_3^{uD} \\ M_{11}^{Tu} & M_{12}^{Tu} & M_{13}^{Tu} & M_{11}^{TT} & M_{12}^{TT} & M_{13}^{TT} & M_1^{T\varphi} & M_1^{TD} \\ M_{21}^{Tu} & M_{22}^{Tu} & M_{23}^{Tu} & M_{21}^{TT} & M_{22}^{TT} & M_{23}^{TT} & M_2^{T\varphi} & M_2^{TD} \\ M_{31}^{Tu} & M_{32}^{Tu} & M_{33}^{Tu} & M_{31}^{TT} & M_{32}^{TT} & M_{33}^{TT} & M_3^{T\varphi} & M_3^{TD} \\ M_1^{\varphi u} & M_2^{\varphi u} & M_3^{\varphi u} & M_1^{\varphi T} & M_2^{\varphi T} & M_3^{\varphi T} & M^{\varphi\varphi} & M^{\varphi D} \\ M_1^{Du} & M_2^{Du} & M_3^{Du} & M_1^{DT} & M_2^{DT} & M_3^{DT} & M^{D\varphi} & M^{DD} \end{pmatrix} \tag{72}$$

The expressions, obtained above, allow to calculate the admittance of the resonator Y which is its main work characteristic.
The zero boundary conditions for T_{1j} and D_1 on the external free lower and upper surfaces of the construction are used for these calculations:

$$T_{11} = 0,\ T_{12} = 0,\ T_{13} = 0,\ D_1 = 0 \quad \text{on free surfaces} \tag{73}$$

The normal components of a stress tensor are equal to zero because lower and upper surfaces are free, the electric displacement is zero because the electric field of the external source is concentrated only between two electrodes (between their inner surfaces).

Let us denote the mechanical displacements and the electric potential on the lower free surface as $u_1^{(1)}, u_2^{(1)}, u_3^{(1)}, \varphi^{(1)}$ and the same values on the upper free surface as $u_1^{(2)}, u_2^{(2)}, u_3^{(2)}, \varphi^{(2)}$. Then with taking into account (73) these values will be connected each other by the transfer matrix $\mathbf{M}_{\mathrm{FE1QE2G}}$ by the following expression:

$$\begin{pmatrix} u_1^{(2)} \\ u_2^{(2)} \\ u_3^{(2)} \\ 0 \\ 0 \\ 0 \\ \varphi^{(2)} \\ 0 \end{pmatrix} = \begin{pmatrix} M_{11}^{uu} & M_{12}^{uu} & M_{13}^{uu} & M_{11}^{uT} & M_{12}^{uT} & M_{13}^{uT} & M_1^{u\varphi} & M_1^{uD} \\ M_{21}^{uu} & M_{22}^{uu} & M_{23}^{uu} & M_{21}^{uT} & M_{22}^{uT} & M_{23}^{uT} & M_2^{u\varphi} & M_2^{uD} \\ M_{31}^{uu} & M_{32}^{uu} & M_{33}^{uu} & M_{31}^{uT} & M_{32}^{uT} & M_{33}^{uT} & M_3^{u\varphi} & M_3^{uD} \\ M_{11}^{Tu} & M_{12}^{Tu} & M_{13}^{Tu} & M_{11}^{TT} & M_{12}^{TT} & M_{13}^{TT} & M_1^{T\varphi} & M_1^{TD} \\ M_{21}^{Tu} & M_{22}^{Tu} & M_{23}^{Tu} & M_{21}^{TT} & M_{22}^{TT} & M_{23}^{TT} & M_2^{T\varphi} & M_2^{TD} \\ M_{31}^{Tu} & M_{32}^{Tu} & M_{33}^{Tu} & M_{31}^{TT} & M_{32}^{TT} & M_{33}^{TT} & M_3^{T\varphi} & M_3^{TD} \\ M_1^{\varphi u} & M_2^{\varphi u} & M_3^{\varphi u} & M_1^{\varphi T} & M_2^{\varphi T} & M_3^{\varphi T} & M^{\varphi\varphi} & M^{\varphi D} \\ M_1^{Du} & M_2^{Du} & M_3^{Du} & M_1^{DT} & M_2^{DT} & M_3^{DT} & M^{D\varphi} & M^{DD} \end{pmatrix} \cdot \begin{pmatrix} u_1^{(1)} \\ u_2^{(1)} \\ u_3^{(1)} \\ 0 \\ 0 \\ 0 \\ \varphi^{(1)} \\ 0 \end{pmatrix} \tag{74}$$

From here we can write for the 4th – 6th rows separately and for the 8th row separately:

$$\begin{pmatrix} 0 \\ 0 \\ 0 \end{pmatrix} = \begin{pmatrix} M_{11}^{Tu} & M_{12}^{Tu} & M_{13}^{Tu} \\ M_{21}^{Tu} & M_{22}^{Tu} & M_{23}^{Tu} \\ M_{31}^{Tu} & M_{32}^{Tu} & M_{33}^{Tu} \end{pmatrix} \cdot \begin{pmatrix} u_1^{(1)} \\ u_2^{(1)} \\ u_3^{(1)} \end{pmatrix} + \begin{pmatrix} M_1^{T\varphi} \\ M_2^{T\varphi} \\ M_3^{T\varphi} \end{pmatrix} \cdot \varphi^{(1)} \quad 0 = (M_1^{Du} \quad M_2^{Du} \quad M_3^{Du}) \cdot \begin{pmatrix} u_1^{(1)} \\ u_2^{(1)} \\ u_3^{(1)} \end{pmatrix} + M^{D\varphi} \cdot \varphi^{(1)} \tag{75}$$

We can obtain the vector $u_1^{(1)}, u_2^{(1)}, u_3^{(1)}$ from the first equation (75) (using the standard inverse matrix designation):

$$\begin{pmatrix} u_1^{(1)} \\ u_2^{(1)} \\ u_3^{(1)} \end{pmatrix} = -\begin{pmatrix} M_{11}^{Tu} & M_{12}^{Tu} & M_{13}^{Tu} \\ M_{21}^{Tu} & M_{22}^{Tu} & M_{23}^{Tu} \\ M_{31}^{Tu} & M_{32}^{Tu} & M_{33}^{Tu} \end{pmatrix}^{-1} \cdot \begin{pmatrix} M_1^{T\varphi} \\ M_2^{T\varphi} \\ M_3^{T\varphi} \end{pmatrix} \cdot \varphi^{(1)} \tag{76}$$

Now we can substitute this into the second equation (75) and obtain:

$$0 = -(M_1^{Du} M_2^{Du} M_3^{Du}) \cdot \begin{pmatrix} M_{11}^{Tu} & M_{12}^{Tu} & M_{13}^{Tu} \\ M_{21}^{Tu} & M_{22}^{Tu} & M_{23}^{Tu} \\ M_{31}^{Tu} & M_{32}^{Tu} & M_{33}^{Tu} \end{pmatrix}^{-1} \cdot \begin{pmatrix} M_1^{T\varphi} \\ M_2^{T\varphi} \\ M_3^{T\varphi} \end{pmatrix} \cdot \varphi^{(1)} + M^{D\varphi} \cdot \varphi^{(1)} \tag{77}$$

In an arbitrary case $\varphi^{(1)} \neq 0$, therefore we obtain from (77) the follow scalar equation:

$$(M_1^{Du} M_2^{Du} M_3^{Du}) \cdot \begin{pmatrix} M_{11}^{Tu} & M_{12}^{Tu} & M_{13}^{Tu} \\ M_{21}^{Tu} & M_{22}^{Tu} & M_{23}^{Tu} \\ M_{31}^{Tu} & M_{32}^{Tu} & M_{33}^{Tu} \end{pmatrix}^{-1} \cdot \begin{pmatrix} M_1^{T\varphi} \\ M_2^{T\varphi} \\ M_3^{T\varphi} \end{pmatrix} - M^{D\varphi} = 0 \tag{78}$$

This is the main equation of the problem. It connects the resonator admittance Y with the frequency ω, because Y value is contained in the transfer matrices of electrodes. We can set the concrete value of ω and calculate from (78) the corresponding value of Y, i.e. we can obtain the frequency response of the resonator - its main work characteristic. Matrix elements in (78) are elements of the total transfer matrix of the whole device - see (72).

In an arbitrary case the equation (78) cannot be solved analytically. The solution can be found only by some numerical method. We used our own algorithm of searching for the global extremum of a function of several variables (Dvoesherstov et. al., 1999). Solution corresponds to the global minimum of the square of the absolute value of the left part of the equation (78). Two arguments of this function are the real and imaginary parts of the admittance Y (for each given frequency).

If there is not any piezoelectric layer in the packets F and G outside the electrodes, the transfer matrices of these packets have the simpler form (62) and the equation (78) can be solved analytically in the following view:

$$\frac{i\omega A}{Y}=(\boldsymbol{M}_Q^{\varphi T}\cdot\boldsymbol{M}_F''+\boldsymbol{M}_Q^{\varphi u})\cdot[\boldsymbol{M}_G''\cdot\boldsymbol{M}_Q^{uu}+\boldsymbol{M}_Q^{TT}\cdot\boldsymbol{M}_F''+\boldsymbol{M}_Q^{Tu}+\boldsymbol{M}_G''\cdot\boldsymbol{M}_Q^{uT}\cdot\boldsymbol{M}_F'']^{-1}\cdot(\boldsymbol{M}_G''\cdot\boldsymbol{M}_Q^{uD}+\boldsymbol{M}_Q^{TD})-\boldsymbol{M}_Q^{\varphi D} \quad (79)$$

Here the compressed form of matrices is used for compactness. For example, $\boldsymbol{M}_Q^{uu}$ means the 3x3 matrix including the first 3 columns and the first 3 rows of the 8x8 matrix, $\boldsymbol{M}_Q^{uT}$ means the 3x3 matrix including the columns 4 - 6 and the first 3 rows of the 8x8 matrix, $\boldsymbol{M}_Q^{\varphi T}$ means the 1x3 matrix including the columns 4 - 6 and the 7th row of the 8x8 matrix, and so on. Index Q means that all these elements are taken from transfer matrix of the Q packet (not for the whole device). $\boldsymbol{M}_F''$ and $\boldsymbol{M}_G''$ are 3x3 matrices, obtained as follows:

$$\boldsymbol{M}_F''=(\boldsymbol{M}_{E1m}\cdot\boldsymbol{M}_F)^{Tu}\cdot[(\boldsymbol{M}_{E1m}\cdot\boldsymbol{M}_F)^{uu}]^{-1} \quad \boldsymbol{M}_G''=[(\boldsymbol{M}_G\cdot\boldsymbol{M}_{E2m})^{TT}]^{-1}\cdot(\boldsymbol{M}_G\cdot\boldsymbol{M}_{E2m})^{Tu} \quad (80)$$

In these expressions the lower indexes F and G also designate the corresponding packets, $\boldsymbol{M}_F$ and $\boldsymbol{M}_G$ are the whole 8x8 matrices of the corresponding packets, $\boldsymbol{M}_{E1m}$ and $\boldsymbol{M}_{E2m}$ - the "mechanical" parts of the electrodes 8x8 matrices and upper indexes uu, Tu, uT and TT means that corresponding 3x3 matrices are taken from whole 8x8 matrices.

Practically all the concrete FBAR devices do not contain piezoelectric layers outside the electrodes, i.e. practically for all these devices the frequency response can be calculated with expressions (79) - (80).

But not only the frequency response can be calculated by the technique, described here. The expression (57) allows to calculate all eight values u_j, T_{1j}, D_1 and φ not only on the second surface of the layer but also for any coordinate x_1 inside the layer, if these eight values are known for the first "input" surface of this layer. These values on the second "output" surface of the first layer can be used as "input" values for the second layer for the same calculations for any coordinate x_1 inside the second layer and so on, i.e. the spatial distribution of all eight values inside the whole multilayer system can be obtained. As was mentioned above, the values on the first "input" surface of the first layer must be known for such calculations (for frequency response calculations all eight values on the first surface of the first layer are not needed).

Four of eight values, namely, T_{1j} and D_1 are known, they are zero - see (73). The absolute value of the electric potential is not essential from point of view of the spatial distribution of all eight values. We can set any (but not zero) value of the electric potential on the first surface of the first layer, for instance $\varphi^{(1)}$ = -1 V. Then we can obtain all three values of the mechanical displacements $u_1^{(1)}, u_2^{(1)}, u_3^{(1)}$ from the equation (76). So all eight values on the first surface of the first layer are determined and the spatial distribution of all these values can be obtained for any multilayer resonator with two electrodes. The admittance Y for given frequency ω must be calculated preliminary, because both these values are needed for the spatial distribution calculation.

The spatial distribution gives a possibility to obtain some information about physical wave processes those take place inside the multilayer structure, in particular - how the Bragg reflector "works".

Fig. 9 shows the frequency response of the membrane type resonator (as in Fig. 3a), obtained by technique, described above. The mass density of all the materials are taken in a form $(1 + i\delta)\rho$, where δ =-0.001 in this case. The frequency response is calculated for two variants of the Al electrode thickness - zero and 0.1 μm.

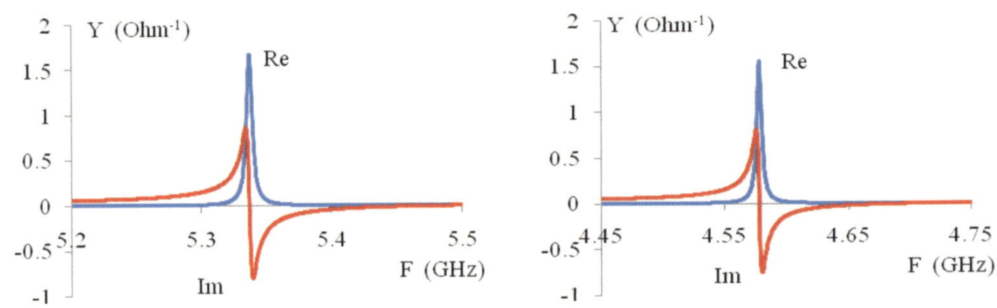

a) b)

Fig. 9. Frequency response of the membrane type resonator. Active layer – AlN, thickness 1 μm. a) – zero electrode thikness, F_{res} = 5.337 GHz, b) – Al electrode thickness 0.1 μm, F_{res} = 4.577 GHz. Electrode area 0.01 mm^2.

Fig. 9 illustrates an influence of the electrode thickness on a resonance frequency (this frequency is obtained directly from a graphic as coordinate of a maximum of a Y real part). The resonance frequency is decreased by the electrodes of a finite thickness, because the whole device with more total thickness corresponds to more half-wavelength. This illustrates Fig. 10 in which the spatial distribution of the T_{11} component of the stress tensor is shown, obtained also by a technique, described above.

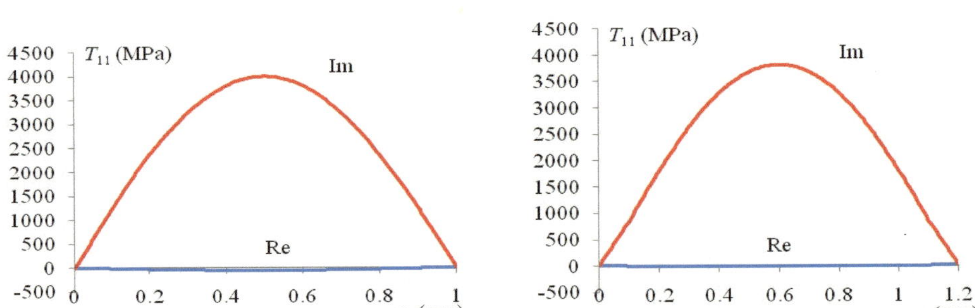

a) F = F_{res} = 5.337 GHz b) F = F_{res} = 4.577 GHz

Fig. 10. Spatial distribution of T_{11} component of the stress tensor for two variants, shown in Fig. 9. F = F_{res} in both cases.

A half-wavelength corresponds to a distance between neighbouring points with zero stress. In a case a) this distance is 1 μm and corresponds to a resonance frequency 5.337 GHz, whereas in a case b) a half-wavelength is equal to 1.2 μm and corresponds to a lower frequency 4.577 GHz. This gives a possibility to control the resonance frequency by changing of the top electrode thickness. For example, Fig. 11 shows dependences of the resonance frequency on a top electrode thickness for two materials of this electrode - Al and Au. The bottom electrode is Al of a thickness 0.1 μm in both cases.

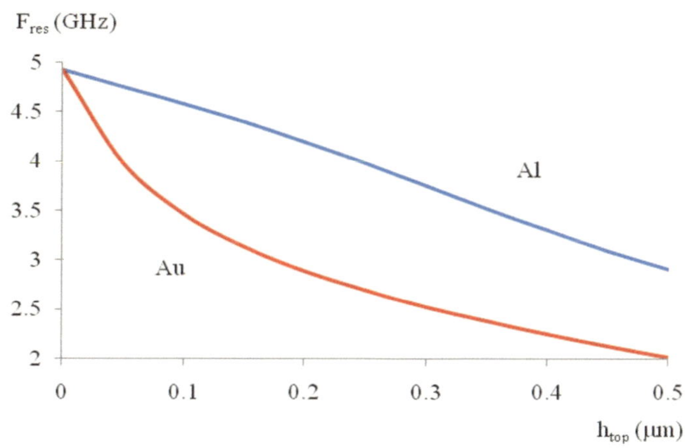

Fig. 11. Dependences of the resonance frequency on the top electrode tickness for Al and Au. The bottom electrode is Al (0.1 μm) in both cases. The thickness of AlN is 1 μm.

For displaying of the possibilities of the described method Fig. 12 shows also the spatial distributions of the longitudinal component of the displacement u_1 and the electric potential φ for the membrane type resonator, corresponding to Figs. 9b and 10b.

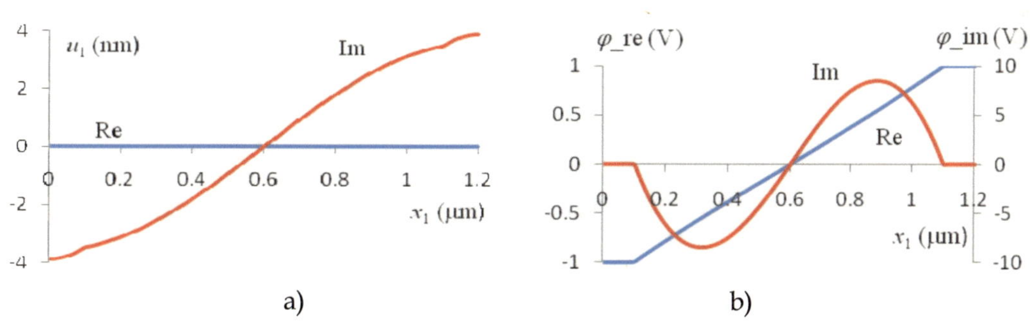

Fig. 12. Spatial distribution of the longitudinal component of the displacement u_1 (a) and the electric potential φ (b) for the membrane type resonator with Al electrodes of finite thickness 0.1 μm.

Distribution of D_1 is not shown here because it is very simple - D_1 = const between the electrodes and equal to zero outside the inner surfaces of the electrodes.

If membrane type resonator is placed on the substrate of not very large thickness, then multiple modes appear, and this resonator can be a multi-frequency resonator, as shown in Fig. 13a.

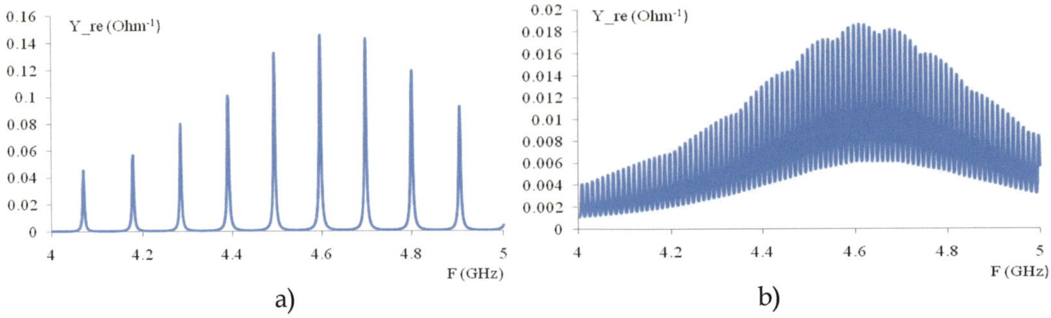

Fig. 13. FBAR membrane type resonator on a Si substrate of thicness 100 μm (a) and 1000 μm (b). Electrodes - Al, thicness 0.1 μm, active layer - AlN, thickness 1 μm.

But if the substrate is too thick, there are too many modes and the resonator transforms from multi-mode actually into a "not any mode" resonator, as one can see in Fig. 13b.

So, the membrane type resonator cannot be placed on the massive substrate directly because of an acoustic interaction with this substrate. One must to provide an acoustic isolation between an active zone of the resonator and a substrate. One of techniques of such isolation is a Bragg reflector between the active zone and the substrate (as shown in Fig. 3b). This reflector contains several pairs of materials with different acoustic properties. The difference of the acoustic properties of two materials in a pair must not be small. Acoustic properties of materials, used for Bragg reflector, are characterized by a value ρV, where ρ is a mass density and V is a velocity. Values ρV are shown in Fig. 14 for some isotropic materials. Material constants are taken from (Ballandras et. al., 1997).

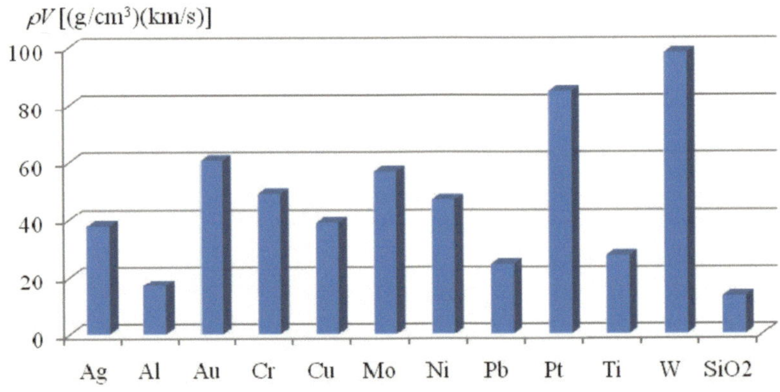

Fig. 14. The value ρV for some isotropic materials.

As one can see in Fig. 14, the best combination for a Bragg reflector is SiO_2/W. A pair Ti/W is good too, and a combination Ti/Mo also can be used successfully (combinations of Au or Pt with other materials also can be not bad, but not cheap).

The thickness of each layer of the reflector must be equal to a quarter-wavelength in a material of the layer for a resonance frequency. As it was mentioned above, the resonance frequency is defined mainly by the active layer thickness and can be adjusted by proper choice of the top electrode thickness.

The computation technique, based on the described here rigorous solution of the wave equations, allows to calculate any bulk wave resonators with any quantity of any layers, including the resonators with Bragg reflector. For example, Fig. 15a shows a frequency response of the resonator, containing an AlN active layer (1 μm), two Al electrodes (both 0.2 μm), three pairs of layers SiO_2/W, and a Si substrate (1000 μm).

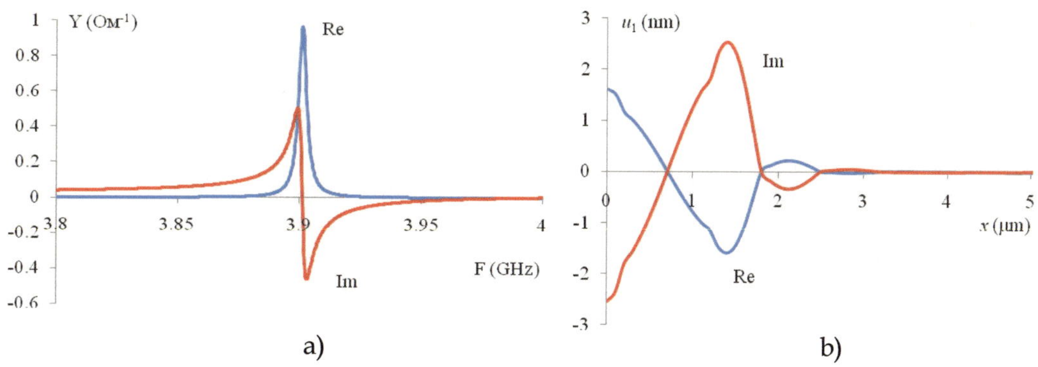

Fig. 15. A frequency response (a) and a distribution of u_1 (b) for a resonator with a Bragg reflector, containing three pairs of layers SiO_2/W.

A thickness of a Bragg reflector layer is 0.38 μm for SiO_2 and 0.33 μm for W (a quarter-wavelength in a corresponding material for a resonance frequency). Fig. 15a shows, that three pairs of SiO_2/W combination is quite enough for full acoustic isolation of an active zone and a substrate. A spatial distribution of a wave amplitude illustrates an influence of the Bragg reflector on a wave propagation, for example, Fig. 15b shows this distribution for a longitudinal component of a mechanical displacement. A coordinate axis x here is directed from a top surface of a top electrode ($x = 0$) towards a substrate. One can see in Fig. 15b that a wave rapidly attenuates in the Bragg reflector and does not reach the substrate. Calculation results show, that the first layer after an electrode must be one with lower value ρV - the SiO_2 layer in this case. In a contrary case a reflection will not take place.

If difference of values ρV of two layers of each pair is not large enough, then three pairs may not be sufficient for effective reflection. For example, calculations show that three or even four pairs of Ti/Mo layers are not sufficient for suppressing the wave in the substrate. Only five pairs give a desired effect in this case and provide results similar shown in Fig. 15 for SiO_2/W layers.

So, the described technique allows to calculate any multilayer FBAR resonators, containing any combinations of any quantity of any layers. The main results of these calculations are a frequency response of a resonator and spatial distributions of physical characteristics of the wave (displacement, stress, electric displacement and potential).

In addition this technique gives a possibility to calculate a thermal sensitivity of the resonator too, i.e. an influence a temperature on a resonance frequency. A resonance frequency always changes in general case when a temperature changes. This change is characterized by a temperature coefficient of a frequency:

$$TCF = \frac{1}{F_r}\frac{dF_r}{dT} \tag{81}$$

Here T is a temperature, F_r is a resonance frequency.
A computation technique, used here, allows to apply this expression for *TCF* calculation directly and to calculate this value by numerical differentiation.
A temperature influence on a resonance frequency is due to three basic causes:

1. A temperature dependence of material constants (stiffness, piezoelectric, dielectric tensors) - TCF_C
2. A temperature dependence of a mass density – TCF_ρ
3. A temperature dependence of a layer thickness – TCF_h

A temperature dependence of material constants is described by temperature coefficients of these constants, a temperature dependence of a mass density is described by three linear expansion coefficients or by a single bulk expansion coefficient, a temperature dependence of a thickness is described by a linear expansion coefficient along a thickness direction. All these coefficients can be found in a literature, for example, for materials, usually used for FBAR resonators, one can see corresponding values in (Ivira et al., 2008).
First we will consider the simplest variant – a membrane type FBAR resonator with a single AlN layer and infinite thin electrodes. For typical values of AlN temperature coefficients we can easily obtain:

$$TCF = TCF_c + TCF_\rho + TCF_h = (-29.639 + 7.343 - 5.268) \cdot 10^{-6}/^oC = -27.564 \cdot 10^{-6}/^oC$$

One can check by a direct calculation, that this result does not depend on a thickness of AlN layer (for this variant with electrodes of finite thickness and for any multilayer structure with layers of finite thickness it is not so). TCF_ρ value is always positive, TCF_h value is always negative. A sign of TCF_c is defined mainly by a sign of temperature coefficients of stiffness constants. If temperature coefficients of stiffness constants are negative (for most materials, including AlN), then TCF_c is negative, if temperature coefficients of some stiffness constants are positive (rare case, for example quartz), then TCF_c can be positive and a total *TCF* can be zero.
For AlN a *TCF* value is always negative. Al electrodes aggravate this position, besause temperature coefficients of Al stiffness constants are negative too. From this point of view Mo electrodes are more preferable, because absolute values of temperature coefficients of its stiffness constants are significantly less than ones for Al (althouth they are also negative). For example, the concrete membrane type resonator Al/AlN/Al with an Al thickness 0.2 μm and an AlN thickness 1.1 μm we can obtain: $TCF = -44.23 \cdot 10^{-6}/^oC$ (F_r = 3.648 GHz), and for Mo/AlN/Mo resonator with the same geometry: $TCF = -33.76 \cdot 10^{-6}/^oC$ (F_r = 2.615 GHz).
For most applications a resonator must be thermostable, i.e. *TCF* must be equal to zero. The single possibility to compensate the negative *TCF* of AlN and of electrodes and to provide a total zero *TCF* is to add some additional layer with positive temperature coefficients of stiffness constants. Such material is, for example SiO_2. Fig. 16 shows dependenses of *TCF* of membrane type resonator with Mo electrodes on a thickness h_t of a SiO_2 layer for two cases: SiO_2 layer is placed together with AlN layer between electrodes (structure Mo/SiO_2/AlN/Mo) and SiO_2 layer is placed outside the electrodes (structure SiO_2/Mo/AlN/Mo). Corresponding dependences of a resonance frequency are presented in Fig. 16 too.
Fig. 16 shows that a SiO_2 layer more effectively influences on both *TCF* and a resonance frequency, when it is placed between electrodes.

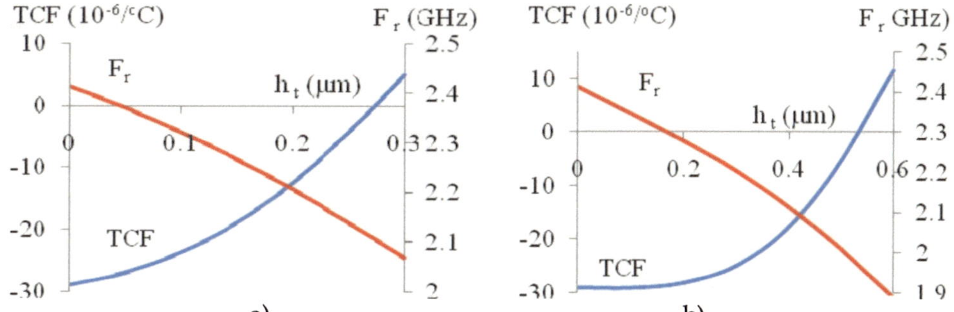

Fig. 16. Dependences of *TCF* and a resonance frequency on a SiO_2 layer thickness h_t for cases, when SiO_2 is placed between electrodes (a) and when SiO_2 is placed outside electrodes (b). A thickness of Mo electrodes is 0.06 μm, a thickness of AlN is 1.9 μm.

Calculations show that a Bragg reflector does not change a resonance frequency of the corresponding membrane type resonator, if a thickness of each layer of the reflector is exactly equal to a quarter-wavelength. But a Bragg reflector influences on a *TCF*. For this reason it is reasonable to choose SiO_2 as one material of a reflector. In this case a thickness of an additional compensating SiO_2 layer can be reduced. For example, a thickness of SiO_2 layer outside electrodes, corresponding to *TCF* = 0, is equal about 0.53 μm for variant, shown in Fig. 16b for membrane type resonator. A resonance frequency is about 2.11 GHz for this case. The Bragg reflector with three pairs of SiO_2/Mo, corresponding this frequency, does not change this frequency, but a *TCF* becomes positive due to SiO_2 material presense in the reflector. One must reduce a thickness of an additional compensating SiO_2 layer to return a TCF to zero. But then a resonance frequency will increase. We must either increase an AlN layer thickness to return a resonance frequency or to change thickness of a Bragg reflector layers to adjust the reflector to a new resonance frequency. In any case several steps of sequential approximation are necessary. The technique, described here, allows to do this without problem. For example, presented in Fig. 16b, full thermocompensation can be obtained for h_t = 0.4 μm (instead of 0.53 μm for membrane type resonator) and for thickness of SiO_2 and Mo layers in a Bragg reflector 0.71 μm and 0.75 μm respectively. The AlN layer thickness remains 1.9 μm and a resonance frequency slightly shifts remaining in the vicinity of 2.1 GHz.

In many cases a presentation of FBAR resonator by means of some equivalent circuit is convenient - see for example (Hara et. al., 2009). The simplest variant of an equivalent circuit is shown in Fig. 17.

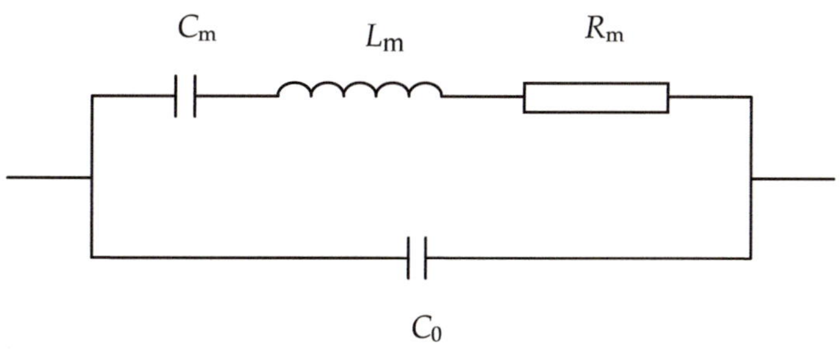

Fig. 17. An equivalent circuit of FBAR resonator.

Hear C_0 is a static capacitance of a resonator – a real physical value, which can be calculated by the geometry of the resonator and the dielectric properties of the layers between the electrodes:

$$C_0 = \left(\frac{1}{\varepsilon_0 A} \sum_{i=1}^{m} \frac{l_i}{\varepsilon_i} \right)^{-1} \tag{82}$$

where ε_i and l_i is a relative dielectric permittivity (element ε_{11} of a tensor) of a layer number i and its thickness, $\varepsilon_0 = 8.854 \cdot 10^{-12}$ F/m – the dielectric constant, A is an area of a resonator electrode, m is a quantity of layers between electrodes.

Values C_m, L_m, and R_m are equivalent dinamic capacitance, inductance and resistance of the resonator – values, which can not be determined from any physical representation – only by comparison with experimental frequency response or with response, obtained by some exact theory. Theory, described here, allows to obtain these values.

An admittance of the equivalent circuit, shown in Fig. 17, can be calculated by following expressions:

$$Y_e = \left(j\omega L_m + \frac{1}{j\omega C_m} + R_m \right)^{-1} + j\omega C_0 = Y_{Rm} + Y_{\mathrm{Im}} + j\omega C_0 \tag{83}$$

Here

$$Y_{Rm} = \frac{R_m}{\left(\frac{1}{\omega C_m} - \omega L_m \right)^2 + R_m^2} \qquad Y_{\mathrm{Im}} = j \frac{\frac{1}{\omega C_m} - \omega L_m}{\left(\frac{1}{\omega C_m} - \omega L_m \right)^2 + R_m^2} \tag{84}$$

Y_{Rm} and Y_{Im} are an active and reactive components of a dinamic admittance of the resonator, $j\omega C_0$ is an admittance of the static capacitance.

Comparison of admittance, calculated by (83) and (84), with admittance, calculated by a rigorous theory, described here, allows to obtain the unique values C_m, L_m, and R_m, which give a frequency response, equivalent to the response, given by the rigorous theory.

The resonance frequency of the equivalent circuit, shown in Fig. 17, is defined as:

$$\omega_r = 2\pi F_r = \frac{1}{\sqrt{L_m C_m}} \tag{85}$$

The value R_m corresponds to a maximum of the active component of the admittance (see (84)):

$$R_m = \frac{1}{(Y_{Rm})_{\max}} \tag{86}$$

We can find a quality-factor from curve of a active component of the admittance:

$$Q = \frac{F_r}{\Delta F} \tag{87}$$

Here ΔF is a full width of the curve at a level 0.5 of a maximum.
Then we can calculate an equivalent dynamic inductance:

$$L_m = \frac{QR_m}{2\pi F_r} \tag{88}$$

At last we can calculate an equivalent dynamic capacitance with help of (85):

$$C_m = \frac{1}{L_m (2\pi F_r)^2} \tag{89}$$

All these calculations the computer program performs automatically and shows obtained results in corresponding windows of the program interface (a program is made in a Borland C++ Builder medium and provides automatic transfer of main results into Excel worksheet). A frequency response, calculated by expressions (83) and (84) with values C_m, L_m, R_m, obtained by such a manner, practically coincides with a frequency response, calculated with rigorous theory, described here (if there is only one resonance peak in a frequency range). In a wide frequency range may be several resonance peaks. In this case one can connect required quantity of C_m, L_m, R_m circuits in parallel (but with only one C_o for all them) in Fig. 17. C_m, L_m, R_m values for every circuit can be determined by comparison with corresponding peak, given by a rigorous theory.

4. Conclusion

General methods of surface and bulk acoustic wave in multilayer structures calculation are described in this chapter. Corresponding equations are formulated. These equations allow to calculate all the main wave propagation characteristics and the device parameters. A phase velocity, an electromechanical coupling coefficient, a temperature coefficient of delay, a power flow angle and others for surface wave devices and a frequency response, a spatial distribution of the wave characteristics, a resonance frequency, a temperature coefficient of frequency, parameters of an equivalent circuit for bulk acoustic resonators are available for calculations by described techniques. Obtained results allow better to understand processes taking place in these devices and to improve their characteristics. Corresponding algorithms and computer programs can be used for design of surface and bulk acoustic wave devices.

5. References

Adler, E.L. (1994). SAW and pseudo-SAW properties using matrix methods, *IEEE Transactions on Electronics, Ferroelectrics, and Frequency Control.*, Vol. 41, No. 6, (November 1994), pp. (876-882).

Ballandras, S.; Gavignet, E., Bigler, E., and Henry, E. (1997). Temperature Derivatives of the Fundamental Elastic Constants of Isotropic Materials, *Appl. Phys. Lett.* Vol. 71, No. 12, (September 1997), pp. (1625-1627).

Campbell, J. J. & Jones, W.R. (1968). A method for estimating optimal crystal cuts and propagation directions for excitation of piezoelectric surface waves, *IEEE Transactions on Sonics and Ultrasonics*, Vol. 15, No. 4, (October 1968), pp. (209-217).

Dvoesherstov, M.Y., Cherednick, V.I., Chirimanov, A.P., Petrov, S.G. (1999). A method of search for SAW and leaky waves based on numerical global multi-variable procedures, *SPIE* Vol. 3900, pp. (290-296).

Hara, M.; Yokoyama, T., Sakashita, T., Ueda, M., and Satoh, Y. (2009). A Study of the Thin Film Bulk Acoustic Resonator Filters in Several Ten GHz band, *IEEE International Ultrasonics Symposium Proceedings*, (2009), pp. (851-854).

Ivira, B.; Benech, P., Fillit, R., Ndagijimana, F., Ancey, P., Parat, G. (2008). Modeling for Temperature Compensation and Temperature Characterizations of BAW Resonators at GHz Frequencies, IEEE *Transactions on Ultrasonics, Ferroelectrics, and Frequency Control*, Vol. 55, No. 2, (February 2008), pp.(421-430).

Kovacs, G., Anhorn, M., Engan, H.E., Visiniti, G. and Ruppel, C.C.W. (1990). Improved Material Constants for $LiNbO_3$ and $LiTaO_3$, *IEEE International Ultrasonics Symposium Proceedings*, (1990), pp. (435-438).

Nowotny, H. & Benes, E. (1987). General one-dimensional treatment of the layered piezoelectric resonator with two electrodes, *Journal of Acoustic. Society of America*, Vol. 82 (2), (August 1987), pp. (513-521).

Shimizu, Y.; Terazaki, A., Sakaue, T. (1976). Temperature dependence of SAW velocity for metal film on α-quartz, *IEEE International Ultrasonics Symposium Proceedings*, (1976), pp. (519-522).

Shimizu, Y. & Yamamoto, Y. (1980). Saw propagation characteristics of complete cut of quartz and new cuts with zero temperature coefficient of delay, *IEEE International Ultrasonics symposium Proceedings*, (1980), pp. (420-423).

4

The Features of Low Frequency Atomic Vibrations and Propagation of Acoustic Waves in Heterogeneous Systems

Alexander Feher[1], Eugen Syrkin[2], Sergey Feodosyev[2], Igor Gospodarev[2], Elena Manzhelii[2], Alexander Kotlar[2] and Kirill Kravchenko[2]

[1]Institute of Physics, Faculty of Science P. J. Šafárik University, Košice

[2]B.I.Verkin Institute for Low Temperature Physics and Engineering NASU, Kharkov

[1]Slovakia

[2]Ukraine

1. Introduction

In recent years, the quasi-particle spectra of various condensed systems, crystalline as well as disordered and amorphous, became also the "object" of applications and technical developments and not only of fundamental research. This led to the interest in the theoretical and experimental study of the quasi-particle spectrum of such compounds, which are among the most popular and advanced structural materials. Most of these substances have heterogeneous structure, which is understood as a strong spatial heterogeneity of the location of different atoms and, consequently, the heterogeneity of local physical properties of the system, and not as the coexistence of different phases (i.e. heterophase). To these structures belong disordered solid solutions, crystals with a large number of atoms per unit cell as well as nanoclusters.

This chapter is devoted to the study of vibration states in heterogeneous structures. In such systems, the crystalline regularity in the arrangement of atoms is either absent or its effect on the physical properties of the systems is weak, affecting substantially the local spectral functions of different atoms forming this structure. This effect is manifested in the behavior of non-additive thermodynamic properties of different atoms (e.g. mean-square amplitudes of atomic displacements) and in the contribution of individual atoms to the additive thermodynamic and kinetic quantities. The most important elementary excitations appearing in crystalline and disordered systems are acoustic phonons. Moreover, in heterogeneous nanostructures the application of the continuum approximation is significantly restricted; therefore we must take into account the discreteness of the lattice. This chapter contains a theoretical analysis at the microscopic level of the behavior of the spectral characteristics of acoustic phonons as well as their manifestations in the low-temperature thermodynamic properties.

The chapter consists of three sections. The first section contains a detailed analysis at the microscopic level of the propagation of acoustic phonons in crystalline solids and disordered solid solutions. We analyze the changes of phonon spectrum of the broken crystal regularity of the arrangement of atoms in the formation of a disordered solid solution with heavy isotope impurities and randomly distributed impurities weakly coupled both with the atoms of the host lattice and among themselves. As is well known, such defects enrich the low-frequency phonon spectrum and lead to a significant change in the low-temperature thermodynamic and kinetic characteristics (see, for example, Kosevich, 1999; Maradudin et al., 1982; Lifshitz, 1952a). In particular, the impurity atoms cause the so-called quasi-localized vibrations (Kagan & Iosilevskij, 1962; Peresada & Tolstoluzhskij 1970, 1977; Cape et al., 1966; Manzhelii et al., 1970). This section analyzes in detail the conditions of the formation and evolution of quasi-localized vibrations with increasing concentration of impurities. It is shown that the quasi-local maximum in the phonon spectrum of the low-frequency zone is formed by vibrations localized on impurity atoms. Rapidly propagating phonons corresponding to the vibrations of the host lattice atoms are scattered by localized vibrations. This scattering forms a kink in the local spectral densities of these atoms that is similar in shape to the first van Hove singularity of a perfect crystal. In the description of the spectral characteristics of the elementary excitations in heterogeneous structures such theoretical methods are necessary which do not involve the translational symmetry of the crystal lattice. In this section we use such a method for computing the local Green's functions and the local and partial spectral densities.

These self-averaging spectral characteristics (Lifshitz et al., 1988) can be determined also for compounds that do not possess regularity in their crystal structure. An effective method for describing disordered systems and calculating their quasi-particle spectra is the method of Jacobi matrices (J -matrices) (Peresada, 1968; Peresada et al., 1975, Haydock et al., 1972). By this method the majority of the calculations in this paper were carried out.

The second section is devoted to the analysis of the reasons for the strong temperature dependence of the Debye temperature $\Theta_D(T)$ under $T \leq 0.1\Theta_D$. The temperature dependence $\Theta_D(T)$ is a solution of the transcendental equation $C_D(T/\Theta_D) = C(T)$, where $C(T)$ is calculated at the microscopic level or experimentally determined and $C_D(T/\Theta_D)$ is the temperature dependence of the Debye heat capacity. It is shown that the reason for the formation of a low-temperature minimum on the dependence $\Theta_D(T)$ are the fast-propagating low-frequency phonons (propagons) (Allen et al., 1999) scattered on the slow quasi-particles. In the case of a defect (random reduction of force constants) the quasi-localized vibrations do not form, but in the ratio of the phonon density of states $\nu(\omega)$ to the square of the frequency a maximum in the propagon zone of the phonon spectrum is formed with increasing concentration of defects. The maxima in the ratio $\nu(\omega)/\omega^2$ are called boson peaks (see, for example, Feher et al., 1994; Gurevich et al., 2003; Schrimacher et al., 1998). They are intensively studied for systems with topological disorder, glasses and such compounds as molecular crystals with rotational degrees of freedom. In this section we analyze the arising of such features in solid solutions with only vibration degrees of freedom. The frequency of the boson peak coincides with the frequency of the quasi-local vibrations corresponding to a weakly bound impurity at concentrations, for which the average distance between the randomly distributed impurity atoms corresponds to the propagon frequency equal to the frequency of quasi-local vibrations. That is, the distance between the impurities (*disorder parameter*) becomes comparable to the wavelength of rapid acoustic phonons with the frequency equal to the quasi-local vibration frequency. This corresponds to the phonon Ioffe-Regel crossover (Klinger & Kosevich, 2001, 2002). It is shown that the temperature-dependence and magnitude of $\Theta_D(T)$ are even more informative than the phonon density of states. It was shown on the example of a $Kr_{1-p}Ar_p$ solid solution that for $p \approx 25\%$ there are no singularities both in the propagon zone of the phonon density of states and in the phonon density relation to the square of frequency. At the same time, the temperature dependence of the relative (compared with pure Kr) changes in the low-temperature heat capacity shows two peaks that can not be explained by the superposition of contributions of isolated impurities, impurity pairs, etc. The reason of this behavior is the scattering of fast propagating phonons, corresponding to the krypton atoms, on significantly slower phonons corresponding to the vibrations of atoms in argon clusters.

Many features of the phonon spectra and vibrational characteristic of disordered heterogeneous structures are also inherent to the crystals with polyatomic unit cells. Third section of this work is devoted to the analysis of phonon spectra and vibrational characteristics of such crystals. The manifestations of the phonon Ioffe-Regel crossover in multilayered regular crystalline structures are analyzed. The presence of the quasi-two-dimensional and quasi-one-dimensional features in the behavior of the vibrational characteristics of multilayer compounds is shown. The macroscopic characteristics of such compounds are derived from the low-dimensional ones. This allows us to describe the vibrational characteristics of such complex compounds in frames of low-dimensional models. The features of the interaction of phonons with a planar defect are investigated using these models. In particular, the resonance effects in the scattering of acoustic waves and the formation of localized and resonance vibrational states in the planar defect are considered. Such effects may lead to singularities in the experimentally observed kinetic characteristics of the grain boundaries. The heat transfer between two different media, on the condition that the Fano resonance occurs, is analyzed.

2. Low-frequency characteristics of the phonon spectra of disordered solid solutions

This chapter is devoted to the study of the propagation of acoustic phonons at different frequencies of quasi-continuous FCC crystal phonon spectrum. We analyze in more detail the analogy of the Van Hove singularity in the phonon spectrum of the perfect crystal with similar features of the phonon spectra of structures with broken regularity in the arrangement of atoms of a crystal. For any solid (both crystal and the one which does not possess the translational symmetry of the atoms arrangement), a low-frequency range exists where the dispersion relation of phonons has the form $\omega(\boldsymbol{k}) = s(\boldsymbol{\kappa})k$ (k is a module of the wave vector $\boldsymbol{k}$, $\boldsymbol{\kappa} \equiv \boldsymbol{k}/k$, and $s(\boldsymbol{\kappa})$ is the velocity of sound). The phonon density of states in this range takes the Debye form $\nu(\omega) \sim \omega^2$. With the increase of the k-value the phonon dispersion relation increasingly deviates from the linear one (frequency ω becomes lower than sk) and the actual density of states deviates upwards from the Debye one. At low frequencies, the sound propagation occurs along all crystallographic directions. With increasing frequency the propagation velocity of acoustic phonons decreases, this decrease being different for different crystallographic directions. In a perfect crystal, when the phonon frequency corresponds to the frequency of the first van Hove singularity $\omega = \omega^*$, the propagation of the transverse sound along one of the crystallographic directions (in the FCC it is the crystal direction ΓL, Fig. 1**a**) ceases and the corresponding group velocity is zero. Phonons with frequencies $\omega < \omega^*$ were named *propagons* and those with higher frequencies are called *diffusons*. With a further frequency increase also the number of directions increases along which the propagation of sound ceases. The highest frequency of the van Hove singularity ($\omega = \omega^{**}$) corresponds to the frequency at which the wavelength of the longest wavelength phonons is smaller than the interatomic distance (Fig. 1**b**). Phonons with $\omega > \omega^{**}$ are almost localized and they were named *locons*, while the frequency interval (ω^{**}, ω_m) is called the *locon band*.

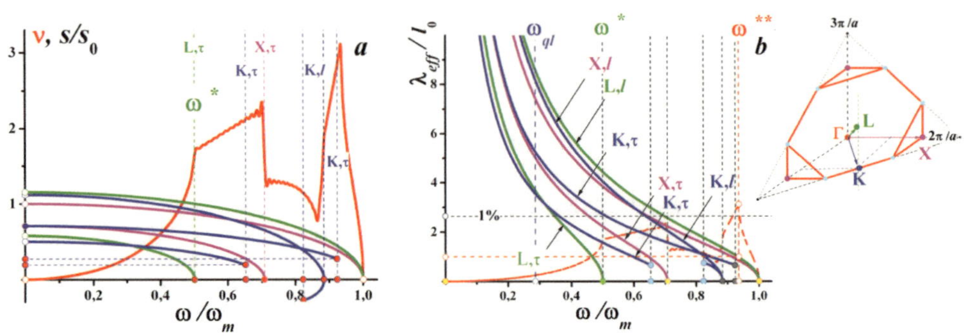

Fig. 1. The phonon density of states (red lines), the frequency dependences of the group velocities of phonon modes (part **a**) and frequency dependences of the values $\lambda_{eff}(\omega)/l_0$ (part **b**) along the main of the highly symmetrical crystallographic directions of FCC crystal with central nearest-neighbors interaction (blue, purple and olive line, depending on the direction). The first octant of the first Brillouin zone of a FCC crystal with indications of considered high-symmetry directions is shown on the right ($l_0 \equiv a/\sqrt{2}$ is the distance between nearest neighbors).

Note that the propagation character of locons and diffusons practically does not differ from the propagation character of optical phonons in a crystal with complex lattice. A translational symmetry disturbance does not lead to a qualitative modification in the nature of acoustic phonons and does not change their classification.

Let in the crystal randomly introduce heavy isotopic substitution impurities or impurities weakly coupled to the atoms of the host lattice and among themselves. Neglecting the interaction between impurities the variations of the phonon spectrum are satisfactorily described by the theory of regular perturbations devised by I.M. Lifshitz (Lifshitz, 1952a). In particular, the formation of the so-called *quasi-local vibrations* (QLV) due to the presence of heavy or weakly bound impurities was predicted (Kagan & Iosilevskij, 1962) and studied in detail both theoretically (see, for example, Peresada & Tolstoluzhskij 1977) and experimentally (see, for example, Cape et al., 1966; Manzhelii et al., 1970). The QLV are manifested by the resonance peaks in the low-frequency part of the phonon spectrum and they contribute essentially to the low-temperature thermodynamic properties. At low impurity atoms concentrations $p \ll 1$, the vibration characteristics of the solid solution can be described within the linear in p approximation:

$$\tilde{\nu}(\omega) = \nu(\omega) + p\sum_i \Delta\rho^{(i)}(\omega) \tag{1}$$

The summation is performed over all cyclic subspaces (Peresada, 1968; Peresada et al.,1975), in which the operator $\hat{\Lambda}$ describing the perturbation of the lattice vibrations by either isolated heavy or weakly coupled impurity is non-zero, $\Delta\rho^{(i)}(\omega)$ is the spectral density change in each of these subspaces, $\tilde{\nu}(\omega)$ and $\nu(\omega)$ are the phonon densities of solid solution and perfect crystal states, respectively. If in each of the cyclic subspaces the operator $\hat{\Lambda}$ induces a regular degenerate operator, then the value $\Delta\rho^{(i)}(\omega)$ can be calculated using the spectral shift function $\xi(\omega)$ (Lifshitz, 1952a). Using the expressions obtained for this function in the J-matrix method (Peresada, 1968; Peresada et al., 1975, Peresada & Tolstoluzhskij 1977), we obtain:

$$\Delta\rho(\omega) = -\frac{d\xi(\omega)}{d\omega} = \frac{\rho^2(\omega)}{\pi^2\rho^2(\omega) + \left[S(\omega) - \mathrm{Re}\,G(\omega)\right]^2} \cdot \frac{d}{d\omega}\left[\frac{S(\omega) - \mathrm{Re}\,G(\omega)}{\rho(\omega)}\right] \tag{2}$$

where the function $S(\omega)$ describes the perturbation by defect and depends on the defect parameters, $G(\omega)$ is the local Green's function of a perfect crystal. If in any cyclic subspace the solution of the equation

$$S(\omega) - \mathrm{Re}\,G(\omega) = 0 \tag{3}$$

is $\omega = \omega_k$, then in the vicinity of this value the expression (2) has a resonant character:

$$-\frac{d\xi}{d\omega} = \frac{2}{\pi}\frac{\Gamma}{4(\omega-\omega_k)^2 + \Gamma^2}; \qquad \Gamma \equiv \frac{\pi\rho(\omega)}{\frac{d}{d\omega}\left[S(\omega) - \mathrm{Re}\,G(\omega)\right]_{\omega=\omega_k}} \quad . \tag{4}$$

The equation (3) formally coincides with the Lifshitz equation which yields (of course for other values of $S(\omega)$) the frequencies of discrete vibrational levels, lying outside the band of quasi-continuous spectrum of the crystal (Lifshitz, 1952a). However, these discrete levels are, in contrast to the values ω_k, the poles of the perturbed local Green's function. The Green's function can not have poles within the quasi-continuous spectrum. The possibility to determine the QLV frequencies using equation (3) arises from the fact that at low frequencies $|\mathrm{Re}\,G_{00}(\omega)| >> \mathrm{Im}\,G_{00}(\omega)$.

Let us analyze the quasi-local oscillations due to the substitution impurity in an FCC crystal with the central interaction of nearest-neighbors. The interaction of the impurity with the host lattice is also considered as a purely central and, therefore, the perturbation caused by such an impurity should be regular and degenerate. Let us consider two cases: the isotopic impurity with a mass four times higher than that of the host lattice (i.e. the mass defect is $\varepsilon \equiv \Delta m/m = 3$) and the impurity atom with a mass equal to the mass of a host lattice atom $\varepsilon = 0$ and coupled to the host lattice four times weaker than are the atoms of the host lattice between each other ($\upsilon \equiv \Delta\alpha/\alpha = -3/4$ is the coupling defect). In the first case, operator $\hat{\Lambda}$ induces a non-zero operator only in the cyclic subspace which is generated by the displacement of the impurity atom. The vectors corresponding to this subspace transform according to the irreducible representation τ_-^5 of the symmetry group of the lattice O_h (the notation of Kovalev, 1961). In the given subspace the spectral density of perfect lattice coincides with its density of states. For an isotopic impurity the function $S(\omega)$ (Peresada & Tolstoluzhskij, 1977) reads:

$$S_{is}(\omega) = -\frac{2}{\omega\varepsilon} \tag{5}$$

In the second case, except the subspace $H^{(\tau_-^5)}$ where the function $S(\omega)$ is

$$S_w^{(\tau_-^5)}(\omega) = \frac{2}{\omega} + \frac{\omega_m^2(1+\upsilon)}{\omega^3\upsilon} \tag{6}$$

the non-zero operators will be operators induced by the operator $\hat{\Lambda}$ in cyclic subspaces transformed according to irreducible representations, τ_+^1, τ_+^3, τ_+^4 and τ_-^4 of the same group O_h. Over all of these four subspaces $S(\omega)=16\omega/(\omega_m^2\upsilon)$. For weakly bound impurity, the function $S_w \le S_{\lim} = -16\omega/\omega_m$ and equation (3) can not have solutions in the propagon zone within these cyclic subspaces. Therefore, for real values of parameter υ equation (3) has a solution in the subspace $H^{(\tau_-^5)}$ only. This solution for both cases shows Fig. 2. The real part of the Green's function (curves 2 in both parts) crosses the dashed curves 3, which represent the equations (5) (part ***a***) and (6) (part ***b***), at points ω_k. This figure also shows the spectral densities $\rho^{(\tau_-^5)}(\omega)$ of the perfect crystal, coinciding with its phonon density of states $\nu(\omega)$ (dashed curves 1), and phonon densities of states of the corresponding solid solutions with concentration $p = 5\%$. This figure shows the phonon density of states (curves 1) for both the heavy isotopes (part ***a***) and weakly bound impurities (part ***b***). Curves 4 show the contributions from impurities and curves 6 those from the matrix lattice. We can see that the maxima formed on the phonon densities $\tilde{\nu}(\omega,p)$ are (curves 1) completely caused by the vibrations of impurity atoms. Let us analyze figures 2 and 3 together. The value of the phonon density of states of a perfect crystal at $\omega=\omega_k$ can not be considered as negligible, since it is comparable to the value of the real part of the Green's function at this frequency ($\nu(\omega_k) \sim 0.1\,\mathrm{Re}\,G(\omega_k)$). Therefore, as is seen from the figures, though the frequencies of the maxima on the curves $\tilde{\nu}(\omega,p)$ and $\nu_{imp}(\omega,p)$ are close to the frequency ω_k, they do not coincide with it (Fig. 2**b**) (especially in the case of a weakly bound impurity). For weakly bound impurity one should expect a higher degree of localization of QLV on impurity atoms. In Fig. 3 the values of $\nu_{imp}(\omega,p)$ are compared with the spectral density of isolated impurity atoms $\tilde{\rho}^{(\tau_-^5)}(\omega) = \frac{2\omega}{\pi}\mathrm{Im}\left(\vec{h}_0^{(\tau_-^5)},\left[\omega^2\hat{I}-\hat{L}-\hat{\Lambda}\right]^{-1}\vec{h}_0^{(\tau_-^5)}\right)$. The function $\nu_{imp}(\omega,p)$ is nonzero only near frequencies ω_{ql}, which are the maxima on curves 5. Therefore the frequency ω_{ql} can be more reasonably than ω_k considered as the frequency of QLV (*quasi-local frequency*).

Therefore, QLV can be represented as waves slowly diverging from the impurity, similar to spherical waves. Fig. 2 also presents (curves 4) the values of $\rho_{01}^{(\tau_-^5)}(\omega)$, i.e. the spectral correlators of displacements of impurity atoms with their first coordination sphere

$$\rho_{01}^{(\tau_-^5)}(\omega) = \frac{2\omega}{\pi}\mathrm{Im}\left(\vec{h}_1^{(\tau_-^5)},\left[\omega^2\hat{I}-\hat{L}-\hat{\Lambda}\right]^{-1}\vec{h}_0^{(\tau_-^5)}\right) = P_1(\omega^2)\rho^{(\tau_-^5)}(\omega), \tag{8}$$

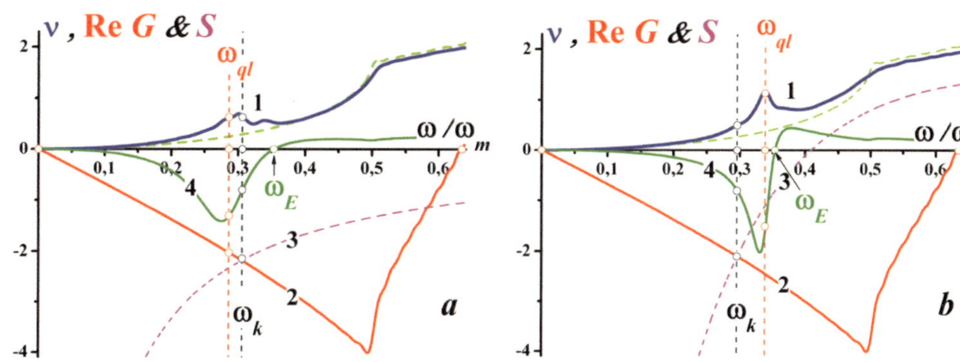

Fig. 2. Phonon densities of disordered solid solutions with impurity concentration 5% (curves 1) and solutions of the equation (3) (intersection of curves 2 and 3) for cases of a heavy isotopic impurity (part **a**) and weakly bound impurities (part **b**). Curves 4 in both parts are spectral correlators of vibrations of the impurity atom with its first coordination sphere.

where $P_1\left(\omega^2\right)$ is the polynomial defined by the recurrence relations for the J-matrix of the perturbed operator $\hat{L}+\hat{\Lambda}$ (Peresada, 1968; Peresada et al.,1975). Spectral correlator $\rho_{01}(\omega)$ vanishes when $\omega=\omega_E$, where ω_E is the Einstein frequency of the correspondent subspace ($\omega_E^2=\int_0^{\omega_m}\omega^2\nu(\omega)d\omega$). Thus, when $\omega=\omega_E$, the correlation with the first coordination sphere is absent, and the close is the frequency ω_{ql} to ω_E, the stronger is the degree of localization of QLV. As it could be seen from Fig. 2 the QLV frequency for weakly bound impurity is nearly three times closer to ω_E than that the heavy isotopic one, and the quasi-local maximum for a weakly bound impurity has a sharper resonance form than the maximum for a heavy isotopic defect.

The QLV are localized near the impurity atoms and their formation is very similar to the occurrence of discrete vibrational levels (local oscillations) outside the continuous spectral band of the host lattice in the presence of light or strongly coupled impurities in a crystal. However, there is an important fundamental difference between the local and quasi-local vibrations, manifested under increasing concentration of impurity atoms. Local vibrations are the poles of the Green's function of the perturbed crystal, and their amplitudes decay exponentially with the distance from the impurity. Being located outside the quasi-continuous spectrum, these vibrations do not interact with the phonon modes of the host lattice. With an increasing concentration of either light or strongly coupled impurities their effect upon phonon spectrum can be determined by taking into account the expansion of concentration (Lifshitz et al., 1988). Thus, the increase of the concentration of light impurities leads to the appearance of sharp resonant peaks in phonon spectrum with frequencies coincident with those of local vibrations of the isolated impurity atom pairs and eventually, regular triangles and tetrahedrons (Kosevich et al., 2007). The QLV are not the poles of the Green's function, they are common non-divergence maxima in the phonon density of states. Though, as is shown in the next section, these peaks are formed by the impurity atoms vibrations which interact with the phonon modes of host lattice. Therefore, at finite (even low enough, about few percents) concentrations of heavy or weakly coupled impurity atoms, the significant modification of the entire phonon spectrum of the crystal occurs,

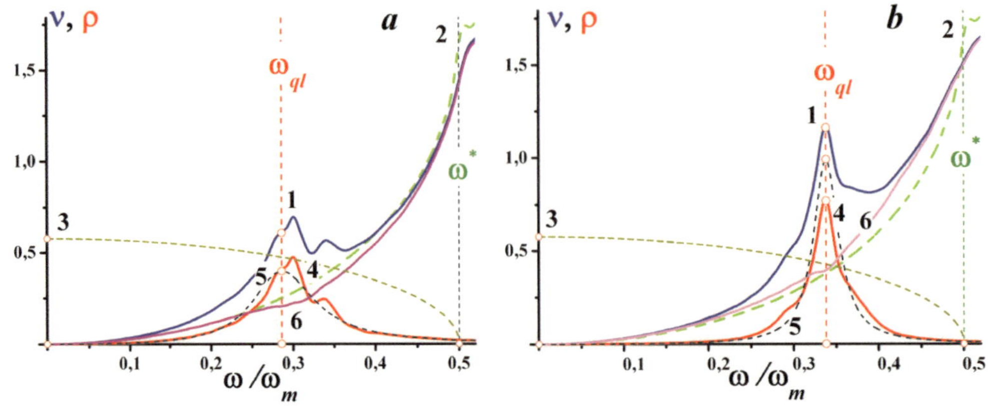

Fig. 3. Phonon densities of disordered solid solutions with impurity concentration 5% (curves 1); from contributions impurity atoms (curves 4) and atoms of the host lattice (curves 6). Part **a** corresponds to the heavy isotopic impurity ($m'/m = 4$), part **b** corresponds to a weakly bound impurity ($\alpha'/\alpha = 1/4$). Curves 2 show the phonon density of the original perfect lattice, curves 3 represent the frequency dependence of the transverse sound velocity along the crystallographic direction *ΓL*, and curves 5 show the spectral density of single isolated impurities, multiplied by the concentration $p = 0.05$.

which can not be described by the expansion of the impurity concentration. Thus the weakening of bonds between the argon impurities in krypton matrix leads to a characteristic "two-extreme" behavior of the temperature dependence of the relative change in the low-temperature heat capacity unexplained by the superposition of contributions of isolated impurities, impurity pairs, triples and etc., without taking into account the restructuring of the entire spectrum (Bagatskii et al., 2007). The restructuring of the phonon spectrum of the crystal and the delocalization of QLV at finite concentrations of impurities in the coherent potential approximation was considered in (Ivanov, 1970; Ivanov & Skripnik, 1994).

The QLV usually occur in the frequency range where corresponding wavelengths of acoustic phonons of the host lattice become comparable to the average distance between the defects (the so-called *disorder parameter*). This is valid even for low concentrations of impurity atoms as is illustrated in Fig. 1***b***. The value $\lambda(\omega_{ql})$ for most phonon modes exceeds the disorder parameter even at $p \approx 1\%$. Therefore an interaction of QLV with rapidly propagating acoustic phonons of the host lattice (*propagons*) appears as the Ioffe-Regel crossover as is shown in (Klinger & Kosevich, 2001, 2002) and can lead to the formation of a boson peak (BP). The BP is an anomalous override of the low-frequency phonon density over the Debye density. The BP was observed in the Raman and Brillouin scattering spectra (Hehlen et al., 2000; Rufflé et al., 2006) and in inelastic neutron scattering experiments (Buchenau et al., 1984) as maxima in the frequency dependence $\nu(\omega)/\omega^2$ or $I(\omega)/\omega^2$ ($I(\omega)$ is the scattering intensity). These peaks appear in the low-frequency region (between 0.5 and 2 THz) of the vibration density of states (Ahmad et al., 1986), i.e. far below the Debye frequency. At the BP frequency the transition occurs from the fast-propagating low-frequency phonons (propagons), with dispersion relation close to the acoustic one, to the so-called *diffusons*, i.e. phonons, whose propagation is hampered by the scattering on localized states (Feher et al., 1994). In the frequency range (ω^*,ω^{**}), the number of localized vibrations increases with the frequency increase (Fig. 1**a**). Therefore, the phonons with frequencies lying in that interval (*diffuson* area) are either diffusons or locons. The similarity of the boson peak in disordered systems (e.g. glasses and substitution solid solutions) to the first van Hove singularity in crystal structures is noted in (Buchenau et al., 2004; Gospodarev et al., 2008). BPs are also observed in polymeric and metallic glasses (Duval et al., 2003; Arai et al., 1999).

As is shown in Fig.4 in the frequency range $[0,\omega^*]$ the vibrations of atoms of the host lattice propagate rapidly and are scattered by the quasi-localized states formed by impurity vibrations. Curves 6 in this figure depict the frequency dependence $\tilde{\nu}(\omega,p)-\nu_{imp}(\omega,p)$. At frequencies $\omega<\omega_{ql}$ there vibrations propagate as plane waves. Corresponding parts of curves 6 are smooth and have parabolic (quasi-Debye) form. At $\omega\approx\omega_{ql}$ a kink similar to the shape of the first van Hove singularity can be seen on curves 6. At this frequency as well as at $\omega=\omega^*$ in the phonon spectrum of a perfect crystal (Fig. 1**a**) there is a sharp change of the average group velocity of phonons. The frequency ω_{ql} is the upper limit of propagon zone of solid solution. This is clearly exhibiting with increasing impurity concentration p. Fig. 4 shows the evolution of the contribution to the phonon density of states by the displacements of impurity atoms (part **a**) and by the displacements of atoms of the host lattice (part **b**) with increasing impurity concentration. Note that both dependences can be determined experimentally (e.g., by the method described in (Fedotov et al., 1993). On both parts of Fig. 4 dashed curves show phonon densities of states of the perfect host lattice. In addition, functions $p\tilde{\rho}^{(\tau_-^5)}(\omega)$ are depicted in Fig. 4**a** by dashed lines. It is seen that at concentrations $p\in[0.1,0.5]$ the values of both $\nu_{imp}(\omega,p)$ and $p\tilde{\rho}^{(\tau_-^5)}(\omega)$ are different from zero in the same frequency range near the quasi-local frequency ω_{ql}. For $\omega<\omega_{ql}$ the frequency dependence takes parabolic form (quasi-Debay form). The frequency dependences $\tilde{\nu}(\omega,p)-\nu_{imp}(\omega,p)$ also have a characteristic kink at $\omega\approx\omega_{ql}$, similar in shape to the first van Hove singularity (observed at all concentrations, even at $p=0.9$). That is at $p\le 0.5$ the quasi-local frequency is an upper bound of the propagon zone for the vibrations of both impurity atoms and atoms of the host lattice.

With increasing concentration ($p\ge 0.25$) a singularity of the kink type begins to form on the function $\nu_{imp}(\omega,p)$ at $\omega\approx\omega_{ql}$. At concentrations $p>0.5$ large enough impurity clusters are formed in the solid solution. There is a short-range order in such clusters and we can identify the different crystallographic directions. The structure consisting of such clusters can already be considered as a structure with topological disorder and for given values of the concentration the upper limit of the propagon zone corresponds to the vanishing of the group velocity of the transversally polarized phonons along the crystallographic direction *ΓL* in impurity clusters. This frequency, as shown in Fig. 4**a**, is lower than ω_{ql}. With increasing *p* it approaches to the value of the frequency of the first van Hove singularity of perfect crystal consisting of heavy impurity atoms $\tilde{\omega}^*$.

Thus the influence of impurity atoms, which are heavy or weakly bound to the atoms of host lattice, on the phonon spectrum and the vibrational characteristics is manifested both in the formation of quasilocal vibrations caused by the vibrations of impurities and in the scattering on these vibrations of fast acoustic phonons generated by atomic vibrations of the

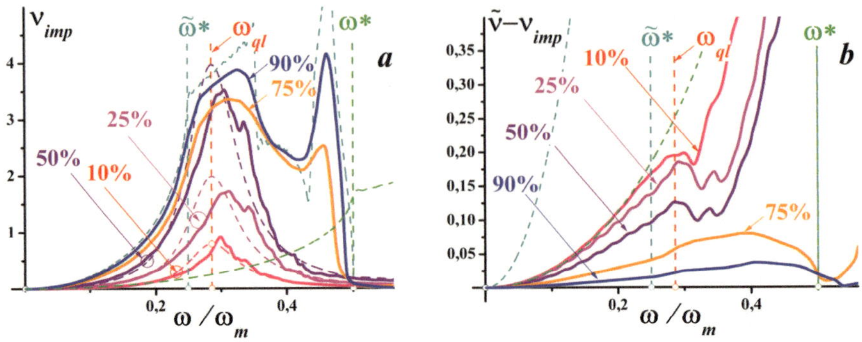

Fig. 4. Part **a** shows the evolution of function $\nu_{imp}(\omega,p)$ with increasing concentration of impurities. Part **b** shows the evolution of function $\tilde{\nu}(\omega,p)-\nu_{imp}(\omega,p)$ with increasing concentration of impurities.

host lattice. Up to concentrations of $p \approx 0.5$ the quasi-local frequency is an upper boundary of the propagon band, i.e. the frequency interval in which the phonons propagate freely in all directions. Further increase in the concentration is accompanied by the shift of the propagon zone upper boundary to the frequency of the first van Hove singularity of the crystal consisting of impurity atoms $\tilde{\omega}^* < \omega_{ql}$. At the same time for the propagation of atomic vibrations of the host lattice the upper boundary of the propagon zone is quasi-local frequency ω_{ql}.

3. Phonon spectra and low-temperature heat capacity of heterogeneous structures with bonds randomly distributed between atoms

The Debye approximation widely used for the description of the thermal properties of solids is based on an approximation of the real vibrational spectrum of the crystal by phonons with acoustic dispersion law. The corresponding density of states is (see, e.g. Kosevich, 1999):

$$\nu_D^{(q)}(\omega) = \frac{q\omega^{\,q-1}}{\omega_D{}^q}. \tag{9}$$

Fig. 5**a** shows the Debye density of states $\nu_D^{(3)}(\omega) = 3\omega^2/\omega_D^3$, defined by (9) at $q=3$ (curve 1), compared with the true density of states of the FCC lattice with central interaction of nearest neighbors (curve 2). It is seen that at $\omega \le 0.25\omega_m$ these curves almost coincide. With the frequency increase a deviation of the phonon density from $\nu_D^{(3)}(\omega)$ occurs. This leads to a deviation of the temperature dependence of the phonon heat capacity from its Debye form $C_D(T)$. Moreover, this deviation is more apparent the lower the frequencies are at which such deviation starts. As a rule, the deviation of the true phonon heat capacity from $C_D(T)$ is described as a temperature dependence of the Debye temperature Θ_D. This dependence can be derived from the transcendental equation

$$C_v(T) = C_D(T) \equiv 3R\left\{D\left(\frac{\Theta_D}{T}\right) - \frac{\Theta_D}{T}D'\left(\frac{\Theta_D}{T}\right)\right\}; \quad D(x) \equiv \frac{3}{x^3}\int_0^x \frac{z^3 dz}{e^z - 1}, \tag{10}$$

where the heat capacity $C_v(T)$ is determined from experiment or microscopic calculation as

$$C_v(T) = 3R \int_0^{\omega_m} \left(\frac{\hbar\omega}{2kT}\right) \cdot \mathrm{sh}^{-2}\left(\frac{\hbar\omega}{2kT}\right) \cdot \nu(\omega) d\omega . \tag{11}$$

Of course, at $\nu(\omega) = \nu_D^{(3)}(\omega)$ the expressions (10) and (11) coincide and $\Theta_D(T) = \Theta_P \equiv \hbar\omega_m/k$, i.e. the Debye temperature does not depend on temperature. At low temperatures ($T << \Theta_P, \Theta_D$) the main contribution to the heat capacity is provided by the long-wavelength phonons with the sound dispersion relation. It seems that the dependence $C_v(T)$ is well described by (10). That is, the Debye temperature should be practically the same as Θ_P. Indeed, as seen from Fig. 5**b** (curve 2), exactly in the temperature range $T \le 0.1\Theta_P$ the dependence $\Theta_D(T)$ is most intense. This is typical for a large number of compounds (Leibfried, 1955). To find the cause of a strong temperature dependence of Θ_D at $T << \Theta_D$ we consider the function $\Theta_D(T)$ for a system for which the phonon density of states is a linear combination of the function $\nu_D^{(3)}(\omega)$ (curve 1 in Fig. 5**a**) and the Einstein density of states $\delta(\omega - \omega^*)$, where ω^* is the frequency of the first van Hove singularity (dashed line 3 in Fig. 5**a**). Curve 3 in Fig. 5**b** shows the $\Theta_D(T)$ in the case when the phonon density of states is $\nu_{appr}(\omega) = \frac{8}{39}\delta(\omega - \omega^*) + \frac{31}{39}\nu_D^{(3)}(\omega)$. The coefficients of this linear combination are selected from the averaging over all the high-symmetry directions in the FCC lattice. As shown in Fig. 5**b**, curve 3 quite satisfactorily coincides with the dependence $\Theta_D(T)$ of the FCC crystal (curve 2). This is manifested in the behavior of $\Theta_D(T)$ at $T \le 0.1\Theta_P$ and in the coincidence of the minima (both in temperature and in magnitude). Thus, one can assert that the dependence $\Theta_D(T)$ at low temperatures is conditioned by the changes in the character of the phonon propagation on the frequency of the first van Hove singularity.

Taking into account the Einstein level tailing can improve the approximation of the $\Theta_D(T)$ function at low temperatures (curves 4).

As mentioned above, the frequency of the first van Hove singularity ω^* is an "interface" frequency between the fast and slow phonons, i.e. between propagons and diffusons. It can be interpreted as the Ioffe-Regel singularity (or its equivalent) in a regular crystal system.

Maxima on the ratio $\nu(\omega)/\omega^2$ can be considered as BPs only when $\omega < \omega^*$, because the maximum on the mentioned ratio, corresponding to the first van Hove singularity, always exists. Within this frequency interval the phonon density can be approximated by a parabola, and its deviation from the Debye density $\nu_D^{(3)}(\omega)$ can be expressed by the frequency dependence of the value ω_D, i.e. writing the phonon density in a form analogous to (9). At q=3 we have

$$\nu(\omega) \equiv \frac{3\omega^2}{\omega_D^3(\omega)} . \tag{12}$$

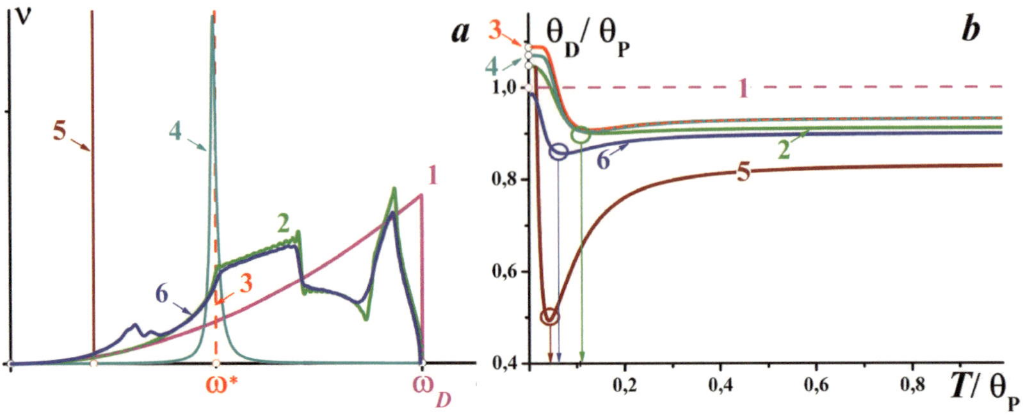

Fig. 5. Relationship of the temperature dependence of the Debye temperature (part **b**) to the character of the long-wavelength phonons propagation in a crystal (part **a**).

Then, using the definition of ω_D, the ratio of the phonon density to the squared frequency can be expressed by the dispersion of sound velocities $s_i(\omega)$

$$\frac{\nu(\omega)}{\omega^2} \equiv \frac{3}{\omega_D^3(\omega)} = \frac{V_0}{6\pi^2}\sum_{i=1}^{3}\frac{1}{s_i^3(\omega)}, \tag{13}$$

where V_0 is the unit cell volume. Thus, the occurrence of the maximum on the ratio $\nu(\omega)/\omega^2$ is caused by the additional dispersion of sound velocities. This dispersion is caused by the heterogeneity of the structure, which is the source of quasi-localized vibrations. Such additional sound velocity dispersion must be manifested in the behavior of the temperature dependence Θ_D. On the curve $\Theta_D(T)$ a low-temperature minimum should appear (see curve 5, Fig. 5**b**), deeper than those on curves 1–4 in Fig. 5**b**. This curve corresponds, in addition to the quasi-localized perturbations on the frequency of the first van Hove singularity in the phonon spectrum with the density of states $\nu_{appr}(\omega)$, to the presence of an additional resonance level with the frequency $\delta(\omega - \omega_D/5)$ (see Fig. 5**a**). Curves 6 in both parts of Fig.5 correspond to the 5% solution of a heavy isotope impurity in the FCC crystal. The formation of the QLV leads to a significant deepening of the $\Theta_D(T)$ low-temperature minimum and to be shifting of its temperature below that of the perfect crystal.

In the first section it was shown that heavy or weakly bound impurities form QLV caused by their motion. On these vibrations the fast acoustic phonons associated with the displacements of atoms of the host lattice are scattered. This leads to kinks in the contribution to the phonon spectral density (see curve 6 in Fig. 3) which are a manifestation of the Ioffe-Regel crossover. On the background of large quasi-local maxima it is difficult to distinguish their influence on the vibrational characteristics of the crystal. The study of this effect is possible in systems in which interatomic interactions are not accompanied by the formation of QLV, or in systems in which the frequencies of QLV lie beyond the propagon zone. Examples of such systems are crystals with weakly bound impurities. Fig. 6 shows the low-frequency parts of the phonon density of states (*a*) and the temperature dependence Θ_D (*b*) for the FCC lattice, in which force constants of impurities (p = 5%) are four and eight times weakened (curves 3 and 4, respectively). Part ***a*** shows the functions $\frac{\nu(\omega)}{4}\left(\frac{\omega_m}{\omega}\right)^2$ (curves 3′ and 4′), for which the deviation of the phonon density of states from the Debye form is more pronounced. Curve 1 corresponds to a perfect crystal. Curve 2 shows the frequency dependence of the group velocity in the direction ΓL (see Fig. 1***a***). Values $\omega_{ql}^{(4)}$ and $\omega_{ql}^{(8)}$ correspond to the frequencies of QLV in a lattice containing an isolated weakly bound impurity ($\alpha'/\alpha = 1/4$ and $\alpha'/\alpha = 1/8$, respectively). As can be seen from the figure, the phonon densities are qualitatively different from the quasi-Debye behavior, starting from the frequencies $\omega = \omega_{ql}^{(4)}$ (curve 3) and $\omega = \omega_{ql}^{(8)}$ (curve 4). In this system the formation of QLV with such frequencies corresponds to the existence of atoms with few weakened force interactions (at least two, along the same line), i.e. to the formation of defect clusters (or impurity molecules). The minimum size of the defect cluster is equal to two interatomic distances and the Ioffe-Regel crossover can occur in a wide range of values (see Fig. 1***b***). Fig. 6***b*** shows that there are notable low-temperature minima on $\Theta_D(T)$ for crystals with impurities (p = 5%) whose force interactions are four and eight times weakened (curves 3 and 4, respectively). These minima points to a slowdown of acoustic phonons due to their localization on the defect clusters and due to the scattering of additional phonons, remaining delocalized on the resulting quasi-localized states.

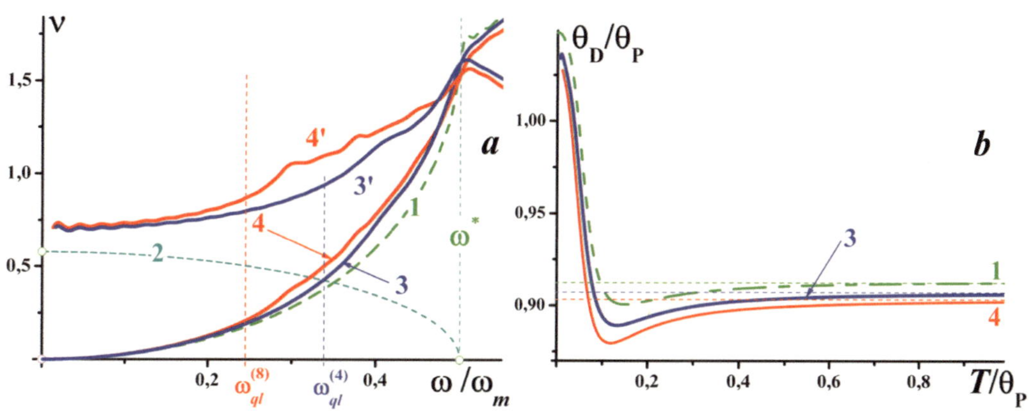

Fig. 6. Low-frequency parts of phonon spectra (part **a**) and temperature dependences Θ_D (part **b**) of FCC crystals with 5% of weakened force interactions

The high sensitivity of the low-temperature heat capacity to the slowing of the long-wavelength phonons is clearly manifested in the case when not only the interaction of impurity atoms with the host lattice is weakened, but also the interaction between substitution impurities in the matrix of the host lattice. An example of such a system is the solid solution $Kr_{1-p}Ar_p$. Krypton and argon are highly soluble in each other and the concentration p can take any value from zero to one. Argon is ≈ 2.09 times lighter than krypton, and the interaction of the impurity of argon with krypton atoms is slightly weaker than the interaction of krypton atoms between each other, so an isolated Ar impurity in the

Kr matrix behaves almost like a light isotope. At the same time, in a krypton matrix the interaction of argon impurities between each other is more than five times weaker than the interaction between the krypton atoms (Bagatskii et al., 2007). Fig. 7**a** shows the phonon densities of states of pure krypton and argon as well as that of the $Kr_{0.756}Ar_{0.244}$ solid solution. At such a concentration there is a sufficient number of isolated impurities and defect clusters with dimensions less than two interatomic distances in the solution (Fig. 7**b**). This leads, in comparison with the pure Kr phonon spectrum, to the increase of the number of high-frequency states in the phonon spectrum of the solution (Bagatskii et al., 1992). In such clusters weakly coupled argon impurities are not created and quasi-local vibrations are not formed. At the same time in such a solution larger defect clusters are formed, which consists of weakly coupled Ar impurities. However, the frequency of QLV formed by these clusters is $\omega_{ql}^{Kr-Ar} \approx 0.86\omega_{Kr}^{*}$, that is (unlike the previous case) slightly less than the frequency of the first van Hove singularity for the Kr lattice. Therefore, neither on the solution phonon density of states nor on its relationship to the square frequency any singularities do appear. Extension of the of quasi-continuous spectrum of the Kr-Ar solution as compared with pure Kr, as seen in Fig. 7**a**, occurs mainly due to the phonons with frequencies in the interval $[\omega^*, \omega^{**}]$ (diffuson zone).

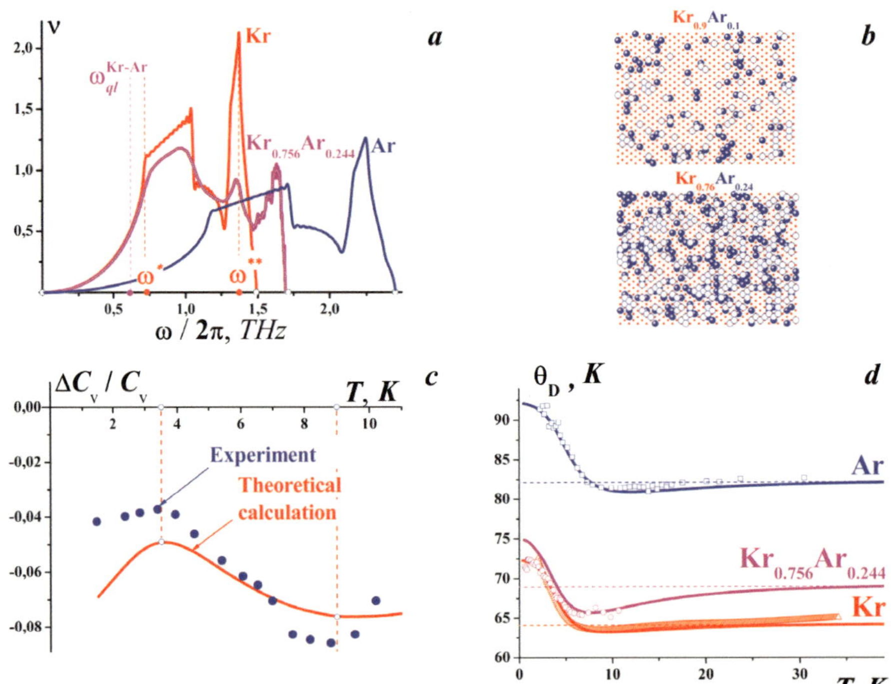

Fig. 7. Phonon densities (**a**) and temperature dependences of the Debye temperature (**d**) of the krypton, argon and the $Kr_{0.756}Ar_{0.244}$ solid solution. Part **b** shows in the [111] plane, some typical configurations of the displacements of argon impurity in the in krypton matrix at $p = 0.1$ and at 0.24 (circles and filled circles correspond to the Ar atoms, lying in different neighboring layers). Part **c** is shows the relative change of the heat capacity.

Note that the phonon densities of states of the solution and of pure krypton are practically the same in the most part of the propagon zone. The redistribution of the phonon frequency leads to a characteristic two-extremum behavior of the temperature dependence of the relative change of the low-temperature heat capacity (Fig. 7**c**), the maximum on which indicates that there is an additional slowing-down of the long-wavelength acoustic phonons on slow phonons, corresponding to the quasi-local vibrations of weakly couple argon atoms. This scattering, as in earlier cases, forms a significant minimum in the temperature dependence of Θ_D. Fig. 7**d** plots the values $\Theta_D(T)$ for pure krypton, argon, and the $Kr_{0.756}Ar_{0.244}$ solution. These dependences are the solutions of the transcendental equation (10) for the heat capacity, calculated theoretically and determined experimentally, see Fig. 7c (Bagatskii et al., 1992). The results of the theoretical calculations show a good agreement with experimentally obtained results, especially near the minimum on $\Theta_D(T)$. This minimum can appear also in the case when the maximum of the ratio $\nu(\omega)/\omega^2$ is not observed.

Thus, the results presented in this section allow us to make the conclusion that both the low temperature heat capacity and the temperature dependence of the value Θ_D are highly sensitive not only to the formation of quasi-localized states, but also to the reduction of the rate of propagation of long-wavelength acoustic phonons due to their scattering on these states. This slowdown is clearly manifested in the frequency range as boson peaks in the ratio $\nu(\omega)/\omega^2$, or as another singularities of the Ioffe-Regel type, but only when certain conditions are fulfilled. They are, according to our analysis:

1. For such defects as local weakening of the interatomic interactions or light weakly bound impurities the QLV scattering frequency must be low enough, and so, in other words, the "power of the defect" should be large enough.
2. Defect cluster should be large enough (at least two atomic distances) which requires a high enough (~ 15-20%) concentration of defects.

4. Low-frequency features of the phonon spectra of layered crystals with complex lattice

As it has been shown in the previous sections the low-frequency region of the phonon density of states of heterogeneous systems differs from the Debye form. This is caused by the formation of the quasi-localized states on the structure heterogeneities and by the scattering of the fast longwavelength acoustic phonons (propagons) on them. However, it is not necessary that these heterogeneities were defects violating the regularity of the crystalline arrangement of atoms. If, in the crystal with polyatomic unit cell the force interaction between atoms of one unit cell is much weaker than the interaction between cells, then optical branches occur in the phonon spectrum of the crystal at the frequencies significantly lower than the compound Debye frequency. These optical branches are inherent to the phonon spectra of many highly anisotropic layered crystals and they may cross the acoustic branches, causing additional features in the propagon area of phonon spectrum (Wakabayashi et al., 1974; Moncton et al., 1975; Syrkin & Feodosyev, 1982). Note that the deviation of the phonon spectrum of such compounds from $\nu_D^{(3)}(\omega)$ at low frequencies may be a manifestation of their quasi-low-dimensional structure as well

(Tarasov, 1950) of the flexure stiffness of single layers (Lifshitz, 1952b). However, the crossing of the low-lying optical modes with the acoustic ones may also occur in systems, in whose propagon zone of the phonon spectrum no quasi-low-dimensional peculiarities and no flexural vibrations are present. These compounds include high-temperature superconductors, dichalcogenides of transition metals, a number of polymers and biopolymers, as well as many other natural and synthesized materials. A distinctive feature of the structure of these substances is the alternation of layers with strong interatomic interactions (covalent or metal) with layers in which atomic interactions are much weaker, e.g. the van der Waals interaction. Since this interaction is weak along all directions, the propagation of the propagons is three-dimensional and can be characterized by the temperature dependence of the Θ_D determined by formulas (10, 11).

Let us examine a simple model of such a structure, i.e. the system based on a FCC crystal lattice and generated by "separating" the atomic layers along the [111] axis into a structure consisting of stacked layers of the closely packed ...A - B - B - A - B - B -... type. To describe the interatomic interaction we shall restrict our attention to the central interaction between nearest neighbors. We assume that the interaction between atoms of the B type (lying in one layer as well as in different layers) is half as strong as the interaction between A type atoms and atoms of different types (we assume these interactions are the same). The phonon spectrum of considered model contains nine branches (three acoustic and six optical) and the optical modes are not separated from the acoustic modes by a gap. The frequencies of all phonons polarized along the [111] axis (axis *c*) lie in the low-frequency region. At $k = 0$ two optical modes have low frequencies corresponding to a change in the topology of the isofrequency surfaces (from closed one to the open one along the *c* axis) both for transverse and longitudinal modes. Thus, these frequencies play the role of the van Hove frequencies ω^* and are shown in Figs. 8**a-d** and 9**a** as vertical dashed lines ω_τ^* and ω_l^*.

Fig. 8 displays the spectral densities corresponding to displacements of A and B atoms in the basal plane *ab* and along the *c* axis (curves 1). The normalization of each spectral density corresponds to its contribution to the total phonon density of states $\nu(\omega)$ presented in Fig. 9**a**:

$$\nu(\omega) = \frac{2}{9} \cdot \rho_{ab}^{(A)}(\omega) + \frac{1}{9} \cdot \rho_c^{(A)}(\omega) + \frac{4}{9} \cdot \rho_{ab}^{(B)}(\omega) + \frac{2}{9} \cdot \rho_c^{(B)}(\omega). \tag{14}$$

Fig. 8 also displays the quantities proportional to the ratio of the corresponding spectral densities to the squared frequency (curves 2). The coefficients of proportionality are chosen so that these curves may be placed in the same coordinate system as the corresponding spectral density. The functions $\rho_c^{(A)}(\omega)$ and $\rho_c^{(B)}(\omega)$ and their ratios to ω^2 have distinct features at ω_l^* as well as at a certain frequency ω_c lying below ω_τ^*. This frequency corresponds to the crossing of the longitudinal acoustic mode, polarized along the *c* axis, with the transversely polarized optical mode propagating in the plane of the layer. The velocity of sound in this acoustic mode is $s_l^{(c)} \sim \sqrt{C_{33}}$ (in the described model the elastic moduli of elasticity C_{ik} satisfy the relations $C_{11} = 2.125 \cdot C_{33} = 3 \cdot C_{66} = 7.5 \cdot C_{44}$). The spectral densities $\rho_{ab}^{(A)}(\omega)$ and $\rho_{ab}^{(B)}(\omega)$ have additional features at frequencies $\omega_{ab}^{(\tau)} < \omega^*$ and $\omega_{ab}^{(l)} \in \left[\omega_\tau^*, \omega_l^*\right]$. These features are related to the crossing of acoustic branches with the low-frequency optical mode which is polarized along the c axis. There are three acoustic waves propagating in the basal plane and differing substantially from one another (longitudinal wave $s_l^{(ab)} \sim \sqrt{C_{11}}$ and two transverse waves). One of the transverse waves is polarized in the basal plane ($s_\tau^{(ab)} \sim \sqrt{C_{66}}$) and another one is polarized along the c axis ($s_n^{(ab)} \sim \sqrt{C_{44}}$). The acoustic modes with sound velocities $s_l^{(ab)}$ and $s_\tau^{(ab)}$ cross the low-frequency optical mode. In this optical mode at $k = 0$ the frequency of the vibrations is $\omega = \omega_l^*$, and at the point K at the boundary of the first Brillouin zone (see Fig 1) the mode joins the slowest acoustic mode, polarized along the c axis. Appreciable dispersion of this optical mode leads to a small value of $\omega_{ab}^{(\tau)}$ ($\omega_{ab}^{(\tau)} < \omega_c$) and to the blurring of the feature near $\omega_{ab}^{(l)}$.

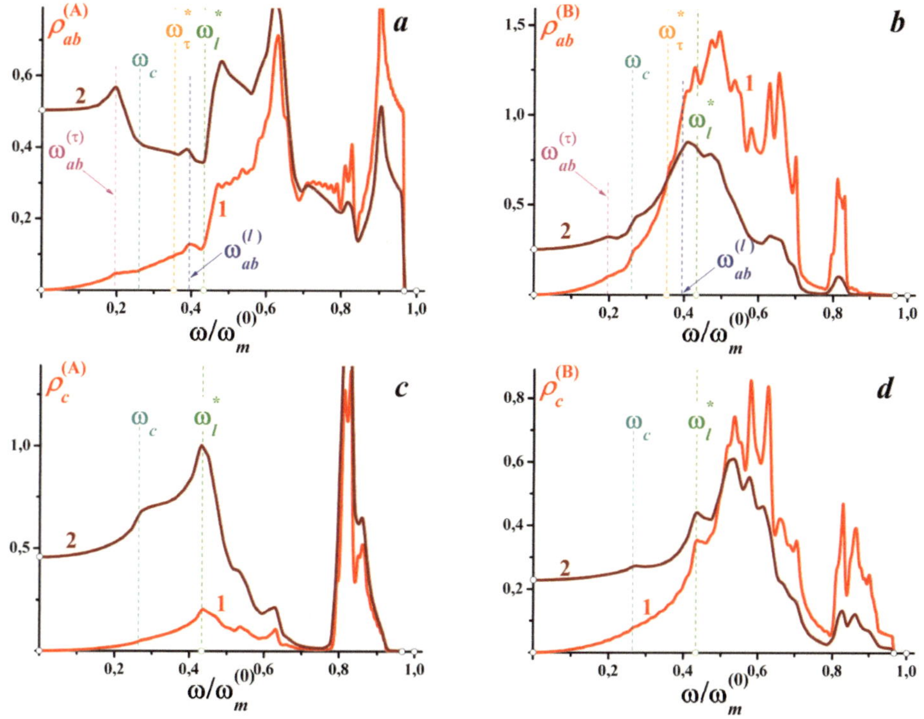

Fig. 8. Spectral densities (curves 1) and their ratio to the squared frequency (curves 2), corresponding to displacements of atoms of different sublattices along different crystallographic directions.

All spectral densities at quite low frequencies are proportional to ω^2, i.e. at low-temperatures the thermodynamic quantities should be determined by an ordinary three-dimensional behavior (see Fig. 8). Fig. 9**b** shows the temperature dependence of the Debye temperature (10, 11) for the considered model. For comparison, on Fig. 9**a** and 9**b** the characteristics of the "initial" FCC lattice is shown (lattice of A type atoms). As a result of the weakening (as compared to the A lattice) of some force bonds the function $\nu(\omega)$ increases at low frequencies (Fig. 9**a**) and therefore Θ_D decreases. The scattering of the propagons on slow optical phonons forms a distinct low-temperature minimum on $\Theta_D(T)$.

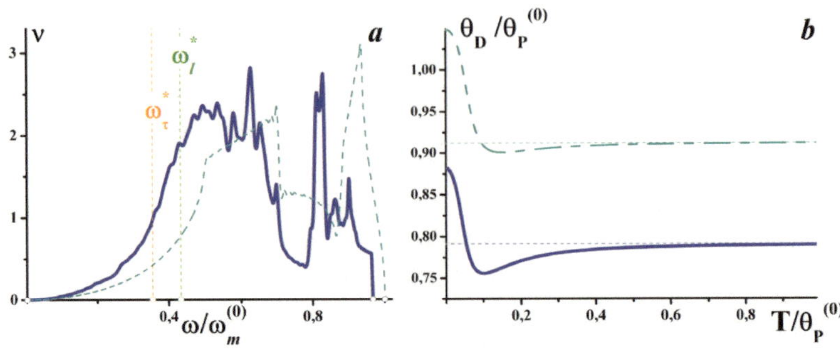

Fig. 9. Phonon density of states (**a**) and temperature dependence of $\Theta_D(T)$ (**b**) of a layered crystal with a three-atom unit cell (solid curves) and analogous characteristics of an ideal FCC lattice with central interaction of the nearest neighbors (dashed curves).

The Ioffe-Regel crossover determined by the intersections of the acoustic branches with the low-lying optical one is clearly apparent on the niobium diselenide phonon spectrum. This compound has a three-layer Se-Nb-Se "sandwich" structure. Fig. 10 (center) shows the dispersion curves of the $NbSe_2$ low-frequency branches (Wakabayashi et al., 1974)]. The low-frequency optical modes Δ_2 and Δ_5 correspond to a weak van der Waals interaction between "sandwiches". They cross at points C2, C3, C4, S1, A1 and A2 with acoustic branches polarized in the plane of layers. The wavelength λ_{eff} (see Sec. 2) corresponding to frequency of each of these crossovers exceeds the thickness h of the "sandwich". The parameter h plays in this case the same role as the distance between impurities in solid solutions, i.e. the condition of the Ioffe-Regel is met. Therefore, for given values of frequency as well as for the van Hove frequencies (points D1, D2 and D4) an abrupt change of the propagon group velocity occurs. This leads to the appearance of peaks on the dependences $\nu(\omega)$ and $\nu(\omega)/\omega^2$ (curves 1 and 2 in Fig. 10**a**) and to the formation of a rather deep low-temperature minimum in the dependence $\Theta_D(T)$ (Fig. 10**b**). For the longitudinal acoustic mode Δ_1 polarized along the c axis at the frequency corresponding to the point of its intersection with the branch Δ_5 (point C1), the value λ_{eff} is less than h . Therefore, at this point the group velocity of phonons does not have a jump and does not change its sign. There are no peculiarities at point C1 on the phonon density of states and on the function $\nu(\omega)/\omega^2$.

Thus, in the crystalline ordered heterogeneous structures the scattering of fast phonons on slow optical ones is possible. This scattering is similar to the scattering of such phonons on quasi-localized vibrations in disordered systems and is completely analogous to that considered in (Klinger & Kosevich, 2001, 2002). It leads to the formation of the same low-frequency peculiarities on the phonon density of states than are those manifested in the behavior of low-temperature vibrational characteristics. The elastic properties of structures discussed in this section differ essentially from the properties of low-dimensional structure. However, at high frequencies (larger than the frequencies of the van Hove singularities, which correspond to the transition from closed to open isofrequency surfaces along the c axis) the phonon density of states exhibits quasi-two dimensional behavior seen on parts ***a*** of Figs. 8, 9 and 10. Such a behavior is inherent to many heterogeneous crystals, in particular high-temperature superconductors (see, e. g., Feodosiev et al., 1995; Gospodarev et al., 1996), as was confirmed experimentally (Eremenko et al., 2006). This allows us to describe the vibrational characteristics of such complex compounds in the frames of low-dimensional models.

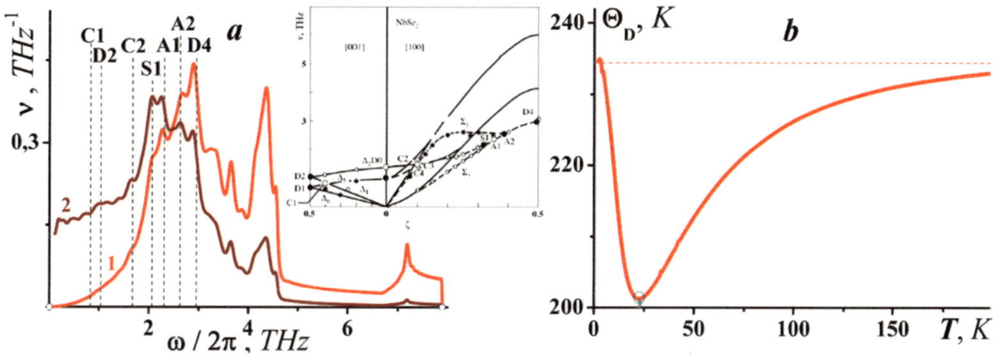

Fig. 10. Vibrational characteristics of $NbSe_2$. Part **a** shows the phonon density of states (curve 1) and ratio $\nu(\omega)/\omega^2$ (curve 2). On the inset the dispersion curves of the low-frequency vibration modes determined by the method of neutron diffraction are shown. Part **b** shows the dependence $\Theta_D(T)$.

The theory developed for the multichannel resonance transport of phonons across the interface between two media (Kosevich Yu. et al., 2008) can be applied to interpret the experimental measurements of the phonon ballistic transport in an Si-Cu point contact (Shkorbatov et al., 1996, 1998). These works revealed for the first time the low temperature quantum ballistic transport of phonons in the temperature region 0.1 - 3 K. Besides, in some works (Shkorbatov et al., 1996, 1998) a reduced point contact heat flux in the regime of the geometric optics was investigated in the temperature interval 3 - 10 K. The results obtained in these works showed that in this temperature interval the reduced heat flow through the point contact is a non-monotonous temperature function and has pronounced peaks at temperatures T_1 = 4.46 K, T_2 = 6.53 K and T_3 = 8.77 K. We suppose that the series of peaks for the reduced heat flow (Shkorbatov et al., 1996, 1998) could be explained by the models represented in Fig.11 **a,b**. These peaks are a result of the resonance transport. In the case of the single-channel resonance transport studied in work (Feher et al., 1992) a model of the narrow resonance peak was applied, meaning the following: the total heat flux $\dot{Q}$ may be written as the sum of the ballistic flux $\dot{Q}_B$ and the resonance heat fluxes $\dot{Q}_R$, $\dot{Q}=\dot{Q}_B+\dot{Q}_R$. Assuming the narrow resonance peak near the frequency ω_0^α we obtain the formula describing the temperature dependence of the heat flux:

$$\dot{Q}(T,T_0)=C\Delta T^4+\sum_\alpha K_\alpha\left[\frac{1}{\exp\left(\hbar\omega_0^\alpha/T\right)-1}-\frac{1}{\exp\left(\hbar\omega_0^\alpha/T_0\right)-1}\right]. \quad (15)$$

To separate the two parts of the total heat flux, its value must be divided by $\Delta T^4=(T^4-T_0^4)$.

This model (using only one frequency) can be fitted to our experimental data with a correlation factor of about 0.95. The resonance frequency ω_0 is connected with T_{max} by the

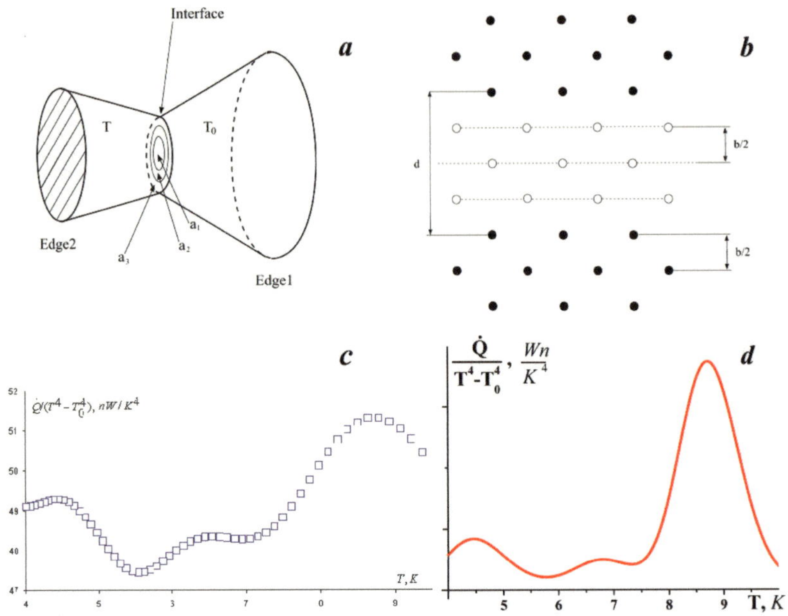

Fig. 11. **a**) Schematic model of a contact. **T** and $\mathbf{T_0}$ are the temperatures of the massive edges of the contact; $\mathbf{a_1}$, $\mathbf{a_2}$, and $\mathbf{a_3}$ are the zones with different composition of the interface layer. **b**) Schematic figure showing an interface between two crystal lattices that contains three intercalate impurity layers. **c)** Experimentally observed temperature dependence of the reduced heat flux through the *Si-Cu* point contact. **d)** Results of a numerical calculation using the considered model.

relation $\hbar\omega_0 = 3.89T_{\max}$. Using the model of the multichannel resonance transport we modified the expression (15) in a following way:

$$\frac{\dot{Q}}{\Delta T^4} = \frac{1}{\Delta T^4}\sum_{n=1}^{3} K_n \left[\left(\exp\left(3.89\frac{T_n}{T}\left(1 + \frac{1}{T_s^2}(T - T_n)^2 \right) \right) - 1 \right)^{-1} - \left(\exp\left(3.89 \cdot \frac{T_n}{T_0} \right) - 1 \right)^{-1} \right] + C \,. \quad (16)$$

The optimal correspondence between the values calculated by this formula and the experimental results was obtained for the following values of parameters:

$$T_0 = 0.15\ K;\ T_1 = 4.46\ K;\ T_2 = 6.8\ K;\ T_3 = 8.71\ K;\ T_s = 1.5\ K\,.$$

$$K_1 = 0.7\ nW;\ K_2 = 2\ nW;\ K_3 = 50\ nW;\ C = 49.55\ nW / K^4\,.$$

The expression (16) takes into account the presence of three channels of the resonance transport as well as (using an additional term containing the intrinsic temperature T_S) the instability of the intermediate layer of weakly bound impurities near the resonance. Results of numerical calculations by formula (16) are given in Fig.11***d***. These results evidence that the proposed model describes in much detail the experimental results presented in Fig. 11***c***. It should be noted that the temperature T_S used in our calculations corresponds to the binding energy of the impurity layer with contact banks. This temperature is by two orders of magnitude lower than the Debye temperature of crystals forming the banks of contacts.

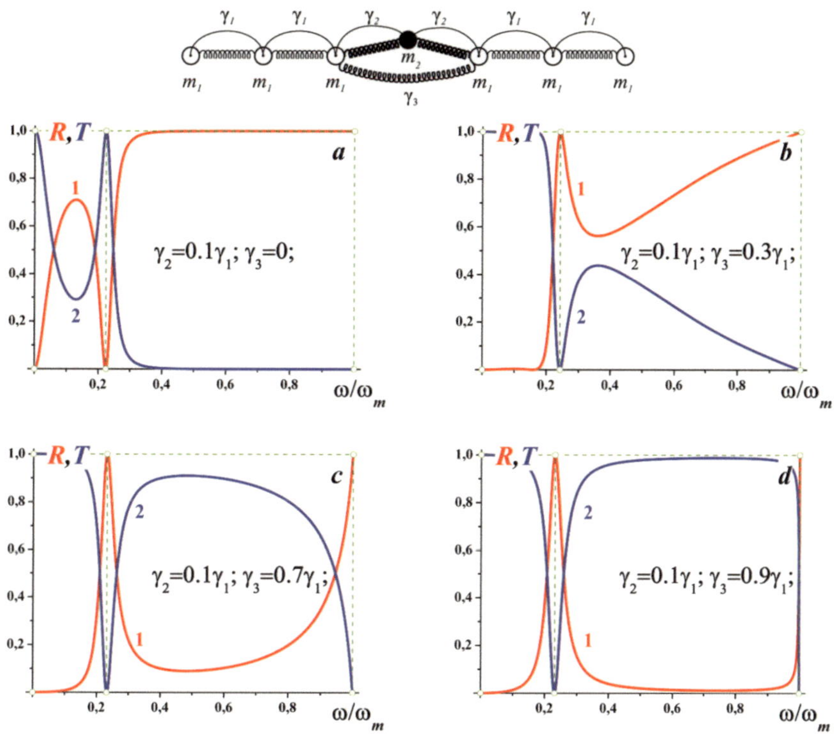

Fig. 12. Coefficients of the phonon energy reflection (curve 1, red line) and transmission (curve 2, blue line) through an impurity atom.

This is in agreement with the fact that the binding constant of the impurity layer with contact banks is by two orders of magnitude lower than the binding constant in crystals forming this contact (Shklyarevskii et al., 1975; Koestler et al., 1986; Lang, 1986). Coefficients K are proportional to the squares of the area of different interface layers. Using the results presented in Fig. 11**d** we can interpret experimental results (Shkorbatov et al., 1996, 1998) presented in Fig. 11**c**.

Finally we consider the resonance reflection and transmission of phonons through an intercalated layer between two semi-infinite crystal lattices. We consider an infinitely long chain which contains a substitution impurity atom weakly coupled to the matrix atoms (see model in Fig. 12). In this system quasi-local (resonance) impurity oscillations emerge with such a frequency, at which the transmission coefficient through the impurity becomes equal to unity (full phonon transmission through the interface, see Fig.12***a***). Let us compare these results with the results received taking into account the force constant γ_3, corresponding to the interaction between non-nearest neighbors. We have shown that if the non-nearest neighbor force constant γ_3 is larger than the weak bounding force constant γ_2 (Kosevich, et al., 2008) (see Fig.12), two frequency regions with enhanced phonon transmission are formed, separated by the frequency region with enhanced phonon reflection. Namely, for $\gamma_3 \approx \gamma_1$ a strong transmission "valley" occurs at the same resonance frequency at which there is a transmission maximum for $\gamma_3 << \gamma_2 < \gamma_1$. Moreover, this transmission minimum occurs on the background of an almost total phonon transmission through the impurity atom due to the strong interaction of matrix atoms through the defect (with force constant $\gamma_3 \approx \gamma_1$). For large values of γ_3, the resonant transmission frequency corresponds to the frequency of total reflection (so-called Fano effect) (Fano, 1961). Such a system permits to make a filter which reflects the phonons in a very narrow region of frequencies (heat transmission is minimum at corresponding temperatures) while the total transmission is observed in other regions of frequencies. It is worth to mention that such an inversion of the transmission and reflection spectra in the two limiting cases is directly related to the Fano-type interference. Similar inversion of the Fano-type transmission and reflection resonances also occurs in sound transmission through two-dimensional periodic arrays of thin-walled hollow cylinders due to the their flexural vibration modes (see Liu, 2000).

5. Acknowledgments

This work was supported by the grants of the Ukrainian Academy of Sciences under the contract No. 4/10-H and by the grant of the Scientific Grant Agency of the Ministry of Education of Slovak Republic and the Slovak Academy of Sciences under No. 1/0159/09.

6. References

Ahmad N., Hutt K.W. & Phillips W.A. (1986) Low-frequency vibrational states in As_2S_3 glasses. *J. Phys. C.: Solid State Phys.* , Vol.19, pp. 3765 - 3773.

Allen P.B., Feldman J.L., Fabian J. & Wooten F. (1999) Diffusons, locons and propagons: character of atomic vibrations in amorphous Si. *Philosophical Magazine B.* Vol. 79, pp. 1715-1731.

Arai M., Inamura Y. & Otomo T. (1999) Novel dynamics of vitreous silica and metallic glass. *Philosophical Magazine B.* Vol. 79, pp. 1733 – 1739.

Bagatskii M.I., Syrkin E.S. & Feodosyev S.B. (1992) Influence of the light impurities on the lattice dynamics of cryocrystals. *Low Temp. Phys.* Vol. 18, pp. 629-635.

Bagatskii M.I., Feodosyev S.B., Gospodarev I.A., Kotlyar O.V., Manzhelii E.V. Nedzvetskiy A.V. & Syrkin E.S. (2007) To the theory of rare gas alloys: heat capacity. *Low Temp. Phys.* Vol. 33, pp. 741-746.

Buchenau U., Nücker N. & Dianoux A. J. (1984) Neutron Scattering Study of the Low-Frequency Vibrations in Vitreous Silica. *Phys. Rev. Lett.,* Vol. 53, pp. 2316 – 2319

Buchenau U., Wisschnewski A., Ohl M. & Fabiani E. (2007) Neutron scattering evidence on the nature of the boson peak. *J. Phys.: Condens. Matter* Vol. 19, pp. 205106.

Cape J.A., Lehman G.M., Johnston W.V. & de Wames R.E. (1966) Calorimetric observation of virtual bound-mode phonon states in dilute Mg-Pb and Mg-Cd alloys. *Phys. Rev. Lett.* Vol. 16, pp. 892 - 895.

Duval E., Saviot L., David L., Etienne S. & Jalbibitem J.F. (2003) Effect of physical aging on the low-frequency vibrational density of states of a glassy polymer. *Europhys. Lett.* , Vol. 63, pp. 778 - 784.

Eremenko V.V., Gospodarev I.A., Ibulaev V.V., Sirenko V.A., Feodosyev S.B. & Shvedun M.Yu. (2006) Anisotropy of $Eu_{1+x}(Ba_{1-y}Ry)_{2-x}Cu3O_{7-d}$ lattice parameters variation with temperature in quasi-harmonic limit. *Low Temp. Phys.* Vol. 32, pp. 1560-1565.

Fano U. (1961) Effects of Configuration Interaction on Intensities and Phase Shifts. *Phys. Rev.* Vol. 124, pp. 1866 – 1878.

Feher A., Stefanyi P., Zaboi R., Shkorbatov A.G. & Sarkisyants T.Z. (1992) Heat conductivity of dielectric- dielectric and metal- dielectric point contacts at low temperatures. *Sov. J. Low Temp. Phys.* Vol. 18, pp. 373 - 375.

Feher A., Yurkin I.M., Deich L.I., Orendach M. & Turyanitsa I.D. (1994) The comparative analysis of some low-frequency vibrational state density models of the amorphous materials – applied to the As_2S_3 glass. *Physica B*, Vol. 194 - 196, pp. 395 - 396.

Fedotov V.K., Kolesnikov A.I., Kulikov V.I., Ponjtovskii V.G., Natkanec I., Mayier Ya. & Kravchik Ya. (1993) The studing of vibration anharmonisms of copper and oxygen atoms in the yttrium ceramics by inelastic neutron scattering. *Fiz. Tv. Tela.*, Vol. 35, pp. 310-319 (in Russian).

Feodosyev S.B., Gospodarev I.A. Kosevich A.M. & Syrkin E.S. (1995) Quasi-Low-Dimensional Effects in Vibrational Characteristics of 3D-Crystals. *Phys. Low-Dim. Struct.* , Vol. 10-11, pp. 209 - 219

Gospodarev I.A., Kosevich A.M., Syrkin E.S. & Feodosyev S.B. (1996) Localized and quasi-low-dimensional phonons in multilayered crystals of HTS type. *Low Temp. Phys.* Vol. 22, pp. 593 - 596

Gospodarev I. A., Grishaev V. I., Kotlyar O. V., Kravchenko K. V., Manzheliĭ E. V., Syrkin E. S. & Feodosyev S. B. (2008) Ioffe_Regel crossover and boson peaks in disordered solid solutions and similar peculiarities in heterogeneous structures. *Low Temp. Phys.* Vol. 34, pp. 829 - 841.

Gurevich V.L., Parshin D.A. & Schrober H.R. (2003) Anharmonicity, vibrational instability, and the Boson peak in glasses. *Phys.Rev. B.*, Vol. 67. pp. 094203-01 - 094203-10.

Haydock R., Heine V. & Kelly M.J. (1972) Electronic structure based on the local atomic environment for tight-binding bands. *Journ. of Phys. C.* Vol. 5, pp. 2845-2858.

Hehlen B., Courtens E., Vacher R., Yamanaka A., Kataoka M. & Inoue K. (2000) Hyper-Raman Scattering Observation of the Boson Peak in Vitreous Silica, *Phys. Rev. Lett.*, Vol. 84, pp. 5355 - 5358

Ivanov M.A. (1970) Dynamics of quasi-local vibrations at a high concentration of impurity centers. *Fiz. Tv. Tela,* Vol. 12, pp. 1895–1905 (in Russian).

Ivanov M.A. & Skripnik Yu. V. (1994) On the convergence and pair corrections to the coherent potential. *Fiz. Tv. Tela*, Vol. 36, pp. 94–105 (in Russian).

Kagan Yu. & Iosilevsky Ya. (1962) Mossbauer effect for an impurity nucleus in a crystal. *Zh. Eksp. Teor. Fiz*. Vol. 44, pp. 259-272 (in Russian).

Klinger M.I. & Kosevich A.M. (2001) Soft-mode-dynamics model of acoustic-like high-frequency excitations in boson-peak spectra of glasses. *Phys. Lett. A*, Vol. 280, pp. 365 - 370.

Klinger M.I. & Kosevich A.M. (2002) Soft-mode dynamics model of boson peak and high frequency sound in glasses: "inelastic" Ioffe-Regel crossover and strong hybridization of excitations. *Phys. Lett. A*, Vol. 295, pp. 311 - 317.

Koestler L., Wurdack S., Dietsche W. & Kinder H. (1986). Phonon Spectroscopy of Adsorbed H_2O Molecules, In: *Phonon scattering in condensed matter V*, Meissner M., Pohl R.O. (Ed), 171 -173, ISBN 0-387-17057-X Springer-Verlag, New York, Berlin, Heidelberg.

Kosevich A.M. (1999). *The Crystal Lattice (Phonons, Solitons, Dislocations)*, WILEY-VCH Verlag Berlin GmBH, Berlin

Kosevich A.M., Gospodarev I.A., Grishaev V.I., Kotlyar O.V., Kruglov V.O., Manzheliĭ E.V., Syrkin E.S. & Feodosyev S.B. (2007) Peculiarities of transformations of local levels into impurity zone on example of disordered solid solutions $Ag_{1-p}Al_p$. *Journal of Experimental and Theoretical Physics,* Vol. 132, pp. 11 - 18.

Kosevich Yu.A., Feher A. & Syrkin E.S. (2008) Resonance absorption, reflection, transmission of phonons and heat transfer through interface between two solids. *Low Temp. Phys.* Vol. 34, pp. 725 - 733.

Kovalev O.V. (1961) *Irreducible Representations of the Space Groups*, Gordon and Breach, New York 1965, Ukrainian Academy of Sciences Press, Kiev.

Lang N.D. (1986) Theory of single-atom imaging in the scanning tunneling microscope. *Phys. Rev. Lett.*, Vol. 56, pp. 1164 - 1167.

Leibfried G. (1955) *Gittertheorie der Mechanischen und thermishen Eigenschaften der Kristalle.* Springer-Verlag, Berlin, Göttingen, Heidelberg.

Lifshitz I.M., Gredeskul S.A. & Pastur L.A. (1988) *Introduction to the Theory of Disordered Systems,* Wiley, New York.

Lifshits I.M. (1952a) About one problem of perturbation theory in quantum statistics. *Usp. Mat. Nauk,* Vol. 7, pp. 171 – 180 (in Russian).

Lifshits I.M. (1952b) About heat properties of chain and layered structures. *Journal of Experimental and Theoretical Physics,* Vol. 22, pp.475 - 478.

Liu Z., Zhang X., Mao Y., Zhu Y.Y., Yang Z., Chan C. T. & Sheng P. (2000) Locally resonant sonic materials. *Science,* Vol. 289, pp. 1734 - 1736.

Manzhelii V.G., Chausov V.P. & Kovalenko S.I. (1970) Heat capacity of solid solutions Ar—Xe in low temperature. *Fiz. Tv. Tela.,* Vol. 12, pp. 2764 - 2766.

Maradudin A.A., Montroll E.W., Weiss G.N. & Ipatova I.P. (1982) *Lattice Dynamics and Models of Interatomic Forces,* Springer Series in Solid State Sciences. Springer-Verlag, Berlin, Heidelberg, New York.

Moncton D.E., Axe J.D. & Di Salvo F.J. (1975) Study of Superlattice Formation in 2*H*-$TaSe_2$ by Neutron Scattering. *Phys. Rev. Lett.* , Vol. 34, pp. 734 - 737.

Peresada V.I. (1968) New computational method in theory of crystal lattice. In *Condensed Matter Physics(in Russian),* 172 - 210, FTINT AN UkrSSR, Kharkov, Ukraine.

Peresada V.I. & Tolstoluzhskij V.P. (1970) *On the influence of impurity atoms on the thermodynamic properties of the face centered cubic lattice.* Preprint (in Russian), FTINT AN UkrSSR (Institute for Low Temperature Physics and Engineering, National Academy of Sciences of Ukraine), Kharkov, Ukraine.

Peresada V.I. & Tolstoluzhski V.P. (1977) Low-temperature heat capacity of FCC crystal containing substitution impurities. *Fiz. Nizk. Temp.* (*Sov. J. Low Temp. Phys.*), Vol. 3, pp. 788 - 800.

Peresada V.I., Afanas'ev V.N., & Borovikov V.S. (1975) Calculation of the distribution function for single-magnon excitation in an antiferromagned. *Fiz. Nizk. Temp. (Sov. J. Low Temp. Phys.),* Vol. 1, pp. 461 - 472.

Rufflé B., Guimbretière G., Courtens E., Vacher R. & Monaco G. (2006) Glass-Specific Behavior in the Damping of Acousticlike Vibrations. *Phys. Rev. Lett.,* Vol. 96, pp. 045502-1 – 045502-4.

Schrimacher W., Diezemann G., & Ganter C. (1998) Harmonic Vibrational Excitations in Disordered Solids and the "Boson Peak". *Phys. Rev. Lett.,* Vol. 81, pp. 136-139.

Shklyarevski O.I, Yanson I.K. & Lysykh A.A. (1975). The inelastic tunneling effect in thin films of solidified gases *Fiz. Nizk. Temp.* (*Sov. J. Low Temp. Phys.*), Vol. 1, pp. 502 - 505.

Shkorbatov A.G., Feher A. & Stefanyi P. (1996) Ballistic Landauer-tipe thermal conductivity of a dielectric point contact. *Physica B,* Vol. 218, pp. 242 - 244.

Shkorbatov A.G., Stefanyi P., Bystrenova E. & Feher A. (1998). Phonon transport in pressure-made point contacts.*J. Phys.: Condens. Matter,* Vol. 10, pp. 8313 - 8326.

Syrkin E.S. & Feodosyev S.B. (1982) The contribution of optical vibrations in the low-temperature heat capacity of layered crystals. *Low Temp. Phys.* Vol. 8, pp. 760 - 764.

Tarasov V.V. (1950) Theory of heat capacity of chain and layered structures. *Russian Journal of Physical Chemistry A,* Vol. 24, pp. 111 -128.

Wakabayashi N., Smith H.G. & Shanks R. (1974) Two dimensional Kohn anomaly in $NbSe_2$. *Phys. Lett. A* , Vol. 5, pp. 367 - 368.

5

Multiple Scattering of Elastic Waves in Granular Media: Theory and Experiments

Leonardo Trujillo[1,2], Franklin Peniche[3] and Xiaoping Jia[4]
[1]*Centro de Física, Instituto Venezolano de Investigaciones Científicas (IVIC), Caracas*
[2]*The Abdus Salam International Centre for Theoretical Physics (ICTP), Trieste*
[3]*Departamento de Física y Electrónica, Universidad de Córdoba, Montería*
[4]*Laboratoire de Physique des Matériaux Divisés et des Interfaces (LPMDI), Université Paris–Est*
[1]*Venezuela,*
[2]*Italy,*
[3]*Colombia,*
[4]*France*

1. Introduction

Granular materials consist of a collection of discrete macroscopic solid particles interacting via repulsive contact forces. Classical examples are sand, powders, sugar, salt and gravel, which range from tens of micrometers to the macroscopic scale. Their physical behaviour involves complex nonlinear phenomena, such as non equilibrium configurations, energy dissipation, nonlinear elastic response, and peculiar flow dynamics (Jaeger et al. (1996)).

We know from *classical* continuum dynamics that when a deformable body is under the action of a uniform external load, the force is transmitted to every point inside the body (Landau & Lifshitz (1999)). Conversely, when we deal with a static packing of granular particles, the way by which the forces are transmitted within the packing remains a complex, and still unresolved problem (Luding (2005)). One important aspect lies on the observation that force networks form the skeleton that carries most of the load in a static granular medium (Majmudar & Behringer (2005)). The mechanical response of granular packings to external perturbations plays a major role in numerous scientific endeavours (such as soil mechanics and geophysics), as well as in industry (oil exploration, structural stability, product formulation in pharmacology and domestics, granular composites, heterogeneous materials, etc.)

Most laboratory experiments have been carried out in two-dimensional disc packings using the photoelastic (i.e., birefringent under strain) discs, which have allowed the finest visualization of the generation and dynamical evolution of force chains (Majmudar & Behringer (2005)). However, real granular materials are optically opaque, the photoelastic technique becomes difficult to practice. New tools such as pulsed ultrasound propagation through granular beds under stress have been recently developed to probe the elastic response of three-dimensional granular packings (Brunet et al. (2008a;b); Jia et al. (1999); Jia (2004); Jia et al (2009); Johnson & Jia (2005); Khidas & Jia (2010)). In particular, by studying the

low-amplitude coherent wave propagation and multiple ultrasound scattering, it is possible to infer many fundamental properties of granular materials such as elastic constants and dissipation mechanisms (Brunet et al. (2008a;b); Jia et al. (1999); Jia (2004); Jia et al (2009)).
In soil mechanics and geophysics, the effective medium approach (a continuum mechanics, long wavelength description) has been commonly used to describe sound propagation in granular media (Duffy & Mindlin (1957); Goddard (1990)). However, the experimental investigation performed by Liu & Nagel (1992) on the sound propagation in glass bead packings under gravity revealed strong hysteretic behaviour and high sensitivity to the arrangement of the particles in the container. The authors interpreted this as being due to sound propagating within the granular medium predominantly along strong force chains.
In recent years, our understanding of wave propagation in granular materials has advanced both experimentally and theoretically, covering topics such as surface elastic waves, booming avalanches (Bonneau et al. (2007; 2008)), earthquake triggering (Johnson & Jia (2005)), and coda-like scattered waves (Jia et al. (1999); Jia (2004); Jia et al (2009)).
The understanding of wave motion in granular media took a major step forward with the experimental observation of the coexistence of a coherent ballistic pulse travelling through an "effective contact medium" and a multiply scattered signal (Jia et al. (1999)). Such a picture is confirmed by the experiments of sound propagation in two dimensional regular lattice of spheres under isotropic compression (Gilles & Coste (2003)), where the transmitted coherent signal remains almost unchanged for different packing realizations followed by an incoherent tail which depends on the specific packing configuration, in agreement with the completely disordered three–dimensional case (Jia et al. (1999); Jia (2004)). Similar results were obtained with extensive numerical simulations for two and three dimensional confined granular systems by Somfai et al. (2005). This reconciles both points of view in terms of classical wave propagation in a random medium: (i) At low frequencies such that the wavelengths are very long compared with the correlation length of force chains or the spacing between them, the granular medium is effectively homogeneous continuum to the propagating wave. In this case, nonlinear effective medium theories based upon the Hertz-Mindlin theory of grain-grain contacts describe correctly the pressure dependence of sound velocity observed if one includes the increasing number of contacts with the external load (Goddard (1990); Makse et al. (1999)). Most standard measurements of acoustic velocities and attenuation focus on the coherent propagation of effective waves, which provide a means for characterizing the large-scale properties of the granular medium, though micro structural features are not readily resolved; (ii) At high frequencies when the wavelength decreases down to the order of the grain size, scattering effects caused by the spatial fluctuations of force chains become very significant and the effective contact medium is no longer a valid description. One can observe that the continuous wave trains in the tail portion of the transmitted temporal signal through the granular packing have a broadband strongly irregular high frequency spectrum ("coda waves") (Jia et al. (1999)). The energy of a propagating wave spreads in many directions, and strong interference effects occur between scattered wave that have travelled different path through the medium, resulting in a complicated pattern of nodes and antinodes (i.e. *acoustic speckles*) (Jia et al. (1999)).
The speckles are highly sensitive to changes in the granular medium, and configuration specific, i.e. fingerprints of the structure of the force chains. If the multiply scattered waves are excited between or during a temporal change in the granular packing, then one can exploit the sensitivity of the waves to quantify structural variations of the contact network. This opens the possibility to engineer a novel and useful method for investigating the complex response

of granular packings under mechanical perturbations. This new technique could be extended to other amorphous materials such as glasses where heterogeneous force chains have been observed (Tanguy et al. (2002)). Before scattered waves are made more quantitative probe, it is essential first to understand the nature of the wave transport in the granular medium.

Quite recently, experimental evidence (Jia (2004); Jia et al (2009)) indicates a well-established diffusive behaviour of the elastic waves transport over long distance scales. These experiments were carried out in confined granular packings under uniaxial loading. The system was excited using high-frequency ultrasonic pulses. A qualitative investigation of the statistical characteristics of scattered waves was performed using the intensity evolution in space and time of a wave train injected into the granular medium. A key observation reported in (Jia (2004)) is that the intensity of scattered waves is very similar to the transmitted pulses of classical waves across strongly scattering random media, in accordance with the diffusive field approximation. A further important observation is that under strong static loading, the normal and shear loads of individual grain contacts exhibit a random distribution. Therefore, the topological disorder of the granular medium induces space fluctuations on both density and elastic stiffness. This opens the possibility to interpret the propagation of ultrasonic waves within granular media in terms of random fluctuations of density and elastic stiffness by employing the same framework used to describe the vibrational properties of heterogeneous materials (Frisch (1968); Sheng (2006); Vitelli et al. (2010)).

The modelling of multiple scattering and the diffusive wave motion in granular media is by no means a simple task. However, the energy envelope of the waveform can be interpreted within the framework of the Radiative Transport Equation (RTE) developed for modelling wave propagation in random media (Ryzhik et al., (1996); Weaver (1990)) The energy density calculated using RTE for multiple isotropic scattering processes converges to the diffusion solution over long distance scales, and describes adequately the transport of elastic waves dominated by shear waves in granular media (Jia (2004); Jia et al (2009)). This development is rather heuristic and it lacks a rigorous basis on the wave equation.

The potential importance of the original calculation (Jia (2004)) stimulated further investigations towards the construction of a self-consistent theory of transport equations for elastic waves in granular media (Trujillo et al. (2010)). The theory is based upon a nonlinear elasticity of granular media developed by Jiang & Liu (2007), which emphasizes the role of intrinsic features of granular dynamics such as volume dilatancy, mechanical yield and anisotropies in the stress distribution. The formalism developed in (Trujillo et al. (2010)) introduces an extension of the Jiang-Liu granular elasticity that includes spatial fluctuations for the elastic moduli and density, providing a characterization at the grain-scale.

In this chapter we give an introduction to elastic wave propagation in confined granular systems under external load. Our analysis of elastic wave scattering is developed from both experimental and theoretical viewpoints. The present systematic description and interpretation of multiple scattering of elastic waves in granular media is based on a synthesis between the experiments carried out by Jia and co-workers from 1999, and the theory constructed by Trujillo, Peniche and Sigalotti in 2010. This chapter is structured in four main parts as follows: In section 2 we elaborate a presentation of the principal experimental outcomes of ultrasound propagation through a granular packing. We start with a brief schema of the laboratory setup and experimental protocol. Then, we present the characteristics of the transmitted signals, observing a *coexistence* of a coherent ballistic and a speckle-like multiply scattered signal. For over long distances scales, the diffusion approximation is shown to describe adequately the transport of elastic waves dominated by shear waves; In section 3 we elaborate a theory for elastic wave propagation in granular media. As granular materials are disordered systems, we have to introduce several concepts that are unknown in the *classical* theory of linear elastic waves in homogeneous solids. After a presentation of the Jiang–Liu model for nonlinear granular elasticity, we provide a rational path for the choice of the spatial variations of the elastic constants. Furthermore we derive the equation of motion for elastic waves in granular media and present the vector-field mathematical formalism; In section 4 the mathematical formulation of the problem leads to a theoretic formalism analogous to the analytical structure of a quantum field theory. Then, introducing the disorder perturbation as a small fluctuation of the time-evolution operator associated to a Schrödinger-like equation, the RTE and the related diffusion equation are constructed. This result provides the theoretical interpretation, derived from first principles, of the intensity of scattered waves propagating through granular packing; In section 5 we summarize the relevant conclusions.

2. Experiments of ultrasound propagation in externally stressed granular packings

This section deals with different experimental aspects of the propagation of acoustic waves in granular media. After a short presentation of the experimental protocol, we describe two observations that demonstrate the presence of multiple scattering in granular packings under stress: (i) the coexistence of a coherent ballistic pulse and a multiply scattered signal; (ii) the intensity of scattered waves is described by the diffuse field approximation.

2.1 Experimental setup and procedure

A schematic diagram of the experimental setup is shown in Fig. 1. The samples consist of random packings of polydisperse glass beads of diameter $d = 0.6$–0.8 mm, confined in a duralumin cylinder of inner diameter 30 mm. The container is closed with two pistons and a normal load P is applied to the granular sample across the top and bottom pistons. To minimize the hysteretic behavior and improve the reproducibility of experiments, one cycle of loading and unloading is performed in the granular packs before the ultrasonic measurements. Statistically independent ensembles of the packing configuration are realized by stirring vigorously glass beads after each measurement and repeating carefully the same loading protocol. The volume fraction of our glass beads packs is 0.63 ± 0.01. The height of the granular sample ranges from $L = 5$ mm to 20 mm. A generating transducer of 30 mm and a small detecting transducer of 2 mm are placed on the axis at the top and the bottom of the cell. Both the source transducer and detector are in direct contact with glass beads. The excitation is realized by using ten-cycle tone burst excitation of 20 μs duration centered at a frequency $f(= \omega/2\pi) = 500$ kHz is applied to the calibrated longitudinal source transducer. This narrow band excitation corresponds to the product of granular skeleton acoustic wave number and bead diameter, $kd = \omega d/v \approx 2.9$, with $v \approx 750$ m/s being a typical sound speed in the solid frame. At such a high frequency, one expects to deal with a strongly scattering medium. The transmitted ultrasonic signals are digitized and signal averaged to improve the signal–to–noise ratio and to permit subsequent analysis of data.

2.2 Transmitted signals: Coherent propagation and codalike multiple scattering

In Fig. 2 we show the transmitted ultrasonic field through a granular packing of thickness $L = 11.4$ mm under axial stress of $P = 0.75$ MPa, a typical pressure at depths of tens of meters in soils due to the weight of the overburden. To ascertain that the ultrasound propagates from one grain to its neighbors only through their mutual contacts and not via air, we have

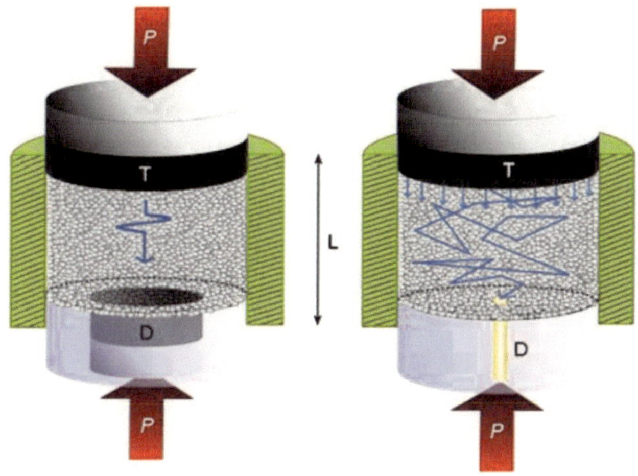

Fig. 1. Multiple scattering of elastic waves in a confined granular packing under stress P. "T" and "D" correspond to a large source transducer and a small detector, respectively. (Reprinted from Jia et al., Chinese Sci. Bull. 54, 4327 (2009))

checked that no ultrasonic signal is detected at vanishing external load. A typical record is presented in Fig. 2 (a), where the transmitted pulse exhibits a primary *low frequency* (LF) coherent component E_P. This coherent pulse E_P well defined at the leading edge of the transmitted signal corresponds to a self–averaging effective wave propagating ballistically at compressional wave velocity $c_P \approx 1000$ m/s, and frequency $f_P \sim 70$ kHz. After the arrival of the coherent pulse, one observes a continuous wave trains in the tail portion of the signal. This wave trains, which are named "coda" (Fehler & Sato (2003)), looks like a random interference pattern having an envelope whose amplitude gradually decreases with increasing time. This coda type *high frequency* (HF) incoherent signal S, with frequency $f_S \sim 500$ kHz, is associated with speckle–like scattered waves by the inhomogeneous distribution of force chains (Jia et al. (1999)). Indeed, at such a high frequency, the acoustic wavelength is comparable to the bead diameter, $\lambda/d \sim 1.5$; thus one expects to encounter strong sound scattering in a granular medium.

The sensitivity of the coherent and incoherent waves to changes in packing configurations is shown in Fig. 2 (b) over 15 independent granular samples. In contrast to the coherent pulse, which is self–averaged and configuration insensitive, the acoustic speckles are configuration specific, and exhibits a fluctuating behavior due to the random phases of the scattered waves through a given contact force network. Hence, an *ensemble average* of configurations can cancel scattered wave signals and leave only the coherent wave E_P. Moreover, another coherent signal noted as E_S survives from this averaging procedure, which propagates ballistically at a shear velocity about $c_S \approx 450$ m/s. The inset of Fig. 2 (b) shows that the use of a transverse transducer as source can lead to a considerable enhancement of this shear wave excitation without ensemble averaging thanks to the temporal and spatial coherence.

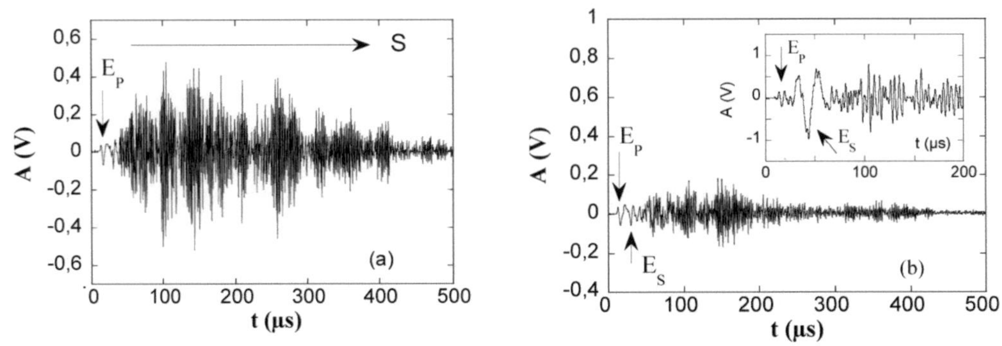

Fig. 2. Transmitted ultrasonic signal through a dry glass beads packing (a) at a given configuration excited by a compressional transducer, and (b) after ensemble averaging over 15 independent configurations. The inset illustrates the transmitted signal at a given configuration using a shear transducer. (Reprinted from Jia, Phys. Rev. Lett. 93, 154303 (2004))

2.3 Transmitted intensity and diffusive wave propagation

Now we present one the most relevant result of this experiment which builds a bridge between experiments and the quest towards a theory of elastic waves propagation in granular media. To investigate quantitatively the statistical characteristics of scattered waves we measure the time–resolved transmitted intensity $I(t)$ through the granular sample. For each configuration we subtract the LF coherent pulses E_P and E_S from the transmitted ultrasonic field by means of a high–pass (HP) filter ($f \geq 300$ kHz) and determine the intensity of the scattered wave by squaring the envelope of the filtered waveform. In the inset of Fig. 3, we present the corresponding average amplitude profile of scattered wave transmission, which rise gradually from an early time value below the noise level to a maximum and decays exponentially at late times. The *ensemble–average* is performed over fifty independent configurations realized according to the same protocol of sample preparation. In Fig. 3, we show that the average transmitted intensity $I(t)$, at $L = 11$ mm, was found to decay exponentially at long times, with the entire time dependence of $I(t)$ being well described by the diffusion model (Jia (2004)). We conclude that multiply scattered ultrasound propagates, for this experiment, in a normal, diffusive way. In what follows, we will provide theoretical foundations to model the propagation of elastic waves in granular packings. In subsection 4.2.1 we derive the analytical expression for the transmitted intensity, Eq. (58). This expression fits the time profile of the average transmitted intensity (solid line in Fig. 3).

3. Theory of elastic wave propagation in compressed granular media

We now proceed to show that a theory for elastic wave propagation in granular media can be constructed based upon a nonlinear elasticity developed by Jiang & Liu (2007), which emphasizes the role of intrinsic features of granular materials such as volume dilatancy, mechanical yield and anisotropies in the stress distribution.

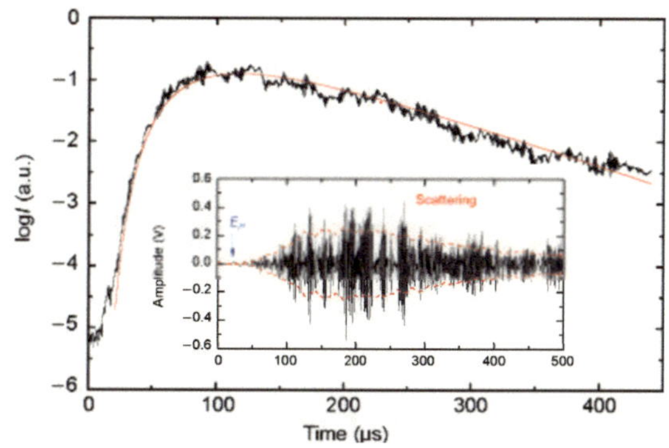

Fig. 3. Averaged time–dependent transmitted intensity $I(t)$ of the scattered waves traveling across a granular sample of height $L = 11$ mm. Solid line corresponds to the theoretical solution (58) with the fitting parameters $D = 0.13\ \mathrm{m}^2/\mathrm{s}$ and $Q^{-1} = 0.005$. Inset: Transmitted ultrasonic signal at a given configuration. Dashed lines correspond to the average amplitude profile. (Reprinted from Jia et al., Chinese Sci. Bull. 54, 4327 (2009))

3.1 Granular elasticity

The rigorous passage from a microscopic to a macroscopic (continuum) mechanical description of granular materials is a challenging task due to the intrinsic disorder of these materials (shape, size, density, contact forces and friction) and the apparent lack of well separated scales between the grain–level dynamics and the entire bulk. An important aspect concerns the anisotropic environment at the particle scale where force chains are clearly evidenced, i.e., chains of contact along which the forces are stronger than the mean interparticle force. The presence of these force chains, implying preferred force paths, has served as empirical argument against an isotropic continuum description of granular matter. However, recent experimental findings on the stress distribution response to localized perturbations have shed some light on the validity of using a continuum elastic theory (e.g., Serero et al. (2001)). On the other hand, Goldhirsch & Goldenberg (2002) showed that exact continuum forms of the balance equations (for mass, momentum and energy) can be established as relations between weighted space (and time) averages. In this framework it is possible to derive exact expressions for the elasticity of disordered solids, such as a granular packing. For granular systems under strong static compression, as is the case for experiments presented in section 2, the theory starts with the assumption of small deformations, i.e., for an infinitesimal deformation the displacements field $\mathbf{u} = \mathbf{r} - \mathbf{r}'$ and their gradients are small compared to unity ($|\mathbf{u}|$ and $|\nabla\mathbf{u}| \ll 1$). This assumption is physically reasonable for small amplitude of wave motion. The displacement associated with deforming the grains, stores energy reversible and maintains a static strain. Sliding and rolling lead to irreversible, plastic process that only heat up the system. So, the total strain tensor ϵ_{ij} may be decomposed into elastic and plastic parts $\epsilon_{ij} = u_{ij} + u^p_{ij}$. In this work we limit our analysis to granular packings under strong static compression. Therefore, the granular energy is a function of the elastic energy alone and we can neglect the plastic strain contribution. Up to linear order, the strain tensor is

$$\epsilon_{ij} \approx u_{ij} = \frac{1}{2}\left(\frac{\partial u_i}{\partial x_j} + \frac{\partial u_j}{\partial x_i}\right).$$

The only way to ensure that the elastic strains are indeed reversible is to specify a strain energy potential $\mathcal{F}(\epsilon_{ij})$ (Helmholtz's free energy), from which the stresses are given as functions of the strains: $\sigma_{ij} = \partial\mathcal{F}(\epsilon_{ij})/\partial\epsilon_{ij}$. A material which possesses such a strain energy function is said to be *hyperelastic*. Jiang and Liu have recently proposed the following free energy strain potential for granular materials

$$\mathcal{F}(\epsilon) = \frac{1}{2}\mathrm{K}\epsilon_{ll}^2 + \mathrm{G}\epsilon_s^2, \tag{1}$$

where K and G are the compressional and shear moduli, respectively, ϵ_{ll} is the trace of the strain tensor ϵ_{ij}, and $\epsilon_s^2 \equiv \epsilon_{ij}^0\epsilon_{ij}^0$, with ϵ_{ij}^0 the traceless part, i.e., $\epsilon_{ij}^0 = \epsilon_{ij} - \epsilon_{ll}\delta_{ij}/3$, and δ_{ij} is the Dirac delta function. As always, summation over repeated indices is assumed implicitly. In the presence of external forces the granular packing deforms leading to a change in the density given by $\delta = 1 - \rho_0/\rho = -\epsilon_{ll}$, where ρ_0 is the density in the absence of external forces. In granular elasticity the elastic moduli are modified to take into account the interaction between grains, which may deform as a result of contact with one another. The elastic moduli K and G are assumed to be proportional to the volume compression ϵ_{ll}, i.e.,

$$\mathrm{K} = \widetilde{\mathrm{K}}\delta^b, \ \mathrm{G} = \widetilde{\mathrm{G}}\delta^a, \tag{2}$$

with $\widetilde{\mathrm{K}}, \widetilde{\mathrm{G}} > 0$ for $\delta \geq 0$ and $\widetilde{\mathrm{K}}, \widetilde{\mathrm{G}} = 0$ for $\delta < 0$ so that the elastic moduli remain finite. The exponents a and b are related to the type of contact between the grains. That is, when $a = b = 0$ linear elasticity is recovered, whereas $a = b = 1/2$ implies Hertz contacts (Landau & Lifshitz (1999)). This formulation provides a much better approximation to granular elastic behavior in which we can specify any type of contact by suitably choosing the exponents a and b.

The free energy strain potential (1) is stable only in the range of strains values that keeps it convex. Therefore, Eq.(1) naturally accounts for unstable configurations of the system, as the yield, which appears as a phase transition on a potential–strain diagram.

If there are no changes in temperature (or an analogous *granular temperature*), the stresses are given as derivatives of the Helmholtz free energy with respect to the strains. The constitutive behavior of a granular packing is completely specified by (1), then we get the following stress-strain relation

$$\sigma_{ij} = -\mathrm{K}\epsilon_{ll}\delta_{ij} + 2\mathrm{G}\epsilon_{ij}^0 - \frac{1}{\delta}\left(\frac{1}{2}b\mathrm{K}\epsilon_{ll}^2 + a\mathrm{G}\epsilon_s^2\right). \tag{3}$$

The above equation contains the stress elements of both the linear and Boussinesq elasticity models as we may see from inspection of the first two terms of the right-hand side of Eq.(3). In this way, the stress is completely defined once we specify a and b together with the stress-dependent elastic moduli, which account for the desired granular behavior. The strain-stress relation (3) also includes the effects of volume dilatancy. These effects are represented by the second term between parentheses on the right–hand side of Eq.(3), where the pure shear stress is proportional to shear strain and the volumetric deformation δ.

3.1.1 Local disorder and randomness

Apart from the validity of continuum elasticity description, granular packings are heterogenous materials due to the intrinsic disorder. In order to capture these heterogeneities we introduce spatially–varying constitutive relations. To do so we need to know how the particle displacements are related to the local disorder in the deformation of the bulk. This can be elucidated with the aid of the stress–strain relation. In particular, we introduce spatial fluctuations for the elastic moduli, i.e., $K(\mathbf{r}) = \lambda(\mathbf{r}) + \frac{2}{3}\mu(\mathbf{r})$ and $G(\mathbf{r}) = \mu(\mathbf{r})$, where $\lambda(\mathbf{r})$ and $\mu(\mathbf{r})$ are the Lamé coefficients and assume that these fluctuations can be described by a random process. Moreover, the stress tensor becomes spatial dependent $\hat{\sigma} \to \hat{\sigma}(\mathbf{r})$.

For the range of frequency used in the experiments, which is below the acoustic resonances of individual glass beads, the granular network can be modeled as an effective random network (Jia et al. (1999); Jia (2004)). Due to the inhomogeneous distribution of individual bead contact forces, this network exhibits spatial fluctuations of both density and elastic modulus, closely analogous to an amorphous solid. However, it is worth emphasizing the peculiar position of the granular medium amongst randomly scattering media in general. In fact, the topologically disordered granular packing plays a twofold role. It builds the medium in which elastic wave propagates, but being random it is also responsible for the disorder effects. There is no separation of the system into a reference medium and scatterers. Here we formulate a heuristic approach to calculate the local spatial variations. Let us recall that in the Jiang-Liu elastic theory, the elastic moduli K and G are strain-dependent functions given by Eq.(2), with $\delta = \delta(\mathbf{r}) = -\mathrm{Tr}[\epsilon_{ij}(\mathbf{r})]$ being the volume compression. Here Tr() denotes the trace. We note that in absence of compression and shear $\delta = 0$ when the grains are in contact. Therefore, we can define the localized compressional fluctuations at position $\mathbf{r}$ as

$$\delta(\mathbf{r}) = \delta_0 + \Delta(\mathbf{r}), \tag{4}$$

where $\delta_0 = \langle\delta(\mathbf{r})\rangle$ is the imposed bulk compression. Angular brackets designate average expectation values with respect to the corresponding probability distribution. Here $\Delta(\mathbf{r})$ is assumed to be a delta correlated Gaussian random process with a zero mean and covariance given by

$$\langle\Delta(\mathbf{r})\Delta(\mathbf{r}')\rangle = \sigma^2\delta(\mathbf{r} - \mathbf{r}'), \tag{5}$$

where σ is the strength of the delta correlated disorder. We assume that $\sigma \sim d$ (Goldenberg et al. (2007)). Our choice of $\delta(\mathbf{r})$ is based on the empirical observation that for isotropically compressed systems the mean normal force (at the grain level) is distributed randomly around an average value with short-range correlation (Majmudar & Behringer (2005)). However, when the system is subjected to an external shear, the force correlations are of much larger and longer range and may be characterized by a power law (Majmudar & Behringer (2005)). Here we restrict only to the case when the system is compressed isotropically and ignore for simplicity the action of external shear. The spatial variations of the local compression can be estimated from the strain field which, up to linear order, can be calculated using the coarse-graining procedure introduced by Goldhirsch & Goldenberg (2002).

The fluctuations in the Lamé coefficients can be expressed in terms of the fluctuating local compression by means of the compressional and shear elastic moduli, i.e.,

$$\lambda(\mathbf{r}) = \widetilde{K}\delta^b(\mathbf{r}) - \frac{2}{3}\widetilde{G}\delta^a(\mathbf{r}),$$
$$\mu(\mathbf{r}) = \widetilde{G}\delta^a(\mathbf{r}).$$

For Hertzian interactions $a = b = 1/2$ and therefore

$$\lambda(\mathbf{r}) = \lambda_0\sqrt{\delta(\mathbf{r})}, \quad \mu(\mathbf{r}) = \mu_0\sqrt{\delta(\mathbf{r})}, \tag{6}$$

where $\lambda_0 = \widetilde{\mathrm{K}} - \frac{2}{3}\widetilde{\mathrm{G}}$ and $\mu_0 = \widetilde{\mathrm{G}}$. The coupling between the elastic moduli through the compression $\delta(\mathbf{r}) = -\mathrm{Tr}[\epsilon_{ij}(\mathbf{r})]$ reduces the number of free parameters needed to characterize the spatial perturbation to three, namely σ, λ_0, and μ_0. This avoids other cross couplings between the elastic moduli (Trégourès & van Tiggelen (2002)) and simplifies the analytical calculations. Finally, let us remark that in the present work the word *random* implies some kind of statistical or ensemble averaging in the theory. Since a granular packing is a nonequilibrium, quenched disordered medium, we do not have access to a true statistical ensemble. Thus, theory and experiment can only be connected through some kind of ergodicity. The equivalence between theoretical and observational averaging is a very difficult task and will not be addressed in this Chapter.

3.2 Equation of motion

Now we are ready to formulate the mathematics of elastic wave propagation in granular media, in a way that is suited to apply the methods of multiple scattering of waves. We start with the equation of motion for the elastic displacement field $\mathbf{u}$ at time t and position $\mathbf{r}$,

$$\rho(\mathbf{r})\frac{\partial^2}{\partial t^2}\mathbf{u}(\mathbf{r},t) = \nabla\cdot\hat{\sigma}(\mathbf{r}) + \mathbf{f}(\mathbf{r},t), \tag{7}$$

where $\rho(\mathbf{r})$ is the local density and $\mathbf{f}(\mathbf{r},t)$ is an external force per unit volume. Mathematically, if $\partial\mathcal{R}$ is the boundary of a region $\mathcal{R}$ occupied by the granular packing, then $\mathbf{u}$ is prescribed on $\partial\mathcal{R}$. This assumption is physically reasonable for the experiments reported in this work where the source and detecting transducers are placed at the surface of the granular medium. From the Jiang-Liu elastic model the spatial dependent stress tensor (3) which, by the Hooke's law, is given in terms of the Lamé coefficients by

$$\begin{aligned}\sigma_{ij}(\mathbf{r}) &= C_{ijkl}(\mathbf{r})\epsilon_{kl}(\mathbf{r}),\\ &= \left[\lambda(\mathbf{r})\delta_{ij}\delta_{kl} + \mu(\mathbf{r})\left(\delta_{ik}\delta_{jl} + \delta_{il}\delta_{jk}\right)\right]\epsilon_{kl}(\mathbf{r}),\\ &= -\left\{\left[1+\frac{b}{2}\right]\lambda(\mathbf{r}) + \frac{1}{3}\left[b-a\right]\mu(\mathbf{r})\right\}\epsilon_{nn}(\mathbf{r})\delta_{ij} - 2\mu(\mathbf{r})\epsilon_{ij}(\mathbf{r}) + \frac{a}{\delta}\epsilon_{lk}(\mathbf{r})\epsilon_{lk}(\mathbf{r})\delta_{ij},\end{aligned} \tag{8}$$

where the second equality applies to an isotropic medium, in which case the fourth-rank stiffness tensor $C_{ijkl}(\mathbf{r})$ can only have two independent contributions, proportional to the Lamé moduli $\lambda(\mathbf{r})$ and $\mu(\mathbf{r})$. Inserting the third equality of Eq.(8) into Eq.(7) and rearranging terms, the equation of motion in index notation is given by

$$\begin{aligned}\rho(\mathbf{r})\partial_t\partial_t u_i(\mathbf{r},t) &= \left\{\left[1+\frac{b}{2}\right]\lambda(\mathbf{r}) + \left[\frac{1}{3}(b-a)+1\right]\mu(\mathbf{r})\right\}\partial_i\partial_k u_k(\mathbf{r},t)\\ &+\mu(\mathbf{r})\partial_j\partial_j u_i(\mathbf{r},t) + \left\{\left[1+\frac{b}{2}\right]\partial_i\lambda(\mathbf{r}) + \frac{1}{3}\left[b-a\right]\partial_i\mu(\mathbf{r})\right\}\partial_k u_k(\mathbf{r},t)\\ &+2\left[\partial_j\mu(\mathbf{r})\right]\epsilon_{ij} - \partial_i\left[\frac{a\mu(\mathbf{r})}{\delta}\epsilon_{lk}\epsilon_{lk}\right] + f_i(\mathbf{r},t),\end{aligned} \tag{9}$$

where the symbols ∂_t and ∂_i are used to denote the partial derivatives with respect to time and space. Most of the existing theoretical and numerical models on granular materials currently use the Hertzian force law because it simulates the nonlinear elastic contacts between grains with fairly good approximation. In what follows, we will restrict ourselves only to the case of Hertz contacts (i.e., $a = b = 1/2$). Then, the dynamics of the displacement fields becomes

$$\rho(\mathbf{r})\partial_t\partial_t u_i(\mathbf{r},t) = \left[\frac{5}{4}\lambda(\mathbf{r}) + \mu(\mathbf{r})\right]\partial_i\partial_k u_k(\mathbf{r},t) + \mu(\mathbf{r})\partial_j\partial_j u_i(\mathbf{r},t) \qquad (10)$$
$$+\frac{5}{4}\partial_i\lambda(\mathbf{r})\partial_k u_k(\mathbf{r},t) + 2\left[\partial_j\mu(\mathbf{r})\right]\epsilon_{ij} - \partial_i\left[\frac{\mu(\mathbf{r})}{2\delta}\epsilon_{lk}\epsilon_{lk}\right] + f_i(\mathbf{r},t),$$

For pedagogical completeness let us remark that setting $a = b = 0$, as it would be appropriate for linear elasticity, in vector notation, Eq. (9) reduces to

$$\rho(\mathbf{r})\frac{\partial^2}{\partial t^2}\mathbf{u}(\mathbf{r},t) = [\lambda(\mathbf{r}) + 2\mu(\mathbf{r})]\nabla[\nabla\cdot\mathbf{u}(\mathbf{r},t)] - \mu(\mathbf{r})\nabla\times\nabla\times\mathbf{u}(\mathbf{r},t) + \nabla\lambda(\mathbf{r})\nabla\cdot\mathbf{u}(\mathbf{r},t)$$
$$+[\nabla\mu(\mathbf{r})]\times[\nabla\times\mathbf{u}(\mathbf{r},t)] + 2[\nabla\mu(\mathbf{r})\cdot\nabla]\mathbf{u}(\mathbf{r},t) + \mathbf{f}(\mathbf{r},t), \qquad (11)$$

this last equality indicates that elastic waves have both dilatational $\nabla\cdot\mathbf{u}(\mathbf{r},t)$ and rotational deformations $\nabla\times\mathbf{u}(\mathbf{r},t)$. For isotropic homogeneous media, the Lamé coefficients $\lambda(\mathbf{r})$ and $\mu(\mathbf{r})$ are independent of $\mathbf{r}$, and the above equation further simplifies to the well–known wave equation

$$\frac{\partial^2}{\partial t^2}\mathbf{u}(\mathbf{r},t) - c_P^2\nabla[\nabla\cdot\mathbf{u}(\mathbf{r},t)] - c_S^2\nabla\times\nabla\times\mathbf{u}(\mathbf{r},t) = \frac{\mathbf{f}(\mathbf{r},t)}{\rho(\mathbf{r})}. \qquad (12)$$

The terms $c_P = \sqrt{(\lambda + 2\mu)/\rho(\mathbf{r})}$ and $c_S = \sqrt{\mu/\rho(\mathbf{r})}$ are the compressional wavespeed, and shear or transverse wavespeed, respectively. This proves that the Jiang-Liu granular elasticity includes the well-known linear elasticity of isotropic and homogeneous elastic solids.

3.3 Total elastic energy

Now we proceed to calculate the total elastic energy for a compressed granular packing. It is well–known that for a deformable granular packing the deformation represented by the strain tensor $\hat{\epsilon}$ is caused by the external forces applied to the packing itself. For simplicity, we shall assume that no appreciable changes on temperature occur within the packing owing to its deformation so that the flux of heat across its boundary $\partial\mathcal{R}$ can be neglected. Moreover, since we assume that the contact between the grains are governed by a Hertzian law, the assumption is also made that they deform the packing at a sufficiently slow rate. The potential energy $\mathcal{U}_P$ due to the displacement field $\mathbf{u}$ is given by the strain energy (Chou & Pagano (1992))

$$\mathcal{U}_P = \frac{1}{2}\int_{\mathcal{R}} d^3\mathbf{r}\left\{\lambda(\mathbf{r})[\nabla\cdot\mathbf{u}]^2 + \mu(\mathbf{r})\mathrm{Tr}\left[\nabla\mathbf{u} + (\nabla\mathbf{u})^T\right]^2\right\},$$
$$= \frac{1}{2}\int_{\mathcal{R}} d^3\mathbf{r}\left\{[\lambda(\mathbf{r}) + 2\mu(\mathbf{r})][\nabla\mathbf{u}]^2 + \mu(\mathbf{r})[\nabla\times\mathbf{u}]^2\right\}, \qquad (13)$$

where in the first equality the superindex ^{T} means transposition. In the second equality we have shown the potential energy in terms of the dilatational and rotational deformations. The different terms in the second equality of (13) represent the compressional $\mathcal{E}_P$ and shear $\mathcal{E}_S$ energy. Note that this relation (13) is strictly valid only when the integral is independent of the path of deformation (Norris & Johnson (1997)). This provides a fairly good approximation provided that the displacements of the grains in the packing are assumed to be small enough. Keeping in mind that the kinetic energy is just the volume integral of the quantity $\frac{1}{2}\rho(\partial \mathbf{u}/\partial t)^2$, we may then evaluate the total energy $\mathcal{E}_T$ of the elastic displacement $\mathbf{u}$ as

$$\mathcal{E}_T = \frac{1}{2}\int_{\mathcal{R}} d^3\mathbf{r}\left\{\rho\left(\frac{\partial \mathbf{u}}{\partial t}\right)^2 + \lambda(\mathbf{r})\left[\nabla\cdot\mathbf{u}\right]^2 + \mu(\mathbf{r})\mathrm{Tr}\left[\nabla\mathbf{u} + (\nabla\mathbf{u})^T\right]^2\right\}, \tag{14}$$

which works for the Jiang–Liu model specialized to Hertzian contact.

In subsection 4.2.2, we will discuss that after several scattering, a stationary regime is set and mode conversion equilibrates: the initial energy spreads with equal probability over all modes. The stabilization of the energy partition ratio $\mathcal{E}_S/\mathcal{E}_P$ is a strong indication that coda waves are made of multiple scattered waves (Hennino et al. (2001); Jia (2004); Jia et al (2009); Papanicolaou et el. (1996); Weaver (1990)).

3.4 Vector–field mathematical formalism

Working with the wave equation (10) for the analysis multiple scattering of elastic waves in granular media is greatly facilitated by introducing an abstract vector space formed by the collection of vector fields

$$\mathbf{\Psi}(\mathbf{r},t) \equiv \begin{pmatrix} -i\sqrt{\frac{\lambda(\mathbf{r})}{2}}\partial_j u_j(\mathbf{r},t) \\ i\sqrt{\frac{\rho(\mathbf{r})}{2}}\partial_t u_j(\mathbf{r},t) \\ -i\sqrt{\mu(\mathbf{r})}\epsilon_{jk}(\mathbf{r},t) \end{pmatrix}, \tag{15}$$

where $i = \sqrt{-1}$ is the imaginary number. This vector has 13 components for the running indexes j and k. The most convenient way to perform the algebra of this vector field is by using the Dirac bra–ket notation of quantum mechanics, i.e., $|\mathbf{\Psi}\rangle \equiv \mathbf{\Psi}$ and, for the complex conjugate field, $\langle\mathbf{\Psi}| \equiv \mathbf{\Psi}^*$. A physical interpretation of vector $\mathbf{\Psi}$ follows by realizing that the Cartesian scalar inner product

$$\langle\mathbf{\Psi}(\mathbf{r},t)|\mathbf{\Psi}(\mathbf{r},t)\rangle \equiv \int d^3\mathbf{r}\mathbf{\Psi}^*(\mathbf{r},t)\cdot\mathbf{\Psi}(\mathbf{r},t), \tag{16}$$

is exactly to the total energy $\mathcal{E}_T$ of the elastic displacement $\mathbf{u}$, defined by Eq.(14). This result suggests that $\mathbf{\Psi}$ can be viewed as a complex amplitude for the elastic energy of the granular media. After some algebra, we show that Eq.(10) for the elastic displacement field $\mathbf{u}$ is indeed equivalent to a Schrödinger-like equation for $\mathbf{\Psi}$, i.e.,

$$i\partial_t|\mathbf{\Psi}(\mathbf{r},t)\rangle = \mathbf{K}\cdot|\mathbf{\Psi}(t)\rangle + |\mathbf{\Psi}_f(t)\rangle, \tag{17}$$

where $|\mathbf{\Psi}_f(t)\rangle$ is the external force field defined by the 13-component vector

$$\mathbf{\Psi}_f(\mathbf{r},t) \equiv \begin{pmatrix} 0 \\ -\frac{1}{\sqrt{\rho(\mathbf{r})}}\mathbf{f}(\mathbf{r},t) \\ (\mathbf{0})_9^T \end{pmatrix}, \tag{18}$$

and $\mathbf{K}$ is the time-evolution operator given by the 13×13 matrix

$$\mathbf{K} \equiv \begin{pmatrix} 0 & \sqrt{\lambda(\mathbf{r})}\mathbf{P}\frac{1}{\sqrt{\rho(\mathbf{r})}} & (\mathbf{0})_9 \\ \frac{5}{4\sqrt{\rho(\mathbf{r})}}\mathbf{P}^T\sqrt{\lambda(\mathbf{r})} & (\mathbf{0})_{3\times 3} & \frac{1}{\sqrt{\rho(\mathbf{r})}}\left[L^l_{jk}(\mathbf{P}^T)\sqrt{2\mu(\mathbf{r})} - \mathbf{P}^T\frac{1}{2\delta}\sqrt{\frac{\mu(\mathbf{r})}{2}}\epsilon_{jk}\right] \\ (\mathbf{0})^T_9 & \sqrt{2\mu(\mathbf{r})}L^{jk}_l(\mathbf{P})\frac{1}{\sqrt{\rho(\mathbf{r})}} & (\mathbf{0})_{9\times 9} \end{pmatrix}, \tag{19}$$

where we have defined the operator $\mathbf{P} = -i\nabla$ and introduced the third-rank tensor $L_{jkl} \equiv 1/2\left(\mathbf{P}_j\delta_{kl} + \mathbf{P}_k\delta_{jl}\right)$. In Eqs. (18) and (19), in order to gain compactness in writing the vector and matrix representations, the notation $(\mathbf{0})^T_9$ is used to signify a column array consisting of nine zeros. Conversely, $(\mathbf{0})_9$ is employed to denote a row array of nine zeros filling the right top part of the matrix. In addition, $(\mathbf{0})_{3\times 3}$ and $(\mathbf{0})_{9\times 9}$ indicate square arrays of 3×3 and 9×9 zeros, respectively.

A similar time-evolution operator to Eq. (19) was previously obtained by Trégourès & van Tiggelen (2002) for elastic wave scattering and transport in heterogeneous media, except for the adding term $\mathbf{P}^T\frac{1}{2\delta}\sqrt{\frac{\mu(\mathbf{r})}{2}}\epsilon_{jk}$ between square brackets in the middle of the right column of Eq. (19). It arises because of the additional term that appears in the Jiang-Liu formulation of the elastic stress [see Eq. (10)] compared to the traditional expression given by Eq. (12). It is this remarkable difference along with the stress-dependent moduli that allow for a theoretical description of granular features such as volume dilatancy, mechanical yield, and anisotropy in the stress distribution, which are always absent in a pure elastic medium under deformation.

4. Multiple scattering, radiative transport and diffusion approximation

In the previous section we have presented the main steps to build up a theory for the propagation of elastic waves in disordered granular packings. Now we proceed to develop the rigorous basis to modeling the multiple scattering and the diffusive wave motion in granular media by employing the same mathematical framework used to describe the vibrational properties of heterogeneous materials (Frisch (1968); Karal & Keller (1964); Ryzhik et al., (1996); Sheng (2006); Weaver (1990)). The inclusion of spatially–varying constitutive relations (i.e., Eqns. (4)–(6)) to capture local disorder in the nonlinear granular elastic theory and the formulation of elastic wave equation in terms of a vector–field formalism, Eq. (17), are both important steps to build up a theory of diffusivity of ultrasound in granular media. In this section, we derive and analyze a radiative transport equation for the energy density of waves in a granular medium. Then, we derive the related diffusion equation and calculate the transmitted intensity by a *plane–wave* pulse.

4.1 Radiative transport and quantum field theory formalism

The theory of radiative transport provides a mathematical framework for studying the propagation of energy throughout a medium under the effects of absorption, emission and scattering processes (e.g., (Ryzhik et al., (1996); Weaver (1990)). The formulation we present here is well known, but most closely follows Frisch (1968); Ryzhik et al., (1996); Trégourès & van Tiggelen (2002); Weaver (1990). As the starting point, we take the Laplace transform of

Eq. (17) to find the solution

$$|\mathbf{\Psi}(\mathbf{z})\rangle = \mathbf{G}(\mathbf{z})\left[i|\mathbf{\Psi}(t=0)\rangle + |\mathbf{\Psi}_{\mathbf{f}}(\mathbf{z})\rangle\right], \tag{20}$$

where $\mathrm{Im}(\mathbf{z}) > 0$, with $\mathbf{z} = \omega + i\epsilon$ and $\epsilon \sim 0$ in order to ensure analyticity for all values of the frequency ω. The operator $\mathbf{G}(\mathbf{z})$ is the Green's function $\mathbf{G}(\mathbf{z}) := [\mathbf{z} - \mathbf{K}]^{-1}$, defined by the equation $[\mathbf{z} - \mathbf{K}]\,\mathbf{G}(\mathbf{z}) = \mathbf{I}\delta(\mathbf{r} - \mathbf{r}')$, where $\mathbf{I}$ is the identity tensor. Physically, it represents the response of the system to the force field for a range of frequencies ω and defines the source for waves at $t = 0$. A clear introduction to Green's function and notation used here is given in the book by Economou (2006). We shall be mainly interested in two average Green's functions: (i) the *configurational averaged* Green's function, related to the mean field; (ii) the covariance between two Green's function, related to the *ensemble–averages intensity*. Mathematical problems of this kind arise in the application of the methods of quantum field theory (QFT) to the statistical theory of waves in random media (Frisch (1968)). In what follows, we derive a multiple scattering formalism for the mean Green's function (analogous to the Dyson equation), and the covariance of the Green's function (analogous to the Bethe–Salpeter equation). The covariance is found to obey an equation of radiative transfer for which a diffusion limit is taken and then compared with the experiments.

4.1.1 Configuration-specific acoustic transmission

A *deterministic* description of the transmitted signal through a granular medium is almost impossible, and would also be of little interest. For example, a fundamental difference between the coherent E and incoherent S signals lies in their sensitivity to changes in packing configurations. This appears when comparing a first signal measured under a static load P with that detected after performing a "loading cycle", i.e., complete unloading, then reloading to the same P level. As illustrated in Fig. 4 S is highly non reproducible, i.e., configuration sensitive. This kind of phenomenon arises in almost every branch of physics that is concerned with systems having a large number of degrees of freedom, such as the many–body problem. It usually does not matter, because only average quantities are of interest. In order to obtain such average equation, one must use a statistical description of both the medium and the wave. To calculate the response of the granular packing to wave propagation we first perform a configurational averaging over random realizations of the disorder contained in the constitutive relations for the elastic moduli and their local fluctuations (see subsection 3.1.1). As the fluctuations in the Lamé coefficients $\lambda(\mathbf{r})$ and $\mu(\mathbf{r})$ can be expressed in terms of the fluctuating local compression (see Eq.(6)), then the operator $\mathbf{K}$ (Eq.(19)) is a stochastic operator.

The mathematical formulation of the problem leads to a partial differential equation whose coefficients are random functions of space. Due to the well–known difficulty to obtaining exact solutions, our goal is to construct a perturbative solution for the ensemble averaged quantities based on the smallness of the random fluctuations of the system. For simplicity, we shall ignore variations of the density and assume that $\rho(\mathbf{r}) \approx \rho_0$, where ρ_0 is a constant reference density. This latter assumption represents a good approximation for systems under strong compression, which is the case for the experiments analyzed here. We then introduce the disorder perturbation as a small fluctuation $\delta\mathbf{K}$ of operator (19) so that

$$\mathbf{K} = \mathbf{K}_0 + \delta\mathbf{K}, \tag{21}$$

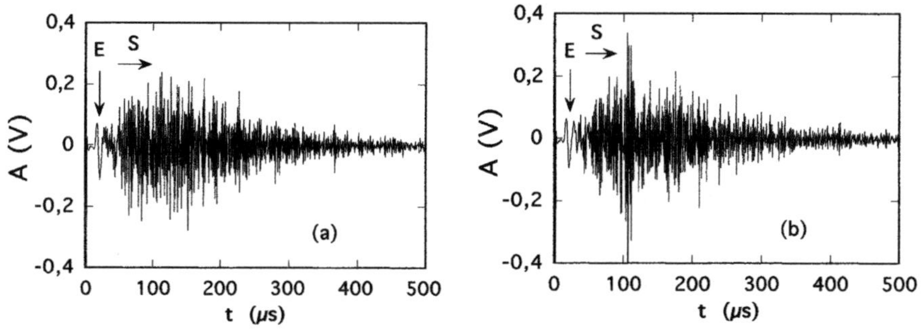

Fig. 4. Transmitted ultrasonic signal through a dry glass beads packing with $d = 0.4 - -0.8$ mm, detected by a transducer of diameter 2 mm and external normal stress $P = 0.75$ MPa: (a) First loading; (b) reloading. (Reprinted from Jia et al., Phys. Rev. Lett. 82, 1863 (1999))

where $\mathbf{K}_0$ is the unperturbed time-evolution operator in the "homogeneous" Jiang-Liu nonlinear elasticity. Using Eq. (19) along with Eqs. (4)–(6), we obtain after some algebraic manipulations the perturbation operator

$$\delta\mathbf{K} = \begin{pmatrix} 0 & \frac{1}{4}\sqrt{\frac{\lambda_0}{\rho_0}}\frac{\Delta(\mathbf{r})}{\delta_0}\mathbf{P} & (\mathbf{0})_9 \\ \frac{1}{4}\sqrt{\frac{\lambda_0}{\rho_0}}\mathbf{P}^t\Delta_1 & (\mathbf{0})_{3\times3} & \frac{1}{2\sqrt{2}}\sqrt{\frac{\mu_0}{\rho_0}}\left[L^l_{jk}(\mathbf{P}^t)\frac{\Delta(\mathbf{r})}{\delta_0} - \frac{\mathbf{P}^t\epsilon^{m*}_{jk}}{\delta_0}\Delta_2\right] \\ (\mathbf{0})^t_9 & \frac{1}{2\sqrt{2}}\sqrt{\frac{\mu_0}{\rho_0}}\frac{\Delta(\mathbf{r})}{\delta_0}L^{jk}_l(\mathbf{P}) & (\mathbf{0})_{9\times9} \end{pmatrix}, \tag{22}$$

where $\Delta_1 = 1 + 5\Delta(\mathbf{r})/(4\delta_0)$ and $\Delta_2 = 1 - \Delta(\mathbf{r})/\delta_0$.

4.1.1.1 The Dyson equation and mode conversion

We may now write the ensemble average Green's function as

$$\langle \mathbf{G}(\omega)\rangle = \left\langle [\omega + i\epsilon - \mathbf{K}]^{-1}\right\rangle = \left[\mathbf{G}_0^{-1}(\omega) - \mathbf{\Sigma}(\omega)\right]^{-1}, \tag{23}$$

where $\mathbf{G}_0(\omega) = [\omega + i\epsilon - \mathbf{K}_0]^{-1}$ is the "retarded" (outgoing) Green's function for the bare medium, i.e., the solution to (20) when $\Delta(\mathbf{r}) = 0$. The second equality is the Dyson equation and $\mathbf{\Sigma}$ denotes the "self–energy" or "mass" operator, in deference to its original definition in the context of quantum field theory (Das (2008)). This equation is exact. An approximation is, however, necessary for the evaluation of $\mathbf{\Sigma}$. The lowest order contribution is calculated under the closure hypothesis of local independence using the method of smoothing perturbation (Frisch (1968)). The expression for $\mathbf{\Sigma}$ is

$$\mathbf{\Sigma}(\omega) \approx \left\langle \delta\mathbf{K}\cdot[\omega + i\epsilon - \mathbf{K}_0]^{-1}\cdot\delta\mathbf{K}\right\rangle \tag{24}$$

The Green's function is calculated by means of a standard expansion in an orthonormal and complete set of its eigenmodes $\mathbf{\Psi}_n$, each with a natural frequency ω_n (Economou (2006)). If the perturbation is weak, we can use first-order perturbation theory (Frisch (1968)) and write the expanded Green's function as

$$\mathbf{G}(\omega) = \sum_n \frac{|\mathbf{\Psi}_n\rangle\langle\mathbf{\Psi}_n|}{\omega - \omega_n - \Sigma_n(\omega)}, \tag{25}$$

with

$$\Sigma_n(\omega) = \sum_m \frac{\langle|\langle\mathbf{\Psi}_n|\delta\mathbf{K}|\mathbf{\Psi}_m\rangle|^2\rangle}{\omega - \omega_m + i\epsilon}. \tag{26}$$

The eigenmodes obey the orthonormality condition $\langle\mathbf{\Psi}_n|\mathbf{\Psi}_m\rangle = \int d^3\mathbf{r}\mathbf{\Psi}_n^* \cdot \mathbf{\Psi}_m = \delta_{nm}$. A straightforward calculation, employing integration by parts, leads to the mode conversion effective cross–section

$$\left\langle|\langle\mathbf{\Psi}_n|\delta\mathbf{K}|\mathbf{\Psi}_m\rangle|^2\right\rangle = \omega^2 \int d^3\mathbf{r}\sigma^2 \left\{ \left| \frac{9\lambda_0}{32\delta_0} (\nabla\cdot\mathbf{u}_n)^* (\nabla\cdot\mathbf{u}_m) + \frac{\mu_0}{2\delta_0}\epsilon_{ji}^{n*}\epsilon_{ij}^{m} \right|^2 \right.$$
$$\left. + \left| \frac{\mu_0}{4\delta_0^2}\epsilon_{ji}^{m*}\epsilon_{ij}^{m} (\nabla\cdot\mathbf{u}_n)^* \right|^2 \right\}, \tag{27}$$

We may now derive an expression for the scattering mean free-time from Eqs. (26) and (27). To do so we first recall that the extinction time of mode n is given by $1/\tau_n = -2\text{Im}\Sigma_n(\omega)$ and replace in Eq. (27) the integers n and m by $i\mathbf{k}_i$ and $j\mathbf{k}_j$, respectively, where i and j are the branch indices obtained from the scattering relations that arise when we solve the eigenvalue problem for a homogeneous and isotropic elastic plate (Trégourès & van Tiggelen (2002)). In this way, mode n corresponds to the mode at frequency ω on the ith branch with wave vector $\mathbf{k}_i$. Similarly, mode m is the mode on the jth branch with wave vector $\mathbf{k}_j$. With the above replacements, the sum Σ_m on the right-hand side of Eq. (26) becomes $\sum_i A \int d^2\hat{\mathbf{k}}_i/(2\pi)^2$. Finally, if we use Eq. (27) into Eq. (26) with the above provisions, we obtain the expression for the scattering mean free-time, or extinction time

$$\frac{1}{\tau_j(\omega)} = \omega^2 \sum_i n_i \int \frac{d^2\hat{\mathbf{k}}_i}{2\pi} W(i\mathbf{k}_i, j\mathbf{k}_j), \tag{28}$$

where

$$W(i\mathbf{k}_i, j\mathbf{k}_j) = \int_0^L dz\sigma^2 \left\{ \left| \frac{9\lambda_0}{32\delta_0} \left(\nabla\cdot\mathbf{u}_{j\mathbf{k}_j}\right)^* \left(\nabla\cdot\mathbf{u}_{i\mathbf{k}_i}\right) + \frac{\mu_0}{2\delta_0}\mathfrak{S}_{j\mathbf{k}_j}^* : \mathfrak{S}_{i\mathbf{k}_i} \right|^2 \right.$$
$$\left. + \left| \frac{\mu_0}{4\delta_0^2}\mathfrak{S}_{i\mathbf{k}_i}^* : \mathfrak{S}_{i\mathbf{k}_i} \left(\nabla\cdot\mathbf{u}_{j\mathbf{k}_j}\right)^* \right|^2 \right\}, \tag{29}$$

is the *mode scattering cross-section* and $n_i(\omega) := k_i(\omega)/v_i$ is the spectral weight per unit surface of mode i at frequency ω in phase space. In Eq.(29) we have made use of the dyadic strain tensor $\mathfrak{S} = 1/2[\nabla\mathbf{u} + (\nabla\mathbf{u})^T]$.

4.1.1.2 The Bethe–Salpeter equation

To track the wave transport behavior after phase coherence is destroyed by disordered scatterings, we must consider the energy density of a pulse which is injected into the granular medium. We start by noting that the wave energy density is proportional to the Green's function squared. Moreover, the evaluation of the ensemble average of two Green's functions requires an equation that relates it to the effect of scattering. The main observable is given by the ensemble-average intensity Green's function $\langle \mathbf{G}(\omega^+) \otimes \mathbf{G}^*(\omega^-) \rangle$, where $\otimes$ denotes the outer product, $\omega^\pm = \omega \pm \Omega/2$, where Ω is a slowly varying envelope frequency, and $\mathbf{G}(\omega^+)$, $\mathbf{G}^*(\omega^-)$ are, respectively, the retarded and the advanced Green's functions. The covariance between these two Green's functions is given by

$$\langle \mathbf{G}(\omega^+) \otimes \mathbf{G}^*(\omega^-) \rangle = \mathbf{G}(\omega^+) \otimes \mathbf{G}^*(\omega^-) + \mathbf{G}(\omega^+) \otimes \mathbf{G}^*(\omega^-) : \mathbf{U} : \langle \mathbf{G}(\omega^+) \otimes \mathbf{G}^*(\omega^-) \rangle \tag{30}$$

The above equation is known as the Bethe-Salpeter equation and is the analog of the Dyson equation for $\mathbf{G}(\omega^+)$. It defines the *irreducible vertex* function $\mathbf{U}$, which is analogous to the self-energy operator $\mathbf{\Sigma}$. This equation can be expanded in the complete base $\mathbf{\Psi}_n$ of the homogeneous case. In this base, we find that $\langle \mathbf{G}(\omega^+) \otimes \mathbf{G}^*(\omega^-) \rangle = \mathcal{L}_{nn'mm'}(\omega,\Omega)$, which defines the object that determines the exact microscopic space-time behavior of the disturbance, where $\mathbf{G}(\omega^+) \otimes \mathbf{G}^*(\omega^-) = G_n(\omega^+)G^*_{n'}(\omega^-)\delta_{nm}\delta_{n'm'}$. The Bethe–Salpeter equation for this object reads

$$\mathcal{L}_{nn'mm'}(\omega,\Omega) = G_n(\omega^+)G^*_{n'}(\omega^-) \left[\delta_{nm}\delta_{n'm'} + \sum_{ll'} \mathrm{U}_{nn'll'}(\omega,\Omega)\mathcal{L}_{ll'mm'}(\omega,\Omega) \right]. \tag{31}$$

Upon introducing $\Delta G_{nn'}(\omega,\Omega) \equiv G_n(\omega^+) - G^*_{n'}(\omega^-)$ and $\Delta\Sigma_{nn'}(\omega,\Omega) \equiv \Sigma_n(\omega^+) - \Sigma^*_{n'}(\omega^-)$ this equation can be rearranged into

$$\begin{aligned} &\left[\Omega - (\omega_n - \omega^*_{n'}) - \Delta\Sigma_{nn'}(\omega,\Omega)\right] \mathcal{L}_{nn'mm'}(\omega,\Omega) = \\ &\Delta G_{nn'}(\omega,\Omega) \left[\delta_{nm}\delta_{n'm'} + \sum_{ll'} \mathrm{U}_{nn'll'}(\omega,\Omega)\mathcal{L}_{ll'mm'}(\omega,\Omega) \right]. \end{aligned} \tag{32}$$

4.1.2 Radiative transport equation

Equation (32) is formally exact and contains all the information required to derive the radiative transport equation (RTE), but approximations are required for the operator $\mathbf{U}$. Using the method of smoothing perturbation, we have that $\mathrm{U}_{nn'll'}(\omega,\Omega) \approx \langle\langle \mathbf{\Psi}_n|\delta K|\mathbf{\Psi}_l\rangle\langle \mathbf{\Psi}_{n'}|\delta K|\mathbf{\Psi}_{l'}\rangle\rangle$. In most cases $\omega >> \Omega$. Therefore, we may neglect Ω in any functional dependence on frequency. The integer index n consists of one discrete branch index j, with the discrete contribution of $\mathbf{k}$ becoming continuous in the limit when $A \to \infty$. In the quasi-two-dimensional approximation we can also neglect all overlaps between the different branches (Trégourès & van Tiggelen (2002)) and use the equivalence $\Delta G_{nn'}(\omega,\Omega) \sim 2\pi i\delta_{nn'}\delta\left[\omega - \omega_n(\mathbf{k})\right]$. As a next step, we need to introduce the following definition for the *specific intensity* $L_{n\mathbf{k}}(\mathbf{q},\Omega)$ of mode $j\mathbf{k}_j$ at frequency ω

$$\sum_{mm'} \mathcal{L}_{nn'mm'}(\omega,\Omega)S_m S^*_{m'} \equiv 2\pi\delta\left[\omega - \omega_n(\mathbf{k})\right]\delta_{nn'}L_{n\mathbf{k}}(\mathbf{q},\Omega), \tag{33}$$

where $S_m(\omega)$ is the source of radiation at frequency ω, which in mode representation can be written as

$$S_m(\omega) = \langle \mathbf{\Psi}_m|\mathbf{\Psi}_f\rangle = \omega \int d^3\mathbf{r} \cdot \mathbf{f}^*(\mathbf{r},\omega) \cdot \mathbf{u}_m(\mathbf{r}), \tag{34}$$

and $\mathbf{q} = \mathbf{k} - \mathbf{k}'$ is the scattering wave vector. If we now multiply Eq. (32) by $S_m S^*_{m'}$ and sum over the integer indices m and m', we obtain that

$$[\Omega - (\omega_n - \omega^*_{n'}) - \Delta\Sigma_{nn'}] \sum_{mm'} \mathcal{L}_{nn'mm'} S_m S^*_{m'} = \Delta G_{nn'} \left[\sum_{mm'} \delta_{nm}\delta_{n'm'} S_m S^*_{m'} \right.$$
$$\left. + \sum_{ll'} U_{nn'll'} \sum_{mm'} \mathcal{L}_{ll'mm'} S_m S^*_{m'} \right]. \quad (35)$$

According to Eq. (33), we may then write

$$2\pi \left[\Omega - (\omega_n - \omega^*_{n'}) - \Delta\Sigma_{nn'}\right] \delta(\omega - \omega_n)\delta_{nn'} L_{n\mathbf{k}_n}(\mathbf{q}, \Omega) = 2\pi i \delta(\omega - \omega_n)\delta_{nn'}$$
$$\times \left[\sum_{mm'} \delta_{nm}\delta_{n'm'} S_m S^*_{m'} + \sum_{ll'} U_{nn'll'} 2\pi\delta(\omega - \omega_l)\delta_{ll'} L_{l\mathbf{k}_l}(\mathbf{q}, \Omega) \right].$$

Substituting the above relation into Eq. (35) and performing the summations over the indices n', m, m', and l', we obtain after some algebraic manipulations that Eq. (35) reduces to

$$\left[-i\Omega - 2\mathrm{Im}(\omega_n) + i\Delta\Sigma_{nn}\right] L_{n\mathbf{k}_n} = |S_n|^2 + \sum_l U_{nnll} 2\pi\delta(\omega - \omega_l) L_{l\mathbf{k}_l}. \quad (36)$$

Note that since the imaginary part of ω_n is small, we can drop the term $2Im(\omega_n)$ on the left-hand side of Eq. (58). Moreover, since $\Omega \ll \omega$ then $\omega^+ \approx \omega^- = \omega$ and hence $\Delta\Sigma_{nn} = -i/\tau_{n\mathbf{k}_n}(\omega)$. On the other hand, recalling that $U_{nnll} = \langle |\langle \Psi_n|\delta K|\Psi_l\rangle|^2 \rangle$, replacing n by j, and making the equivalence $\sum_l \longrightarrow \sum_{j'} A(2\pi)^{-2} \int d^2\mathbf{k}_{j'}$, Eq. (36) becomes

$$\left[-i\Omega + \frac{1}{\tau_{j\mathbf{k}_j}}\right] L_{j\mathbf{k}_j} = |S_{j\mathbf{k}_j}|^2 + \sum_{j'} \int \frac{d^2\mathbf{k}_{j'}}{2\pi} \left\langle |\langle \Psi_j|\delta K|\Psi_{j'}\rangle|^2 \right\rangle \delta(\omega - \omega_{j'}) L_{j'\mathbf{k}'_j}$$
$$= |S_{j\mathbf{k}_j}|^2 + \omega^2 \sum_{j'} \int \frac{d^2\hat{\mathbf{k}}_{j'}}{2\pi} W(j\mathbf{k}_j, j'\mathbf{k}_{j'}) L_{j'\mathbf{k}'_j} n_{j'}. \quad (37)$$

Finally, if we turn out to the real space-time domain by taking the inverse Fourier transform of Eq. (37), it follows that

$$\left(\partial_t + \mathbf{v}_j \cdot \nabla + \frac{1}{\tau_{j\mathbf{k}_j}}\right) L_{j\mathbf{k}_j}(\mathbf{x}, t) = |S_{j\mathbf{k}_j}(\omega)|^2 \delta(\mathbf{x})\delta(t)$$
$$+ \omega^2 \sum_{j'} \int \frac{d^2\hat{\mathbf{k}}_{j'}}{2\pi} W(j\mathbf{k}_j, j'\mathbf{k}_{j'}) L_{j'\mathbf{k}'_j}(\mathbf{x}, t) n_{j'}. \quad (38)$$

This is the desired RTE. The first two terms between brackets on the left-hand side of Eq. (38) define the mobile operator $d/dt = \partial_t + \mathbf{v}_j \cdot \nabla$, where ∂_t is the Lagrangian time derivative and $\mathbf{v}_j \cdot \nabla$ is a hydrodynamic convective flow term, while the $1/\tau_{j\mathbf{k}_j}$-term comes from the average amplitude and represents the loss of energy (extinction). The second term on the right-hand side of Eq. (38) contains crucial new information. It represents the scattered intensity from all directions $\mathbf{k}'$ into the direction $\mathbf{k}$. The object $W(j\mathbf{k}_j, j'\mathbf{k}_{j'})$ is the rigorous theoretical microscopic building block for scattering processes in the granular medium. The first term is a source term that shows up from the initial value problem. The physical interpretation of Eq. (38) can therefore be summarized in the following statement:

$$\left(\partial_t + \mathbf{v}_j \cdot \nabla + \text{losses}\right) L_{j\mathbf{k}_j}(\mathbf{x}, t) = \text{source} + \text{scattering}, \tag{39}$$

which mathematically describes the phenomenon of multiple scattering of elastic waves in granular media. This completes our derivation of the transport equation for the propagation of elastic waves in these systems.

Remark: For granular media the contribution to the loss of energy due to absorption must be included in the extinction time $1/\tau_j$. We refer the reader to Brunet et al. (2008b) for a recent discussion on the mechanisms for wave absorption. Whereas in the context of the nonlinear elastic theory employed in the present analysis intrinsic attenuation is not explicitly considered (similar to the "classical" elastic theory), its effects can be easily accounted for by letting the total extinction time be the sum of two terms: $1/\tau_j = 1/\tau_j^s + 1/\tau_j^a$, where $1/\tau_j^a$ is the extinction-time due to absorption. A rigorous calculation of this term would demand modifying the scattering cross-section (Papanicolaou et el. (1996)), implying that the non-linear elastic theory should be extended to account for inelastic contributions. In this chapter we do not go further on this way and keep the inclusion of the extinction time due to absorption at a heuristic level.

4.2 Diffusion equation

Now we derive the form of Eq. (38) in the diffusion limit and solve it to study the diffusive behavior of elastic wave propagation in granular media. Integrating Eq. (38) over $\hat{\mathbf{k}}$ and performing some rearrangements we obtain the equation

$$\partial_t U_i + \nabla \cdot \mathbf{J}_i = n_i \int \frac{d^2\hat{\mathbf{k}}}{2\pi} |S_{i\mathbf{k}_i}|^2 \delta(t)\delta(\mathbf{x}) - \frac{1}{\tau^a_{i\mathbf{k}_i}} U_i - \sum_j C_{ij} U_j, \tag{40}$$

where

$$U_i := \int \frac{d^2\hat{\mathbf{k}}}{2\pi} \delta(\omega - \omega_{i\mathbf{k}}) L_{i\mathbf{k}} = n_i \int \frac{d^2\hat{\mathbf{k}}}{2\pi} L_{i\mathbf{k}_i}, \tag{41}$$

is the spectral energy density (or fluence rate) U_i,

$$\mathbf{J}_i := \int \frac{d^2\hat{\mathbf{k}}}{2\pi} \delta(\omega - \omega_{i\mathbf{k}}) \mathbf{v}_i L_{i\mathbf{k}} = n_i \int \frac{d^2\hat{\mathbf{k}}}{2\pi} \mathbf{v}_i L_{i\mathbf{k}_i}, \tag{42}$$

is the current density (or energy flux) $\mathbf{J}_i$, and

$$C_{ij} := \frac{\delta_{ij}}{\tau^s_{i\mathbf{k}_i}} - \omega^2 n_i \int \frac{d^2\hat{\mathbf{k}}_j}{2\pi} W(i\mathbf{k}_i, j\mathbf{k}_j), \tag{43}$$

is the *mode conversion matrix* C_{ij}.

The diffusion approximation is basically a first-order approximation to Eq. (38) with respect to the angular dependence. This approximation assumes that wave propagation occurs in a medium in which very few absorption events take place compared to the number of scattering events and therefore the radiance will be nearly isotropic. Under these assumptions the fractional change of the current density remains small and the radiance can be approximated by the series expansion $L_{i\mathbf{k}}(\mathbf{q},\Omega) \simeq \frac{1}{n_i}U_i(\mathbf{q},\Omega) + \frac{2}{n_i v_i^2}\mathbf{v}_i \cdot \mathbf{J}_i(\mathbf{q},\Omega) + \cdots$, where the zeroth-order term contains the spectral energy density and the first-order one involves the dot product between the flow velocity and the current density; the latter quantity being the vector counterpart of the fluence rate pointing in the prevalent direction of the energy flow. Replacing this series approximation into Eq. (38) produces the equation

$$\frac{1}{n_i}\left[\partial_t U_i + \mathbf{v}_i \cdot \nabla U_i + \frac{2}{\mathbf{v}_i^2}\mathbf{v}_i \cdot \partial_t \mathbf{J}_i + 2\mathbf{v}_i \cdot \nabla \mathbf{J}_i + \left(\frac{1}{\tau^s_{i\mathbf{k}_i}} + \frac{1}{\tau^a_{i\mathbf{k}_i}}\right)\left(U_i + \frac{2}{v_i^2}\mathbf{v}_i \cdot \mathbf{J}_i\right)\right] \approx$$

$$|S_{i\mathbf{k}}(\omega)|^2\delta(t)\delta(\mathbf{x}) + \omega^2 \sum_j \int \frac{d^2\hat{\mathbf{k}}_j}{2\pi} W(i\mathbf{k}_i, j\mathbf{k}_j)\left(U_i + \frac{2}{v_i^2}\mathbf{v}_i \cdot \mathbf{J}_i\right). \tag{44}$$

From the above assumptions we can make the following approximations: $\partial_t U_i \to 0$ and $\frac{d}{dt}\mathbf{J}_i = \mathbf{v}_i \cdot \partial_t \mathbf{J}_i + \mathbf{v}_i \cdot \nabla \mathbf{J}_i \to 0$. Moreover, we can also neglect the contribution of $1/\tau^a_{i\mathbf{k}_i}$. The absorption term modifies the solution of the scattering cross-section making it to decay exponentially, with a decay rate that vanishes when $\tau^a_{i\mathbf{k}_i} \to \infty$ (Papanicolaou et el. (1996)). Furthermore, noting that in the diffusive regime $U_j/n_j \approx U_i/n_i$, the above equation can be manipulated and put into the more convenient form

$$2\sum_j \left(\frac{\delta_{ij}}{\mathbf{v}_i^2 \tau^s_{i\mathbf{k}_i}} - n_i\omega^2 \int \frac{d^2\hat{\mathbf{k}}_j}{2\pi} W(i\mathbf{k}_i, j\mathbf{k}_j)\frac{\mathbf{v}_i \cdot \mathbf{v}_j}{\mathbf{v}_i^2 \mathbf{v}_j^2}\right)\mathbf{J}_j \approx -\nabla U_i. \tag{45}$$

It is evident from this equation that we can define the *diffusion matrix* as

$$(\mathbf{D}^{-1})_{ij} := 2\left(\frac{\delta_{ij}}{\mathbf{v}_i^2 \tau^s_{i\mathbf{k}_i}} - n_i\omega^2 \int \frac{d^2\hat{\mathbf{k}}_j}{2\pi} W(i\mathbf{k}_i, j\mathbf{k}_j)\frac{\mathbf{v}_i \cdot \mathbf{v}_j}{\mathbf{v}_i^2 \mathbf{v}_j^2}\right), \tag{46}$$

which allows us to express the current density $\mathbf{J}_i$ as a generalized Fick's Law:

$$\mathbf{J}_i = -\sum_j D_{ij} \nabla U_j. \tag{47}$$

A generalized diffusion equation then follows by combining the continuity-like equation (40) with the Fick's law (47), which reads

$$\partial_t U_i - \nabla \cdot \left(\sum_j D_{ij} \nabla U_j\right) = S_i(\omega)\delta(t)\delta(\mathbf{x}) - \sum_j C_{ij} U_j - \frac{1}{\tau^a_{i\mathbf{k}_i}} U_i, \tag{48}$$

where the source $S_i(\omega)$ is defined by the integral $S_i(\omega) = n_i \int \frac{d^2\hat{\mathbf{k}}}{2\pi}|S_{i\mathbf{k}_i}(\omega)|^2$. At this point it is a simple matter to derive the diffusion equation for the total energy density $U = \sum_i U_i$. Summing all terms in Eq. (48) over the index i, introducing the definitions: $S(\omega) = \sum_i S_i(\omega)$

for the total source along with $D(\omega) := \sum_{ij} D_{ij}(\omega)n_j / \sum_j n_j$, for the total diffusion coefficient, and $\xi := \frac{1}{\tau_a} = \sum_i \frac{n_i}{\tau_{i\mathbf{k}_i^a}} / \sum_i n_i$, for the total absorption rate, and noting that

$$\frac{\sum_{ij} C_{ij}(\omega)n_j}{\sum_j n_j} = \frac{\sum_i \left[\sum_j \frac{\delta_{ij}n_j}{\tau^s_{i\mathbf{k}_i}} - n_i\omega^2 \sum_j n_j \int \frac{d^2\hat{\mathbf{k}}_j}{2\pi} W(i\mathbf{k}_i, j\mathbf{k}_j)\right]}{\sum_j n_j} = 0,$$

where we have made use of Eq. (43), we finally obtain the time-dependent equation

$$\partial_t U - D(\omega)\nabla^2 U + \frac{1}{\tau_a} U = S(\omega)\delta(t)\delta(\mathbf{x}), \tag{49}$$

which describes the diffusive propagation of elastic waves.

4.2.1 Transmitted intensity

In section 2.3, Fig.3 we showed that the averaged transmitted intensity $I(t)$ decays exponentially at long times. This picture is reminiscent of the diffusively transmitted pulses of classical waves through strongly scattering random media (Sheng (2006); Snieder & Page (2007); Tourin et al. (2000)). This is the main result of the present work, which stimulated the construction of the theory for elastic wave propagation in granular media presented above. Now we conclude our analysis with the derivation of the mathematical formula for the transmitted intensity $I(t)$, corroborating that it fix very well with the experimental data.
In the experiment the perturbation source and the measuring transducer were placed at the axisymmetric surfaces and the energy density was measured on the axis of the cylinder. We can make use of Eq. (49) to calculate the analytical expression for the transmitted flux. In order to keep the problem mathematically tractable we assume that the horizontal spatial domain is of infinite extent (i.e., $-\infty < x < \infty$ and $-\infty < y < \infty$), while in the z-direction the spatial domain is limited by the interval $(0 < z < L)$. The former assumption is valid for not too long time scales and for a depth smaller than half of the container diameter. With the use of Cartesian coordinates, a solution to Eq. (49) can be readily found by separation of variables with appropriate boundary conditions at the bottom ($z = 0$) and top of the cylinder ($z = L$). The separation of variables is obtained by guessing a solution of the form $U(x, y, z, t) = U^{\mathbf{x}}(\mathbf{x}, t)U^z(z, t)$. It is not difficult to show that if the surface of the cylinder is brought to infinity, Eq. (49) satisfies the solution for an infinite medium

$$U^{\mathbf{x}}(\mathbf{x}, t) = \frac{S(\omega)}{4\pi D(\omega)t} \exp\left[-\frac{\mathbf{x}^2}{4D(\omega)t}\right] \exp\left(-\frac{t}{\tau_a}\right). \tag{50}$$

It is well known that for vanishing or total internal reflection the Dirichlet or the Neumann boundary conditions apply, respectively, for any function obeying a diffusion equation with open boundaries. In the case of granular packings we need to take into account the internal reflections. In this way, there will be some incoming flux due to the reflection at the boundaries and appropriate boundary conditions will require introducing a reflection coefficient R, which is defined as the ratio of the incoming flux to the outgoing flux at the boundaries (Sheng (2006)). Mixed boundary conditions are implemented for the z-coordinate, which in terms of the mean free path l^* are simply (Sheng (2006)):

$$U^z - c\partial_z U^z = 0 \text{ at } z = 0, \tag{51}$$

$$U^z + c\partial_z U^z = 0 \text{ at } z = L, \tag{52}$$

where the coefficient $c \equiv \frac{2l^*}{3}\frac{1+R}{1-R}$. Therefore, the solution for U^z reads

$$U^z = \sum_{n=1}^{\infty} Z_n(z) Z_n(z_0) \exp\left(-D(\omega)\alpha_n^2 t\right), \tag{53}$$

where

$$Z_n(z) = \frac{\sin(\alpha_n z) + \kappa\beta_n \cos(\alpha_n z)}{\sqrt{\frac{L}{2}}\left(1 + 2\kappa + \kappa^2\beta_n^2\right)}, \tag{54}$$

with $\alpha_n = \beta_n / L$, $\kappa = c/L$, and the discrete values of β_n determined by the roots of $\tan\beta_n = 2\beta_n\kappa/(\beta_n^2\kappa^2 - 1)$.

Finally, using Eqs. (50) and (53) we can ensemble the solution for the total energy density

$$U = \frac{S}{4\pi Dt} e^{\left(-\mathbf{x}^2/4Dt\right)} e^{-t/\tau_a} \sum_{n=1}^{\infty} C_n \left[\sin(\alpha_n z) + \kappa\beta_n \cos(\alpha_n z)\right] e^{-D\alpha_n^2 t}, \tag{55}$$

where

$$C_n \equiv \frac{2\left[\sin(\alpha_n z_0) + \kappa\beta_n \cos(\alpha_n z_0)\right]}{L\left(1 + 2\kappa + \kappa^2\beta_n^2\right)}. \tag{56}$$

The total transmitted flux at the top wall of the cylinder can be readily calculated by taking the z-derivative of E as defined by Eq. (55) and by evaluating the result at $z = L$ to give

$$\begin{aligned} I(\mathbf{x}, z, t) &= -D\partial_z U|_{z=L} \\ &= \frac{S}{2\pi L^2 t} e^{\left(-\mathbf{x}^2/4Dt\right)} e^{-t/\tau_a} \sum_{n=1}^{\infty} \alpha_n C_n \left[\kappa\beta_n \sin(\beta_n) - \cos(\beta_n)\right] e^{-D\alpha_n^2 t}. \end{aligned} \tag{57}$$

If, as mentioned by Jia (2004), the reflectivity of the wall is very high, then $R \approx 1$. In the full reflection limit the following limits can be verified: $\kappa \to \infty$, $\tan\beta \to 0 \implies \beta_n = n\pi$ for all $n = 0, 1, 2, \ldots$, $\lim_{\kappa\to\infty} C_n = 0$, and $\lim_{\kappa\to\infty} \kappa C_n = (2/\beta_n L)\cos(\alpha_n z_0)$. For a plane-wave source we need to integrate Eq. (57) over $\mathbf{x} = (x, y)$ to obtain

$$I(t) = \frac{vS(\omega)}{2L} \exp\left(-\frac{t}{\tau_a}\right) \sum_{n=1}^{\infty} (-1)^n \cos\left(\frac{n\pi z_0}{L}\right) \exp\left(-\frac{D(\omega)(n\pi)^2}{L^2} t\right), \tag{58}$$

where v is the energy transport velocity and $z_0 \approx l^*$. This equation tells us that the flux transmitted to the detector behaves as $I(t) = vU/4$, when $R \approx 1$. This result provides the theoretical interpretation of the acoustic coda in the context of the present radiative transport theory and assesses the validity of the diffusion approximation for a high-albedo (predominantly scattering) medium as may be the case of granular packings.

4.2.2 Energy partitioning

In section 3.3 we have shown that the total energy $\mathcal{E}_T$ is given by Eq. (14), and that the Cartesian scalar inner product of the vector field $\boldsymbol{\Psi}$ is exactly to the total energy. In the diffusion limit, the conversion between compressional $\mathcal{E}_P$ and shear $\mathcal{E}_S$ energies equilibrates in a universal way, independent of the details of the scattering processes and of the nature of the excitation source. The energy ratio is governed by the *equipartition of energy law*, $K = \mathcal{E}_S/\mathcal{E}_P = 2(c_P/c_S)^3$, where the factor 2 is due to the polarization of the shear waves (Jia et al (2009); Papanicolaou et el. (1996); Ryzhik et al., (1996); Weaver (1990)). For typical values of $c_P/c_S \geq \sqrt{3}$, the equipartition law predicts the energy ratio $K \geq 10$. This shows that in the diffusive regime the shear waves dominate in the scattering wave field, which is observed in seismological data (Hennino et al. (2001); Papanicolaou et el. (1996)). The diffusion coefficient D is a weighted mean of the individual diffusion coefficients of the compressional wave D_P and shear wave D_S: $D = (D_P + D_S)/(1 + K)$. With the weights $K \geq 10$ the diffusion coefficient is approximated to $D \approx D_S = c_S l_S^*/3$. This demonstration confirms the applicability of the diffusion equation for describing the multiple scattering of elastic waves (Jia (2004)).

5. Conclusions

In summary, the experiments presented in this chapter permit one to bridge between two apparently disconnected approaches to acoustic propagation in granular media, namely, the effective medium approach (Duffy & Mindlin (1957); Goddard (1990)), and the extreme configuration sensitive effects (Liu & Nagel (1992)). This unified picture is evidenced in fig.2 with the coexistence of a coherent ballistic pulse E_P and a multiply scattered signal S. The coherent signal was shown to be independent of the packing topological configuration, whereas the coda-like portion of the signal behaves like a fingerprint of the topological configuration as showed in fig.2 b and fig.4.

The experimental confirmation of the applicability of the diffusion approximation to describe the multiple scattering of elastic waves through a compressed granular medium, was decisive to guide the construction of the theoretical model for elastic waves propagation. We have shown that the nonlinear elastic theory proposed by Jiang & Liu (2007) can be used to derive a time-evolution equation for the displacement field. Introducing spatial variations into the elastic coefficients λ and μ, we were able to describe the disorder due to the inhomogeneous force networks. The link between the local disorder expressed through the constitutive relations, and the continuum granular elastic theory, permit us to put together within a single theoretical framework the micro-macro description of a granular packing.

The mathematical formulation of the problem leads to a vector-field theoretic formalism analogous to the analytical structure of a quantum field theory, in which the total energy satisfies a Schrödinger-like equation. Then, introducing the disorder perturbation as a small fluctuation of the time-evolution operator associated to the Schrödinger-like equation, the RTE and the related diffusion equation have been constructed. We have shown that the temporal evolution for the averaged transmitted intensity $I(t)$, Eq.58, fits very well with the experimental data presented in fig.3, providing the theoretical interpretation of the intensity of scattered waves propagating through a granular packing. This opens new theoretical perspectives in this interdisciplinary field, where useful concepts coming from different areas of physics (quantum field theory, statistical mechanics, and condensed-matter physics) are

now merging together as an organic outgrowth of an attempt to describe wave motion and classical fields of a stochastic character.
As perspectives for future research let us mention the study of the evolution of the wave transport behavior in a more tenuous granular network when the applied stress is decreased. The disordered nature of a granular packing has a strong effect on the displacements and forces of in individual realizations, which depends on the intensity of the external loads. A hot topic is the study of acoustic probing to the *jamming transition* in granular media (Vitelli et al. (2010)). This is related to anisotropic effects and the emergence of non–affine deformations of the granular packing. It is necessary a systematic study of the transport properties of elastic waves between the different regimes of external load: strong compression $\leftrightarrow$ weak compression $\leftrightarrow$ zero compression. The last one is related to the behavior of waves at the free surface of the granular packing (Bonneau et al. (2007; 2008); Gusev et al. (2006)). The propagation of sound at the surface of sand is related to the localization of preys by scorpions (Brownell (1977)) and the spontaneous emission of sound by sand avalanches (the so–called song of dunes) (Bonneau et al. (2007)).
We believe that the experiments presented in this chapter point out to the considerable interest in acoustic probing as a tool for studying of the mechanical properties of confined granular media. Clearly, before this can be undertaken, one should study in detail the sensitivity of the acoustic response to configurational variations. On the other hand, the present theory represents a powerful tool to understand complex granular media as, for example, sedimentary rocks whose geometrical configuration is affected by deposition ambients, sediments, accommodation phase, lithostatic overburden, etc. This explains why anisotropy is always present and characterization is so difficult. Therefore, the study of acoustic waves in such complex media gives useful information to sedimentologists. It can also be applied to important oil industry issues such as hole stability in wells. Important geotechnical applications involve accurate seismic migration, seismo-creep motions, and friction dynamics. Finally, let us mention the similarity between the scattering of elastic waves in granular media with the seismic wave propagation in the crust of Earth and Moon (Dainty & Toksöz (1981); Hennino et al. (2001); Snieder & Page (2007)). In particular, the late-arriving coda waves in the lunar seismograms bear a striking resemblance to the multiple scattering of elastic waves in the dry granular packing. Some features of the laboratory experiments may be used to explain some seismic observations in the high-frequency coda of local earthquakes in rocky soils and the granular medium may be useful as model system for the characterization of seismic sources.

6. References

Bonneau, L., Andreotti, B. & Clément, E. (2007). Surface elastic waves in granular media under gravity and their relation to booming avalanches. *Phys. Rev. E* Vol. 75, No. 1 (January 2007), 016602

Bonneau, L., Andreotti, B. & Clément, E. (2008). Evidence of Rayleigh-Hertz surface waves and shear stiffness anomaly in granular media. *Phys. Rev. Lett.* Vol. 101, No. 11 (September 2008), 118001

Browneel, P. H. (1977). Compressional and surface waves in sand: Used by desert scorpions to locate prey. *Science* Vol. 197, No. 4302 (July 1977) 479–482

Brunet, Th, Jia, X. & Johnson, P. A. (2008). Transitional nonlinear elastic behaviour in dense granular media *Geophys. Res. Lett.* Vol. 35 No. 19 (October 2008) L19308

Brunet, Th., Jia, X. & Mills, P. (2008). Mechanisms for acoustic absorption in dry and weakly wet granular media. *Phys. Rev. Lett.* Vol. 101, No. 13, (September 2008), 138001

Chou, P. C. & Pagano, N. J. (1992). *Elasticity tensor dyadic and engineering approaches*, Dover Publications, New York, ISBN 0-486-66958-0

Dainty, A. M. & Toksöz, M. N. (1981). Seismic coda on the Earth and the Moon: A comparison. *Phys. Earth and Planet Inter.* Vol. 26, No. 4 (September 1981) 250-260

Das, A. (2008), *Lectures on quantum field theory*, World Scientific Publishing Co. Singapore, ISBN-10-981-283-285-8

Duffy, J. & Mindlin, R. D. (1957) Stress-strain relation and vibrations of granular medium. *J. Appl. Mech. Trans. ASME* Vol.7, 585–593

Economou, E. N. (2006). *Green's functions in quantum physics*, Springer, Berlin, ISSN: 0171–1873

Fehler, M. & Sato, H. (2003). Coda. *Pure appl. geophys.* Vol. 160, No. 3-4 (March 2003) 541–554

Frisch, U. (1968). *Wave propagation in random media*, In: *Probabilistic methods in applied mathematics*, A. T. Bharucha–Reid, (Ed.), 75–198, Academic Press, New York

Gilles, B. & Coste, C. (2003). Low–frequency behavior of beads constrained on a lattice. *Phys. Rev. Lett.* Vol. 90, No. 17 (May 2003) 174302

Goddard, J. D. (1990). Nonlinear Elasticity and pressure-dependent wave speeds in granular media. *Proc. R. Soc. London A* Vol. 430, No. 1878 (July 1990) 105-131

Goldenberg, C, Tanguy, A. & Barrat, J.–L. (2007). Particle displacements in the elastic deformation of amorphous materials: local fluctuations vs. non-affine field. *Europhys. Lett.*, Vol. 80, No. 1 (October 2007) 16003

Goldhirsch, I. & Goldenberg, C. (2002). On the microscopic foundations of elasticity. *Eur. Phys. J. E*, Vol. 9, No. 3 (December 2002) 245-251

Gusev, V. E., Aleshin, V. & Tournat, V. (2006). Acoustic waves in an elastic channel near the free surface of granular media. *Phys. Rev. Lett.* Vol. 96, No. 21 (June 2006) 214301

Hennino, R., Trégourès, N. P., Shapiro, N. M, Margerin, L., Campillo, M., van Tiggelen, B. A. & Weaver, R. L. (2001). Observation of equipartition of seismic waves. *Phys. Rev. Lett.* Vol. 86, No. 15 (April 2001) 3447-3450

Jaeger, H.M., Nagel, S.R. & Behringer, R. P. (1996). Granular solids, liquids, and gases. *Rev. Mod. Phys.* Vol. 68, No. 4 (October 1996) 1259-1273

Jia, X., Caroli, C. & Velicky, B. (1999). Ultrasound propagation in externally stressed granular media. *Phys. Rev. Lett*, Vol. 82, No. 9 (March 1999), 1863–1866

Jia, X. (2004). Codalike multiple scattering of elastic waves in dense granular media. *Phys. Rev. Lett.*, Vol. 93, No. 15 (October 2004), 154303

Jia, X., Laurent, J., Khidas, Y. & Langlois, V. (2009). Sound scattering in dense granular media. *Chinese Sci. Bull.*, Vol. 54, No. 23 (December 2009), 4327–4336

Jiang, Y. & Liu, M. (2007). A brief review of "granular elasticity": Why and how far is sand elastic? *Eur. Phys. J. E*, Vol. 22, No. 3 (February 2007) 255–260

Johnson, P. A. & Jia, X. (2005) Nonlinear dynamics, granular media and dynamic earthquake triggering. *Nature* Vol. 437, No. 7060 (October 2005), 871-874

Karal, F. C. & Keller, J. B. (1964). Elastic, electromagnetic, and other waves in a random medium. *J. Math. Phys.*, Vol. 5, No. 4 (April 1964) 537–547

Khidas, Y. & Jia, X. (2010) Anisotropic nonlinear elasticity in a spherical-bead pack: Influence of the fabric anisotropy. *Phys. Rev. E* Vol. 81, No. 2 (February 2010) 021303

Landau, L. D. & Lifshitz, E. M. (1999). *Theory of Elasticity*, Butterworth-Heinemann, Oxford, ISBN-10: 075062633X

Liu, C-h. & Nagel, S. R. (1992). Sound in sand. *Phys. Rev. Lett.* Vol. 68, No. 15 (April 1992), 2301-2304

Luding, S. (2005) Granular media: Information propagation. *Nature* Vol. 435, No. 7039 (May 2005) 159–160

Majmudar, T. S. & Behringer, R. P. (2005). Contact force measurements and stress–induced anisotropy in granular materials. *Nature*, Vol. 435, No. 7045 (June 2005)1079–1082

Makse, H. A., Gland, N., Johnson, D. L. & Schwartz, L. M. (1999). Why Effective medium theory fails in granular materials. *Phys. Rev. Lett.* Vol. 83, No. 24 (December 1999), 5070–5073

Norris, A. N. & Johnson, D. L. (1997). Nonlinear elasticity of granular media. *ASME J. App. Mech.*, Vol. 64, No. 1 (March 1997) 39–49

Papanicolaou, G. C., Ryzhik, L. V. & Keller, J. B. (1996). Stability of the P–to–S energy ratio in the diffusive regime. *Bull. Seism. Soc. Am.*, Vol. 86, No. 4 (August 1996) 1107–1115

Ryzhik, L., Papanicolaou, G. & Keller, J. B. (1996) Transport equations for elastic and other waves in random media. *Wave Motion*, Vol. 24, No. 4 (December 1996) 327–370

Serero, D.; Rydellet, G.; Clément E. & Levine, D. (2001). Stress response function of a granular layer: Quantitative comparison between experiments and isotropic elasticity. *Eur. Phys. J. E*, Vol. 6, No. 2, (October 2001) 169–179

Sheng, P. (2006). *Introduction to wave scattering, localization and mesoscopic phenomena*, 2nd edition, Springer–Verlag, Berlin, Heidelberg, ISSN: 0933-033X

Snieder, R. & Page, J. (2007). Multiple scattering in evolving media. *Physics Today* Vol. 60, No. 5 (May 2007) 49–55

Somfai, E., Roux, J. N., Snoeijer, J. H., van Hecke, M. & van Saarloos, W. (2005). Elastic wave propagation in confined granular systems. *Phys. Rev. E* Vol. 72, No. 2, (August 2005) 021301

Tanguy, A., Wittmer, J. P., Leonforte, F. & Barrat, J. -L. (2002). Continuum limit of amorphous elastic bodies: A finite–size study of low–frequency harmonic vibrations. *Phys. Rev. B* Vol. 66, No. 17 (November 2002) 174205

Tourin, A., Fink, M. & Derode, A. (2000). Multiple scattering of sound. *Waves in Random Media* Vol. 10, No. 4 (October 2000) R31-R60

Trégourès, N. P. & van Tiggelen, B. A. (2002). Quasi–two–dimensional transfer of elastic waves. *Phys. Rev. E*, Vol. 66, No. 3 (September 2002), 036601

Trujillo, L., Peniche, F. & Sigalotti, L. Di G. (2010). Derivation of a Schrödinger-like equation for elastic waves in granular media. *Granular Matter*, Vol. 12, No. 4 (July 2010), 417–436

Vitelli, V., Xu, N., Wyart, M., Liu, A. J. & Nagel, S. R. (2010). Heat transport in model jammed solids. *Phys. Rev. E*, Vol. 81, No. 2 (February 2010), 021301

Weaver, R. L. (1990). Diffusivity of ultrasound in polycrystal. *J. Mech. Phys. Solids*, Vol. 38, No. 1, 55–86

6

Interface Waves

Hefeng Dong and Jens M. Hovem
Norwegian University of Science and Technology
Norway

1. Introduction

The word *acoustics* originates from the Greek word meaning "to listen." The original meaning concerned only hearing and sound perception. The word has gradually attained an extended meaning and, in addition to its original sense, is now commonly used for almost everything connected with rapidly varying mechanical vibrations, from noise to seismic and sonar systems, to ultrasound in medical diagnosis and materials technology. An important technical application of acoustics is related to undersea activities, where acoustic waves are used in much the same way that radar and electromagnetic waves are used on land and in the air—for the detection and location of objects, and for communications. The reason that acoustic rather than electromagnetic waves are used in seawater is simple: electromagnetic waves are strongly attenuated in salt water and would, therefore, have too short a range to be useful for most applications.

The objective of this chapter is to give an introduction to interface waves and the use of the interface waves to estimate shear wave (also called S-wave) speed in the sediments. Knowledge of the S-wave speed profile of seabed sediments is important for seafloor geotechnical applications, since S-wave speed provides a good indicator of sediment rigidity, as well as for sediment characterization, seismic exploration, and geohazard assessment. In addition, for environments with high seabed S-wave speeds, S-wave conversion from the compressional wave (also called P-wave) at the seafloor can represent an important ocean acoustic loss mechanism which must be accounted for in propagation modelling and sonar performance predictions. This chapter serves as a basic introduction to acoustic remote sensing of the seabed's structure and composition. In addition to the basic concepts, the chapter also presents technical subjects such as experimental set up for excitation and recording of the interface waves and techniques for using interface waves to estimate the seabed geoacoustic parameters. Particular attention is devoted to an understanding and an explanation of the experimental problems involved with the generation, reception and processing of interface waves.

The chapter is organized as follows. Section 2 introduces acoustic wave propagating in fluids and gases and elastic wave propagating in solid media which support both P-wave and S-wave. Then polarization of S-waves is discussed. Section 3 is devoted to introduce interface waves and their properties. Section 4 presents techniques for using interface waves to estimate the seabed geoacoustic parameters for applications of geotechnical engineering in offshore construction and geohazard investigation. Different signal processing methods for extracting the dispersion curves of the interface waves and inversion schemes are presented. Examples for the inversion are illustrated. Section 5 contains the conclusions.

2. Acoustic and elastic waves

Acoustic waves are mechanical vibrations. When an acoustic wave passes through a substance, it causes local changes in the density that is related to local displacements of mass about the rest positions of the particles in the medium. This displacement leads to the formation of forces that act to restore the density to the equilibrium state, and move the particles back to their rest positions. The medium may be a gas, a fluid, or a solid material.
The basic equations of acoustics are obtained by considering the equations for an inviscid and compressible fluid. In the following these equations are expressed with the notation that p is the pressure, ρ is density and $\mathbf{u}$ is particle displacement. The particle velocity is the derivative of the displacement with respect to time $\mathbf{v} = \dot{\mathbf{u}}$.
The acoustic wave equation for fluids and gases is derived by the application of three simple principles.

- The momentum equation also known as Euler's equation
- The continuity equation, or conservation of mass
- The equation of state: the relationship between changes in pressure and density or volume

Euler's equation is expressed by

$$\rho\left[\frac{\partial \mathbf{v}}{\partial t} + \mathbf{v}\cdot\nabla\mathbf{v}\right] = -\nabla p \,, \tag{1}$$

which is an extension of Newton's second law that states that force equals the product of mass and acceleration. The extension is the second left-hand term in Equation (1) which represents the change in velocity with position for a given time instant, while the first term describes the change with time at a given position. The conservation of mass implies that the net changes in the mass, which result from its flow through the element, must be equal to the changes in the density of the mass of the element. This is expressed by the **continuity equation**

$$\frac{\partial \rho}{\partial t} = -\nabla\cdot(\rho\mathbf{v}) \,. \tag{2}$$

An **equation of state** is required to give a relationship between a change in density and a change in pressure taking into consideration the existing thermodynamic conditions. Assuming that the passage of an acoustic wave is nearly an adiabatic and reversible process the equation of stat may be formulated as pressure as a function of density:

$$p = p(\rho) \,. \tag{3}$$

Equations (1), (2), and (3) are all nonlinear. Applying linearization to these equations and combining them the acoustic wave equation can be obtained

$$\nabla^2 p - \frac{1}{c^2}\frac{\partial^2 p}{\partial t^2} = 0 \,, \tag{4}$$

where ∇^2 is Laplace operator and c is the sound speed at the ambient conditions, which is defined as:

$$c = \sqrt{\frac{K}{\rho}} \, . \tag{5}$$

Thus the sound speed is given by the square root of the ratio between volume stiffness or bulk modulus K, which has the same dimension as pressure expressed in N/m^2 or in pascal (Pa) and density, and the dimension of density is kg/m^3. Both the volume stiffness and density are properties of the medium, and therefore depend on external conditions such as pressure and temperature. Therefore the sound speed is a local parameter, which may vary with the location, for instance, when the sound speed varies with the depth in the water. Equation (4) gives the wave equation for sound pressure. After linearization, the particle velocity is obtained from Newton's second law

$$\frac{\partial \dot{\mathbf{u}}}{\partial t} = -\frac{1}{\rho} \nabla p \, , \tag{6}$$

and the particle displacement satisfies the wave equation

$$\nabla(\nabla \cdot \mathbf{u}) - \frac{1}{c^2} \frac{\partial^2 \mathbf{u}}{\partial t^2} = 0 \, . \tag{7}$$

It is often convenient to describe the particle displacement by a scalar variable as

$$\mathbf{u} = \nabla \phi \, , \tag{8}$$

ϕ is the displacement potential, which also satisfies the wave equation:

$$\nabla^2 \phi - \frac{1}{c^2} \frac{\partial^2 \phi}{\partial t^2} = 0 \, . \tag{9}$$

The sound pressure can be expressed by the displacement potential

$$p = -\rho \frac{\partial^2 \phi}{\partial t^2} \, . \tag{10}$$

By Fourier transformation, the wave equation is transformed from time domain to frequency domain:

$$\Phi(\mathbf{r}, \omega) = \int_{-\infty}^{+\infty} \phi(\mathbf{r}, t) \exp(i\omega t) dt \, , \tag{11}$$

and back to time domain by the inverse transformation

$$\phi(\mathbf{r}, t) = \frac{1}{2\pi} \int_{-\infty}^{+\infty} \Phi(\mathbf{r}, \omega) \exp(-i\omega t) d\omega \, . \tag{12}$$

The wave equation for the displacement potential may be expressed in frequency domain as:

$$\left[\nabla^2 + \kappa^2(\mathbf{r})\right] \Phi(\mathbf{r}, \omega) = 0 \, , \tag{13}$$

where the wave number $\kappa(\mathbf{r})$ is defined as

$$\kappa(\mathbf{r}) = \frac{\omega}{c(\mathbf{r})} . \tag{14}$$

Equation (13) is the Helmholtz equation, which is often easier to solve than the corresponding wave equation in time domain.
A fluid medium can only support pressure or compressional waves also called P-waves or longitudinal waves with particle displacement in the direction of the wave propagation. A solid medium can in addition also support transverse waves or S-waves with particle displacement perpendicular to the direction of wave propagation. The wave equation in solid medium is given as:

$$\rho \frac{\partial^2 \mathbf{u}}{\partial t^2} = (\lambda + 2\mu)\nabla(\nabla \cdot \mathbf{u}) + \mu \nabla^2 \mathbf{u} . \tag{15}$$

In this wave equation, λ and μ are Lamé elasticity coefficients, ρ is the density of the medium, and $\mathbf{u}$ is the particle displacement vector with components u_x, u_y and u_z. It is often convenient to recast equation (14) expressing the particle displacement vector by two potential functions, a scalar potential ϕ and a vector potential $\mathbf{\Psi}$. The particle displacement vector is then expressed as:

$$\mathbf{u} = \nabla\phi + \nabla \times \mathbf{\Psi} . \tag{16}$$

Inserting equation (16) into equation (15) yields

$$\rho\left(\nabla \frac{\partial \phi^2}{\partial t^2} + \nabla \times \frac{\partial^2 \mathbf{\Psi}}{\partial t^2} \right) = (\lambda + \mu)\nabla\left[\nabla \cdot \nabla\phi + \nabla \cdot (\nabla \times \mathbf{\Psi})\right] \\ + \mu \nabla^2 (\nabla\phi + \nabla \times \mathbf{\Psi}) \quad . \tag{17}$$

By definition, $\nabla \cdot (\nabla \times \mathbf{\Psi}) = 0$. In equation (17), the terms containing ϕ and $\mathbf{\Psi}$ are independently selected to satisfy the respective parts of equation (17). This results in the following two wave equations:

$$\rho \frac{\partial^2 \phi}{\partial t^2} = (\lambda + 2\mu)\nabla^2 \phi , \tag{18}$$

$$\rho \frac{\partial^2 \mathbf{\Psi}}{\partial t^2} = \mu \nabla^2 \mathbf{\Psi} . \tag{19}$$

From Equation (18), we observe that the scalar potential ϕ propagates at a speed, called P-wave speed c_p, defined as:

$$c_p = \sqrt{\frac{(\lambda + 2\mu)}{\rho}} = \sqrt{\frac{H}{\rho}} . \tag{20}$$

The vector potential $\mathbf{\Psi}$ of equation (19) propagates with the S-wave speed c_s, defined as:

$$c_s = \sqrt{\frac{\mu}{\rho}} \,. \tag{21}$$

The ratio between the two wave speeds defined by equations (20) and (21) is given by the Poisson ratio ν as:

$$\frac{c_s}{c_p} = \sqrt{\frac{1-2\nu}{2(1-\nu)}} \,. \tag{22}$$

After inserting the two wave speeds into equations (18) and (19), respectively, the two wave equations are rewritten as

$$\nabla^2\phi = \frac{1}{c_p^2}\frac{\partial^2\phi}{\partial t^2} \,, \tag{23}$$

$$\nabla^2\mathbf{\Psi} = \frac{1}{c_s^2}\frac{\partial^2\mathbf{\Psi}}{\partial t^2} \,. \tag{24}$$

Equations (23) and (24) are the two wave equations relevant to acoustic-seismic wave propagation in an isotropic elastic medium. In a boundless, non-absorbing, homogeneous and isotropic solid these two types of body waves propagate independently of each other with speeds given by (20) and (21), respectively. In inhomogeneous media with space-dependent parameters, for instance at an interface between two different media, conversions between P-wave and S-wave take place, and vice versa.

In many applications we are only interested in a two-dimensional case in which the particle movements are in the x-z plane and where there is no y-plane dependency. S-waves that are polarized so that the particle movement is in the x-z plane are called vertically polarized S-waves or SV waves. In general, S-waves are both vertically and horizontally polarized. The horizontal polarized S-waves are also called SH waves. However, in most underwater acoustic applications, we only need to consider vertically polarized S-waves since these are the waves that may be excited in the bottom by a normal volume source in the water column.

An incident P-wave in a fluid medium at an interface between the fluid and a solid medium generates a reflected P-wave in the fluid and two transmitted waves: one P-wave and one S-wave. An incident P-wave at an interface between two solid media generates reflected P-wave and S-wave in the incident medium and transmitted P-wave and S-wave in the second medium. In any case the reflected and transmitted waves are determined by the boundary conditions, which require that the normal stress, normal particle displacement, tangential stress, and tangential particle displacement are continuous at the interface. In the fluid, the tangential stress is zero and there is no constraint on the tangential particle displacement.

3. Interface waves

In this section we introduce interface waves and their properties (Rauch, 1980). The simplest type of interface wave is the well-known Rayleigh wave, which can propagate along a free surface of a solid medium and has a penetration depth of about one wavelength of the Rayleigh wave. A Scholte wave is another wave of the same type that can propagate at a fluid/solid interface and its decay inside the solid is comparable with that of the Rayleigh wave. The penetration depth in the fluid remains small when the adjacent solid is very soft, that is when the S-wave speed in the solid is smaller than the sound speed in the fluid. This is the situation for most water/unconsolidated-sediment combinations. But the penetration depth can be much larger if the S-wave speed in the solid is larger than the sound speed in the fluid, as is normally the case for all water/rock combinations. The most complicated type of interface wave is the well-known Stoneley wave, which can occur at the interface between two solid media for only limited combinations of parameters. Its penetration depth into each of the solid media is similar to that of the Rayleigh wave. The existence of the interface waves discussed above requires that at least one of the two media is a solid while the other medium may be a vacuum, air, a fluid or a solid. Love wave is another type of interface wave which is related to SH wave polarized parallel to a given interface and propagates within solid layers. It is guided by a free surface or a fluid/solid interface (Love, 1926; Sato, 1954).

3.1 Scholte wave

To give some insight into the physics of the interface problem we give a brief mathematical description of a Scholte wave propagating along the interface between two homogeneous, isotropic and non-dissipative half-spaces. The results give an idea of the pertinent propagation mechanism. We consider the situation depicted in Figure 1, where the water ($z<0$) has the sound speed c_0 and density ρ_0. The sea bottom is considered as a solid medium ($z>0$) with P- and S- wave speeds c_{p1} and c_{s1}, and its density is ρ_1.

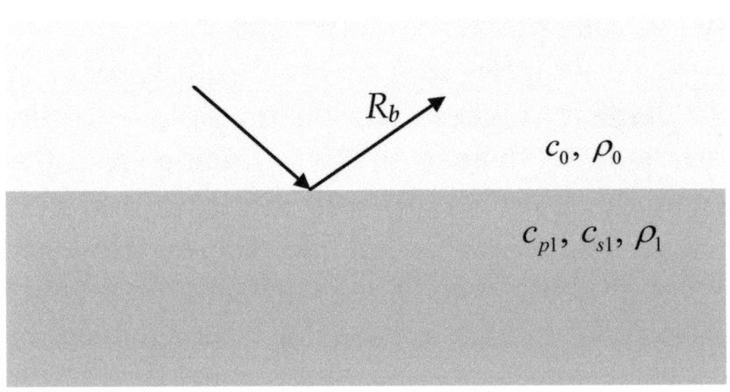

Fig. 1. Wave propagation in a half-space water column over a homogeneous half-space solid bottom; R_b is the reflection coefficient of the bottom.

Since the water depth is infinite there is no reflection from the sea surface. The reflected acoustic wave field is determined by the reflection coefficient at an interface between the water and the solid half-space and given as an integral over horizontal wavenumber k (Hovem, 2011):

$$\Phi_R(x,z,\omega) = \int_{-\infty}^{\infty} \frac{S(\omega)}{4\pi i\gamma_{p0}} R_b(k)\exp\left(-i\gamma_{p0}(z+z_s)\right)\exp(ikx)dk\,, \tag{25}$$

where $\Phi_R(x,z,\omega)$ is the reflected wave field due to a point source with frequency ω and source strength $S(\omega)$ at depth z_s, γ_{p0} is the vertical wave number and $R_b(k)$ is reflection coefficient.

Consider a plane, monochromatic wave of angular frequency $\omega = 2\pi f$ propagating in the $+x$ direction – the problem becomes two-dimensional (no y-coordinate dependency). Therefore the particle displacement has only two components $\mathbf{u} = (u_x, u_z)$ and the vector potential has only one component $\boldsymbol{\psi} = (0, \psi, 0)$. The two components of the particle displacement in equation (16) are then defined as:

$$u_x = \frac{\partial \phi}{\partial x} - \frac{\partial \psi}{\partial z}, \tag{26a}$$

$$u_z = \frac{\partial \phi}{\partial z} + \frac{\partial \psi}{\partial x}. \tag{26b}$$

The components of the stress expressed by the potentials are

$$\sigma_{xx0} = \sigma_{zz0} = -p = \lambda_0 \left(\frac{\partial^2 \phi_0}{\partial x^2} + \frac{\partial^2 \phi_0}{\partial z^2} \right), \tag{27a}$$

$$\sigma_{xz0} = 0 \tag{27b}$$

in the water, and

$$\sigma_{xx1} = (\lambda_1 + 2\mu_1) \left(\frac{\partial^2 \phi_1}{\partial x^2} + \frac{\partial^2 \phi_1}{\partial z^2} \right) - 2\mu_1 \left(\frac{\partial^2 \phi_1}{\partial z^2} + \frac{\partial^2 \psi_1}{\partial x \partial z} \right) \tag{28a}$$

$$\sigma_{xz1} = \mu_1 \left(2\frac{\partial^2 \phi_1}{\partial x \partial z} + \frac{\partial^2 \psi_1}{\partial x^2} - \frac{\partial^2 \psi_1}{\partial z^2} \right), \tag{28b}$$

$$\sigma_{zz1} = (\lambda_1 + 2\mu_1) \left(\frac{\partial^2 \phi_1}{\partial x^2} + \frac{\partial^2 \phi_1}{\partial z^2} \right) - 2\mu_1 \left(\frac{\partial^2 \phi_1}{\partial x^2} - \frac{\partial^2 \psi_1}{\partial x \partial z} \right) \tag{28c}$$

in the bottom. The boundary conditions at the interface between the water and the solid bottom at $z = 0$ are

$$u_{z0} = u_{z1} \tag{29a}$$

$$-p = \sigma_{zz1}. \tag{29b}$$

$$0 = \sigma_{xz1} \tag{29c}$$

Assuming the displacement potentials of the form:

$$\phi_0 = A\exp(\alpha_{p0} z)\exp[i(kx - \omega t)] \; (z \le 0), \tag{30}$$

$$\phi_1 = B\exp(-\alpha_{p1} z)\exp[i(kx - \omega t)] \; (z \ge 0) \tag{31a}$$

$$\psi_1 = C\exp(-\alpha_{s1} z)\exp[i(kx - \omega t)] \; (z \ge 0). \tag{31b}$$

The potentials have to fulfil the wave equations:

$$\nabla^2\phi_0 + \kappa_0^2\phi_0 = 0\ . \tag{32}$$

in the water, and

$$\nabla^2\phi_1 + \kappa_{p1}^2\phi_1 = 0\ , \tag{33}$$

$$\nabla^2\psi_1 + \kappa_{s1}^2\psi_1 = 0\ , \tag{34}$$

in the solid bottom. Since the horizontal wave number, *k*, is the same for all waves at the interface, the vertical wave numbers describing the vertical decays of the fields have to be

$$\alpha_{p0} = -i\gamma_{p0} = \sqrt{k^2 - \kappa_{p0}^2} \tag{35a}$$

$$\alpha_{p1} = -i\gamma_{p1} = \sqrt{k^2 - \kappa_{p1}^2}\ , \tag{35b}$$

$$\alpha_{s1} = -i\gamma_{s1} = \sqrt{k^2 - \kappa_{s1}^2} \tag{35c}$$

where

$$k = \frac{\omega}{v_p},\ \kappa_{p0} = \frac{\omega}{c_{p0}},\ \kappa_{p1} = \frac{\omega}{c_{p1}},\ \kappa_{s1} = \frac{\omega}{c_{s1}}\ . \tag{36}$$

are the horizontal wave number, the wave numbers for the P-wave in the water and the P- and S-waves in the bottom, respectively, and v_p is the phase speed. The use of the boundary conditions Equation (29) leads to a set of three equations for the amplitudes A, B, and C.

$$\begin{pmatrix} 0 & 2ik\alpha_{p1} & (k^2+\alpha_{s1}^2) \\ \lambda_0(k^2-\alpha_{p0}^2) & (\lambda_1+2\mu_1)\alpha_{p1}^2 - \lambda_1 k^2 & -2i\mu_1 k\alpha_{s1} \\ \alpha_{p0} & \alpha_{p1} & -ik \end{pmatrix} \begin{pmatrix} A \\ B \\ C \end{pmatrix} = \begin{pmatrix} 0 \\ 0 \\ 0 \end{pmatrix}. \tag{37}$$

The relationship between these amplitudes are given by

$$B = \frac{\left(k^2+\alpha_{s1}^2\right)}{\kappa_{s1}^2}\frac{\alpha_{p0}}{\alpha_{p1}}A \tag{38a}$$

$$C = \frac{-2ik\alpha_{p0}}{\kappa_{s1}^2}A \tag{38b}$$

The set of homogeneous, linear equations (37) has a non-trivial solution only if the coefficient determinant is vanishing, which results in:

$$4\frac{\alpha_{p1}\alpha_{s1}}{k^2} - \left(2 - \frac{\omega^2}{k^2 c_{s1}^2}\right)^2 = \frac{\rho_0}{\rho_1}\left(\frac{\omega}{kc_{s1}}\right)^4 \frac{\alpha_{p1}}{\alpha_{p0}}\ . \tag{39}$$

Inserting equations (35) and (36) into equation (39), we get the expression for the phase speed of the Scholte wave

$$4\sqrt{1-\left(\frac{v_p}{c_{p1}}\right)^2}\sqrt{1-\left(\frac{v_p}{c_{s1}}\right)^2}-\left(2-\frac{v_p^2}{c_{s1}^2}\right)^2=\frac{\rho_0}{\rho_1}\left(\frac{v_p}{c_{s1}}\right)^4\frac{\sqrt{1-\left(\frac{v_p}{c_{p1}}\right)^2}}{\sqrt{1-\left(\frac{v_p}{c_{p0}}\right)^2}}. \tag{40}$$

Equation (40) has always one positive real root, which is the Scholte wave $v_p = v_{Sch}$ and can be found numerically.

In the general situation with finite water depth, D, the sound propagates as in a waveguide by reflections from both the sea surface and the bottom. The sound field in a waveguide is given by an integral over horizontal wave numbers (Equation (17.66) in Hovem, 2011). The solution of this integral is approximately found by using the residue technique as the sum of the residues at the poles of the integrand. The poles are given by the zeros of the denominator of the integrand

$$1-R_bR_s\exp(-2\alpha_{p0}D)=0\ , \tag{41}$$

where R_s and R_b are reflection coefficients of the sea surface and the bottom, respectively. Assume that at the sea surface R_s= -1 and the poles are given as the solution to

$$1+R_b\exp(-2\alpha_{p0}D)=0\ . \tag{42}$$

Using the expression of the reflection coefficient of the bottom (Equation (15.42) in Hovem, 2011) we can get

$$4\frac{\alpha_{p1}\alpha_{s1}}{k^2}-\left(2-\frac{\omega^2}{k^2c_{s1}^2}\right)^2=\frac{\rho_0}{\rho_1}\frac{\alpha_{p1}}{\alpha_{p0}}\left(\frac{\omega}{kc_{s1}}\right)^4\tanh\left(\alpha_{p0}D\right)\ . \tag{43}$$

By applying equations (35) and (36), this expression can be transformed into

$$\begin{aligned}&4\sqrt{1-\left(\frac{v_p}{c_{p1}}\right)^2}\sqrt{1-\left(\frac{v_p}{c_{s1}}\right)^2}-\left(2-\frac{v_p^2}{c_{s1}^2}\right)^2\\&=\frac{\rho_0}{\rho_1}\left(\frac{v_p}{c_{s1}}\right)^4\frac{\sqrt{1-\left(\frac{v_p}{c_{p1}}\right)^2}}{\sqrt{1-\left(\frac{v_p}{c_{p0}}\right)^2}}\tanh\left(\frac{\omega D}{v_p}\sqrt{1-\left(\frac{v_p}{c_{p0}}\right)^2}\right).\end{aligned} \tag{44}$$

Equation (44) is the dispersion equation for the case with finite water depth and we see that, when $D\rightarrow\infty$, this expression becomes identical to the expression of equation (40) for the

infinite water depth. The dispersion equation gives the phase speed as function of frequency for given media parameters and layer thickness.
The dispersion equation can be solved numerically. While the phase speed v_p is found as the numerical solution of Equation (44), the group speed v_g can be found by differentiation - that is, by taking the derivative - and is expressed as:

$$v_g = \frac{d\omega}{dk} = \frac{v_p}{1 - \frac{\omega}{v_p}\frac{dv_p}{d\omega}} . \tag{45}$$

Figure 2 shows an example of such a numerical solution using the geoacoustic parameter values ρ_0 =1000 kg/m^3, ρ_1 =2000 kg/m^3, c_{p0}=1500 m/s, c_{p1}=2500 m/s, and c_{s1}=400 m/s. Since the frequency appears only in a product with the water depth D, the speed must be a function of the product of f and D. Notice that the phase speed of the interface wave is slightly lower than the S-wave speed, and that the phase speed decreases slightly with increasing frequency. This means that the interface wave is dispersive in the general case. In the limiting case, where the water depth is infinite, the phase speed of the interface wave is approximately 90% of the S-wave speed in the bottom, while with a zero water depth the speed is somewhat higher, about 95% of the bottom S-wave speed.

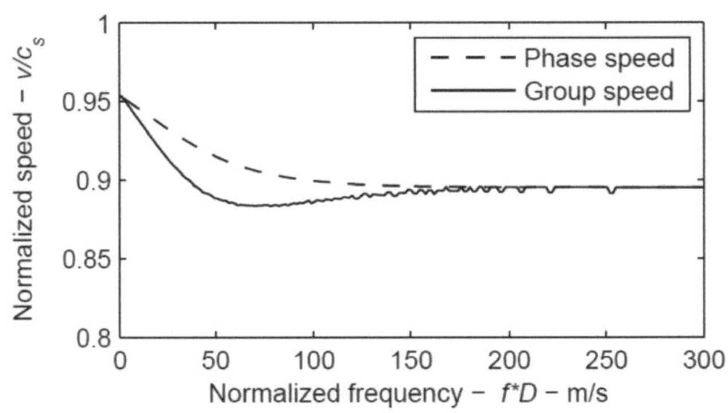

Fig. 2. The phase and group speeds of an interface wave, relative to the S-wave speed, expressed as a function of the frequency-thickness product f^*D for the numerical values given in the text.

The interface wave at a boundary between vacuum or air and a solid is called Rayleigh wave. The equation for the Rayleigh wave speed, v_R, is obtained from Equation (40) by setting $\rho_0 = 0$ or from Equation (44) by setting D = 0, resulting in the dispersion equation:

$$4\sqrt{1-\left(\frac{v_p}{c_{p1}}\right)^2}\sqrt{1-\left(\frac{v_p}{c_{s1}}\right)^2} - \left(2-\frac{v_p^2}{c_{s1}^2}\right)^2 = 0 . \tag{46}$$

Equation (46) has always one positive real solution $v_p = v_R < c_{s1}$ (Rauch, 1980; Brekhovskikh, 1960). The phase speed of the Rayleigh wave is frequency-independent and can be approximated to high accuracy by a simple formula (Rauch, 1980):

$$v_R = \frac{0.87 + 1.12\nu}{1+\nu} c_{s1} \,. \tag{47}$$

where ν is the Poisson's ratio. The phase speed of the Rayleigh is approximately 95% of the S-wave speed, as can be seen in Figure 2. Thus, in a solid medium a measurement of the Rayleigh wave speed may also give an accurate measure of S-wave speed. The phase speed v_p of the Scholte wave is also frequency-independent, and always slightly smaller than the lowest speed occurring in any of the two bordering media i.e. $v_p < \min(c_{p0}, c_{s1})$.

With the use of hydrophones in the water or geophones on or in the bottom, one can detect and record the sound pressure within the water mass and the components of the particle displacement in the solid bottom. We will now determine the real components of the displacement vector of the Scholte wave. Using equations (26), (30), (31) and (36), the displacement components are expressed as

$$u_{x0} = \hat{u}_{x0}(k,z)\sin(kx - \omega t),\ (z \le 0) \tag{48a}$$
$$u_{z0} = \hat{u}_{z0}(k,z)\cos(kx - \omega t),\ (z \le 0)\ , \tag{48b}$$

where

$$\hat{u}_{x0}(k,z) = -kA\exp\left(\alpha_{p0}z\right) \tag{49a}$$
$$\hat{u}_{z0}(k,z) = -\alpha_{p0}A\exp\left(\alpha_{p0}z\right)' \tag{49b}$$

in the water, and

$$u_{x1} = \hat{u}_{x1}(k,z)\sin(kx - \omega t),\ (z \ge 0) \tag{50a}$$
$$u_{z1} = \hat{u}_{z1}(k,z)\cos(kx - \omega t),\ (z \ge 0)' \tag{50b}$$

where

$$\hat{u}_{x1}(k,z) = \frac{\alpha_{p0}kA}{\kappa_{s1}^2\alpha_{p1}}\left\{2\alpha_{p1}\alpha_{s1}\exp(-\alpha_{s1}z) - (k^2 + \alpha_{s1}^2)\exp(-\alpha_{p1}z)\right\} \tag{51a}$$
$$\hat{u}_{z1}(k,z) = \frac{\alpha_{p0}A}{\kappa_{s1}^2}\left\{2k^2\exp(-\alpha_{s1}z) - (k^2 + \alpha_{s1}^2)\exp(-\alpha_{p1}z)\right\}\ , \tag{51b}$$

in the bottom. Equations (48) - (51) are parametric representations of ellipses having their main axes parallel to the axes of the coordinate system. With increasing distance from the interface, the displacement amplitudes $\hat{u}_{x0}$, $\hat{u}_{z0}$ and $\hat{u}_{z1}$ decrease exponentially without changing sign. The horizontal displacement in the bottom, $\hat{u}_{x1}$, shows the same asymptotic behaviour, but with the sign changing at the depth of about one-tenth of the Scholte wavelength. Figure 3 plots the particle displacements as a function of depth relative to the Scholte wavelength λ_{Sch} in the water column (left panel) and the particle orbits (right panels) for a typical water/sediment interface. The same parameters are used as in Figure 2 and the frequency-thickness product $f^*D = 200$. The penetration depth in the water is about one-half of the Scholte wavelength. The right panels plot the particle movement at depth $z = 0.01\lambda_{Sch}$

(bottom right), $0.09\lambda_{Sch}$ (middle right) and $0.5\lambda_{Sch}$ (upper right) above the water/sediment interface. At all depths the particles follow retrograde elliptical movements. The ellipses are close to circular in this case since the eccentricity is close to zero. For harder sediment, the ellipses are more elongated. Figure 4 shows the same plots as in Figure 3 but for the particle displacements in the bottom. The penetration depth in the solid is larger than the wavelength of the Scholte wave. At depth $z = 0.01\lambda_{Sch}$ (upper right) the particles follow a retrograde elliptical movements, while at depth $z = 0.09\lambda_{Sch}$ (middle right) the particle movement follows a vertical line, and at depth $z = 0.5\lambda_{Sch}$ (middle right) the particle movement is a prograde ellipse.

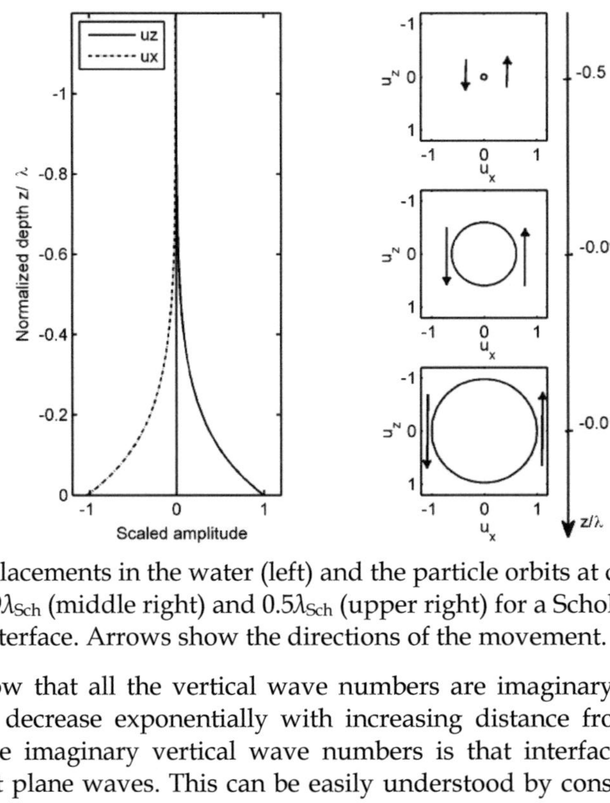

Fig. 3. Particle displacements in the water (left) and the particle orbits at depth $z = 0.01\lambda_{Sch}$ (bottom right), $0.09\lambda_{Sch}$ (middle right) and $0.5\lambda_{Sch}$ (upper right) for a Scholte wave at a water/sediment interface. Arrows show the directions of the movement.

Equations (35) show that all the vertical wave numbers are imaginary, and therefore the signal amplitudes decrease exponentially with increasing distance from the interface. A consequence of the imaginary vertical wave numbers is that interface waves cannot be excited by incident plane waves. This can be easily understood by considering the grazing angle of the wave in the uppermost medium. This angle is expressed as:

$$\cos\theta_0 = \frac{k}{\omega/c_{p0}} = \frac{c_{p0}}{v_p} > 1. \tag{52}$$

Equation (52) means that the angle θ_0 must be imaginary and, consequently, cannot be the incident angle of a propagating plane wave. However, the interface waves can be excited by a point source close to the interface, that is, as a near-field effect.

The interface waves are confined to a narrow stratum close to the interface, which means that they have cylindrical propagation loss (i.e., $1/r$) rather than spherical spreading loss (i.e., $1/r^2$), as would be true of waves from a point source located in a medium of infinite extent. Cylindrical spreading loss indicates that, once an interface wave is excited, it is likely

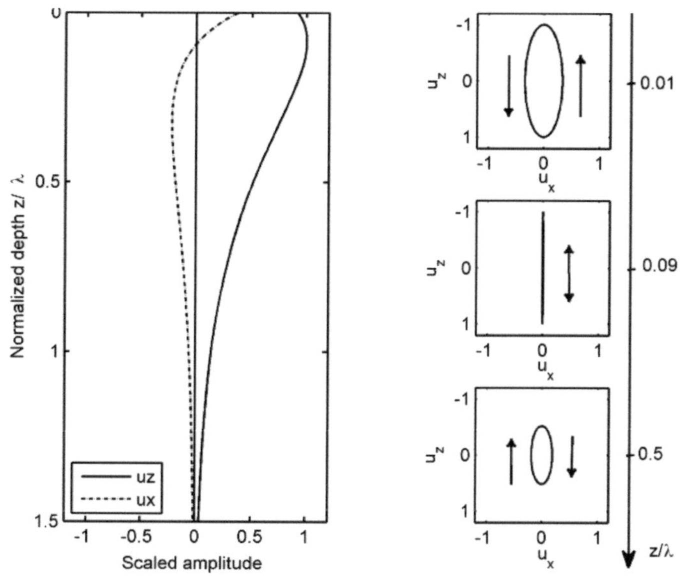

Fig. 4. Particle displacements in the bottom (left) and the particle orbits at depth $z = 0.01\lambda_{Sch}$ (upper right), $0.09\lambda_{Sch}$ (middle right) and $0.5\lambda_{Sch}$ (bottom right) for a Scholte wave at a water/sediment interface. Arrows show the directions of the movement.

to dominate other waves that experience spherical spreading at long distances. This effect is familiar from earthquakes, where exactly this kind of interface wave, the Rayleigh wave, often causes the greatest damage.

4. Applications of interface waves

Knowledge of S-wave speed is important for many applications in underwater acoustics and ocean sciences. In shallow waters the bottom reflection loss, caused by absorption and shear wave conversion, represents a dominating limitation to low frequency sonar performance. For construction works in water, geohazard assessment and geotechnical studies the rigidity of the seabed is an important parameter (Smith, 1986; Bryan & Stoll, 1988; Richardson et al., 1991; Stoll & Batista, 1994; Dong et al., 2006, WILKEN et al., 2008; Hovem et al., 1991).

In some cases the S-wave speed and other geoacoustic properties can be acquired by *in-situ* measurement, or by taking samples of the bottom material with subsequent measurement in laboratories. In practice this direct approach is often not sufficient and has to be supplemented by information acquired by remote measurement techniques in order to obtain the necessary area coverage and the depth resolution.

The next section presents a convenient and cost-effective method for how the S-wave speed as function of depth in the bottom can be determined from measurements of the dispersion properties of the seismo-acoustic interface waves (Caiti et al., 1994; Jensen & Schmidt, 1986; Rauch, 1980).

First the experimental set up for interface wave excitation and reception is presented. Data processing for interface wave visualization is given. Then the methods for time-frequency analysis are introduced. The different inversion approaches are discussed. All the presented methods are applied to some real data collected in underwater and seismic experiments.

4.1 Experimental setup and data collection

In conventional underwater experiments both the source and receiver array are deployed in the water column. In order to excite and receive interface waves in underwater environment the source and receivers should be located close, less than one wavelength of the interface wave, to the bottom. The interface waves can be recorded both by hydrophones, which measure the acoustic pressure, and 3-axis geophones measuring the particle velocity components. In most cases an array of sensors, hydrophones and geophones are used. The spacing between the sensors is required to be smaller than the smallest wavelength of the interface waves in order to fulfil the sampling theorem for obtaining the phase speed dispersion. Low frequency sources should be used in order to excite the low frequency components of the interface waves since the lower frequency components penetrate deeper into the sediments and can provide shear information of the deeper layers. The recording time should be long enough to record the slow and dispersive interface waves. Due to the strong reverberation background and ocean noise the seismic interface waves may be too weak to be observed even if excited. In order to enhance the visualization of interface waves one needs to pre-process the data. The procedure includes three-step: low pass filtering for reducing noise and high-frequency pulses, time-variable gain, and correction of geometrical spreading (Allnor, 2000).

Figure 5 illustrates an experimental setup for excitation and reception of interface wave from a practical case in a shallow water (18 m depth) environment. Small explosive charges were used as sound sources and the signals were received at a 24-hydrophone array positioned on the seafloor; the hydrophones were spaced 1.5 m apart at a distance of 77 – 111.5 m from the source.

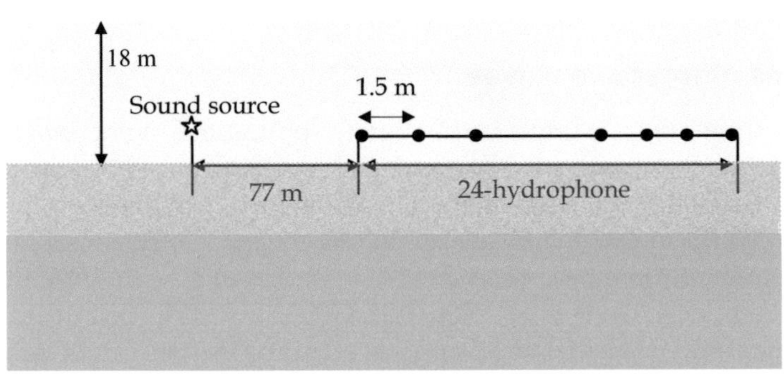

Fig. 5. Experimental setup for excitation and reception of interface waves by a 24-hydrophone array situated on the seafloor.

The 24 signals received by the hydrophone array are plotted in Figure 6. The left panel shows the raw data with the full frequency bandwidth. The middle panel shows the zoomed version of the same traces for the first 0.5 s. The first arrivals are a mixture of refracted and direct waves. In the right panel the raw data have been low pass filtered, which brings out the interface waves. In this case the interface waves appear in the 1.0 - 2.5 s time interval illustrated by the two thick lines. The slopes of the lines with respect to time axis give the speeds of the interface waves in the range of 40 m/s – 100 m/s with the higher-frequency components traveling slower than the lower-frequency components. This indicates that the S-wave speed varies with depth in the seafloor.

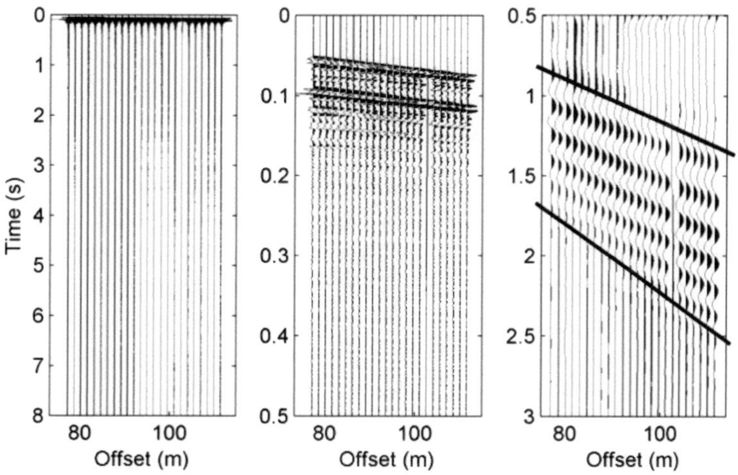

Fig. 6. Recorded and processed data of the 24-hydrophone array. Left panel: the raw data with full bandwidth; Middle panel: zoomed version of the raw data in a time window of 0.0 - 0.5 s. Right panel: low pass filtered data in a time window of 0.5 - 3.0 s.

4.2 Dispersion analysis

There are two classes of methods used for time-frequency analysis to extract the dispersion curve of the interface waves: single-sensor method and multi-sensor method (Dong et al., 2006). Single-sensor method, which can be used to study S-wave speed variations as function of distance (Kritski, 2002), estimates group speed dispersion of one trace at a time from

$$v_g = \frac{d\omega}{dk(\omega)}, \tag{53}$$

where v_g is group speed, ω angular frequency, and $k(\omega)$ wavenumber. This method requires the distance between the source and receiver to be known. The Gabor matrix (Dziewonski, 1969) is the classical method that applies multiple filters to single-sensor data for estimating group-speed dispersion curves. The Wavelet transform (Mallat, 1998) is a more recent method that uses multiple filters with continuously varying filter bandwidth to give a high-resolution group-speed dispersion curves and improved discrimination of the different modes. The sharpest images of dispersion curves are usually found with multi-sensor method (Frivik, 1998 & Land, 1987), which estimates phase-speed dispersion using multiple traces and the expression is given by

$$v_p = \frac{\omega}{k(\omega)}. \tag{54}$$

This method assumes constant seabed parameters over the length of the array. Conventionally, two types of multi-sensor processing methods are used for extracting phase-speed dispersion curves: frequency wavenumber (*f-k*) spectrum and slowness-frequency (*p-ω*) transform methods (McMechan, 1981). The former method requires regular spatial sampling, while the latter can be used with irregular spacing.

Alternatively, the Principal Components method (Allnor, 2000), uses high-resolution beamforming and the Prony method to determine the locations of the spectral lines corresponding to the interface mode in the wavenumber spectra. These wavenumber estimates are then transformed to phase speed estimates at each frequency using the known spacing between multiple sensors.

The low pass filtered data in the right panel in Figure 6 is analyzed by applying Wavelet transform to each trace to obtain the dispersion of group speed. The dispersion of trace number 10 is illustrated by a contour plot in Figure 7. The dispersion data are obtained by picking the maximum values along the each contour as indicated by circles. Only one mode, fundamental mode, is found in this case within the frequency range of 2.5 Hz - 10.0 Hz. The corresponding group speed is in the range of 50 m/s - 90 m/s, which gives a wavelength of 5.0 m - 36 m approximately. After each trace is processed, the dispersion curves of the group speed are averaged to obtain a "mean group speed", which is subsequently used as measured data to an inversion algorithm to estimate S-wave speed profile.

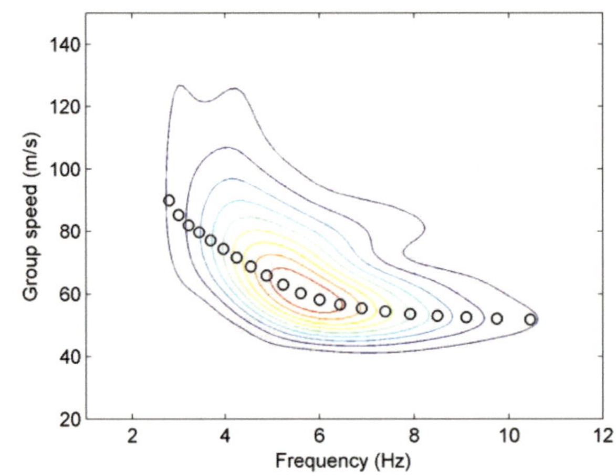

Fig. 7. Dispersion analysis showing estimated group speed as function of frequency in the form of a contour map of the time frequency analysis results. The circles are sampling of the data.

4.3 Inversion methods

The inverse problem can be qualitatively defined as: Given the dispersion data of the interface waves, determine the geoacoustic model of the seafloor that will predict the same dispersion curves. In a more formal way, the objective is to find a set of geoacoustic parameters **m** such that, given a known relation **T** between geoacoustic properties and dispersion data **d**,

$$\mathbf{T}(\mathbf{m}) = \mathbf{d} \,. \tag{55}$$

In general, this problem is nonlinear but we present only a linearized inversion scheme: the Singular Value Decomposition (SVD) of linear system (Caiti et al., 1996). The seafloor model is discretized in m layers, each characterized by thickness h_i, density ρ_i, P-wave speed c_{pi}, and S-wave speed c_{si}. The first simplifying assumption is that the seafloor is considered to be horizontally homogeneous, so that the geoacoustic parameters are only a function of depth in the sediment. The second simplifying assumption is that the dispersion of the interface wave at the water-sediment interface is only a function of S-wave speed of the bottom materials and the layering. The other geoacoustic properties are fixed and not changed during the inversion procedure since the dispersion is not sensitive to these parameters. These assumptions reduce the number of parameters to be estimated and the computational effort needed, but do not seriously affect the accuracy of the estimates.

The actual computation of the predicted dispersion of phase/group speed is performed with a standard Thomson-Haskell integration scheme (Haskell, 1953), which has the advantage of being fast and economical in terms of computer usage. However, different codes can be used to generate predictions without affecting the structure of the inversion algorithm. With the assumptions the model generates the dispersion of phase/group speed $\mathbf{v}_p \in \mathbf{R}^n$ as function of the S-wave speed $\mathbf{c}_s \in \mathbf{R}^m$:

$$\mathbf{T}\mathbf{c}_s = \mathbf{v}_p \,, \tag{56}$$

where Jacobian $\mathbf{T} \in \mathbf{R}^n \times \mathbf{R}^m$. Depending on the system represented by equation (55) is over- or underdetermined, its solution may not exist or may not be unique. So it is customary to look for a solution of (56) in the least square sense; that is, a vector $\mathbf{c_s}$ that minimizes $\left\|\mathbf{T}\mathbf{c}_s - \mathbf{v}_p\right\|^2$. Consider the most common case where $m < n$; that is, we have more data than parameters to be estimated. The least-square solution is found by solving the normal equation:

$$\mathbf{c}_s = (\mathbf{T}^T\mathbf{T})^{-1}\mathbf{T}^T\mathbf{v}_p \,. \tag{57}$$

Here $\mathbf{T}^T$ is the transpose conjugate of matrix **T**. By using the SVD to the rectangular matrix **T** the solution can be expressed as:

$$\mathbf{c}_s = \mathbf{W}\boldsymbol{\Sigma}^{-1}\mathbf{U}^T\mathbf{v}_p \,, \tag{58}$$

$$\mathbf{c}_s = \sum_{i=1}^{m} \frac{(\mathbf{u}_i^T \mathbf{v}_p)}{\sigma_i}\mathbf{w}_i = \sum_{i=1}^{m} \frac{\alpha_i}{\sigma_i}\mathbf{w}_i \,. \tag{59}$$

In equations (57), (58) and (59) $\mathbf{T}^T = \mathbf{W}[\boldsymbol{\Sigma}\ \mathbf{O}]\mathbf{U}^T$, **U** and **W** are unitary orthogonal matrices with dimension ($n\times n$) and ($m\times m$) respectively and $\boldsymbol{\Sigma}$ is a square diagonal matrix of dimension m, with diagonal entries σ_i called singular values of **T** with $\sigma_1>\sigma_2>\ldots>\sigma_m$; **O** is a zero matrix with dimension ($m\times(n-m)$); $\mathbf{u}_i$ is the ith column of **U** and $\mathbf{w}_j$ the jth column of **W**.

Since the matrix $\boldsymbol{\Sigma}$ is ill conditioned in the numerical solution of this inverse problem a technique called regularization is used to deal with the ill conditioning (Tikhonov & Arsenin, 1977). The regularized solution is given by:

$$\tilde{\mathbf{c}}_s = (\mathbf{T}^T\mathbf{T} + \lambda\mathbf{H}^T\mathbf{H})^{-1}\mathbf{T}^T\mathbf{v}_p \,. \tag{60}$$

H with dimension ($m\times m$) is a generic operator that embeds the *a priori* constraints imposed on the solution and regularization parameter $\lambda > 0$. The detailed discussion on regularization can be found in (Caiti et al., 1994). The regularized solution is given by

$$\tilde{\mathbf{c}}_s = T^{\dagger}\mathbf{v}_p \,, \tag{61}$$

with

$$\mathbf{T}^{\dagger} = \mathbf{W}(\boldsymbol{\Sigma} + \boldsymbol{\Sigma}^{-1}\lambda(\mathbf{HW})^T(\mathbf{HW}))^{-1}\mathbf{U}^T \,. \tag{62}$$

The inversion scheme described above is used to estimate S-wave speed profile by inverting the group-speed dispersion data shown in Figure 7. A 6-layered model with equal thickness is assumed to represent the structure of the bottom. The layer thickness, P-wave speeds and densities are kept constant during iterations, but the regularization parameter is adjustable. The inversion results are illustrated in Figure 8. The upper left panel plots the measured (circles) and predicted (solid line) group speed dispersion data. The measured data and predicted dispersion curve agree very well. The eigenvalues and eigenvectors of the Jacobian matrix **T** are plotted in the upper right and bottom right panels respectively. The eigenvalues to the left of the vertical line are larger than the value of the regularization parameter λ (the vertical line). The corresponding eigenvectors marked with black shading constitute the S-wave speed profile. The eigenvectors marked with gray shading give no contribution to the estimated S-wave speed since their eigenvalues are smaller than the regularization parameter. The bottom left panel presents the estimated S-wave speed versus depth (thick line) with error estimates (thin line). The error estimate was generated assuming an uncertainty of 15m/s in the group speed picked from Figure 7.

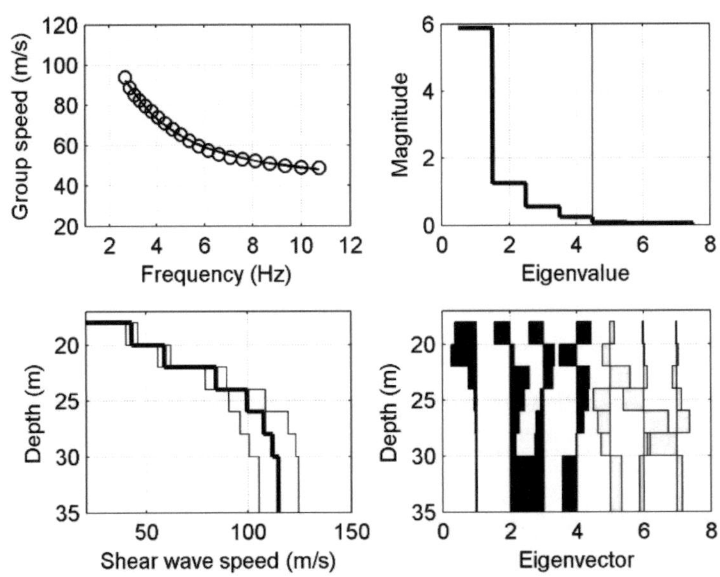

Fig. 8. Inversion results. Top left: measured (circles) and predicted (solid line) group speed dispersion; Top right: eigenvalues of matrix T and the value of the regularization parameter (vertical line). Bottom right: eigenvectors; Bottom left: estimated S-wave speed (thick line) and error estimates (thin line).

The estimated S-wave speed is 45 m/s in the top layer and increases to 115 m/s in the depth of 15 m below the seafloor, which corresponds to one-half of the longest wavelength at 3 Hz.

The errors are smaller in the top layer than that in the deeper layer. This can be explained by the eigenvalues and the behaviors of the corresponding eigenvectors. The eigenvectors with larger eigenvalues give better resolution, but penetrate only to very shallower depth, while the eigenvectors with smaller eigenvalues can penetrate deeper depth, but give relatively poor resolution.

Finally, we present another example to demonstrate the techniques for estimating S-wave speed profiles from measured dispersion curves of interface waves (Dong et al., 2006). The data of this example were collected in a marine seismic survey at a location where the water depth is 70 m. Multicomponent ocean bottom seismometers with 3-axis geophone and a hydrophone were used for the recording. The geophone measured the particle velocity components just below the water-sediment interface. The hydrophones were mounted just above the interface, and measured the acoustic pressure in the water. The receiver spacing was 28 m and the distance from the source to the nearest receiver was 1274 m. A set of data containing 52 receivers with vertical, v_z, and inline, v_x, components of the particle velocity are shown in the left two panels in Figure 9. In order to enhance the interface waves the recorded data are processed by low-pass filtering, time-variable gain and correction of geometrical spreading (Allnor, 2000). The processed data are plotted in the two right panels in Figure 9 where the slow and dispersive interface waves are clearly observed. The thick lines bracket the arrivals of the interface waves. The slopes of the lines with respect to the time-axis define the speeds of the interface waves. In this case the speeds appear to be in the range of 290 m/s - 600 m/s for the v_z component and 390 m/s - 660 m/s for the v_x component. The higher speed of v_x component is a consequence of the fact that the v_x component has weaker fundamental mode and stronger higher-order mode than v_z component, as can be observed in Figure 10.

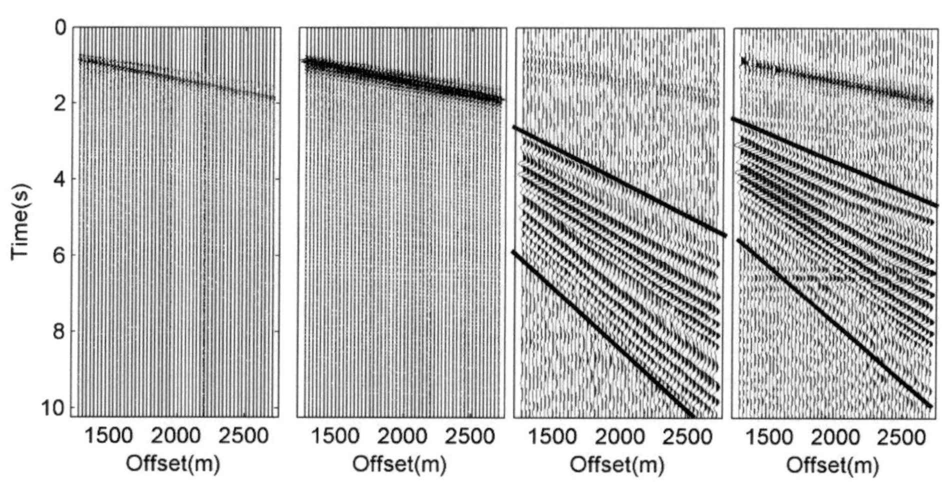

Fig. 9. Raw and processed data. From the left to the right: v_z and v_x components of raw and processed data. The thick lines in the processed data illustrate the arrivals of the interface waves and the slopes of the lines indicate the speed range of the interface waves.

The Principal Components method is applied to the processed data to obtain the phase speed dispersion. The extracted dispersion data of v_z (blue dots) and v_x (red dots) are plotted in Figure 10. The advantage by using multi-component data is that one can identify and separate different modes and obtain higher resolution. By combining both v_z and v_x dispersion data the final dispersion data are extracted and denoted by circles. There are four modes identified, but only the first two modes are used in the inversion algorithm for estimating the S-wave speed. Figure 10 shows that the lower frequency components of the higher-order mode have higher phase speed and therefore longer wavelength than that the higher frequency components of the lower-order mode have. In this case the phase speed of the first-order mode at 2 Hz is 550 m/s, which gives a wavelength of 270 m. A 12-layered model is assumed to represent the structure of the bottom with layer thickness increasing logarithmically with increasing depth. The layer thickness, P-wave speeds and densities are kept constant during iterations, but the regularization parameter is adjustable.

The inversion results are illustrated in Figure 11. The left panel shows the measured phase speed dispersion data (circles) and the predicted (solid line) phase speed dispersion curve. The right panel presents the estimated S-wave speed versus depth (thick line) with error estimates (thin line). The error estimates were generated assuming an uncertainty of 15m/s in the selection of phase speed from Figure 10. The match between the predicted and measured dispersion data is quite good for both the fundamental and the first-order modes. The estimated S-wave speed is 237 m/s in the top layer and increases up to 590 m/s in the depth of 250 m below the seafloor, which is approximately one of the longest wavelength at the frequency of 2.0 Hz. The results from the both examples indicate that the Scholte wave sensitivity to S-wave speed versus depth using multiple modes is larger than that using only fundamental mode.

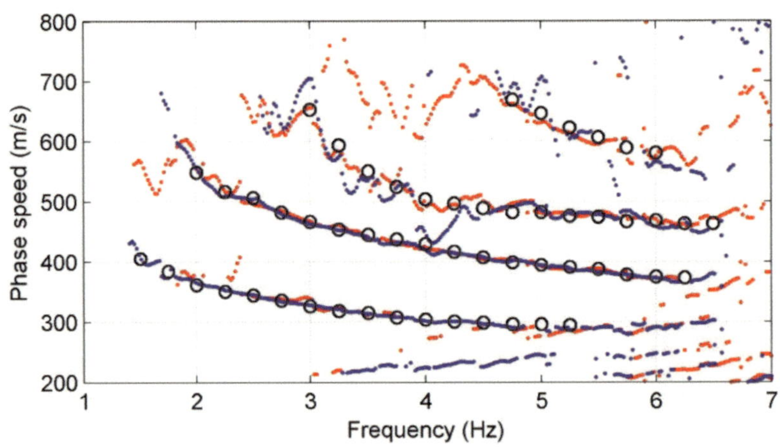

Fig. 10. Phase-speed dispersion of v_z (blue) and v_x (red) components. The circles are the sampling of the data.

Over the years considerable effort has been applied to interface-wave measurement, data processing, and inversion for ocean acoustics applications (Rauch, 1980; Hovem et al., 1991; Richardson, 1991; Caiti et al., 1994; Frivik et al., 1997; Allnor, 2000; Godin & Chapman, 2001; Chapman & Godin, 2001; Dong et al, 2006; Dong et al., 2010). Nonlinear inversion gives both quantitative uncertainty estimation and rigorous estimation of the data error statistics and of an appropriate model parameterization, and is not discussed here. The work on nonlinear inversion can be found in Ivansson et al. (1994), Ohta et al. (2008) and Dong & Dosso (2011). More recently Vanneste et al. (2011) and Socco et al. (2011) used a shear source deployed on

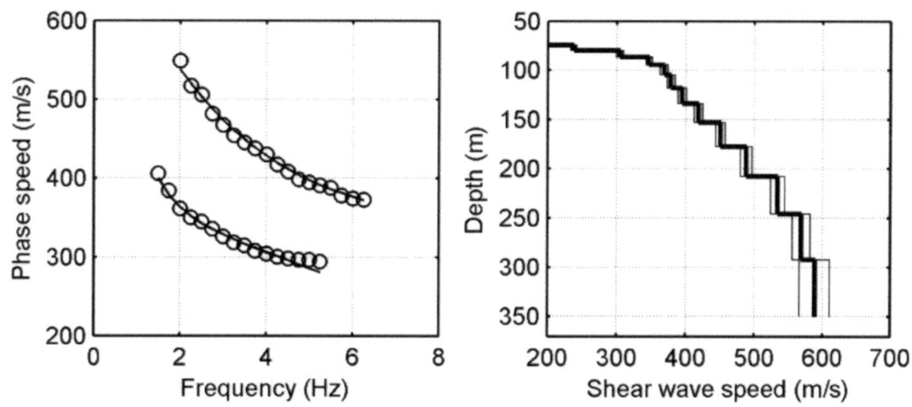

Fig. 11. Inversion results. Left: measured (circles) and predicted (solid line) phase speed dispersion data; Right: estimated S-wave speed versus depth (thick line) and the error estimates (thin line).

the seafloor to generate both vertical and horizontal shear waves in the seafloor. This enabled to measure both Scholte and Love waves and to inverse S-wave speed profile jointly, thereby obtaining information on anisotropy in the subsurface. Another and entirely different approach is based on using ocean ambient noise recorded by ocean bottom cable to extract information on the ocean subsurface. This approach has attracted much attention as being both economical and environmental friendly (Carbone et al., 1998; Shapiro et al., 2005; Bensen et al., 2007; Gerstoft et al., 2008; Bussat & Kugler, 2009; Dong et al., 2010).

5. Conclusions

In this chapter after briefly introducing acoustic and elastic waves, their wave equations and propagation, a detailed presentation on interface waves and their properties is given. The experimental set up for excitation and reception of interface waves are discussed. The techniques for using interface waves to estimate the seabed geoacoustic parameters are introduced and discussed including signal processing for extracting dispersion of the interface waves, and inversion scheme for estimating S-wave speed profile in the sediments. Examples with both hydrophone data and ocean bottom multicomponent data are analyzed to validate the procedures. The study and approaches presented in this chapter provide alternative and supplementary means to estimate the S-wave structure that is valuable for seafloor geotechnical engineering, geohazard assessment, seismic inversion and evaluation of sonar performance.

The work presented in this chapter is resulted from the authors' number of years of teaching and research on underwater acoustics at the Norwegian University of Science and Technology.

6. Acknowledgment

The authors would like to give thanks to Professor N. Ross Chapman, Professor Stan E. Dosso at the University of Victoria and our earlier colleague Dr. Rune Allnor for helpful discussions and collaboration.

7. References

Allnor, R. (2000). *Seismo-Acoustic Remote Sensing of Shear Wave Velocities in Shallow Marine Sediments,* PhD. Thesis, No. 420006, Norwegian University of Science and Technology, Trondheim, Norway

Brekhovskikh, L. M. (1960). *Waves in Layered media,* Academic Press, New York, N. Y. USA

Bryan, G. M. & Stoll, R. D. (1988). The Dynamic Shear Modulus of Marine Sediments, *J. Acoust. Soc. Am.,* Vol. 83, pp. 2159-2164

Bucker, H. P. (1970). Sound Propagation in a Channel with Lossy Boundaries, *J. Acoust. Soc. Am.* Vol. 48, pp. 1187-1194

Bussat, S. & Kugler, S. (2009). Recording Noise - Estimating Shear-Wave Velocities: Feasibility of Offshore Ambient- Noise Surface Wave Tomography (ANSWT) on a Reservior Scale, SEG Expanded Abstracts, pp. 1627-1631

Caiti, A., Akal, T. & Stoll, R.D. (1994). Estimation of Shear Wave Velocity in Shallow Marine Sediments, *IEEE J. Ocean. Eng.,* Vol. 19, pp. 58-72

Carbone, N. M., Deane, G. B. & Buckingham, M. J. (1998). Estimating the compressional and shear wave speeds of a shallow water seabed from the vertical coherence of ambient noise in the water column, *J. Acoust. Soc. Am.,* Vol. 103(2), pp. 801-813

Chapman, D. M. F. & Godin, O. A. (2001). Dispersion of Interface Waves in Sediments with Power-Law Shear Speed Profile. II. Experimental Observations and Seismo-Acoustic Inversion, *J. Acoust. Soc. Am.,* Vol. 110, pp. 1908-1916

Dong, H., Allouche, N., Drijkoningen, G. G. & Versteeg, W. (2010). Estimation of Shear-Wave Velocity in Shallow Marine Environment, *Proc. of the 10th European Conference on Underwater Acoustics,* Vol. 1, pp. 175-180, Akal, T (Ed.)

Dong, H, Hovem, J. M. & Kristensen, Å. (2006). Estimation of Shear Wave Velocity in Shallow Marine Sediment by Multi-Component Seismic Data: a Case Study, *Procs. of the 8th European Conference on Underwater Acoustics,* Vol. 2, pp. 497-502, Jesus, S. M. & Rodringuez, O. C. (Ed.)

Dong, H., Hovem, J. M. & Frivik, S. A. (2006). Estimation of Shear Wave Velocity in Seafloor Sediment by Seismo-Acoustic Interface Waves: a Case Study for Geotechnical Application, In: *Theoretical and Computational Acoustics,* pp. 33-43, World Scientific

Dong, H., Liu, L., Thompson, M. & Morton, A. (2010). Estimation of Seismic Interface Wave Dispersion Using Ambient Noise Recorded by Ocean Bottom Cable, *Proc. of the 10th European Conference on Underwater Acoustics,* Vol. **1**, pp. 183-188, Akal, T. (Ed.)

Dong, H. & Dosso S. E. (2011). Bayesian Inversion of Interface-Wave Dispersion for Seabed Shear-Wave Speed Profiles, *IEEE J. Ocean. Eng.,* Vol. 36 (1), pp. 1-11

Dziewonski, A.S., Bloch, S. & Landisman, M. A. (1969). A technique for the analysis of Transient Seismic Signals, *Bull. Seismol. Soc. Am.* Vol. 59, pp. 427-444

Frivik, S. A., Allnor, R. & Hovem, J. M. (1997). Estimation of Shear Wave Properties in the Upper Sea-Bed Using Seismo-Acoustical Interface Waves, In: *High Frequency Acoustics in Shallow Water,* pp. 155-162, N. G. Pace, E. Pouliquen, O Bergem, & Lyons, A. P. (Ed.), NATO SACLANT Undersea Research Centre, Italy

Gerstoft, P., Hodgkiss, W. S., Siderius, M., Huang, C. H. & Harrison, C. H. (2008). Passive Fathometer Processing, *J. Acoust. Soc. Am.,* Vol. 123, No. 3, pp. 1297-1305

Godin, O. A. & Chapman, D. M. F. (2001). Dispersion of Interface Waves in Sediments with Power-Law Shear Speed Profile. I. Exact and Approximate Analytic Results, *J. Acoust. Soc. Am.,* Vol. 110, pp. 1890-1907

Haskell, N. A. (1953). Dispersion of Surface Waves on Multilayered Media, *Bullet. Seismolog. Soc. of America,* Vol. 43, pp. 17-34

Hovem, J. M. (2011). *Marine Acoustics: The Physics of Sound in Underwater Environments,* Peninsula Publishing, ISBN 978-0-932146-65-6, Los Altos Hills, California, USA.

Hovem, J. M., Richardson, M. D. & Stoll, R. D. (1991), *Shear Waves in Marine Sediments,* Dordrecht: Kluwer Academic

Ivansson, S., Moren, P. & Westerlin, V. (1994). Hydroacoustical Experiments for the Determination of Sediment Properties, *Proc. IEEE Oceans94,* Vol. III, pp. 207-212

Jensen, F. B., & Schmidt, H. (1986). Shear Properties of Ocean Sediments Determined from Numerical Modelling of Scholte Wave Data, In: *Ocean Seismo-Acoustics, Low Frequency Underwater Acoustics,* Akal, T. & Berkson, J. M. (Ed.), pp. 683-692, Plenum Press, New York, USA

Kritski, A., Yuen, D.A. and Vincent, A. P. (2002). Properties of Near Surface Sediments from Wavelet Correlation Analysis, *Geophysical Research letters,* Vol. 29, pp. 1922-1925

Land, S. W., Kurkjian, A. L., McClellan, J. H., Morris, C. F. & Parks, T. W. (1987). Estimating Slowness Dispersion from Arrays of Sonic Logging Waveforms", *Geophysics,* Vol. 52 No. 4, pp. 530-544

Love, A. E. H. (1926). *Some Problems of Geodynamics,* 2nd ed. Cambridge University Press, London

Mallat, S. (1998). *A Wavelet Tour of Signal Processing,* Academic Press, USA

McMechan, G. A. & Yedlin, M. J. (1981). Analysis of Dispersion Waves by Wave-Field Transformation, *Geophysics,* Vol. 46, pp. 869-874

Ohta, K., Matsumoto, S., Okabe, K., Asano, K. & Kanamori, Y. (2008). Estimation of Shear-Wave Speed in Ocean-Bottom Sediment Using Electromagnetic Induction Source, *IEEE J. Ocean. Eng.,* Vol. 33, pp. 233-239

Rauch, D. (1980). *Seismic Interface Waves in Coastal Waters: A Review,* SACLANTCEN Report SR-42, NATO SACLANTASW Research Centre, Italy

Richardson, M. D., Muzi, E., Miaschi, B. & Turgutcan, F. (1991). Shear Wave Velocity Gradients in Near-Surface Marine Sediments, In: *Shear Waves in Marine Sediments,* Hovem, J. M., Richardson, M. D. & Stoll, R. D., (Ed.), pp. 295-304, Dordrecht: Kluwer Academic

Shapiro, N. M., Campillo, M., Stehly, L. & Ritzwoller, M. H. (2005). High Resolution Surface Wave Tomography from Ambient Seismic Noise, Science, Vol. 307, pp. 1165-1167

Smith, D. T. (1986). Geotechnical Characteristics of the Seabed Related to Seismo-Acoustics, In: *Ocean Seismo-Acoustics,* T. Akal and J. M. Berkson, (Ed.), PP. 483-500, Plenum, New York, USA

Socco, V. L., Boiero, D., Maraschini, M., Vanneste, M., Madshus, C., Westerdahl, H., Duffaut, K. & Skomedal, E. (2011). On the use of the Norwegian Geotechnical Institute's prototype seabed-coupled shear wave vibrator for shallow soil characterization – II. Joint inversion of multimodal *Love* and *Scholte* surface waves, *Geophys. J. Int.,* Vol. 185 (1), pp. 237-252

Stato, Y. (1954). Study on Surface waves XI. Definition and Classification of Surface Waves, *Bulletin, Earthquake Research Institute,* Tokyo University, Vol. 32, pp. 161-167

Stoll, R. D. & Batista, e. (1994). New Tools for Studying Seafloor Geotechnical Properties, *J. Acoust. Soc. Am.,* Vol. 96, pp. 2937-2944

Tikhonov, A. N. & Arsenin,V. Y. (1977). *Solution of Ill-Posed Problems,* New York Wiley, USA

Viktirov, I. A. (1967). *Rayleigh and Lamb Waves*, Plenum Press, New York, N. Y. USA
Vanneste, M., Madshus, C., Socco, V.L., Maraschini, M., Sparrevik, P. M., Westerdahl, H., Duffaut, K., Skomedal, E. & Bjørnarå, T. I. (2011). On the use of the Norwegian Geotechnical Institute's prototype seabed-coupled shear wave vibrator for shallow soil characterization - I. Acquisition and processing of multimodal surface waves, *Geophys. J. Int.*, Vol. 185 (1), pp. 221-236
Westwood, E. K.; Tindle, C. T. & Chapman, N. R. (1996). A Normal Mode Model for Acousto-Elastic Ocean Environments, *J. Acoust. Soc. Am.* Vol. 100, pp. 3631-3645
Wilken, D., Wolz, S., Muller, C. & Rabbel, W. (2008). FINOSEIS: A New Approach to Offshore-Building Foundation Soil Analysis Using High Resolution Seismic Reflection and Scholte-Wave Dispersion Analysis, *J. Applied Geophysics*, Vol. 68, pp. 117-123

7

Acoustic Properties of the Globular Photonic Crystals

N. F. Bunkin[1] and V. S. Gorelik[2]
[1]*A.M.Prokhorov General Physics Institute, Russian Academy of Sciences,*
[2]*Lebedev Physical Institute, Russian Academy of Sciences,*
Moscow,
Russia

1. Introduction

Modern technologies allow us to construct new nanomaterials with a periodic superstructure. In particular, the increasing interest has been recently shown in the so-called photonic (PTC) [1 - 4] and phononic (PNC) [5] crystals. In a case of PTC its structure is characterized by the refractive index, which periodically varies in space; the spatial period essentially exceeds the atomic sizes. PNCs are characterized by spatial periodic modulations of the sound velocity caused by the presence of the periodically settled elements of various materials (metals, polymers etc.) inside the sample. It is worth noting that PTC can at the same time be treated as a version of PNC.

PTCs and PNCs can be realized as one, two and three-dimensional structures. Among a wide variety of PTCs the special attention of researchers is paid to the crystal structures, whose lattice period is comparable with a wavelength of electromagnetic wave in the visible range. The periodicity of such PTC-structure results in presence of the so-called band-gaps located in the visible spectral range, i.e. in the spectral areas, where the electromagnetic waves can penetrate inside the sample only to a near-surface region with thickness of about a wavelength of light (0.4-0.8 microns). If the frequency of an electromagnetic wave is close to the band-gap edge, the group velocity of an electromagnetic wave drastically decreases, which results in a sharp increase of spectral density of electromagnetic radiation [2]. The numerous works are devoted to study of such effect for the visible range [2 - 4]. Other interesting feature of PTC consists in the existence of spectral bands characterized by a negative value of the effective refractive index, when the directions of phase and group velocities of an electromagnetic wave appear to be opposite to one another. At last we shall note that effective rest mass of photons in PTC is non-zero and can accept both positive and negative values. The absolute value of the effective rest mass of the photons in PTC is equal to $m = \frac{\hbar\omega}{c^2} \approx 10^{-36}$ kg.

The properties of acoustic waves in PNC are in many respects similar to the properties of electromagnetic waves in PTC. In the given work the review of characteristic properties of acoustic waves in PNC in comparison with the corresponding properties of electromagnetic waves in PTC is given. In particular, the problems of finding the form of dispersion dependences $\omega(k)$ for acoustic waves together with the dispersion dependences of their group velocities and effective mass of the corresponding acoustic phonons are solved. The results of the theoretical analysis and the data of experimental studies of the optical and acoustic phenomena in PTC and PNC, including the studies of spectra of non-elastic scattering of light together with the experiments to observe the stimulated light scattering accompanying by the coherent oscillations of globules are reported.

1.1 Theory of dispersion of electromagnetic and acoustic waves in one-dimensional PTC/PNC

The one-dimensional dielectric medium with two alternating layers (see. Fig. 1) can be considered as a one-dimensional PTC. At the same time, such medium can either be regarded as a one-dimensional PNC characterized by specified propagation velocities of acoustic waves in each of layers. At the first stage, let us consider the dispersion law for electromagnetic waves on the basis of the theory developed earlier [5 - 7]. According to the technique described in detail in Ref. [5], in order to obtain the dispersion relation, we used the plane monochromatic wave approximation with allowance for the boundary conditions at the edges of the layers (see Fig. 1).

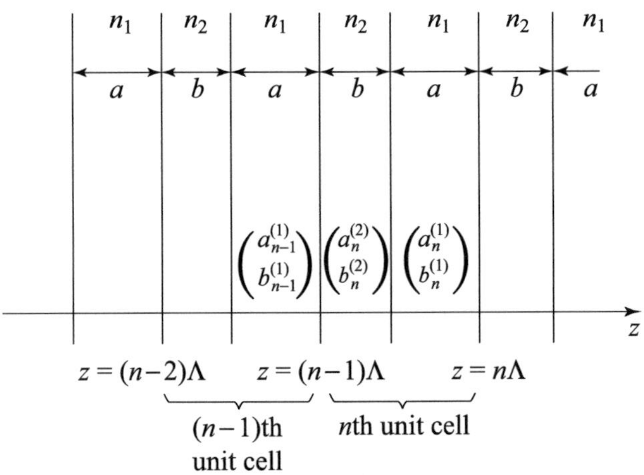

Fig. 1. Schematic of periodic layered medium and plane wave amplitudes corresponding to the n-th unit cell and its neighboring layers [5]

The periodic layered medium under study consists of two various substances with the following structure of the refractive index:

$$n(z) = \begin{cases} n_2, & 0 < z < b, \\ n_1, & b < z < \Lambda, \end{cases} \tag{1}$$

With making allowance for the periodicity of the refractive index, we arrive at:

$$n(z) = n(z + \Lambda). \tag{2}$$

Here the z-axis is perpendicular to the boundaries of layers, while Λ is the spatial period of the superstructure. The general solution to the wave equation for the electric field vector can be sought for in the form

$$\mathbf{E}(\mathbf{r},t)=\mathbf{E}_0(z)\exp\left[i\left(\omega t-k_y y\right)\right]. \tag{3}$$

Here it is assumed that the wave propagates in (yz) plane, whereas k_y is the vector component that remains constant during the propagation through the medium. The electric field strength within each homogeneous layer can be represented as a sum of the incident and reflected plane waves. Complex amplitudes of these two waves are components of the column vector. Thus, the electric field in α-th layer ($\alpha = 1, 2$) of the n-th unit cell (see. Fig. 1) can be written in the form of the column vector

$$\begin{pmatrix} a_n^{(\alpha)} \\ b_n^{(\alpha)} \end{pmatrix},\ \alpha=1,2. \tag{4}$$

The distribution of the electric field strength in the layer under consideration can be represented as

$$E(y,z)=\left\{a_n^{(\alpha)}\exp\left[-ik_{\alpha z}(z-n\Lambda)\right]+b_n^{(\alpha)}\exp\left[-ik_{\alpha z}(z-n\Lambda)\right]\right\}\exp\left(-ik_y y\right), \tag{5}$$

where

$$k_{\alpha z}=\sqrt{\left(\frac{n_\alpha\omega}{c}\right)^2-k_y^2},\ \alpha=1,2. \tag{6}$$

The column vectors are related to each other by the conditions of continuity at the interfaces. As a consequence, only one vector (or two components of different vectors) can be chosen arbitrarily. For TE-waves (vector E is perpendicular to the yz plane), the condition for the continuity of the components E_x and H_y ($H_y \sim \partial E_x/\partial z$) [6, 7] at the interfaces $z = (n-1)\Lambda$ and $z = (n-1)\Lambda + b$ (see Fig. 1) leads to the following equations:

$$\begin{gathered} a_{n-1}+b_{n-1}=e^{ik_{2z}\Lambda}c_n+e^{-ik_{2z}\Lambda}d_n,\ ik_{1z}\left(a_{n-1}-b_{n-1}\right)=ik_{2z}\left(e^{ik_{2z}\Lambda}c_n-e^{-ik_{2z}\Lambda}d_n\right), \\ e^{ik_{2z}a}c_n+e^{-ik_{2z}a}d_n=e^{ik_{1z}a}a_n+e^{-ik_{1z}a}b_n,\ ik_{2z}\left(e^{ik_{2z}a}c_n+e^{-ik_{2z}a}d_n\right)=ik_{1z}\left(e^{ik_{1z}a}a_n+e^{-ik_{1z}a}b_n\right). \end{gathered} \tag{7}$$

These four equations can be written as a system of two matrix equations:

$$\begin{pmatrix} 1 & 1 \\ 1 & -1 \end{pmatrix}\cdot\begin{pmatrix} a_{n-1} \\ b_{n-1} \end{pmatrix}=\begin{pmatrix} \exp(ik_{2z}\Lambda) & \exp(-ik_{2z}\Lambda) \\ \frac{k_{2z}}{k_{1z}}\exp(ik_{2z}\Lambda) & -\frac{k_{2z}}{k_{1z}}\exp(-ik_{2z}\Lambda) \end{pmatrix}\cdot\begin{pmatrix} c_n \\ d_n \end{pmatrix}, \tag{8}$$

$$\begin{pmatrix} \exp(ik_{2z}a) & \exp(-ik_{2z}a) \\ \exp(ik_{2z}a) & -\exp(-ik_{2z}a) \end{pmatrix}\cdot\begin{pmatrix} c_n \\ d_n \end{pmatrix}=\begin{pmatrix} \exp(ik_{1z}a) & \exp(-ik_{1z}a) \\ \frac{k_{1z}}{k_{2z}}\exp(ik_{1z}a) & -\frac{k_{1z}}{k_{2z}}\exp(-ik_{1z}a) \end{pmatrix}\cdot\begin{pmatrix} a_n \\ b_n \end{pmatrix}, \tag{9}$$

where

$$a_n \equiv a_n^{(1)},\ b_n \equiv b_n^{(1)},\ c_n \equiv a_n^{(2)},\ d_n \equiv b_n^{(2)}. \tag{10}$$

Eliminating the column vector $(c_n, d_n)^T$ from this system, we obtain the matrix equation

$$\begin{pmatrix} a_{n-1} \\ b_{n-1} \end{pmatrix} = \begin{pmatrix} A & B \\ C & D \end{pmatrix} \begin{pmatrix} a_n \\ b_n \end{pmatrix}. \tag{11}$$

The matrix elements in this equation are:

$$\begin{aligned} A &= \exp(ik_{1z}a)\cdot\left[\cos k_{2z}b + \frac{1}{2}i\left(\frac{k_{2z}}{k_{1z}} + \frac{k_{1z}}{k_{2z}}\right)\sin k_{2z}b\right],\ B = \exp(-ik_{1z}a)\cdot\left[\frac{1}{2}i\left(\frac{k_{2z}}{k_{1z}} - \frac{k_{1z}}{k_{2z}}\right)\sin k_{2z}b\right], \\ C &= \exp(ik_{1z}a)\cdot\left[-\frac{1}{2}i\left(\frac{k_{2z}}{k_{1z}} - \frac{k_{1z}}{k_{2z}}\right)\sin k_{2z}b\right],\ D = \exp(-ik_{1z}a)\cdot\left[\cos k_{2z}b - \frac{1}{2}i\left(\frac{k_{2z}}{k_{1z}} + \frac{k_{1z}}{k_{2z}}\right)\sin k_{2z}b\right]. \end{aligned} \tag{12}$$

Since the matrix (11) relates amplitudes of the field of two equivalent layers with identical refractive indices, it is unimodular, i.e.,

$$AD - BC = 1 \tag{13}$$

As was pointed out above, only one column vector is independent. For this vector one can choose, for instance, the column vector for layer 1 in the zero unit cell. The remaining column vectors of the equivalent layers are connected with the vector for the zero unit cell by the relation

$$\begin{pmatrix} a_0 \\ b_0 \end{pmatrix} = \begin{pmatrix} A & B \\ C & D \end{pmatrix}^n \begin{pmatrix} a_n \\ b_n \end{pmatrix}. \tag{14}$$

It follows from here that

$$\begin{pmatrix} a_n \\ b_n \end{pmatrix} = \begin{pmatrix} A & B \\ C & D \end{pmatrix}^{-n} \begin{pmatrix} a_0 \\ b_0 \end{pmatrix}, \tag{15}$$

or, in view of (14)

$$\begin{pmatrix} a_n \\ b_n \end{pmatrix} = \begin{pmatrix} D & -B \\ -C & A \end{pmatrix}^n \begin{pmatrix} a_0 \\ b_0 \end{pmatrix}. \tag{16}$$

A periodic layered medium is equivalent to a one-dimensional PTC that is invariant under translations to the lattice constant. The lattice translation operator T is defined by the expression

$$Tz = z - l\Lambda,\ l \in Z.$$

Thus, we arrive at

$$T\mathbf{E}(z) = \mathbf{E}(T^{-1}z) = \mathbf{E}(z + l\Lambda). \tag{17}$$

According to the Bloch theorem [6, 7], the vector of the electric field of the normal mode in the layered periodic medium has the form:

$$\mathbf{E} = \mathbf{E}_K(z)\exp(-iKz)\exp\left[i(\omega t - k_y y)\right], \tag{18}$$

where $\mathbf{E}_K(z)$ is the periodic function with the period Λ, i.e.,

$$\mathbf{E}_K(z) = \mathbf{E}_K(z+\Lambda). \tag{19}$$

Using the column vector representation and expression (5), the periodicity condition (19) for the Bloch wave can be written as:

$$\begin{pmatrix} a_n \\ b_n \end{pmatrix} = \exp(-iK\Lambda)\begin{pmatrix} a_{n-1} \\ b_{n-1} \end{pmatrix}. \tag{20}$$

As follows from Eqns. (11) and (20), the column vector of the Bloch wave obeys the eigenvalue equation:

$$\begin{pmatrix} A & B \\ C & D \end{pmatrix}\begin{pmatrix} a_n \\ b_n \end{pmatrix} = \exp(iK\Lambda)\begin{pmatrix} a_n \\ b_n \end{pmatrix}. \tag{21}$$

Thus, the phase factor is the eigenvalue of the translation matrix ($A\ B\ C\ D$) and satisfies the characteristic equation

$$\det\begin{pmatrix} A-\exp(iK\Lambda) & B \\ C & D-\exp(iK\Lambda) \end{pmatrix} = 0.$$

The solution to this equation has the form

$$\exp(iK\Lambda) = \frac{1}{2}(A+D) \pm \sqrt{\frac{1}{2}(A+D)^2 - 1}. \tag{22}$$

Eigenvectors corresponding to these eigenvalues are solutions to Eqn. (21), and accurate to an arbitrary constant they can be represented in the form

$$\begin{pmatrix} a_0 \\ b_0 \end{pmatrix} = \begin{pmatrix} B \\ \exp(iK\Lambda) - A \end{pmatrix} \tag{23}$$

According to (20), the corresponding column eigenvector for the n-th unit cell is

$$\begin{pmatrix} a_n \\ b_n \end{pmatrix} = \exp(-inK\Lambda)\begin{pmatrix} B \\ \exp(iK\Lambda) - A \end{pmatrix}. \tag{24}$$

The Bloch waves obtained from (23) and (24) can be considered as eigenvectors of the translation matrix with the eigenvalues $\exp(iK\Lambda)$, given by Eqn. (22); this equation results in the dispersion relation of the kind:

$$K\left(k_y,\omega\right)=\frac{1}{\Lambda}\arccos\left(\frac{A+D}{2}\right). \tag{25}$$

The modes, in which $|A + D|/2 < 1$, correspond to the real K. If $|A + D|/2 < 1$, the relation $K = m\pi/\lambda + iK_m$ takes place, i.e., the imaginary part in the wave vector K is non-zero, and the wave is damped. Thus the so-called band-gap opens. The frequencies corresponding to the band-gap boundaries are found from the condition $|A+D|/2=1$.

At normal incidence (k_y =0), the dispersion dependence $\omega(K)$ has, according to (25), the following form:

$$\cos(K\Lambda)=\cos(k_1 a)\cos(k_2 b)-\frac{1}{2}\left(\frac{n_2}{n_1}+\frac{n_1}{n_2}\right)\sin(k_1 a)\sin(k_2 b). \tag{26}$$

The quantities in Eqn. (26) have the following physical meaning: $i = 1$ is the subscript related to the first medium, while $i = 2$ is the subscript related to the second medium; $n_1 = n_1(\omega)$ is the refractive index of the first medium, and $n_2 = n_2(\omega)$ is the refractive index of the second medium; $a=(1-\eta)\Lambda$, $b=\eta\Lambda$, η is the content of the second medium in the layered PTC; $k_i(\omega)=\dfrac{\omega\cdot n_i}{C_0}$ is the wave vector in the i-th medium, and $C_0 = 3\cdot 10^8$ m/s is the velocity of light in vacuum.

As the first approximation we can assume that the refractive index values of both layers n_1 and n_2 are the constant values and do not depend on the electromagnetic radiation frequency. At the next stage we shall take into account their dependence on the frequency (or on the wavelength of a radiation illuminating the PTC). For example, let us consider a one-dimensional PTC, one layer of which is amorphous quartz (SiO_2), and second layer is atmospheric air, for which the refractive index is ~ 1. Additionally, we will consider that the refractive index of SiO_2 is not dependent upon the wavelength, i.e. $n_1 = 1.47$. Finally, we will use the following values of the parameters a_1 and a_2: specifically, $a_1 = 136$ nm, and $a_2 = 48$ nm. In this case the dispersion law for the electromagnetic waves in PTC can be obtained by using Eqn. (26). The results of numerical simulation of the dispersion curves $\omega(k)$ (these curves were taken from our study [8]) are plotted in Fig. 2 by the bold lines. As is seen from this Figure, we can discern in the spectrum three dispersion branches $\omega_j(k)$, $(j=1,2,3)$, where the first one is related to the low-frequency range, including the infra-red spectral area, the second one is related to the visible spectral range, and the third one is related to the near ultra-violet spectrum. The additional branches are plotted in view of anomalous dispersion of the refractive index at approaching the electronic absorption band in SiO_2. In Fig. 2 two photonic band-gaps, where the first one corresponds to the Brillouin zone boundary ($k=\dfrac{\pi}{\Lambda}$), and the second one is related to the Brillouin zone center (k = 0), are seen. The second dispersion branch is related to the case, where the directions of the phase and group velocities are opposite with respect to one another, i.e. to the case of negative effective refractive index $(n<0)$.

As was found in study [8], the good enough description of the dispersion curves (see the gray curves (2) in Fig. 2, which were plotted on the basis of the exact theory given by Eqn. (26)), can be carried out by using the approximated formulas of the kind:

$$\omega = 2\frac{C_0}{m_1\Lambda}\sin\left(\frac{k\Lambda}{2}\right), \text{ - the first dirpersion branch,} \tag{27}$$

$$\omega^2 = \omega_2^2 - \left(2\frac{C_0}{m_2\Lambda}\right)^2 \sin^2\left(\frac{k\Lambda}{2}\right) \text{ - the second dirpersion branch,} \tag{28}$$

$$\omega^2 = \omega_3^2 + \left(2\frac{C_0}{m_3\Lambda}\right)^2 \sin^2\left(\frac{k\Lambda}{2}\right) \text{ - the third dirpersion branch.} \tag{29}$$

Here ω_1 and ω_2 are the frequencies corresponding to the edges of the second order band-gap, m_1, m_2, and m_3 are the effective refractive indices for the first, the second and the third dispersion branches accordingly. The following values for the frequencies and refractive indices were taken: ω_2 = 3.62·10^{15} radian/s, ω_2 = 3.62·10^{15} radian/s, ω_3=4.14·10^{15} radian/s, and m_1 = 0,94; m_2 = 0.555; m_3 = 0.435. Besides, the opportunity of approximating the dispersion curves close to the center of the Brillouin zone by the so-called quasi-relativistic formulas (30) – (33) was considered:

$$\omega = \frac{C_0}{m_1}k \text{ - the first dirpersion branch,} \tag{30}$$

$$\omega^2 = \omega_2^2 - \left(\frac{C_0}{m_2}\right)^2 k^2 \text{ - the second dirpersion branch,} \tag{31}$$

$$\omega^2 = \omega_3^2 + \left(\frac{C_0}{m_3}\right)^2 k^2 \text{ - the third dirpersion branch,} \tag{32}$$

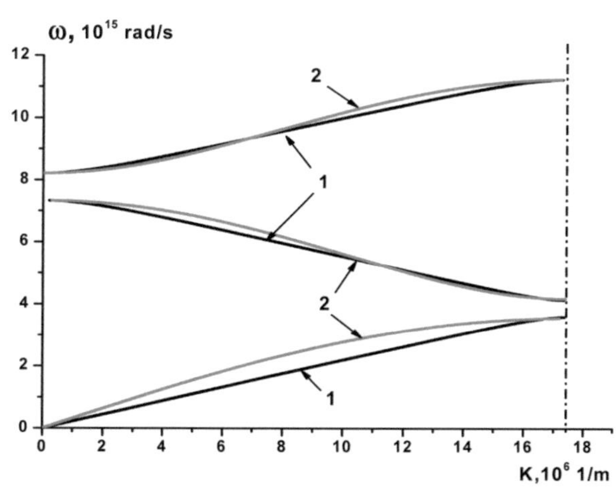

Fig. 2. The dispersion law $\omega(k)$ for a one-dimensional PTC; (1) - the results of calculation of the dispersion dependence $\omega(k)$ according to (26), (2) – the results of calculation of the dispersion dependence $\omega(k)$ with the help of sinusoidal approximation.

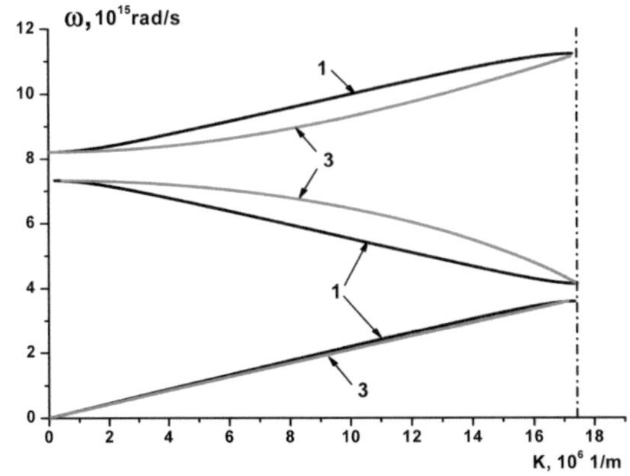

Fig. 3. The dispersion law $\omega(k)$ for a one-dimensional PTC; (1) - the results of calculation of the dispersion dependence $\omega(k)$ according to (26), (3) – the results of calculation of the dispersion dependence $\omega(k)$ with the help of quasi-relativistic approximation.

As is seen from Fig. 3, the satisfactory agreement between the curves 1 and 3 takes place only at small values of a wave vector (i.e., close to the center of the Brillouin zone). Nonetheless, this approach allows us to estimate the effective photonic mass basing on the equality (here m_0 and E_0 are the effective rest mass and the rest energy of the photon.)

$$m = \frac{1}{\frac{d^2E}{dp^2}} = \frac{\hbar}{\frac{d^2\omega}{dk^2}}; \quad |m_0| = \frac{E(0)}{C^2} = \frac{\hbar\omega(0)}{C^2}. \tag{33}$$

For the second dispersion branch the effective rest mass of the photon appears to be negative and equal to $m_2 = -\frac{\hbar\omega_1}{\left(\frac{C_0^2}{m_2^2}\right)} = -0{,}13\cdot 10^{-35}$ kg. Accordingly for the third branch we obtain: $m_3 = \frac{\hbar\omega_2}{\left(\frac{C_0^2}{m_3^2}\right)} = 0{,}09\cdot 10^{-35}$ kg. Thus, the effective rest mass of a photon inside the PTC appears to be non-zero and can acquire both positive and negative magnitudes.

If one takes into account the dispersion of refractive index for the layers, forming PTC, the dispersion law $\omega(k)$ can also be received from numeric solution to Eqn. (26). For example, let us consider PTC, where the first medium is SiO_2, whereas the second medium is atmospheric air or water. The refractive index of air is assumed to be equal to unity. Thus for the dependence of the SiO_2 refractive index versus a wavelength we can use the formula:

$$n_1^2 - 1 = \frac{0{,}6962\lambda^2}{\lambda^2 - 0{,}0684^2} + \frac{0{,}4079\lambda^2}{\lambda^2 - 0{,}1162^2} + \frac{0{,}8975\lambda^2}{\lambda^2 - 9{,}896^2},$$

while the same dependence for water is given by the formula:

$$n_2^2 - 1 = \frac{5{,}667\cdot 10^{-1}\lambda^2}{\lambda^2 - 5{,}084\cdot 10^{-3}} + \frac{1{,}732\cdot 10^{-1}\lambda^2}{\lambda^2 - 1{,}818\cdot 10^{-2}} + \frac{2{,}096\cdot 10^{-2}\lambda^2}{\lambda^2 - 2{,}625\cdot 10^{-2}} + \frac{1{,}125\cdot 10^{-1}\lambda^2}{\lambda^2 - 1{,}074\cdot 10^{1}}$$

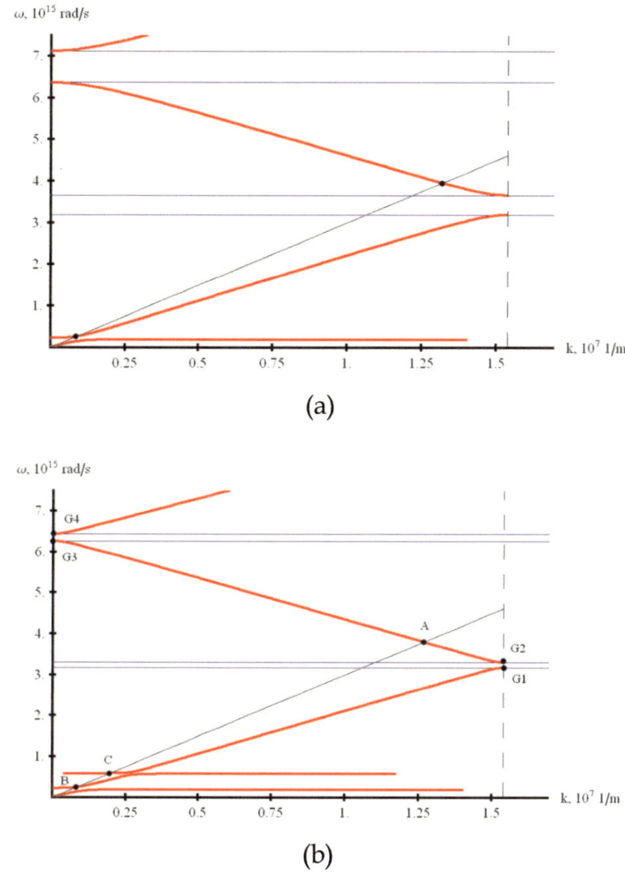

Fig. 4. The calculated dispersion curves; (a) - the initial PTC; (b) - the PTC, filled with water. The band-gap boundaries (thin solid lines), the edge of the first Brillouin zone together with the straight line indicating the dispersion law for a light wave in vacuum ($\omega = C_0 \cdot k$) are indicated.

The calculated dispersion curves for these cases are shown in Fig. 4 (a) and (b). As is seen in this Figure, implantation of water instead of air in PTC results in decreasing the optical contrast and, accordingly, in reducing the band-gap width for the visible and ultraviolet spectral ranges. Accounting for the dispersion of the refractive index for SiO_2 results in occurrence of the additional dispersion branch in the infrared spectral range; this branch is related to the polariton curve, stimulated by the polar vibrations, e.g., the vibrations along the bond Si-O in the microstructure of quartz. In Fig. 4 the points of intersection of the straight line, corresponding to the light wave (this line is set by the formula $\omega = C_0 \cdot k$, for which the effective refractive index is equal to unity) are marked. Thus according to the known Fresnel formulas the reflectance of a light wave from the PTC interface approaches zero, and the material should become absolutely transparent (provided that the absorption is absent).

1.2 Calculation of the dispersion characteristics for the one-dimensional PNC

As was already noted, PNC can either be considered as PTC with making allowance for the fact that the sonic wave velocities depend upon the type of material of a layer. Using the optical-acoustic analogy for describing the dispersion of acoustic waves in PNC and basing upon Eqn. (26), we obtain the following dispersion equation for the acoustic wave propagating in PNC along the crystallographic direction (111):

$$\cos(ka) = \cos(k_1 a_1) \cdot \cos(k_2 a_2) - \frac{1}{2} \frac{V_1^2 + V_2^2}{V_1 \cdot V_2} \sin(k_1 a_1) \cdot \sin(k_2 a_2). \quad (34)$$

The quantities entering into (34) have the following physical meaning: $i = 1$ is the subscript for SiO_2 (opal matrix); $i = 2$ is the subscript for the layer, filled with a metal or liquid; V_1 is the velocity of acoustic waves in opal; V_2 is the velocity of sound in the medium that fills the pores in the opal (see table 1); $\eta = 0{,}26$ is the effective sample porosity coefficient, $D = 220$ nm is the diameter of the quartz globules; $a = D\sqrt{2/3}$ is the period of the structure of the opal samples under investigation; $a_1 = (1 - \eta)a$, $a_2 = \eta a$; ω_i is the cyclic frequency of the acoustic wave; $k_i(\omega) = \omega / v_i$ is the wave vector in the i-th medium.

Material	Transverse wave velocity, km·s^{-1}	Longitudinal wave velocity, km·s^{-1}
Opal	3.3	5.3
Air	0.1	0.3
Water	0.6	1.5
Gold	1.2	3.2

Table 1. Velocities of longitudinal and transverse acoustic waves

Based on the numerical analysis of (34), we constructed the dispersion dependences $\omega(k)$ for different branches in the acoustic region of the spectrum. The numerically simulated acoustic properties of various PNCs are shown in Fig. 5.

The abscissas are the wave vector values, scaled in m^{-1}, and the ordinates are the cyclic frequencies (rad·s^{-1}); the solid lines indicate longitudinal waves and the dashed lines indicate transverse waves. Fig. 5 (a) corresponds to the initial (unfilled) opal containing air in its pores, whereas Fig. 5 (b) shows the dispersion dependence $\omega(k)$ for the sample with water in its pores, and Fig. 5 (c) displays acoustic branches of PTC with nanoparticles of gold. As is seen from these graphs, both in PTC and PNC the band-gaps are being formed, whose location and bandwidth depend on the spatial period of the superstructure and the type of material. The dispersion dependence of quasi-particles traveling along the crystalline lattice can be found from the known relation [6, 7]:

$$V_{gr}(\omega) = \frac{d\omega(k)}{d\,k} = \frac{1}{\dfrac{dk(\omega)}{d\omega}}. \quad (35)$$

The corresponding dependences of group velocities are shown in Fig. 6. Figures 6 (a) – (c) correspond to the initial PNC, to the crystal filled with water, and to the opal filled with gold, respectively. Note that the group velocity of the acoustic waves becomes zero at the boundaries of the band-gaps. Besides, at the band-gap boundaries the group velocity of the acoustic phonons approaches zero, which corresponds to "stopping" of phonons at the corresponding frequencies. Such phenomenon is quite similar to "stopping" of photons, related to the band-gap boundaries of PNC.

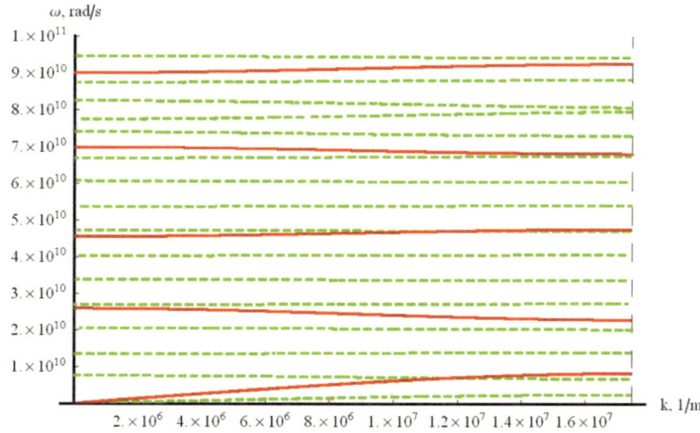

(a)

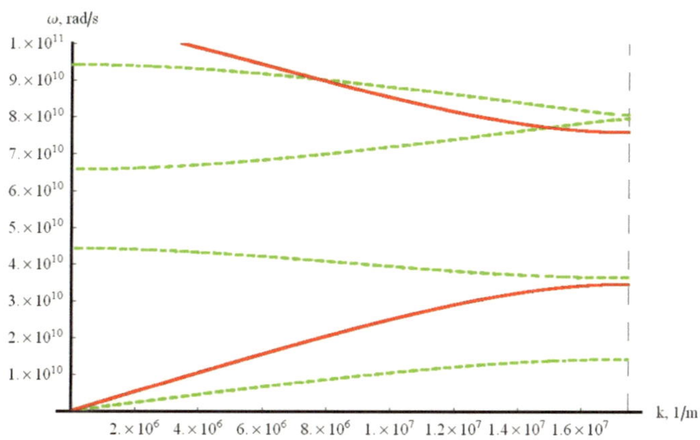

(b)

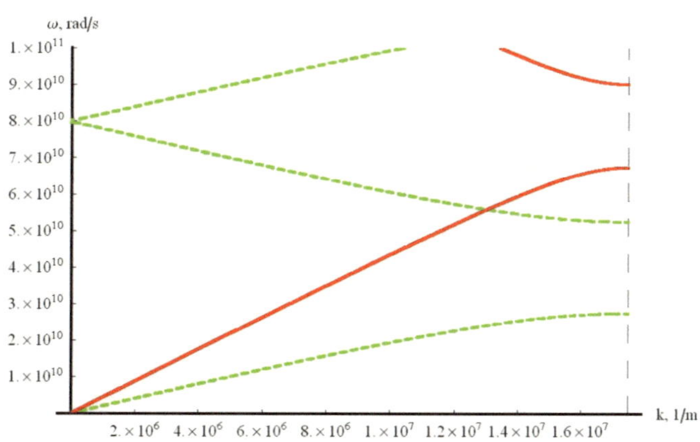

(c)

Fig. 5. Dispersion dependence $\omega(k)$ for three types of PNCs: (a) the initial PNC, (b) the PNC, filled with H_2O, (c) the PNC with Au nanoparticles. Solid and dashed curves correspond to longitudinal and transverse waves, respectively.

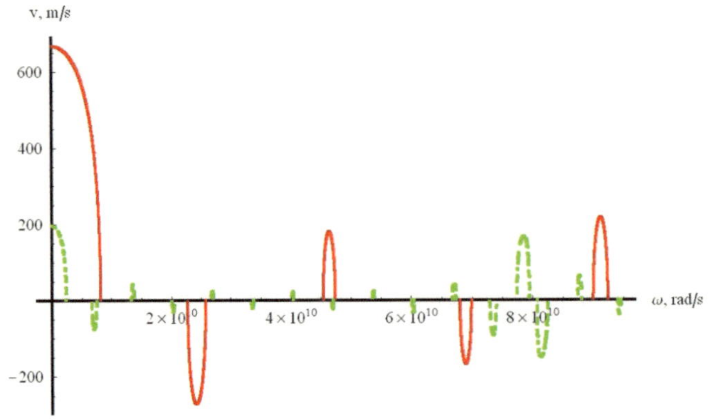

(a)

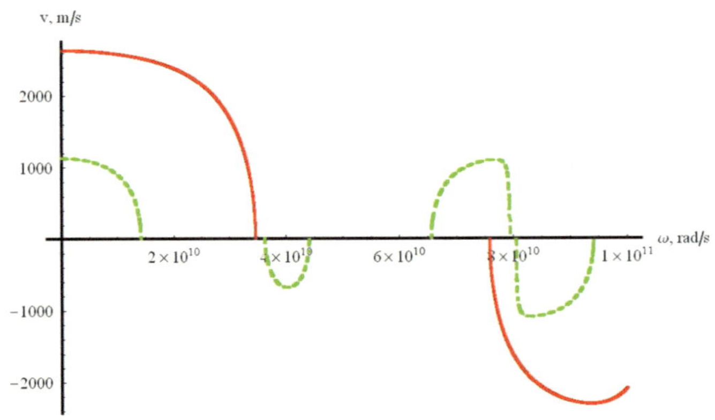

(b)

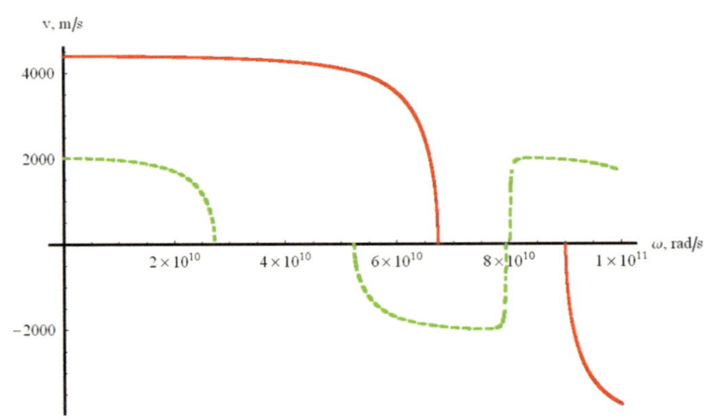

(c)

Fig. 6. Group velocity of phonons in the investigated samples: (a) the initial PNC, (b) the PNC filled with water, (c) the PNC filled with gold. Solid and dashed curves correspond to longitudinal and transverse waves, respectively.

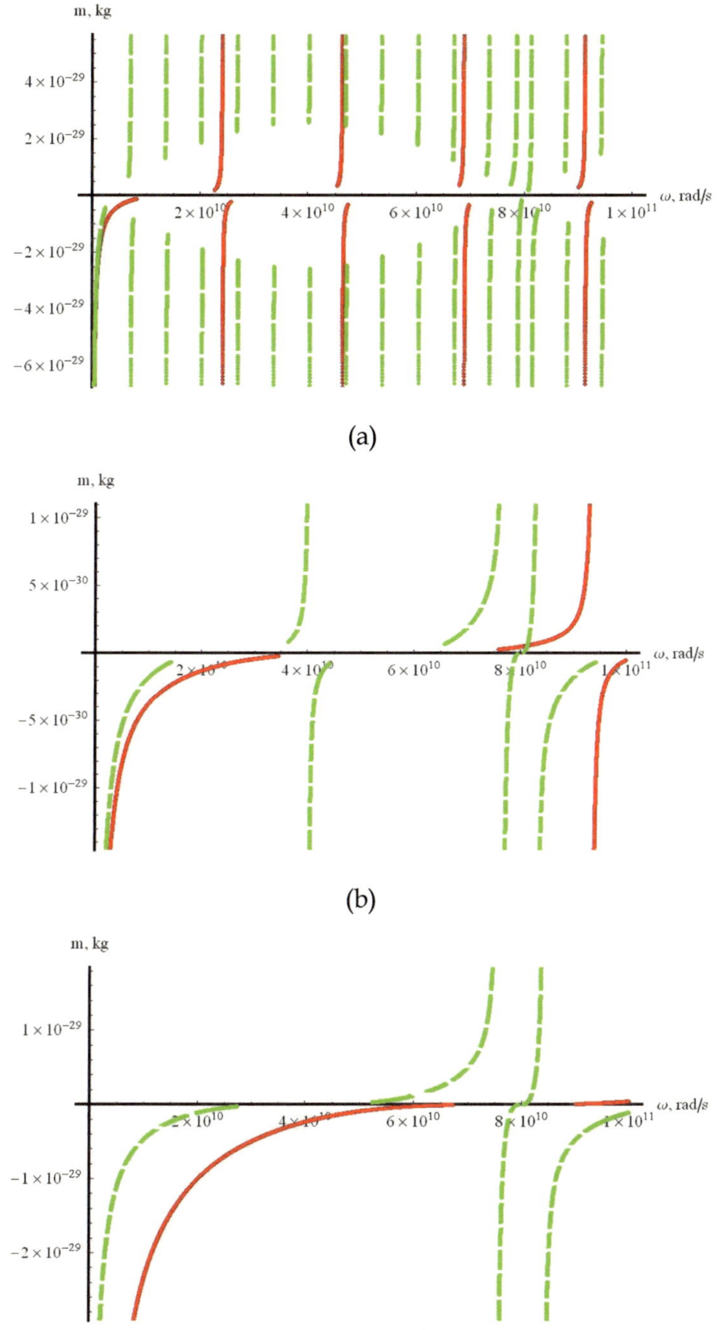

Fig. 7. Mass of acoustic quasi-particles for different types of opal: (a) initial opal, (b) opal with water, (c) opal with gold. Solid and dashed curves correspond to longitudinal and transverse waves, respectively.

According to the general definition of the effective mass of a quasi-particle [6, 7], the effective mass of acoustic phonons can be calculated by the formula

$$m(\omega) = \hbar\left[\frac{d^2\omega(k)}{dk^2}\right]^{-1} = \frac{\hbar}{V_{gr}(\omega)\left[dV_{gr}(\omega)/d\omega\right]}. \tag{36}$$

This effective mass is related to slow acoustic waves and is many orders of magnitude smaller than the mass of photons in PNC, and can be estimated from the relation $|m_0| = \frac{\hbar\omega(0)}{S^2}$, where S is corresponding sonic velocity. In particular, the effective mass of the transverse acoustic phonons related to the second dispersion branch of PNC containing the atmospheric air (see Fig. 6) is equal to $m_0 = -24\cdot10^{-30}$ kg; for PNC containing water we have $m_0 = -3{,}64\cdot10^{-30}$ kg; and for PNC containing gold we obtain $m_0 = -6{,}94\cdot10^{-30}$ kg. Accordingly for the third dispersion branch the effective mass appears to be positive and slightly exceeds (by the absolute value) the indicated above values of the effective rest mass of phonons.
Summarizing, in PNC the acoustic phonons possess by the rest mass; the phonon rest mass by its absolute value is 5 - 6 orders of magnitude less than the effective rest mass of photons in PTC, and can be both positive and negative.

1.3 Structure and the techniques of preparation of the globular photonic crystals

The important example of the three-dimensional PTC (PNC) is the so-called globular photonic crystal composed of densely packed balls (globules) as the face-centered cubic crystal lattice. The diameter of the globules is slightly changed within the whole structure of a crystal. Depending on the technological process this diameter can vary within the range of 200 - 1000 nm. To the present time the globular photonic crystals composed of the balls of synthetic opal (SiO_2), titanium oxide (TiO_2), and Polystyrene are known. There exist the voids (pores) between the globules of a photonic crystal, which can be filled with some foreign additives. For example, it is possible to implant into the pores of a globular phoptonic crystal some liquids, which moisturize the globule interface, and solid dielectrics, including piezoelectrics and ferroelectrics. Besides, it is possible to implant magnetic

Fig. 8. Samples of 3D-PTC, obtained from the synthetic opals under different technological conditions.

materials, semiconductors, metals and superconductors. Thus, we have a wide opportunity to create new materials of a hybrid-like type: dielectric-ferroelectric, dielectric-magnetic, dielectric-metal etc. We also can control the dielectric, acoustic and galvanic properties of such hybrid materials by changing the diameter of globules.
Some samples of three-dimensional PTCs under study are illustrated in the photo, see Fig. 8. The white large sample (at the foot of the photo) was annealed in the atmospheric air at the temperature of 600 C. Color (green and blue) samples were annealed in the atmosphere of argon. During the process of growth and annealing these samples were saturated by carbon as the result of destruction of organic molecules, which were initially (in the trace amounts) located in the samples.

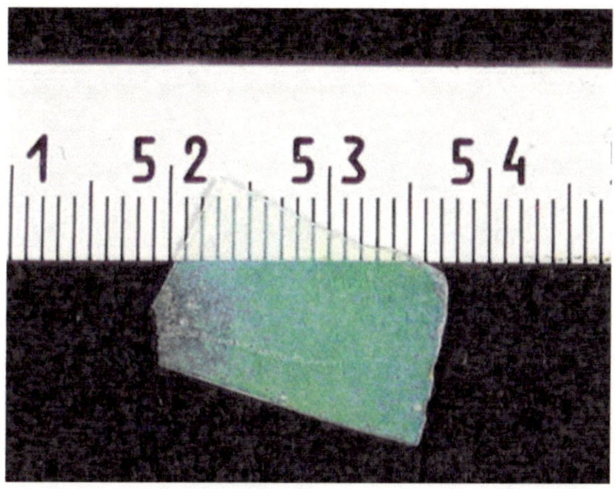

(a)

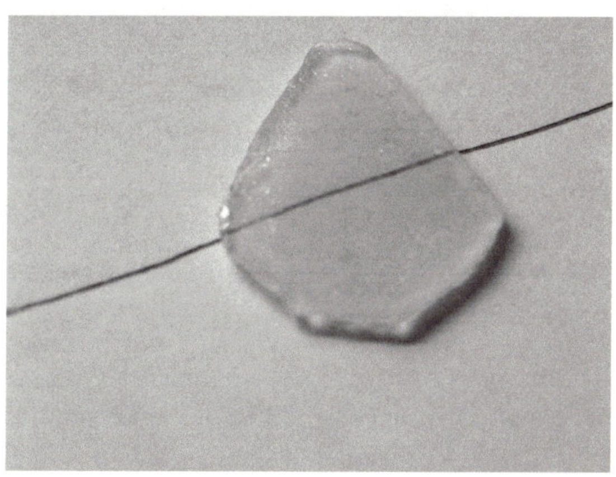

(b)

Fig. 9. PTC, transparent in the visible spectral range; (a) - PTC, containing the quantum dots. This sample was filled by ZrO_2 nanoparticles and then was subjected to annealing at high temperature (up to 1200 C); as the result, the sample became transparent as the size of implanted inclusions of ZrO_2 nanoparticles was essentially less than the photonic crystal lattice constant and a wavelength in the visible range. (b) - PTC, filled with glycerol-water mixture.

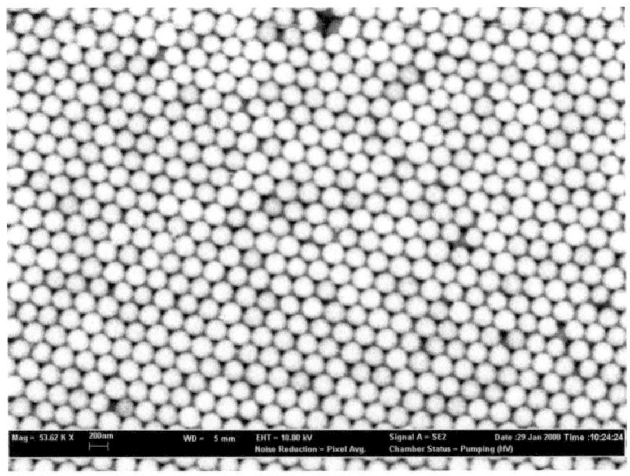

(a)

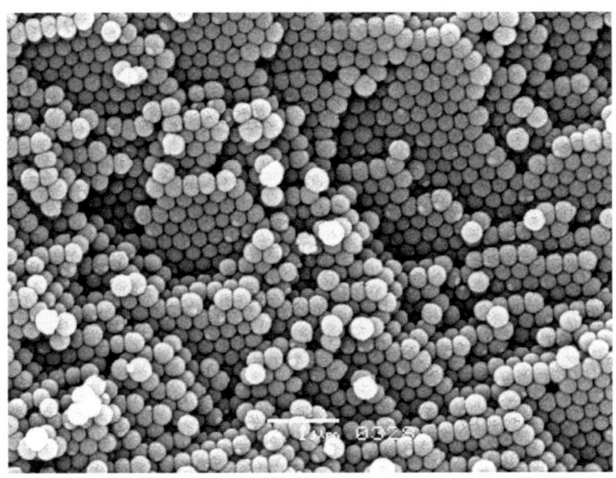

(b)

Fig. 10. The images of (111) surfaces for two ((a) and (b)) investigated synthetic opals, obtained with the help of electronic microscope.

Electronic images of the globular PTC surface (111) for two investigated samples are shown in Fig. 10 (a) and (b). We can see that the nanostructure of sample in Fig.10 (a) is close to the ideal one. In the case of the second sample (Fig. 10 (b)) there exist numerous defects arisen due to certain disordering processes. Initial synthetic opals have been filled with some organic (Stilbene, glycerol, acetone, nitrobenzene) or inorganic (sodium nitrite, sulfur, ZrO_2) chemicals. At the certain concentration of glycerol-water mixture its refractive index appeared to be very close to that for a quarts globule. In this way almost transparent 3D-PTC have been obtained (see Fig. 9 (b)).

The processes of the opal sample processing are shown in Fig. 11 (a, b). We have implanted nanoparticles of some metals (Au, Ag, Ga) into the photonic crystal pores localized between the globules. The sample was filled with ZrO_2 nanoparticles and then was subjected to annealing at high temperature (up to 1200 C); as the result, the sample became transparent as the size of implanted inclusions of ZrO_2 nanoparticles was essentially less than the photonic crystal lattice constant and the visible range wavelength. Accordingly, such spatial arrangement of inclusions can be described as the array of spatially ordered quantum dots in the transparent crystal of quartz. The schematic nanostructure of such quantum dots in PTC is illustrated in Fig. 11 c.

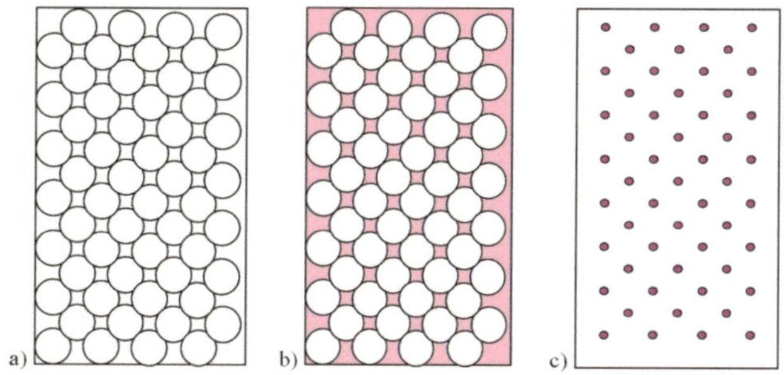

Fig. 11. Structures of 3D-PTC filled by dielectrics or metals; (a) - initial synthetic opal, (b) - opal, filled with some substance, (c) – result of the high temperature annealing of the sample, containing the particles of ZrO_2, whose melting temperature is higher than that for quartz.

1.4 Optical properties of the globular PTC

In what follows we will analyze the optical and acoustic properties of globular PTC; it is clear that we can describe the both properties in the framework of the same approach. This is why the following considerations basically repeat the models applied above, but now we should bear in mind that we deal with the three-dimensional periodic medium. Assuming that the light wave is directed along the (111) vector in a crystal, it is still possible to use the approximation of effective one-dimensional model of the layered PTC [6, 7]. In this case the dispersion law of the globular PTC on the basis of the synthetic opal, whose pores are filled with atmospheric air, is given by the following formula, which is quite similar to Eqn. (34) for the dispersion law of acoustic waves in the layered PTC:

$$\cos k_1 a_1 \cdot \cos k_2 a_2 - \frac{1}{2}\frac{\varepsilon_1 + \varepsilon_2}{\sqrt{\varepsilon_1 \cdot \varepsilon_2}} \sin k_1 a_1 \cdot \sin k_2 a_2 = \cos ka. \tag{37}$$

The parameters here are the following: ε_1 is the dielectric permittivity of quartz (naturally for the oprtical range of frequencies); ε_2 is the dielectric permittivity of air, $a_1 = (1-\eta)a$, $a_2 = \eta a$, where η is the effective sample porosity, $a = D\sqrt{2/3}$ is the period of the structure of the sample, D is the effective diameter of quartz globule, ω is the cyclic frequency of the electromagnetic wave, $k_i(\omega) = \frac{\omega}{c_0}\sqrt{\varepsilon_i}$ is the wave vector in SiO_2 ($i = 1$) and in the air ($i = 2$).

In Fig. 12 the dispersion dependence $\omega(k)$ for the incident (along the direction (111)) electromagnetic wave in the globular PTC, whose pores filled with atmospheric air, and the effective globule diameter is $D = 225$ nm. The Fig. 13 illustrates the two-branch dependence $\omega(k)$ for the globular PTC filled with the liquid having the refractive index close to that for SiO_2. As is seen from the graphs, in that case the band-gap width approaches zero.

Figs. 14 and 15 illustrate the dispersion law $\omega(k)$ of electromagnetic waves for the globular PTC, filled with the dielectric or metal accordingly. Figs. 16 and 17 show the character of changing the dispersion law owing to the occurrence of the low and high frequency

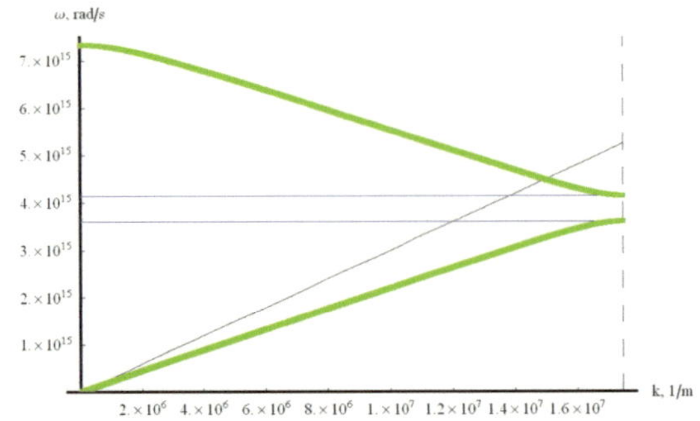

Fig. 12. The dispersion curves ω(*k*) for the first two branches of the globular PTC filled with air. The straight line obeys the dispersion law in vacuum.

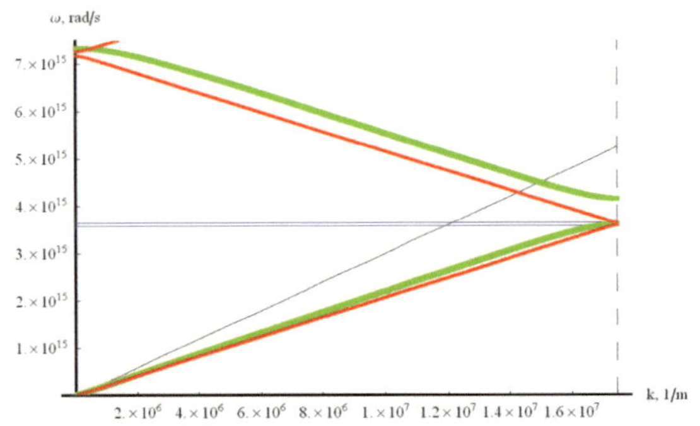

Fig. 13. The dispersion curves *ω*(*k*) for the first two branches of the globular PTC, filled with water. The upper curve corresponds to the initial (free of water) crystal, the lower curve corresponds to the crystal, whose pores contain a liquid with the refractive index close to that for quartz.

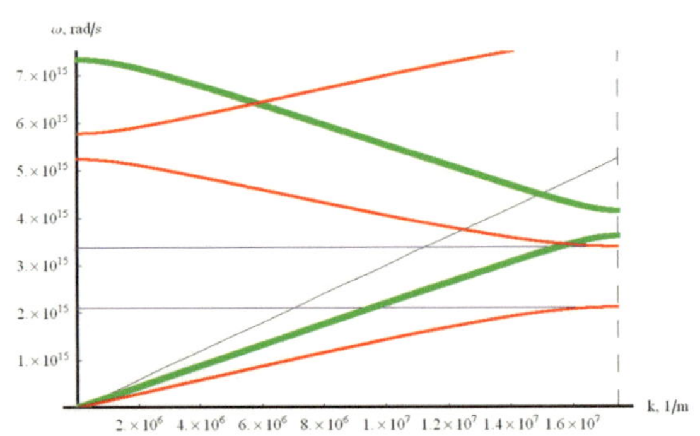

Fig. 14. The dispersion curves *ω*(*k*) for the first two branches of the globular PTC, filled with the dielectric.

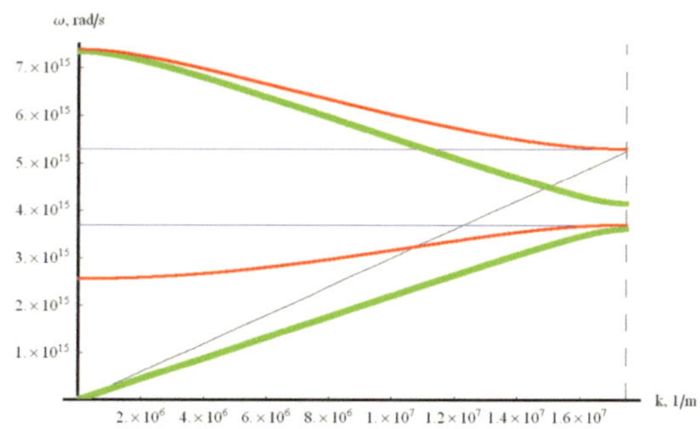

Fig. 15. The dispersion curves $\omega(k)$ for the first two branches of the globular PTC, filled with the metal.

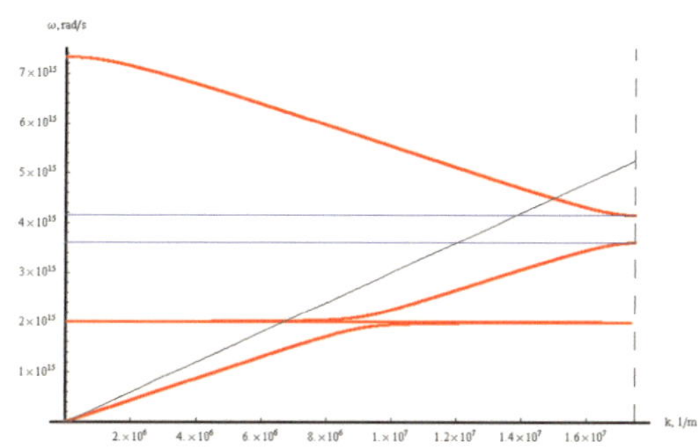

Fig. 16. The dispersion curves $\omega(k)$ for the first two branches of the globular PTC for the case, where the low-frequency resonance exists.

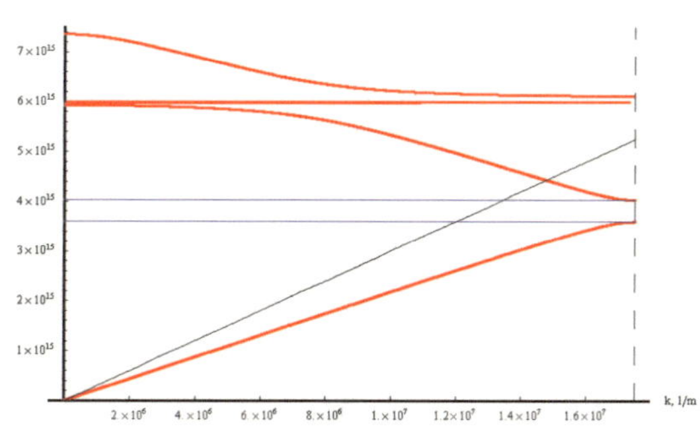

Fig. 17. The dispersion curves $\omega(k)$ for the first two branches of the globular PTC for the case, where the high-frequency resonance exists.

resonances accordingly; the resonances arise due to adding the certain substance in the pores. As is seen from Figs. 13 - 15, the implantation of dielectrics, whose refractive index exceeds that of quartz, into the pores of the globular PTC results in changing the width of the band-gap and its shifting to lower frequencies. At the same time, at implanting metal into these pores the band-gap shifts to higher frequencies, see Fig. 15. If one implants the substance, characterizing by the presence of resonances close to the band-gap spectrum, the dispersion curves $\omega(k)$ drastically change; it becomes possible that new band-gaps are being formed, and this process is essentially dependent on the resonant frequencies of the implanted substance, see Figs. 16, 17.

The implantation of various chemicals into the globular PTC was carried out by various techniques: among these was impregnation by a liquid wetting quartz, saturation of the crystal matrix by solutions of various salts with subsequent annealing, and also some laser methods including ablation. To analyze the spectra of reflectance of incident broadband electromagnetic radiation from the globular PTC interface, whose pores contain various substances, the experimental setup (see Fig. 18) was designed; its characteristics are described in Ref. [9]. In this setup the radiation of halogen or deuterium lamp (14) was directed with the help of an optical fiber probe perpendicular to the crystal interface (3). The optical fiber diameter was 100 μm, and the spatial resolution of the setup was on the level of 0.2 mm. With the help of another optical wave-guide the oppositely reflected radiation was

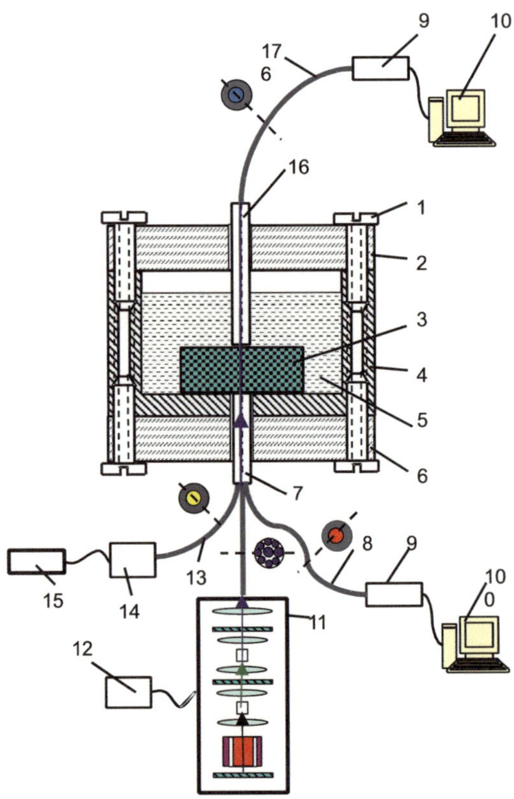

Fig. 18. The schematic of the experimental setup for analyzing the spectra of radiation reflected from the PTC interface; (1) - screws; (2) - the top Teflon cover-sheet; (3) – the PTC; (4) – the cell; (5) – the liquid sample; (6) - the bottom Teflon cover-sheet; (7) – the optical fiber probe; (8) – the wave-guide; (9) – the mini-spectromemer; (10) – the computer; (11) – the YAG:Nd^{3+} - laser; (12) - the power supply unit for the wave-guides; (13) – the wave-guide; (14) – the halogen lamp; (15) – the power supply unit for the lamp; (16) – the optical fiber probe for investigating the transmission spectra; (17) – the wave-guide.

input to a mini-spectrometer FSD-8, where the reflectance spectra in the range of 200 - 1000 nm were processed in the real time. The spectral resolution of the reflectance spectra was ≤ 1 nm. Using the laser radiation (pulse repeating YAG:Nd^{3+} laser with the possibility of doubling or quadrupling the frequency of the radiation) allowed us to carry out additional implantation of dielectrics or metals into the pores of the crystal with the simultaneous controlling the spectrum of the band-gap (this spectrum depends on the type and amount of the implanted substance). Using the additional optical fiber probe (16) allowed us to analyze the transmission spectrum with the help of second mini-spectrometer (9). The experimental data were input to the analog-to-digital converter of the computer (10) for the final processing.

In Fig. 19 the reflectance spectra of the globular PTC with various globule diameter and containing the atmospheric air (curve 1 in Fig. 19 (a) - (c)), and water (curve 2 in Fig. 8 (a) - (c)) are given. It is seen that at increase of the globule diameter, and at implantation of water into the pores the reflectance peak corresponding to the band-gap is shifted to higher frequencies. This experimental result is in agreement with formulas (38) and (39), which are relevant for the PTC model in question:

$$\lambda_{\max} = 2\sqrt{\frac{2}{3}}D\sqrt{n_{eff}{}^{2} - \sin^{2}\theta}, \tag{38}$$

$$n_{eff} = \sqrt{n_1^2\beta + n_2^2(1-\beta)}. \tag{39}$$

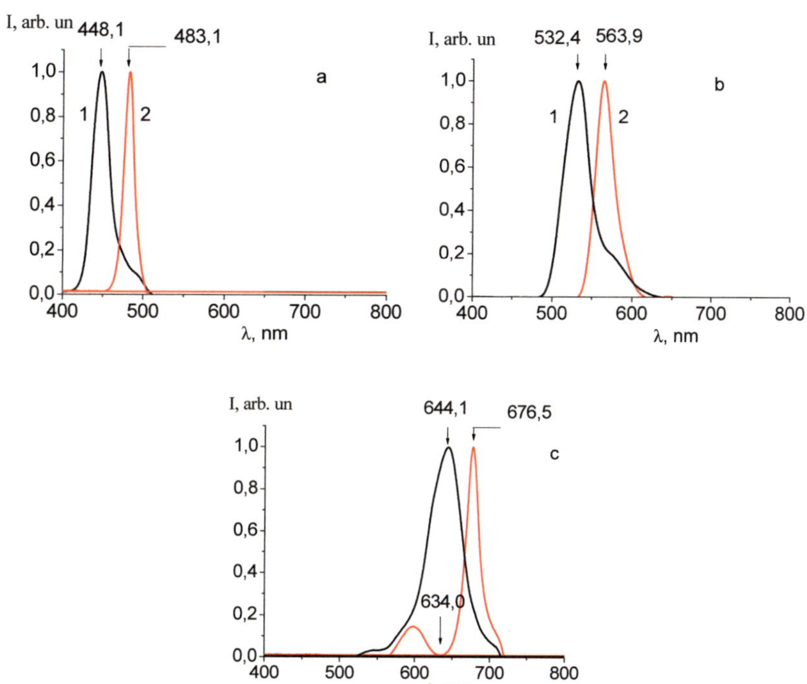

Fig. 19. The spectra of radiation reflected from (111) interface of the globular PTC with various globule diameters: D = 200 (a), 240 (b) and 290 nm (c).

Here θ is the angle of the radiation incidence onto the interface (111) of the PTC, D is the globule diameter, and n_1, n_2 are the refractive indices of SiO_2 and an implanted substance respectively.

As is seen in Fig. 19, the impregnation of the crystal matrix by water results in narrowing the band-gap. This is in conformity with the optical contrast decrease at approaching the refractive indices n_2 and n_1 to one another, see Eqn. (40) for the band-gap width.

$$\Delta\lambda = \lambda_{max} \frac{4}{\pi} \frac{|n_2 - n_1|}{(n_2 + n_1)}. \tag{40}$$

Fig. 20 illustrates the reflectance spectrum for the first and the second band-gap. According to Eqn. (38) the frequency of the reflectance spectral maximum should belong to the visible range, and for the second band-gap that frequency should be duplicated. As is seen in this Figure, the additional reflectance peak is indeed observed in the near ultra-violet range. The curve (1) in this Figure characterizes the parameters of the second band-gap. It is noteworthy that spectral boundaries of this band-gap are shifted towards larger wavelengths. This result is due to the growth of refractive index of SiO_2 in the ultra-violet spectral range.

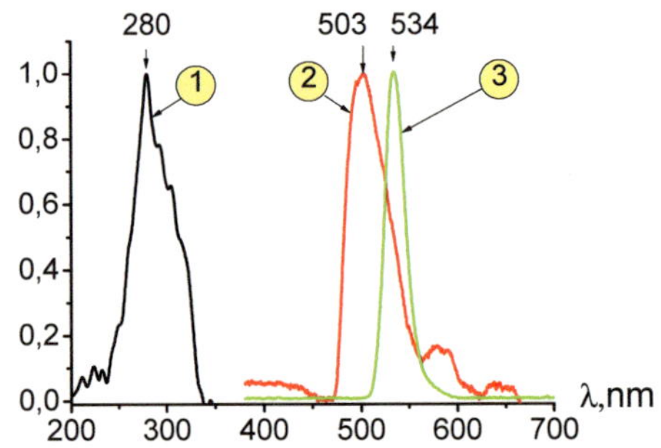

Fig. 20. The reflectance spectra of the globular PTC, filled with air (curve (2)) and water (curves (1) and (3)). The curves (2) and (3) are related to using the halogen lamp with a broad bandwidth in the visible range. The curve (1) is related to using the deuterium lamp with a broad bandwidth in the ultra-violet range.

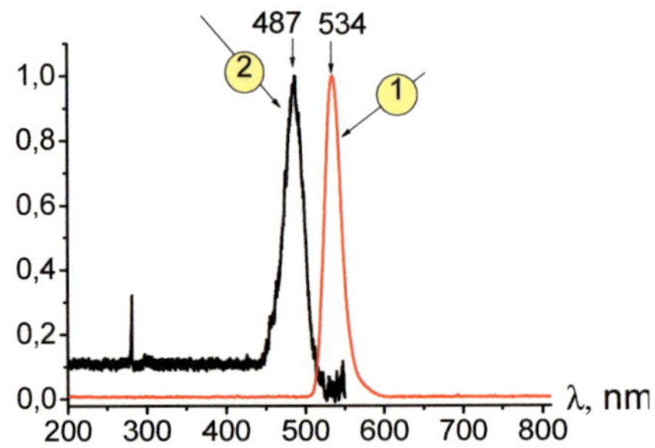

Fig. 21. The reflectance spectrum for the initial PTC (the curve 2), and the PTC doped with the nanoparticles of gold (the curve 1).

Fig. 21 shows the reflectance spectra for the same geometry of an incident light (the radiation is reflected from (111) surface of the crystal), but now the golden particles are implanted into the crystal pores by the technique of laser ablation. As is seen in this Figure, the implantation of metal into the pores results in shifting the band-gap to higher frequencies, which is due to the fact that the real part of the metal dielectric permittivity in the range of optical frequencies is negative.

1.5 Acoustic properties of globular PNC

Basing on the classical Lamb model, in Refs. [6, 7] the theory of natural oscillations (modes) of isolated isotropic spherical globules was developed. In this theory the existence of two kinds of globular oscillations (modes), characterized by the subscripts l and n, was predicted. For describing these modes the following dimensionless values were introduced:

$$\xi_{nl} = \frac{\pi v_{nl} D}{V_L}, \quad \eta_{nl} = \frac{\pi v_{nl} D}{V_T}. \tag{41}$$

Here V_L and V_T are the velocities of longitudinal and transverse acoustic waves accordingly, D is the diameter of globules, v_{nl} are the corresponding frequencies in Hz. The equation for the eigenvalues ξ_{nl} and η_{nl} related to the oscillating modes, which are induced in a sphere, has the form:

$$\begin{aligned} &2\left[\eta^2 + (l-1)(l+2)\left(\frac{\eta j_{l+1}(\eta)}{j_l(\eta)} - (l+1)\right)\right]\frac{\xi j_{l+1}(\xi)}{j_l(\xi)} - \frac{1}{2}\eta^4 + \\ &+(l-1)(2l+1)\eta^2 + \left[\eta^2 - 2l(l-1)(l+2)\right]\frac{\eta j_{l+1}(\eta)}{j_l(\eta)} = 0 \end{aligned} \tag{42}$$

where η and ξ are the corresponding eigenvalues, and $j_l(\eta)$ is spherical first order Bessel function. The solution to this equation gives the following relationship between the frequencies:

$$v_{nl} = \frac{v_0(n,l)}{D}, \tag{43}$$

where $v_0(n, l)$ is some function, dependent upon the numbers n and l.

The modes characterized by even numbers n and l are the Raman-active ones, and thus can contribute to the spectra of two-photon light scattering (by contrast to the libration modes, which cannot be displayed in the two-photon processes due the rules of selection). The equation (43) was analyzed in Refs. [6, 7] for the spherical globules made of quartz; the velocities V_L = 5279 m/s and V_T =3344 m/s for the longitudinal and transverse sound velocities in the amorphous quartz were substituted in the corresponding equations. The calculated values of the frequencies in the GHz frequency range for some globular modes are the following:

$$v_{10} = 2.617/D = 0.44 \text{ cm}^{-1}, \; v_{20} = 4.017/D = 0.68 \text{ cm}^{-1}, \tag{44}$$

where D = 200 nm, which is in a good conformity with the experimental data, see below. Thus, in the case of the opal matrixes the nano-sized spherical globules play a role of vibrating molecules. The standing waves are induced in each globule of the crystal. The pulsating modes arising in the PNC globules are related to the movements, resulting in the change of the globule material density. This is why the vibrating excitation of one particular globule can transfer to another globule; accordingly the excitation wave of the globules can travel along the crystal. As is known, it is possible to observe various kinds of non-elastic scattering in medium, e.g., the Raman scattering, the Brillouin scattering, the Bragg scattering etc. In the case of Raman scattering the oscillatory quanta corresponding to the molecular vibrations are excited (or damped). Thus if we deal with PNC, the globules with the size of several hundred of nanometers play a role of vibrating molecules. Accordingly, the non-elastic scattering of light caused by the excitations of radial vibrations of the globules was termed as Globular Scattering (GS) of light. At low intensities of incident radiation this scattering is of spontaneous character. In Ref. [10] the spectra of spontaneous GS in the synthetic opals were for the first time observed at irradiation of a CW Ar^{++} - laser with the wavelength of 514.5 nm in the back-scattering geometry. For such measurements the synthetic opals having the effective sphere diameter D = 204, 237, 284 and 340 nm were used.

The GS spectrum investigated in this work consisted of six well-pronounced Stokes and anti-Stokes spectral peaks, whose frequencies could be associated with the resonant globular modes belonging to the range of 7 - 27 GHz. The presence of the anti-Stokes satellites is explained by a high "population density" of low vibration states at room temperature. As was found out, the frequencies and relative intensities of the satellites do not depend on the polarization and the angle of incidence of the radiation. Besides, these parameters did not change at rotating the sample around the normal axis in the point of incidence of laser radiation. In Ref. [10] the dependence of frequency of various acoustic modes upon the sphere diameter was studied. As against to the spontaneous Brillouin scattering, GS can be observed both in the "forward" and "backward" geometry. The frequency shift for GS appears to be essentially smaller than that for Raman scattering caused by the molecular vibrations.

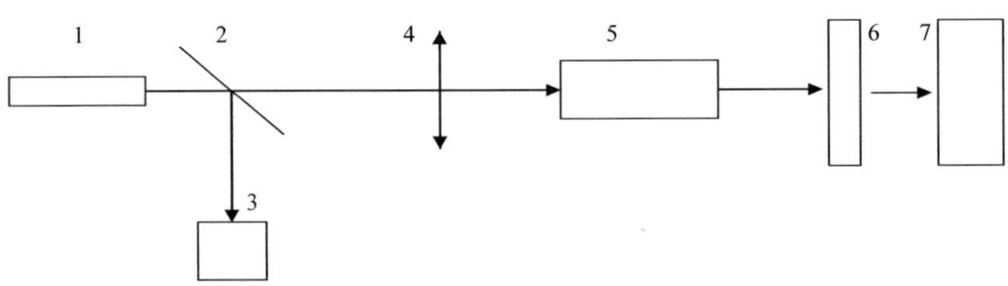

Fig. 22. The schematic of experimental setup for observing the Stimulated Globular Scattering (SGS) in the "forward" geometry; 1 – Ruby laser, 2 – half-transparent mirror, 3 – power meter, 4 - focusing system, 5 – the sample under study, 6 - the Fabri-Perot interferometer, 7 – mini-spectrometer

The experiments to observe the Stimulated Globular Scattering (SGS) in the PTC were first described in [6]. The schematic of experimental setup for observing this scattering in the "forward" and "backward" geometry is illustrated in Figs. 22 and 23 accordingly. Here the pulsed Ruby laser with the wavelength of 694.3 nm, the bandwidth 0.015 cm^{-1}, the pulsewidth of 20 ns, and the pulse energy of 0.4 J was used. The laser radiation was directed with the help of focusing lens system 4 (Fig. 22) or 6 (Fig. 23) onto the PTC sample mounted on a copper cooler and placed into a basin made of a foam plastic. We used the lenses of

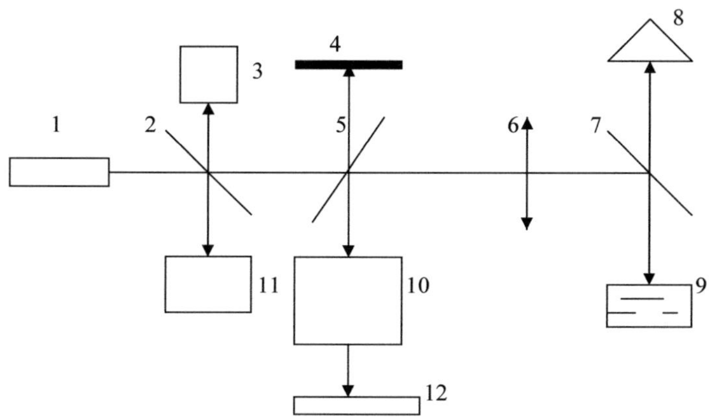

Fig. 23. The schematic of experimental setup for observing the Stimulated Globular Scattering (SGS) in the "backward" geometry; 1 - Ruby laser, 2, 5, 7 - half-transparent mirrors, 3, 8, 11 - power meters, 4 - the mirror, which can be removed (optional), 6 - focusing system, 9 - the sample under study, 10 - the Fabri-Perot interferometer, 12 – mini-spectrometer.

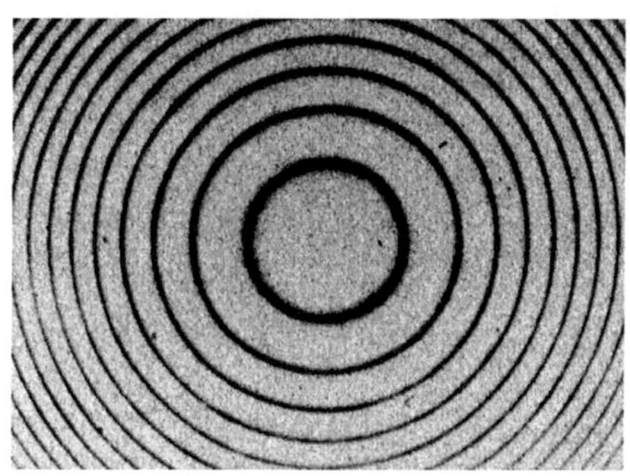

(a)

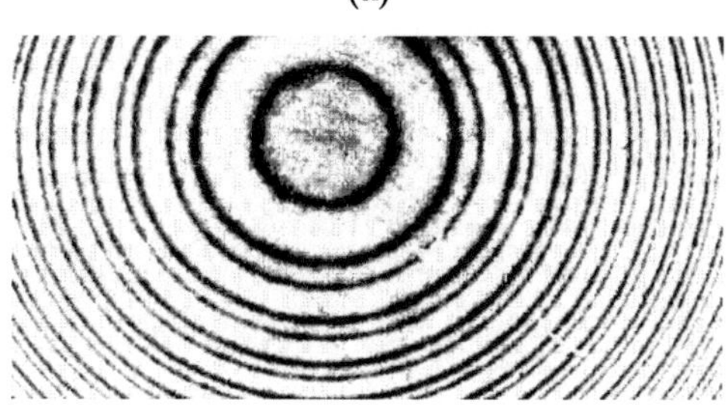

(b)

Fig. 24. The interferograms, obtained with the help of the Fabri-Perot interferometer, relating to the incident radiation spectrum of the Ruby laser (λ = 694.3 nm), case (a), and to the spectrum of SGS in the "backward" geometry, case (b). In the second photo (case (b)) the system of double rings corresponds to the incident wave (the rings of smaller radius; the same rings can be found in the first photo, case (a)), and its scattering Stokes satellite (the rings of greater radius). In this particular case the free spectral range of the interferometer was equal to 0.833 cm^{-1}.

different focal lengths: 50, 90 and 150 mm. Thus it was possible to perform the measurements for various intensities of the radiation as well as for various electromagnetic field distributions inside the sample. The PTC samples under study were manufactured of thin (their thickness was of 2 - 4 mm) synthetic opal plates with the interface corresponding to the (111) crystal plane. The laser radiation was focused normally to the crystal interface. The scattered radiation in the "forward" geometry (Fig. 22) was analyzed in the incident wave direction with the help of the Fabri-Perot interferometer and a mini-spectrometer 7. In case of the "backward" geometry (Fig. 23) the scattered radiation was analyzed in the opposite (with respect to the incident wave) direction with the help of the Fabri-Perot interferometer 10 and a mini-spectrometer 12. The mirror 4 (Fig. 23) was used for comparing the spectra of incident Ruby laser radiation and the scattered radiation.

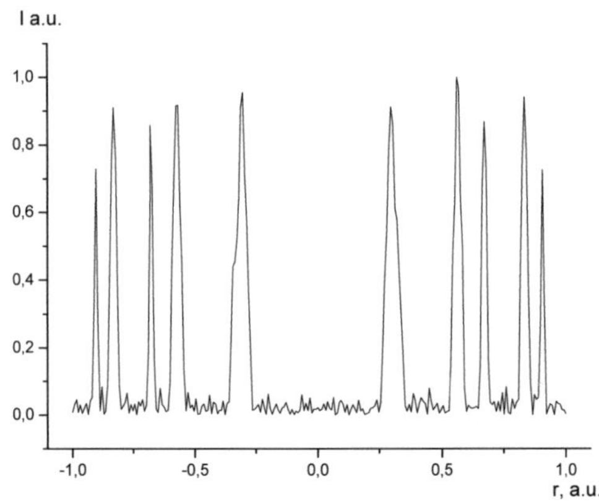

Fig. 25. The Stimulated Globular Scattering (SGS) spectrum in the PTC, obtained in the "backward" geometry. The free spectral range of the Fabri-Perot interferometer was of 0.833 cм$^{-1}$. The broad lines are related to the incident laser radiation, while the narrow ones – to the Stokes satellites of the SGS.

The taking of the scattered radiation spectra was carried out at the room temperature, and also at cooling the samples up to the temperature of nitrogen boiling (78 K). In the latter case copper cooler with samples was placed into a foam plastic basin filled with the liquid nitrogen, while the crystal interface (111) was always above the level of the boiling nitrogen. For researching the spectra the Fabri-Perot interferometers with various (from 0.42 to 1.67 cm^{-1}) free spectral ranges were used. The detectors 3 (Fig. 22), and 3, 8, 11 (Fig. 23) for measuring the energy of laser radiation and the signal of the SGS in the "forward" and "backward" geometry accordingly were applied.

In Fig. 24 (a) and (b) the interferograms related to the setup illustrated in Fig. 23 are given. Fig. 24 (a) shows the spectrum of incident Ruby laser radiation obtained at blocking the scattering signal by turning the half-transparent mirror 5 to the corresponding angle. In this case the spectral pattern had the form of the pattern of single interference rings, whose bandwidth was controlled by incident radiation bandwidth, which was on the level of 0.015 cm^{-1}. Figs. 24 (b) and 25 are related to the experiment, where the scattered radiation was studied in the "backward" geometry. The patterns of double interference rings related to the incident radiation (the rings of smaller radius), and the Stokes satellites related to the SGS (the rings of greater radius) are clearly seen. The frequency shift in this case was about 0.44

cm^{-1}. The Stokes signal intensity appeared to be comparable with the incident radiation intensity. In the case of missing the mirror 4 (see Fig. 23), the spectrum would contain only the pattern of single rings caused by the SGS only, as in this case the incident radiation strongly diverges after its reflecting from the sample interface, and this signal does not input to the detector 12. We plotted the dependence of the frequency shift *f* for the first Stokes component of the SGS as the function of the inverse diameter (1/*D*) of the globules. It occurred that such dependence is close to the linear one.

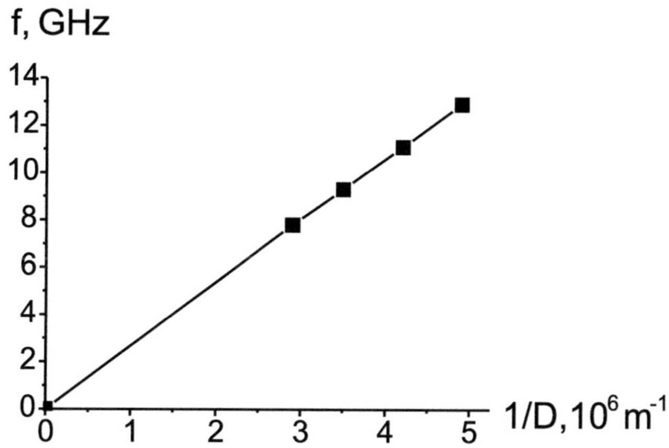

Fig. 26. Dependence of the frequency shift of frequency for the first Stokes component of the SGS versus the inverse diameter of the globules.

The explorations of spectra of the SGS were also performed for the opal matrices filled with liquids of various refractive index n: water (n = 1.333), acetone (n = 1.359), ethanol (n = 1.362), glycerol (n = 1.470), toluene (n = 1.497), benzene (n = 1.501) and nitrobenzene (n = 1.553). Thus, the phase contrast (the value of h = n/nSiO2, i.e. the ratio of refractive index of the liquid to that for quartz) changed in the range from 0.91 to 1.06. For example, impregnation of the opal matrix by acetone sharply decreases the phase contrast, and the sample becomes almost transparent. It provides an opportunity to observe the SGS in the scheme of the "forward" geometry of scattering. Similar to the case of the "backward" geometry, at the "forward" geometry of scattering the pattern of double rings related to the incident radiation (the rings of smaller diameter) and the Stokes component of the SGS (the rings of greater diameter) were seen as well. Due to the transparency of the sample treated by acetone it was possible to observe sufficiently intense Stokes signal of the scattering in the incident wave direction (the "forward" geometry). The Stokes shift in this case was on the level of 0.4 cm-1. Note that the SGS was observed with the liquids pointed above both for the "forward" and "backward" geometry. In the case of "backward" geometry the frequency shift of about 0.4 cm-1 was observed for the incident wave intensity at the level of 0.12 GW/cm2 for the opal matrices, filled with ethanol and acetone. The increase in the incident wave intensity up to 0.21 GW/cm2 resulted in occurrence of the second Stokes components having the frequency shift of about 0.65 cm-1 for acetone, and 0.63 cm-1 for ethanol. At the same time, for the "forward" geometry (and at the room temperature) only one Stokes component with the frequency shift of 0.4 cm-1 was observed both for acetone, and for ethanol.

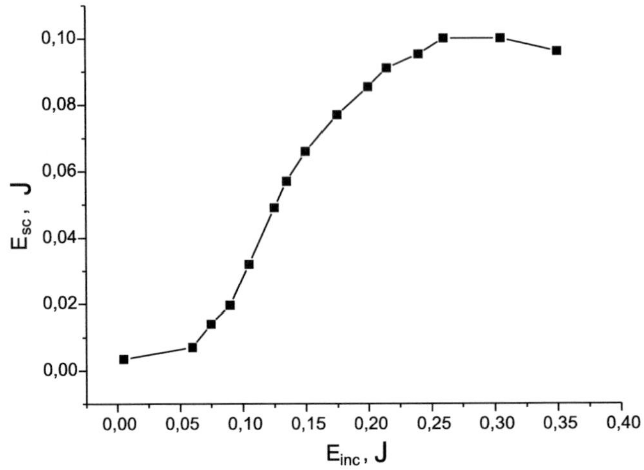

Fig. 27. Dependence of the energy (E_{sc}) of the Stimulated Globular Scattering versus the energy of the incident wave E_l for the synthetic opals with the globule diameter of 245 nm.

The geometry of experiment	The frequency shift, cm^{-1}	The number of the Stokes components
The "forward" geometry, the opal matrix is not saturated by any liquid	0.44	1
The "backward" geometry, the opal matrix is saturated by acetone	0.40 0.65	2
The "forward" geometry, the opal matrix is saturated by acetone	0.40	1
The "backward" geometry, the opal matrix is saturated by ethanol	0.39 0.63	2
The "forward" geometry, the opal matrix is saturated by ethanol	0.37	1

Table 2. The values of the frequency shifts and the number of the Stokes components for various geometries of experiment.

At the decrease of temperature up to 78 K the threshold for the SGS was three times reduced, and the number of observable Stokes components was increased. In the Table 2 the characteristic parameters of the Stokes components, including the frequency shift and the number of the Stokes component observed in the particular geometry of experiment are summarized.

As follows from this table, the values of the frequency shifts for the SGS appear close to those for the Stimulated Brillouin Scattering, observed earlier in the same liquid samples. The threshold for occurrence of SGS and the number of the Stokes components depend on the globule size, the substance in the crystal pores, the energy of incident laser radiation, and the temperature. In Fig. 27 the dependence of energy of the SGS versus the incident light energy is plotted for the liquid nitrogen temperature and for the globule diameter equal to 245 nm in case of the "backward" geometry. As is seen from the graph, at increasing the incident wave energy the SGS signal approaches the saturation level; note that the similar effect has been observed earlier for the Stimulated Raman Scattering. It is thus possible to explain the observed effect by increasing the efficiency of competing nonlinear processes at the growth of intensity of the incident radiation. The highest transformation factor of the incident radiation energy to the SGS energy obtained in our experiments was on the level of 60 %. As the quantum energy of the incident radiation is by four orders of magnitude higher than the energy of the corresponding radial vibrations of the globules, the intensity of acoustic phonons, generated by the globular vibrations should be approximately equal to 10^3 W/cm^2. The lowest threshold for the SGS was realized for the so-called pulsating globular modes, the excitation of which should be accompanied by the oscillations of material density of the globules; this process is expected to have a high Q-factor. The corresponding acoustic waves have a scalar nature, since they are induced by radial vibrations of globules, and their propagation is not associated with any specific direction. In other words, these acoustic waves essentially differ from the acoustic waves of the vector type, i.e. from the transverse or longitudinal phonons. The wave equation for such scalar acoustic waves is analogous to the Klein - Gordon equation, describing the behavior of particles with the non-zero rest mass in the field theory:

$$\left(\Delta - \frac{1}{S^2}\frac{\partial^2}{\partial t^2}\right)u(\mathbf{r},t) = \frac{\omega_0^2}{S^2}u(\mathbf{r},t), \tag{45}$$

where S has the sense of a group velocity of the wave at high values of the wave vector **k**, $\omega_0 = 2\pi c\nu_0 \sim 1/D$, ν_0 is the corresponding magnitude of frequency of the pulsating mode, $u(\mathbf{r},t) = u_0\exp(i\mathbf{kr} - \omega t)$ is the scalar wave function describing the propagation of a pulsating perturbation in space. According to (45), the dispersion law $\omega(k)$ for the scalar acoustic wave has the form:

$$\omega^2 = \omega_0^2 + S^2k^2. \tag{46}$$

As was revealed in the experiments (see Fig. 26), the frequency shift of the first Stokes component of the SGS is inversely proportional to the globule diameter D, which is in conformity with Eqn. (46). For the "forward" geometry of experiment the wave vector of the scalar acoustic wave is equal to

$$k = k_0 - k',$$ (47)

where k_0 and k' are wave vectors of the incident radiation and the SGS wave. If we deal with the "backward" geometry, we arrive at

$$k = k_0 + k'.$$ (48)

Thus, analyzing the spectra of the scattered radiation stimulated by the Ruby laser pulses in the PTC, we established that the spectral pattern is the set of double interference rings. These rings are related to the incident laser radiation and the Stokes components of the SGS. The intensity of the Stokes components is of the same order of magnitude as the incident wave intensity. This is an additional argument in favor of our statement, that the observed phenomenon is associated with the stimulated (not spontaneous) scattering.

2. The conclusion

Summarizing, we shown that implantation of various dielectrics, whose refractive index exceeds that for quartz, into the pores of synthetic opals results in shifting the band-gap of the PTC to lower frequencies. At the same time, at implantation of metals into these pores the band-gap is shifted to higher frequencies. Finally, implantation of various substances having the additional resonant absorption lines belonging to the band-gap of the crystal results in occurrence of additional sharp peaks of reflectance either in the long-wave or the short-wave areas of spectrum. The studies of characteristics of the PTC by the technique of reflectance spectroscopy of the band-gaps have revealed the areas of abnormal increase of density the photon states in the globular crystal; these areas are localized close to the band-gap boundaries. The analysis of such areas can allow us to obtain the lasing in the synthetic opals filled with active media. Besides, we expect that in the framework of this approach it will be possible to realize various nonlinear processes, including the Stimulated Raman Scattering of light, [9], Stimulated Globular Scattering of light [6, 9], generation of optical harmonics, parametrical generation [7] and the afterglow phenomena [11, 12].

On the basis of the results reported here we can make the following conclusions about the dynamics of acoustic phonons in the PTC.

1. The spectrum of acoustic phonons in the PTC contains the allowed and forbidden zones in the GHz frequency range, similar to the allowed and forbidden zones for the photons.
2. The group velocities of the acoustic phonons close to the forbidden zones (band-gaps) boundaries sharply decrease.
3. The effective mass of acoustic phonons close to the band-gap boundaries has an abnormally low magnitude, comparable to the mass of electron.

Let us also note that in the globular PTC a new type of standing acoustic elementary waves is possible [6, 7]. These standing acoustic waves are induced in the globules and can be considered as the coupled states of pairs of the acoustic phonons - the so-termed bi-phonons. As was obtained in the experiments [6, 7], such bi-phonons can be induced by the incident optical radiation, and the interaction between the optical wave and the bi-phonons leads to a new type of the stimulated light scattering – the SGS.

Thus, the globular photonic crystal, being irradiated by powerful enough laser light can be generator of monochromatic acoustic waves in the GHz spectral range; the frequency of such waves should depend on the globule parameters and the type of a substance implanted into the crystal pores.

3. Acknowledgement

The given work was supported by the Russian Foundation for Basic Researches, Grants Nos. 08-02-00114, 10-02-00293, 10-02-90042-Bel, 10-02-90404-Ukr, and by the Presidium of Russian Academy of Sciences, Program for Basic Researches No. 21.

4. References

[1] Yablonovich, E. (1987). Inhibited Spontaneous Emission in Solid-State Physics and Electronics, *Physical Review Letters*, Vol. 58, No. 20, pp. 2059 - 2062.

[2] John, S. (1987). Strong Localization of Photons in Certain Disordered Dielectric Superlattices, *Physical Review Letters*, Vol. 58, No. 23, pp. 2486-2489.

[3] Gorelik, V.S., Zlobina, L.I., Troitskii, O.A., *et al.* (2008). LED-excited emission of opal loaded with silver nanoparticles, *Inorganic Materials*, Vol. 44, No. 1, pp. 58 - 61.

[4] Goncharov, A.P., Gorelik, V.S. (2007). Emission of opal photonic crystals under pulsed laser excitation, *Inorganic Materials*, Vol. 43, No. 4, pp. 386 - 391.

[5] Bunkin, N.F., Gorelik, V.S., Filatov, V.V. (2010). Acoustic properties of globular photonic crystals based on synthetic opals, *Physics of Wave Phenomena*, Vol. 18, No. 2, pp. 90 - 95.

[6] Gorelik, V.S. (2007). Optics of globular photonic crystals, *Quantum Electronics*, Vol. 37, No. 5, pp. 409-432.

[7] Gorelik, V.S. (2008). Optics of globular photonic crystals, *Laser Physics*, Vol. 18, No. 12, pp. 1479-1500.

[8] Voshchinskii, Yu.A., Gorelik, V.S. (2011). Dispersion law in photonic crystals in sinusoidal and quasi-relativistic approximation, *Inorganic Materials*, Vol. 47, No. 2, pp. 148 - 151.

[9] Gorelik, V.S. (2010). Linear and nonlinear optical phenomena in nanostructured photonic crystals, filled by dielectrics or metals, *European Journal – Applied Physics*, Vol. 49, No. 3, 3307.

[10] Kuok, M.H., Lim, H.S., Ng, S.C., Liu, N.N., Wang, Z.K. (2003). Brillouin study of the quantization of acoustic modes in nanospheres, *Physical Review Letters*, Vol. 90, No. 25, 255502.

[11] Gorelik, V.S., Kudryavtseva, A.D., Tareeva, M.V., *et al.* (2006). Spectral characteristics of the radiation of artificial opal crystals in the presence of the photonic flame effect, *JETP Letters*, Vol. 84, No. 9, pp. 485-488.

[12] Gorelik, V.S., Esakov, A.A., Zasavitskii, I.I., (2010), Low-temperature persistent afterglow in opal photonic crystals under pulsed UV excitation, *Inorganic Materials*, Vol. 46, No. 6, pp. 639-643.

[13] Gorelik, V.S., Yurasov, N.I., Gryaznov, V.V., *et al.*, (2009), Optical properties of three-dimensional magnetic opal photonic crystals, *Inorganic Materials*, Vol. 45, No. 9, pp. 1013-1017.

Part 2

Acoustic Waves in Fluids

8

A Fourth-Order Runge-Kutta Method with Low Numerical Dispersion for Simulating 3D Wave Propagation

Dinghui Yang, Xiao Ma, Shan Chen and Meixia Wang
Department of Mathematical Sciences, Tsinghua University, Beijing, China

1. Introduction

The numerical solutions of the acoustic-wave equation via finite-differences, finite-elements, and other related numerical techniques are valuable tools for the simulation of wave propagation. Many numerical methods of modeling waves propagating in various different media have been proposed in past three decades (Kosloff & Baysal, 1982; Booth & Crampin, 1983; Virieux, 1986; Dablain, 1986; Chen, 1993; Carcione, 1996; Blanch & Robertsson, 1997; Komatitsch & Vilotte, 1998; Carcione & Helle, 1999; Carcione et al., 1999; Moczo et al., 2000, Yang et al., 2002, 2006, 2007; many others). These modeling techniques for the 1D and 2D cases are typically used as support for a sound interpretation when dealing with complex geology, or as a benchmark for testing processing algorithms, or used in more or less automatic inversion procedure by perturbation of the parameters characterizing the elastic medium until the synthetic records fit the observed real data. In these methods, the finite-difference (FD) methods were leader and popularly used in Acoustics, Geophysics, and so on due to their simplicity for computer codes.

However, it is well-known that the conventional finite-difference (FD) methods for solving the acoustic wave equation often suffer from serious numerical dispersion when too few grid points per wavelength are used or when the models have large velocity contrasts, or artefacts caused by the source at grid points (Fei & Larner 1995, Yang et al., 2002). Roughly speaking, numerical dispersion is an unphysical phenomenon caused by discretizing the wave equation (Sei & Symes, 1995; Yang et al., 2002). Such a phenomenon makes the wave's velocity frequency dependent. More high-order or accurate FD operators have been developed to minimize the dispersion errors, and those modified FD schemes greatly improved the computational accuracy compared to the conventional operators. For example, the staggered-grid FD method with local operators (Virieux, 1986; Fornberg, 1990; Igel et al., 1995) is an efficient and convenient scheme which improves the local accuracy and has better stability without increasing computation cost and memory usage compared to the conventional second-order FD method. However, the staggered-grid (SG) method still suffers from the numerical dispersion when too few sampling points per minimum wavelength are used and may result in the numerical anisotropy and induce additional numerical errors (Virieux, 1986; Igel et al., 1995). Dablain (1986) developed a series of high-order FD schemes for solving the acoustic wave equation, which greatly improved the computational accuracy. But these high-order schemes also can not cure the numerical dispersion effectively when coarse grids are used, and they usually involve in more grids in a spatial direction than low-order schemes (Yang et al., 2006). For example, the tenth-order compact FD scheme (e.g., Wang et al., 2002), which usually uses more grids than low order schemes, also suffers from numerical dispersion. The demand for more grids in high-order FD methods prevents the algorithms from efficient parallel implementation and artificial boundary treatment. The flux-corrected transport (FCT) technique was suggested for eliminating the numerical dispersion (Fei & Larner 1995, Zhang et al., 1999, Yang et al, 2002; Zheng et al., 2006), but the FCT method can hardly recover the resolution lost by numerical dispersion when the spatial sampling becomes too coarse (Yang et al., 2002). On the other hand, waves have inherent dispersions as they propagate in porous media with fluids. This implies that two kinds of dispersions (numerical dispersion and physical dispersion) might occur simultaneously in wave fields if the conventional FD methods are used to compute the wave fields in a porous medium. In such a case, it is not a good idea to use the FCT technique to eliminate the numerical dispersions because we do not know how to choose the proper control parameters used in the FCT method for suppressing the numerical dispersions (Yang et al., 2006). The pseudo-spectral method (PSM) is attractive as the space operators are exact up to the Nyquist frequency, but it requires the Fourier transform (FFT) of wave-field to be made, which is computationally expensive for 3D anisotropic models and has the difficulties of handling non-periodic boundary conditions and the non-locality on memory access of the FFT, which makes the parallel implementation of the algorithms and boundary treatments less efficient (Mizutani et al., 2000). Meanwhile, it also suffers from numerical dispersion in the time direction, and its numerical dispersion is serious as the Courant number, defined by $\alpha = c_0 \Delta t / \Delta x$ (Dablain 1986; Sei & Symes, 1995), is large, i.e. as the time increment is large (Yang et al., 2006).

Another easy way to deal with the numerical dispersion is to use fine grids to increase spatial samples per wavelength. For example, a spatial sampling rate of more than 20 points per shortest wavelength is needed when a second-order FD scheme is used to obtain reliable results (Holberg, 1987), whereas a fourth-order scheme seems to produce accurate results at ten grid points per shortest wavelength. Dablain (1986) states that eight and four grid points at the Nyquist frequency are required to eliminate numerical dispersion using second-order and fourth-order FD methods, respectively. More grid points per wavelength mean more computational cost and storage. It is not advisable to apply these techniques in large-scale computation, especially for a large scale 3D simulation of seismic wave propagation because of an intensive use of Central Processing Unit (CPU) time and the requirement of a large amount of direct-access memory. Fortunately, with the rapid development of computer performance and the birth of parallel technology in past several decades, 3D wave simulation through using different numerical methods on a large scale or high frequencies becomes affordable, and the study of 3D numerical techniques has been a hot spot and rapidly developed because of its applying to practical issues in the fields of Acoustics and Geophysics.

Recently, the so-called nearly analytic discrete (NAD) method and optimal NAD (ONAD) (Yang et al., 2006) suggested by Yang et al. (2003) for acoustic and elastic equations, which was initially developed by Konddoh et al. (1994) for solving parabolic and hyperbolic equations, is another effective method for decreasing the numerical dispersion. The method, based on the truncated Taylor expansion and the local interpolation compensation for the truncated Taylor series, uses the wave displacement-, velocity- and their gradient fields to restructure the wave displacement-fields. On the basis of such a structure, the NAD and ONAD methods can greatly increase the computational efficiency and save the memory storage. However, the NAD method has only second-order time accuracy. The ONAD method is effective in solving the acoustic and elastic wave equations for a single-phase medium, and it can not be applied to a two-phase porous wave equations such as Biot's porous wave equations (Biot, 1956a, b), because these equations include the particle velocity $\partial U/\partial t$ (U is the wave displacement) and the ONAD method does not compute this field. More recently, the NAD and ONAD methods were also extended to solve the Biot poroelastic equations (Yang et al., 2007a) and the three-dimensional anisotropic wave equations (Yang et al., 2007b).

The main purpose of this chapter is to develop a new 3D numerical method to effectively suppress the numerical dispersion caused by the discretization of the acoustic- and elastic-wave equations through using both the local spatial difference-operator and the fourth-order Runge-Kutta (RK) method so that the numerical technique developed in this chapter has rapid computational speed and can save the memory storage. For to do this, we first transform the original wave equations into a system of first-order partial differential equations with respect to time t, then we use the local high-order interpolation of the wave displacement, the particle velocity, and their gradients to approximate the high-order spatial derivatives, which effectively converts the wave equation to a system of semi-discrete ordinary differential equations (ODEs). Finally, we use the fourth-order RK method to solve the semi-discrete ODEs, and change the 4-stage RK formula to 2-stage scheme resulting that the modified 3D RK algorithm can save the memory storage. Based on such a structure, this method has fourth-order accuracy both in time and space, and it can be directly extended to solve the two-phase porous wave equations including the particle velocity $\partial U/\partial t$ (Biot, 1956a,b) because of simultaneously obtaining the velocity fields when computing the displacement fields.

To demonstrate the numerical behavior of this new method, in this chapter we provide the theoretical study on the properties of the 3D RK method: such as stability criteria, theoretical error, numerical dispersion, and computational efficiency, and compare the numerical error of the 3D RK with those of the second-order conventional FD scheme and the fourth-order LWC method for the 3D initial value problem of acoustic wave equation. Meanwhile, we also compare the numerical solutions computed by the 3D RK with the analytical solutions, and present some wave-field modeling results of this method against those of some high-order FD schemes including the SG and LWC methods for the acoustic case. Besides, we also present the synthetic seismograms in the 3D three-layer isotropic medium and the wave field snapshots in the 3D two-layer medium and the 3D transversely isotropic medium with a vertical symmetry axis (VTI). All these promising numerical results illustrate that the 3D RK can suppress effectively the numerical dispersion caused by discretizing the wave equations when too few sampling points per minimum wavelength are used or models have large velocity contrasts between adjacent layers, further resulting in both increasing the computational efficiency and saving the memory storage when big grids are used. These numerical results also imply that simultaneously using both the wave displacement and its gradients to approximate the high-order spatial derivatives is important for both reducing the numerical dispersion and compensating the important wave field information included in the displacement and particle velocity gradients.

2. Fourth-order RK method for solving ODEs

2.1 Basic RK algorithm

Consider the following ordinary differential equation

$$\frac{du}{dt} = L(u). \tag{1}$$

Where, u is an unknown function of time t, and L is a known operator with respect to u at each spatial point (i, j, k) for the 3D case. Equation (1) can be solved as an ordinary equation using the following fourth-order Runge-Kutta method

$$\begin{cases} u^{(1)} = u^n + \frac{1}{2}\Delta t L(u^n), \\ u^{(2)} = u^n + \frac{1}{2}\Delta t L(u^{(1)}), \\ u^{(3)} = u^n + \Delta t L(u^{(2)}), \\ u^{n+1} = \frac{1}{3}\left(-u^n + u^{(1)} + 2u^{(2)} + u^{(3)}\right) + \frac{1}{6}\Delta t L(u^{(3)}). \end{cases} \tag{2}$$

Where, Δt is the temporal increment, $u^n = u(n\Delta t)$, and $u^{(1)}$, $u^{(2)}$ and $u^{(3)}$ are the intermediate variables. Equation (2) shows that the RK algorithm needs to store these three intermediate variables at each time advancing step, so the storage required for computer code is very large for 3D problems. To save storage, we can equivalently change it into the following two-stage scheme

$$\begin{cases} u^* = u^n + \frac{1}{2}\Delta t L(u^n) + \frac{1}{4}\Delta t^2 L^2(u^n), \\ u^{n+1} = \frac{1}{3}\left(u^n + 2u^*\right) + \frac{1}{3}\Delta t L(u^n) + \frac{1}{3}\Delta t L(u^*) + \frac{1}{6}\Delta t^2 L^2(u^*). \end{cases} \tag{3}$$

Where $L^2 = L \cdot L$. Algorithm (3) uses only one intermediate variable u^*, resulting in that the modified two-stage RK used in this chapter can effectively save the computer memory in the 3D wave propagation modeling.

2.2 Transformations of 3D wave equations

In a 3D anisotropic medium, the wave equations, describing the elastic wave propagation, are written as

$$\frac{\partial \sigma_{ij}}{\partial x_j} + f_i = \rho \frac{\partial^2 u_i}{\partial t^2}, \tag{4a}$$

$$\sigma_{ij} = \frac{1}{2} c_{ijkl} \left(\frac{\partial u_k}{\partial x_l} + \frac{\partial u_l}{\partial x_k} \right), \tag{4b}$$

where subscripts i, j, k and l take the values of 1, 2, 3, $\rho=\rho(x,y,z)$ is the density, u_i and f_i denote the displacement component and the force-source component in the i-th direction,

and x_1, x_2 and x_3 are x, y, and z directions, respectively. σ_{ij} are the second-order stress tensors, c_{ijkl} are the fourth-order tensors of elastic constants which satisfy the symmetrical conditions $c_{ijkl} = c_{jikl} = c_{ijlk} = c_{klij}$, and may be up to 21 independent elastic constants for a 3D anisotropic case. Specially, for the isotropic and transversely isotropic case, the 21 independent elastic constants are reduced to two Lamé constants (λ and μ) and five constants (c_{11}, c_{13}, c_{33}, c_{44}, and c_{66}), respectively.
To demonstrate our present RK method, we transform equation (4) to the following vector equation using the stress-strain relation (4b)

$$\rho\frac{\partial^2 U}{\partial t^2} = D\cdot U + f. \tag{5}$$

Where $U = (u_1, u_2, u_3)^T$, $f = (f_1, f_2, f_3)^T$, D is a second-order partial differential operator with respect to space coordinates. For instance, for a transversely isotropic homogenous case, the partial differential operator can be written as follows

$$D = \begin{bmatrix} c_{11}\frac{\partial^2}{\partial x^2} + c_{66}\frac{\partial^2}{\partial y^2} + c_{55}\frac{\partial^2}{\partial z^2} & (c_{12}+c_{66})\frac{\partial^2}{\partial x\partial y} & (c_{13}+c_{55})\frac{\partial^2}{\partial x\partial z} \\ (c_{12}+c_{66})\frac{\partial^2}{\partial x\partial y} & c_{66}\frac{\partial^2}{\partial x^2} + c_{22}\frac{\partial^2}{\partial y^2} + c_{44}\frac{\partial^2}{\partial z^2} & (c_{23}+c_{44})\frac{\partial^2 u_z}{\partial y\partial z} \\ (c_{13}+c_{55})\frac{\partial^2}{\partial x\partial z} & (c_{23}+c_{44})\frac{\partial^2}{\partial y\partial z} & c_{55}\frac{\partial^2}{\partial x^2} + c_{44}\frac{\partial^2}{\partial y^2} + c_{33}\frac{\partial^2}{\partial z^2} \end{bmatrix}.$$

Let $w_i = \partial u_i / \partial t, i = 1,2,3$, and $W = (w_1, w_2, w_3)^T$, then equation (5) can be rewritten as

$$\begin{aligned} \frac{\partial U}{\partial t} &= W, \\ \frac{\partial W}{\partial t} &= \frac{1}{\rho}D\cdot U + \frac{1}{\rho}f. \end{aligned} \tag{6}$$

Define $V = (U, W)^T$, then equation (6) can be further written as

$$\frac{\partial V}{\partial t} = L\cdot V + F, \tag{7}$$

where $L = \begin{bmatrix} 0 & I_{3\times 3} \\ \frac{1}{\rho}D & 0 \end{bmatrix}, F = \begin{bmatrix} 0 \\ \frac{1}{\rho}f \end{bmatrix}$, $I_{3\times 3}$ is the third-order unit operator.

Define the following vectors and operator matrix:

$$\overline{V} = [V, \frac{\partial V}{\partial x}, \frac{\partial V}{\partial y}, \frac{\partial V}{\partial z}]^T,$$

$$\overline{F} = [F, \frac{\partial F}{\partial x}, \frac{\partial F}{\partial y}, \frac{\partial F}{\partial z}]^T,$$

and

$$\overline{L} = \begin{pmatrix} L & 0 & 0 & 0 \\ 0 & L & 0 & 0 \\ 0 & 0 & L & 0 \\ 0 & 0 & 0 & L \end{pmatrix}.$$

With the previous three definitions, in a homogeneous medium, we have the following equation:

$$\frac{\partial \overline{V}}{\partial t} = \overline{L} \cdot \overline{V} + \overline{F}. \tag{8}$$

2.3 3D fourth-order RK algorithm

We suppose that equation (8) is a semi-discrete equation, on the right–hand side of which the high-order spatial derivatives are explicitly approximated by the local interpolation method (Yang et al., 2010). Under such an assumption, Equation (8) is converted to a system of semi-discrete ODEs with respect to variable $\overline{V}$, and can be solved by the fourth-order RK method (formula (3)). In other words, we can apply formula (3) to solve the semi-discrete ODEs (8) as follows

$$\overline{\mathbf{V}}^*_{i,j,k} = \overline{\mathbf{V}}^n_{i,j,k} + \frac{1}{2}\Delta t \overline{\mathbf{L}}\overline{\mathbf{V}}^n_{i,j,k} + \frac{1}{4}\Delta t^2 \overline{\mathbf{L}}^2 \overline{\mathbf{V}}^n_{i,j,k}, \tag{9a}$$

$$\overline{\mathbf{V}}^{n+1}_{i,j,k} = \frac{1}{3}\left(\overline{\mathbf{V}}^n_{i,j,k} + 2\overline{\mathbf{V}}^*_{i,j,k}\right) + \frac{1}{3}\Delta t \overline{\mathbf{L}}\overline{\mathbf{V}}^n_{i,j,k} + \frac{1}{3}\Delta t \overline{\mathbf{L}}\overline{\mathbf{V}}^*_{i,j,k} + \frac{1}{6}\Delta t^2 \overline{\mathbf{L}}^2 \overline{\mathbf{V}}^*_{i,j,k}. \tag{9b}$$

Where $\overline{V}^n_{i,j,k} = \overline{V}(n\Delta t, i\Delta x, j\Delta y, k\Delta z)$ and the differential operators can be written as

$$\begin{aligned} \overline{L} &= Diag(L,L,L,L) \\ &= Diag\left[\begin{pmatrix} 0 & I_{3\times3} \\ \frac{1}{\rho}D & 0 \end{pmatrix}, \begin{pmatrix} 0 & I_{3\times3} \\ \frac{1}{\rho}D & 0 \end{pmatrix}, \begin{pmatrix} 0 & I_{3\times3} \\ \frac{1}{\rho}D & 0 \end{pmatrix}, \begin{pmatrix} 0 & I_{3\times3} \\ \frac{1}{\rho}D & 0 \end{pmatrix}\right]. \end{aligned} \tag{10a}$$

$$\begin{aligned} \overline{L}^2 &= Diag(L^2, L^2, L^2, L^2) \\ &= Diag\left[\frac{1}{\rho}D, \frac{1}{\rho}D, \frac{1}{\rho}D, \frac{1}{\rho}D, \frac{1}{\rho}D, \frac{1}{\rho}D, \frac{1}{\rho}D, \frac{1}{\rho}D\right]. \end{aligned} \tag{10b}$$

From equation (9) and definitions of $\overline{L}$ and $\overline{L}^2$, we know that the calculations of $\overline{V}^*_{i,j,k}$ and $\overline{V}^{n+1}_{i,j,k}$ only involve in the second- and third-order spatial derivatives of the displacement U

and the particle velocity W, so we can compute these derivatives using equations (A3)-(A7) (see Appendix A).

3. Error analysis and stability conditions

In this section, we investigate the stability criteria and theoretical error of the two-stage RK scheme, and compare the numerical error of the 3D RK with those of the second-order conventional FD scheme and the fourth-order LWC method (Dablain, 1986) for the 3D initially value problem of acoustic wave equation.

3.1 Stability conditions

In order to keep numerical calculation stable, we must consider how to choose the appropriate time and the space grid sizes, Δt and h. As we know, mathematically, the Courant number defined by $\alpha = c_0 \Delta t / h$ gives the relationship among the acoustic velocity c_0 and the two grid sizes, we need to determine the range of α. Following the Fourier analysis (Richtmyer & Morton, 1967; Yang et al., 2006, 2010), after some mathematical derivations (see Appendix B for detail), we obtain the stability conditions for solving 1D, 2D, and 3D acoustic equation as follows:

$$\text{1D case: } \Delta t \le \alpha_{\max} \frac{h}{c_0} \approx 0.730 \frac{h}{c_0}, \tag{11}$$

$$\text{2D case: } \Delta t \le \alpha_{\max} \frac{h}{c_0} \approx 0.707 \frac{h}{c_0}, \tag{12}$$

$$\text{3D case: } \Delta t \le \alpha_{\max} \frac{h}{c_0} \approx 0.577 \frac{h}{c_0}. \tag{13}$$

Where, $\alpha_{\max}$ is the maximum value of the Courant number, $\Delta x = h$ for the 1D case, $\Delta x = \Delta z = h$ for the 2D case, and $\Delta x = \Delta y = \Delta z = h$ for the 3D case.

When the RK method is applied to solve the 3D elastic wave equations, we estimate that the temporal grid size should satisfy the following stability condition,

$$\Delta t \le \Delta t_{\max} \approx 0.577 \frac{h}{c_{\max}}, \tag{14}$$

where $\Delta t_{\max}$ is the maximum temporal increment that keeps the 3D RK method stable and $c_{\max}$ is the maximum P-wave velocity.

The stability condition for a heterogeneous medium can not be directly determined, but it could be approximated by using a local homogenization theory. Equations (11)-(14) are approximately correct for a heterogeneous medium if the maximal values of the wave velocities c_0 and $c_{\max}$ are used.

3.2 Error

To better understand the 3D RK method, we investigate its accuracy both theoretically and numerically, and we also compare it with the fourth-order LWC method (Dablain, 1986) and the second-order conventional FD method (Kelly et al., 1976).

3.2.1 Theoretical error

Using the Taylor series expansion, we find that the errors for the spatial derivatives $(\partial^{q+l+m}U / \partial x^k \partial y^l \partial z^m)_{i,j,k}^n$ $(2 \le q+l+m \le 3)$ are fourth order (i.e. $O(\Delta x^4 + \Delta y^4 + \Delta z^4)$), which results from the local interpolation as formulated in equations (A3)-(A7) in Appendix A. This conclusion is consistent with that given by Yang et al. (2007). Because the fourth-order Runge-Kutta method is used to solve the ODEs in equation (8), the temporal error, caused by the discretization of the temporal derivative, is in the order of $O(\Delta t^4)$. Therefore, we conclude that the error introduced by the two-stage RK scheme (9) is in the order of $O(\Delta t^4 + \Delta x^4 + \Delta y^4 + \Delta z^4)$. In other words, the 3D RK method suggested in this chapter has fourth-order accuracy in both time and space.

3.2.2 Numerical errors

In order to investigate the numerical error of the two-stage RK method proposed in this chapter, we consider the following 3D initial value problem:

$$\begin{cases} \dfrac{\partial^2 u}{\partial x^2} + \dfrac{\partial^2 u}{\partial y^2} + \dfrac{\partial^2 u}{\partial z^2} = \dfrac{1}{c_0^2} \dfrac{\partial^2 u}{\partial t^2}, \\ u(0,x,y,z) = \cos\left[-\dfrac{2\pi f_0}{c_0}(l_0 x + m_0 z + n_0 z) \right], \\ \dfrac{\partial}{\partial t} u(0,x,y,z) = -2\pi f_0 \sin\left[-\dfrac{2\pi f_0}{c_0}(l_0 x + m_0 z + n_0 z) \right], \end{cases} \tag{15}$$

where c_0 is the velocity of the plane wave, f_0 is the frequency, and (l_0, m_0, n_0) is the incident direction at t=0.

Obviously, the analytical solution for the initial problem (15) is given by

$$u(t,x,y,z) = \cos\left[2\pi f_0 \left(t - \frac{x}{c_0} l_0 - \frac{y}{c_0} m_0 - \frac{z}{c_0} n_0 \right) \right]. \tag{16}$$

For comparison, we also use the second-order FD method and the so-called LWC (fourth-order compact scheme (Dablain, 1986)) to solve the initial problem (15).

In the first numerical example, we choose the number of grid points $N = 100$, the frequency f_0=15Hz, the wave velocity c_0=2.5km/s, and $(l_0, m_0, n_0) = (\frac{1}{\sqrt{3}}, \frac{1}{\sqrt{3}}, \frac{1}{\sqrt{3}})$. The relative error ($E_r$) is the ratio of the RMS of the residual ($u_{j,m,l}^n - u(t_n, x_j, y_m, z_l)$) and the RMS of the exact solution $u(t_n, x_j, y_m, z_l)$. Its explicit definition is as follows:

$$E_r(\%) = \left\{ \frac{1}{\sum_{j=1}^{N} \sum_{m=1}^{N} \sum_{l=1}^{N} [u(t_n, x_j, y_m, z_l)]^2} \sum_{j=1}^{N} \sum_{m=1}^{N} \sum_{l=1}^{N} [u_{j,m,l}^n - u(t_n, x_j, y_m, z_l)]^2 \right\}^{\frac{1}{2}} \times 100. \tag{17}$$

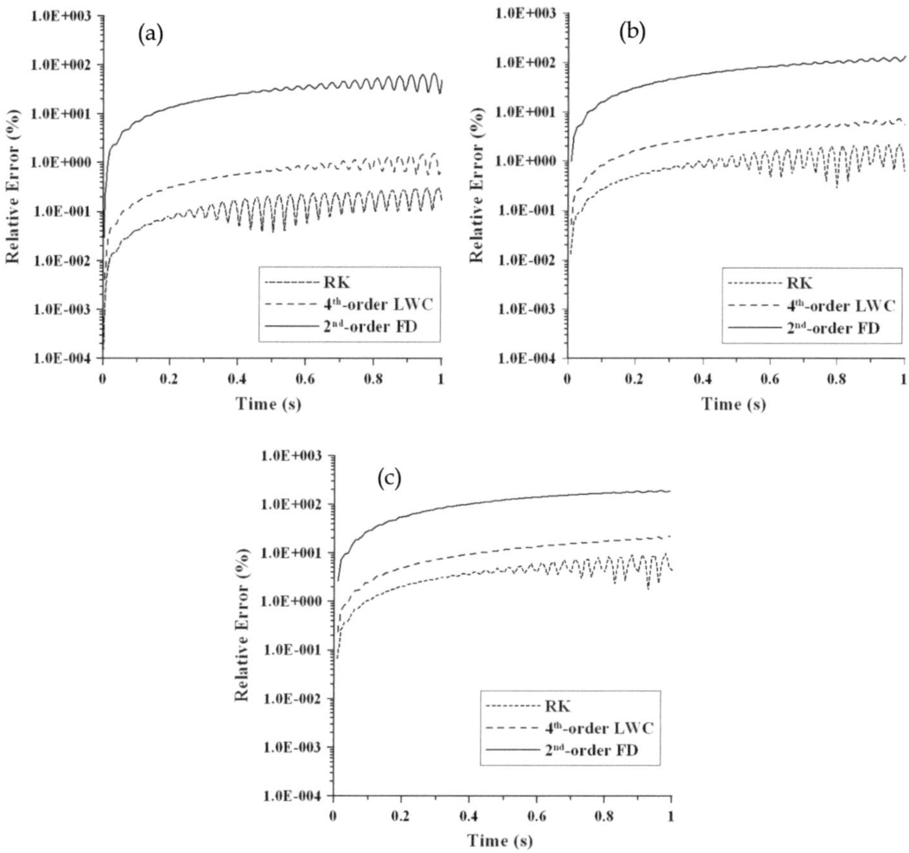

Fig. 1. The relative errors of the second-order FD, the fourth-order LWC, and the RK methods measured by E_r (formula (17)) are shown in a semilog scale for the 3D initial-value problem (15). The spatial and temporal step sizes are chosen by (a) $h=\Delta x=\Delta y=\Delta z=20$m and $\Delta t=5\times10^{-4}$s, (b) $h=\Delta x=\Delta y=\Delta z=30$m and $\Delta t=8\times10^{-4}$s, and (c) $h=\Delta x=\Delta y=\Delta z=40$m and $\Delta t=1\times10^{-3}$s, respectively.

Figures 1(a)-(c) show the computational results of the relative error E_r at different times for cases of different spatial and time increments, where three lines of E_r for the second-order FD method (line —), the fourth-order LWC (line - - - -), and the RK (line -----) are shown in a semi-log scale. In these figures, the maximum relative errors for different cases are listed in Table 1. From these error curves and Table 1 ($\Delta x = \Delta y = \Delta z = h$), we find that E_r increases corresponding to the increase in the time and/or spatial increments for all the three methods. As Figure 1 illustrated, the two-stage RK has the highest numerical accuracy among all three methods

3.3 Convergence order

In this subsection, we discuss the convergence order of the WRK method. In this case, we similarly consider the 3D initial problem (15), and choose the computational domain as $0 \le x \le 1$ km, $0 \le y \le 1$ km, $0 \le z \le 1$ km and the propagation time T =1.0 sec. The same computational parameters are chosen as those used in subsection 3.2.2. In Table 2, we show

Method		2nd-order FD	4th-order LWC	RK
Case 1：	h=20 m $\Delta t=5\times10^{-4}$ s	1.550	2.088	0.306
Case 2：	h=30 m $\Delta t=8\times10^{-4}$ s	7.260	3.963	2.231
Case 3：	h=40 m $\Delta t=1\times10^{-3}$ s	22.298	15.715	9.949

Table 1. Comparisons of maximum relative errors of the three methods in three cases.

the numerical errors of the variable u. For the fixed spatial grid size $h=\Delta x=\Delta y=\Delta z$, the error of the numerical solution u_h with respect to the exact solution u is measured in the discrete L^1, L^2 norms

$$E_{L^m}=\left\|u_h-u\right\|_{L^m}=\left(h^3\sum_{i=1}^{N}\sum_{j=1}^{N}\sum_{k=1}^{N}\left|u_h(x_i,y_j,z_k,T)-u(x_i,y_j,z_k,T)\right|^m\right)^{\frac{1}{m}},\quad m=1,2 \tag{18}$$

h	E_{L^1}	E_{L^2}	O_{L^1}	O_{L^2}
5.000E-02	3.382E-02	5.948E-02	—	—
4.000E-02	2.073E-02	3.317E-02	2.195	2.617
2.500E-02	3.903E-03	6.190E-03	3.552	3.572
2.000E-02	1.422E-03	2.150E-03	4.524	4.738
1.000E-02	4.298E-05	6.367E-05	5.049	5.078

Table 2. Numerical errors and convergence orders of the 3D two-stage RK method.

So if we choose two different spatial steps h^{s-1} and h^s for the same computational domain, we can use (18) to get two L^k errors $E_{L^k}^{s-1}$ and $E_{L^k}^{s}$. Then the orders of numerical convergence can be defined by Dumbser et al. (2007)

$$O_{L^k}=\log\left(\frac{E_{L^k}^{s}}{E_{L^k}^{s-1}}\right)\Big/\log\left(\frac{h^s}{h^{s-1}}\right),\qquad k=1,2. \tag{19}$$

Table 2 shows the numerical errors and the convergence orders, measured by equations (18) and (19), respectively. In Table 2 the first column shows the spatial increment h, and the following four columns show L^1 and L^2 errors and their corresponding to convergence orders O_{L^1} and O_{L^2}. From Table 2 we can find that the errors E_{L^1} and E_{L^2} decrease as the spatial grid size h decreases, which implies that the 3D two-order RK method is convergent.

4. Numerical dispersion and efficiency

As we all know, the numerical dispersion or grid dispersion, which is caused by approximating the continuous wave equation by a discrete finite difference equation, is the major artifact when we use finite difference schemes to model acoustic and elastic wavefields, further resulting in the low computational efficiency of numerical methods. This numerical artifact causes the phase speed to become a function of spatial and time increments. The relative computational merit of most discretization schemes hinges on their ability to minimize this effect. In this section, following the analysis methods presented in Vichnevetsky (1979), Dablain (1986), and Yang et al. (2006), we investigate the dispersion relation between grid dispersion and spatial steps with the RK and the computational efficiencies for different numerical methods through numerical experiments. For comparison, we also present the dispersion results of the fourth-order SG method (Moczo et al., 2000).

4.1 Numerical dispersion

Following the dispersion analysis developed by Moczo et al. (2000) and and Yang et al. (2006), we provide a detailed numerical dispersion analysis with the RK for the 3D case in Appendix C, and compare it with the fourth-order SG method (Moczo et al., 2000). To check the effect of wave-propagation direction on the numerical dispersion, we have chosen different azimuths for two Courant numbers of $\alpha = 0.1$ and 0.3.

Figure 2 shows the dispersion relations as a function of the sampling rate S_p defined by $S_p = h/\lambda$ (Moczo et al., 2000) with h being the grid spacing and λ the wavelength. The curves correspond to different propagation directions. The results plotted in Figure 2(a) and 2(b) are computed by the dispersion relation (C4) given in Appendix C with Courant numbers of 0.1 and 0.3, respectively. Figures 2 and 3 show that the maximum phase velocity error does not exceed 11%, even if there are only 2 grid points per minimum wavelength (S_p=0.5). For a sampling rate of S_p=0.2 the numerical velocity is very close to the actual phase velocity. These Figures also shows that the dispersion curves differ for different propagation directions.

Figure 3 shows the numerical dispersion curves computed by 3D fourth-order SG using the numerical relation (C5) given in Appendix C under the same condition. In contrast with the curves in Figure 2 computed by the RK, the numerical dispersion as derived by the fourth-order SG clearly changes for different propagation directions. It is very clear that the

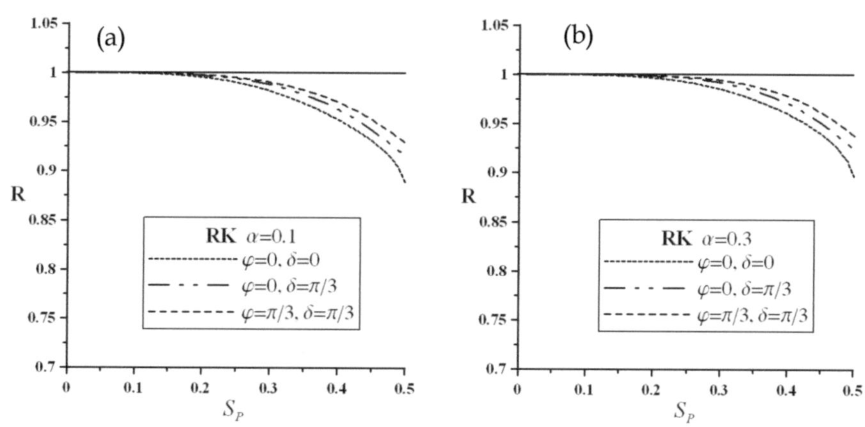

Fig. 2. The dispersion relation of RK method for the Courant number (a) $\alpha = 0.1$ and (b) $\alpha = 0.3$, in which φ is the wave propagating angle to the z-axis, and δ is the propagating angle of the wave projection in the xy plane to the x-axis.

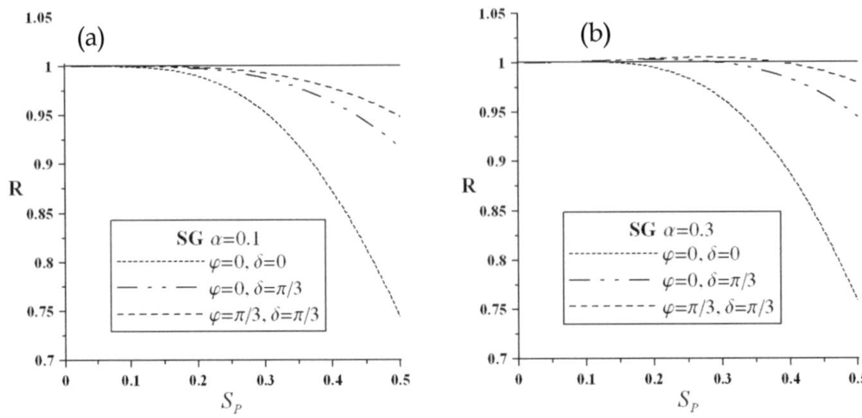

Fig. 3. The dispersion relation of the fourth-order SG method (Moczo et al., 2000) for the Courant number (a) $\alpha = 0.1$ and (b) $\alpha = 0.3$, in which φ is the wave propagating angle to the z-axis, and δ is the propagating angle of the wave projection in the xy plane to the x-axis.

numerical dispersion computed by the fourth-order SG is more serious compared with that of RK. For example, the maximum dispersion error calculated with the latter method is less than 11% (Figure 2a), while the same error calculated with the former one is greater than 26% (Figure 3a). To limit the dispersion error of the phase velocity under 8% (the maximum dispersion error by RK shown in Figure 2a), about 3 grid points per minimum wavelength are required when using fourth-order SG, opposite to only 2.1 grid points per wavelength with RK. Meanwhile, from Figure 2(a) we can observe that the numerical dispersion curves of the RK in different propagation directions are close to each other. It means that the RK has small numerical dispersion anisotropy. In contrast, from Figure 3(a) and 3(b) we can see that the difference of numerical dispersion curves in different propagation directions is very large, implying that the SG has larger numerical dispersion anisotropy than that of the RK.

After comparing Figure 2 computed by the RK with Figure 3 computed by the SG, we conclude that the RK offers smaller numerical dispersion than the SG for the same spatial sampling increment. We will verify this conclusion later via new experiments.

4.2 Computational efficiency

In this subsection, we further investigate the numerical dispersion and computational efficiency of the RK through wave-field modeling, and compare our method with the fourth-order LWC (Dablain, 1986) and the fourth-order SG method. Under this case of our consideration, we choose the following 3D acoustic wave equation

$$\frac{\partial^2 u}{\partial t^2} = c_0^2 \left(\frac{\partial^2 u}{\partial x^2} + \frac{\partial^2 u}{\partial y^2} + \frac{\partial^2 u}{\partial z^2}\right) + f \,, \tag{20}$$

where c_0 is the acoustic velocity. In our present numerical experiment, we choose c_0=4 km/s. The computational domain is $0 \le x \le 5$ km, $0 \le y \le 5$ km, and $0 \le z \le 5$ km, and the number of grid points is 200×200×200. The source is a Ricker wavelet with a peak frequency of f_0 = 37 Hz. The time variation of the source function is

$$f(t) = -5.76 f_0^2 \left[1 - 16(0.6 f_0 - 1)^2\right] \exp\left[-8(0.6 f_0 - 1)^2\right] \tag{21}$$

The force-source included in equation (20) is located at the centre point of the computational domain, and $\partial f/\partial x$ and $\partial f/\partial z$ are set to be zero in this example and other experiments in the following section. The spatial and temporal increments are chosen by $h=\Delta x=\Delta y=\Delta z=25$ m and $\Delta t=1.5\times10^{-3}$s, respectively. The coarse spatial increment of h=25 m is chosen so that we test the effects of sampling rate on the numerical dispersion. A receiver R is placed at the grid point (x_R, y_R, z_R)=(3.575 km, 2.5 km, 2.5 km) to record the waveforms generated by three methods.

Following Dablain's definition (Dablain, 1986), we take the Nyquist frequency of the source to be twice the dominant frequency in this study. The rule of thumb in numerical methods for choosing an appropriate spatial step based on the Nyquist frequency can be written as

$$\Delta x = \frac{v_{\min}}{f_N \cdot G}, \tag{22}$$

where $v_{\min}$ denotes the minimum wave-velocity, f_N is the Nyquist frequency, and G denotes the number of gridpoints needed to cover the Nyquist frequency for non-dispersive propagation (Dablain, 1986). In this case chosen that implies a Nyquist frequency of 74 Hz and the number of gridpoints at Nyquist is about 2.2 in our present numerical experiment.

Figures 4, 5, and 6 show the wave-field snapshots at t=0.5 sec on a coarse grid of $\Delta x=\Delta y=\Delta z=25$ m (G≈2.2), generated by the RK (Fig. 6), the fourth-order LWC (Dablain, 1986) (Fig. 7), and the fourth-order SG (Moczo et al., 2000) (Fig. 8), where Figures (a), (b), and (c) shown in these Figures show the wave-field snapshots in the xy, xz, and yz planes, respectively. Figures 7 and 8 show the wave-field snapshots at t=0.5 sec for the same Courant number ($\alpha = 0.24$), generated by the fourth-order LWC (Fig. 7) and the fourth-order SG (Fig. 8) on a fine grid ($\Delta x=\Delta y=\Delta z=8.3$ m) so that the numerical dispersions caused by the fourth-order LWC and the fourth-order SG are eliminated. We can see that the wavefronts of seismic waves shown in Figures 4-6, simulated by the three methods, are nearly identical. However, the result generated by the RK (Fig. 4) shows much less numerical dispersion even though the space increment is very large, whereas the fourth-order LWC and the fourth-order SG suffer from serious numerical dispersions (see Figs. 7, 8). Comparison between Figure 6 and Figures 7 and 8 demonstrates that the RK on a coarse grid can provide the similar accuracy as those of the

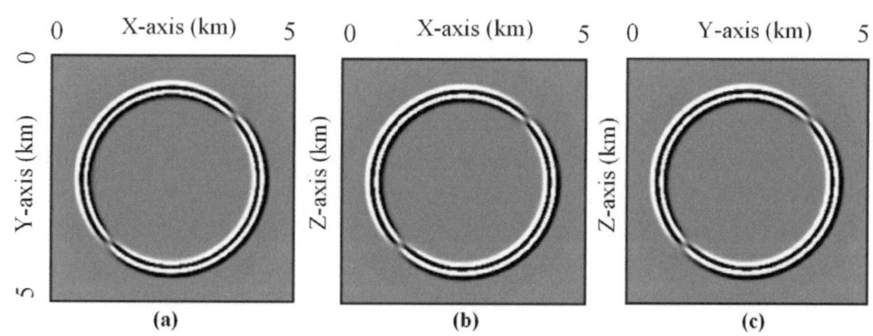

Fig. 4. Snapshots of acoustic wave fields at time 0.5 sec on the coarse grid ($\Delta x=\Delta y=\Delta z=25$m) in the xy (a), xz (b), and yz (c) planes, respectively, computed by the 3D RK method.

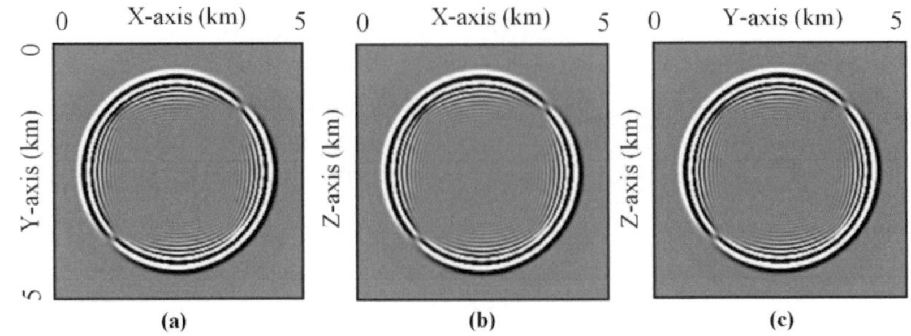

Fig. 5. Snapshots of acoustic wave fields at time 0.5 sec on the coarse grid (Δx=Δy=Δz=25 m) in the *xy* (a), *xz* (b) and *yz* (c) planes, respectively, generated by the fourth-order LWC method.

fourth-order LWC and the fourth-order SG on a fine grid for the same Courant number. But the computational cost of the RK is quite different from the other two methods. For example, it took the RK about 15.3 min to generate Figure 4, whereas the fourth-order LWC and the fourth-order SG took about 50.8 min and 50.6 min to generate Figure 5 and Figure 6, respectively. This suggests that the computational speed of the RK is roughly 3.3 times of the fourth-order LWC and the fourth-order SG to achieve the same accuracy. Thus we can conclude the 3D RK can save the computational cost by using coarse grids to simulate wave propagation in large scale models. The results in Figures 4-8 were computed on a parallel computation with 40 processors and using the message passing interface (MPI).

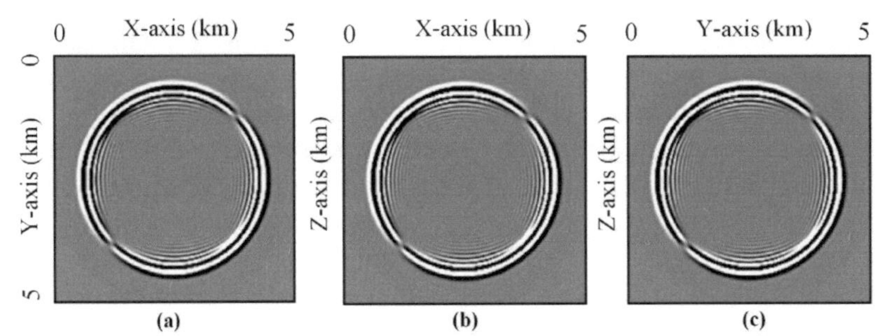

Fig. 6. Snapshots of acoustic wave fields at time 0.5 sec on the coarse grid (Δx=Δy=Δz=25 m) in the *xy* (a), *xz* (b) and *yz* (c) planes, respectively, generated by the fourth-order SG method.

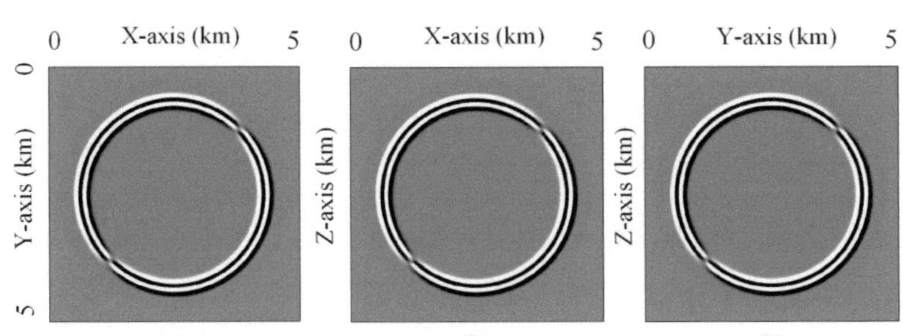

Fig. 7. Snapshots of acoustic wave fields at time 0.5 sec on the fine grid (Δx=Δy=Δz=8.3 m) in the *xy* (a), *xz* (b) and *yz* (c) planes, respectively, generated by the fourth-order LWC method.

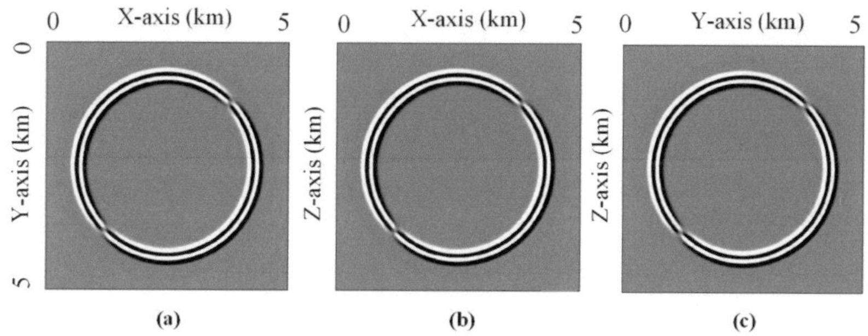

Fig. 8. Snapshots of acoustic wave fields at time 0.5 sec on the fine grid (Δx=Δy=Δz=8.3 m) in the *xy* (a), *xz* (b) and *yz* (c) planes, respectively, generated by the fourth-order SG method.

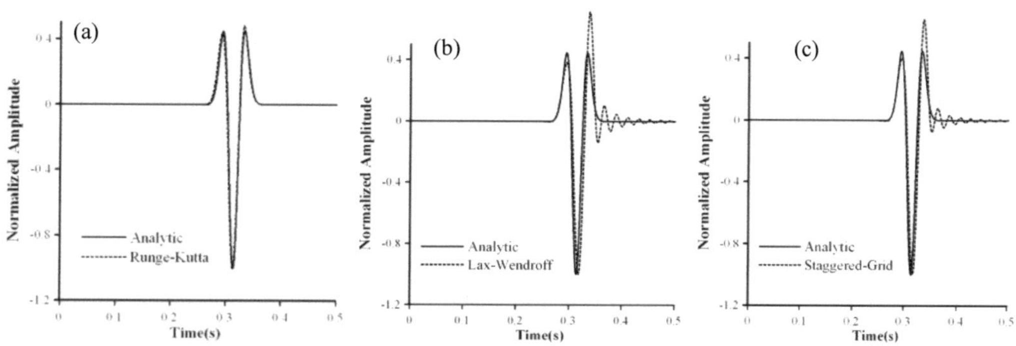

Fig. 9. Comparions of the analytic solution computed by the Cagniard–de Hoop method (de Hoop, 1960) with waveforms generated by (a) the RK, (b) the fourth-order LWC, and (c) the fourth-order SG, respectively.

Note that the memory required by RK is also different from those of the fourth-order LWC and the fourth-order SG methods. The RK needs 20 arrays to hold the wave fields at each time step, and the number of grid points for each array is 200×200×200 on a coarse grid for generating Figure 4. Even though the fourth-order LWC needs only eight arrays to store the wave displacement and the fourth-order SG needs nine arrays to store the wave displacement and the stress at each grid point to generate a comparable result, the two methods require much finer grid sampling. For example, the number of grid points of each array for generating Figures 7 and 8 goes up to 600×600×600 for both the fourth-order LWC and the fourth-order SG. Therefore, the overall memory required by the RK takes only about 31.3% of that needed by the fourth-order LWC and about 27.8% of that of the fourth-order SG.

Now we compare the accuracy of the waveforms at receiver R (3.575 km, 2.5 km, 2.5 km), generated by the RK, the fourth-order LWC, and the fourth-order SG, respectively. Figure 9 shows the waveforms of the analytic solution (solid lines) computed by the Cagniard–de Hoop method (Aki and Richards, 1980) and the numerical solutions (dashed line) computed by three numerical methods on the coarse grid (Δx=Δy=Δz=25 m). Figure 9(a) shows that the waveforms calculated by the 3D RK and the Cagniard-de Hoop method (solid line) are in good overall agreement even on the coarse grid (Δx=Δy=Δz=25 m). In contrast, the results in Figures 9(b) and 9(c), calculated by the fourth-order LWC and the SG methods, respectively, show serious numerical dispersion following the peak wave as contrasted to the analytic solution (solid line). It illustrates that the 3D two-stage RK is accurate in wave-field modeling for the acoustic propagation modeling and it can provide very accurate results even when coarse grids are chosen.

5. Wavefield modelling

In this section, we present the performance of the two-stage RK in the 3D acoustic and elastic cases and compare against the so-called LWC method (Dablain 1986) through wavefield modelling and synthetic seismograms. In particular, we use the RK to simulate the acoustic and elastic waves propagating in the 3D multilayer acoustic medium, two-layer elastic medium, and the transversely isotropic medium with a vertical symmetry axis (VTI).

5.1 Multilayer acoustic model

In this experiment, we consider a special multilayer isotropic medium model, shown in Figure 10. Speaking in detail, when $0 \le y \le 1.5$ km, the model is consisted of three layer media where the acoustic velocities are chosen by 2 km/s, 3 km/s, and 4 km/s, corresponding to the top, middle and bottom layers, respectively, whereas the model is a two layer media with acoustic velocities of 2 km/s and 3 km/s as 1.5 km$<y\le$3 km. The computational domain is $0 \le x \le 3$ km, $0 \le y \le 3$ km, $0 \le z \le 1.8$ km. We choose the spatial increments$\Delta x = \Delta y = \Delta z = 15$ m, the temporal increment $\Delta t = 0.8$ ms. The source of the Ricker wavelet with a peak frequency of $f_0 = 30$ Hz is located at coordinate (x_s, y_s, z_s)=(1.5 km, 1.5 km, 0.015 km), and the expression is the same as equation (21). The perfectly matched layer (PML) absorbing boundary condition suggested by Dimitri and Jeroen (2003) is used in the experiment.

Figure 11, generated by the RK, shows the synthetic seismograms recorded by 201 receivers on the surface spreading respectively along the two lines of y=1.5 km (Fig. 11a) and x=1.5 km (Fig. 11b) shown in Figure 10. In Figure 11, the reflected waves from the inner interfaces are very clear. We can identify the medium structure from the reflected curve wave shown in Figure 11. In this experiment, we use the stiff boundary condition at the free surface because the source is located at the surface. This experiment also illustrates that it is efficient for the RK to combine with the PML absorbing boundary condition (Dimitri and Jeroen, 2003).

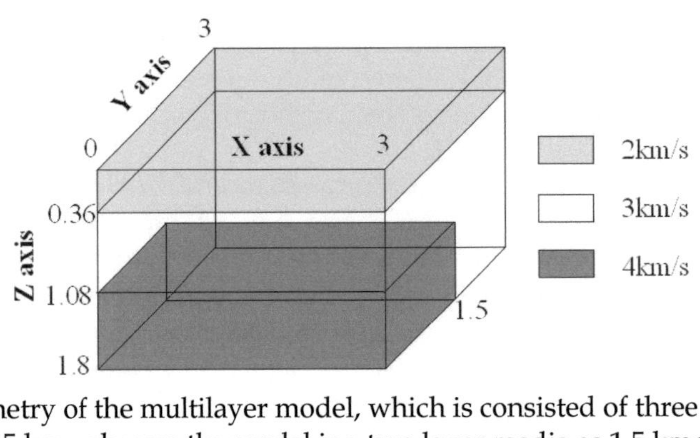

Fig. 10. The geometry of the multilayer model, which is consisted of three layer media in the domain of $0 \le y \le 1.5$ km, whereas the model is a two layer media as 1.5 km$<y\le$3 km.

Figure 12 shows the synthetic VSP seismograms recorded in the wells, generated by the RK. In Figure 14(a), the receivers are spread from receiver R_1 (x, y, z)= (1.8 km, 1.8 km, 0) to receiver R_N (x, y, z)= (1.8 km, 1.8 km, 1.8 km) spaced 0.015 km apart, and from receiver R_1 (x, y, z)= (1.2 km, 1.2 km, 0) to receiver R_N (x, y, z)= (1.2 km, 1.2 km, 1.8 km) in Figure 12(b). From Figure 12 we can see that the VSP seismograms are very clean and have no grid dispersions while the model velocity contrasts between adjacent layers (layers 1 and 2, layers 2 and 3) are about 50% and 33%, respectively. We can also observe the difference of two seismic records shown in Figure 12(a) and 12(b) from different wells.

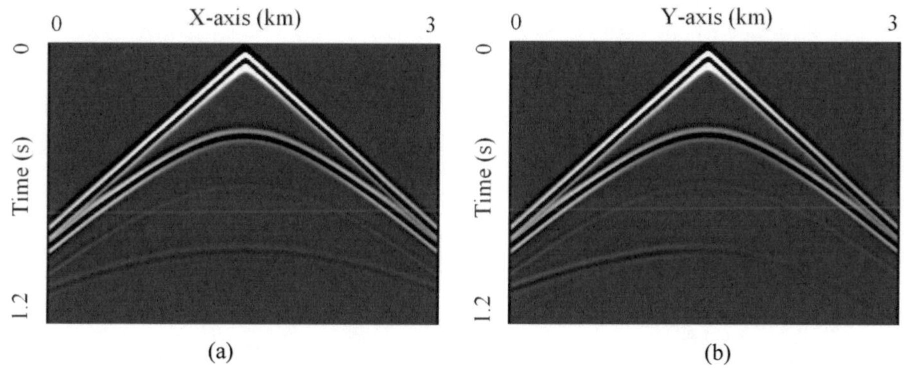

Fig. 11. Synthetic seismograms recorded by 201 receivers on the surface spreading (a) from x=0 to 3 km spaced 0.015 km apart along the line of y=1.5 km, and (b) from y=0 to 3 km spaced 0.015 km apart along the line of x=1.5 km, respectively generated by the RK for the multilayer geological model shown in Figure 14.

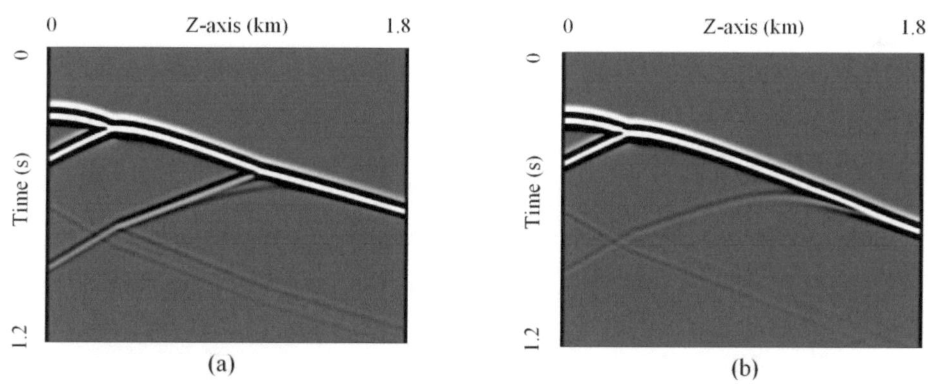

Fig. 12. Synthetic VSP seismograms recorded by 121 receivers in wells spreading (a) from receiver R_1 (x, y, z)= (1.8 km, 1.8 km, 0) to receiver R_N (x, y, z)= (1.8 km, 1.8 km, 1.8 km) , and (b) from receiver R_1 (x, y, z) =(1.2 km, 1.2 km, 0) to receiver R_N (x, y, z) =(1.2 km, 1.2 km, 1.8 km).

5.2 Two-layered elastic wave modeling

Subsurface structures have interfaces where velocities and density are discontinuous, and some of the interfaces may have strong velocity contrasts. Some FD methods, such as conventional FD (Kelly et al., 1976), LWC method (Dablain 1986), often suffer from serious numerical dispersion when the models have large velocity contrast between adjacent layers. So we consider a two-layer medium with inner interface to investigate the validity of the 3D RK in multilayer elastic model. In the two-layer model, the Lamé constants are given as λ_1=1.5 GPa, μ_1=2.5 GPa and ρ_1=1.5g/cm^3, λ_2=11.0 GPa, μ_2=15.0 GPa and ρ_2=2.0g/cm^3, corresponding to the P- and S-wave velocities of 2.082 km/s and 1.291 km/s in the top layer medium, and 4.528 km/s and 2.739 km/s in the bottom medium. The computational domain is $0 \le x \le 4$ km, $0 \le y \le 4$ km, and $0 \le z \le 4$ km. We choose the spatial increments h=Δx=Δy=Δz=20 m and the temporal increment Δt=1.5 ms. The source of the Ricker wavelet with a peak frequency of $f_0 = 20$ Hz is located at $(x_s, y_s, z_s) = (2$ km$, 2$ km$, 1.92$ km$)$, and the source function is the same as equation (21). The three force-source components, corresponding to f_1, f_2, and f_3 included in equation (4a), are chosen by f_1=f_2=f_3=$f(t)$. The horizontal inner interface is located at the depth z=2.4 km. In this experiment, we use similarly the PML absorbing boundary condition presented in Dimitri and Jeroen (2003).

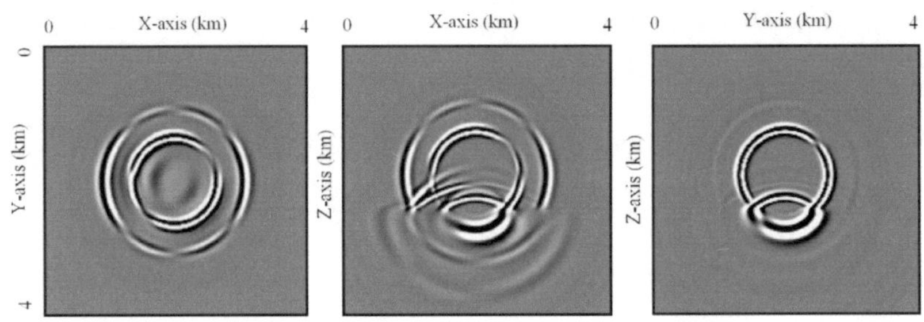

(a) (b) (c)

Fig. 13. Snapshots of the seismic wave fields at time 0.6 sec for the u_1 component in the two-layer isotropic medium, generated by the RK, for (a) the xy plane, (b) the xz plane, and (c) the yz plane.

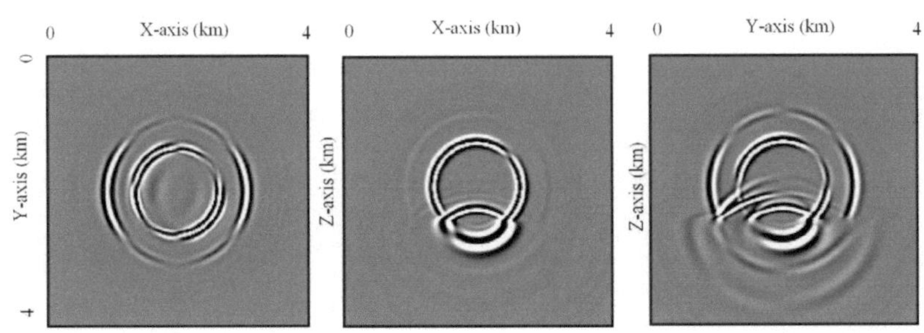

(a) (b) (c)

Fig. 14. Snapshots of the seismic wave fields at time 0.6 sec for the u_2 component in the two-layer isotropic medium, generated by the RK, for (a) the xy plane, (b) the xz plane, and (c) the yz plane.

Figures 13-15 show the wavefield snapshots of the three displacement-components (u_1, u_2, and u_3) at t =0.6 sec on the coarse increments (Δx=Δy=Δz=20 m) for the two-layer elastic model, generated by the RK. Figures 16-17 and Figures 18-19 show the wavefield snapshots of the horizontal and vertical displacement-components (u_1 and u_3) at t =0.6 sec for the same grid

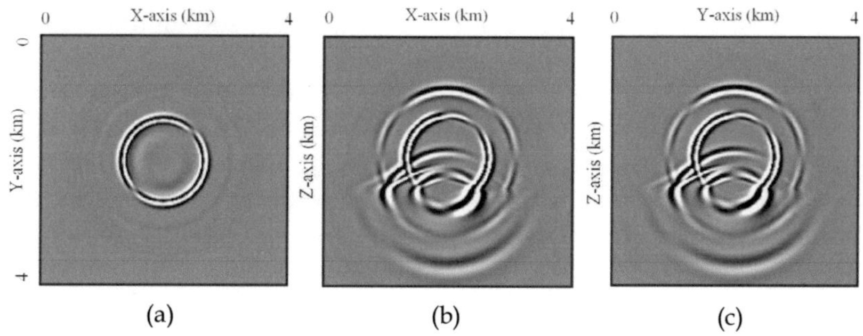

Fig. 15. Snapshots of the seismic wave fields at time 0.6 sec for the u_3 component in the two-layer isotropic medium, generated by the RK, for (a) the *xy* plane, (b) the *xz* plane, and (c) the *yz* plane.

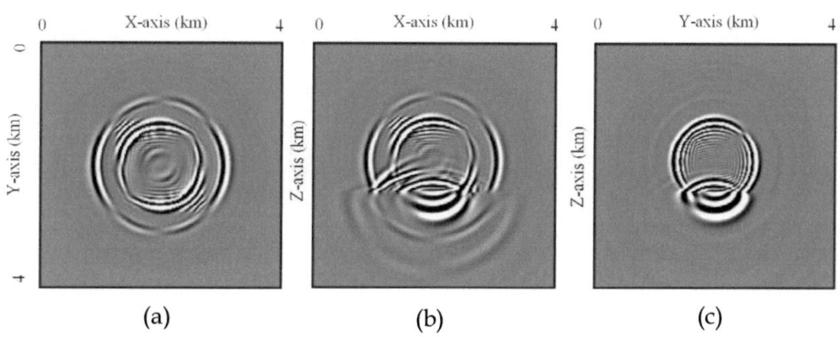

Fig. 16. Snapshots of the seismic wave fields at time 0.6 sec for the u_1 component in the two-layer isotropic medium, generated by the fourth-order LWC, for (a) the *xy* plane, (b) the *xz* plane, and (c) the *yz* plane.

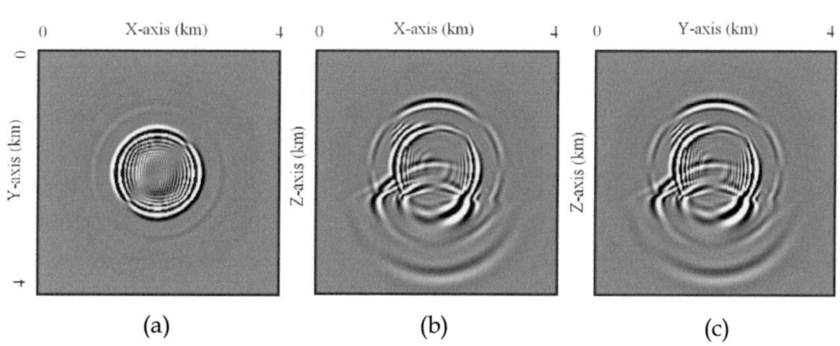

Fig. 17. Snapshots of the seismic wave fields at time 0.6 sec for the u_3 component in the two-layer isotropic medium, generated by the fourth-order LWC, for (a) the *xy* plane, (b) the *xz* plane, and (c) the *yz* plane.

increments and same model, generated by the fourth-order LWC and fourth-order SG methods, respectively. Four snapshots such as Figure 13(b) in the *xz* plane for the u_1 component Figure 14(c) in the *yz* plane for the u_2 component, and Figures 15(b) and 15(c) in the *xz* and *yz* planes for the u_3 component show numerous phases such as direct *P* wave, direct *S* wave, and their reflected, transmitted and converted phases from the inner interface. In Figures 13(c), 14(b), and 15(a), the snapshots in the *yz, xz,* and *xy* planes, corresponding to three displacement-components u_1, u_2, and u_3, respectively, show a very weak P wave and a strong S wave. The wavefield snapshots (Figs. 13-15) also show that the RK has no visible numerical dispersions even if the space increment is chosen 20 m without any additional treatments for the two-layer elastic model with a large velocity contrasts of 2.18 times between the top and bottom layer media, whereas the fourth-order LWC and the fourth-order SG suffer from substantial numerical dispersion for the same computational conditions (see Figs. 16-19).

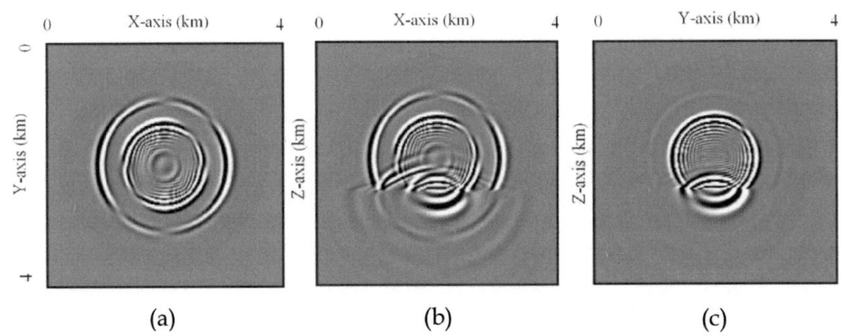

Fig. 18. Snapshots of the seismic wave fields at time 0.6 sec for the u_1 component in the two-layer isotropic medium, generated by the fourth-order SG, for (a) the *xy* plane, (b) the *xz* plane, and (c) the *yz* plane.

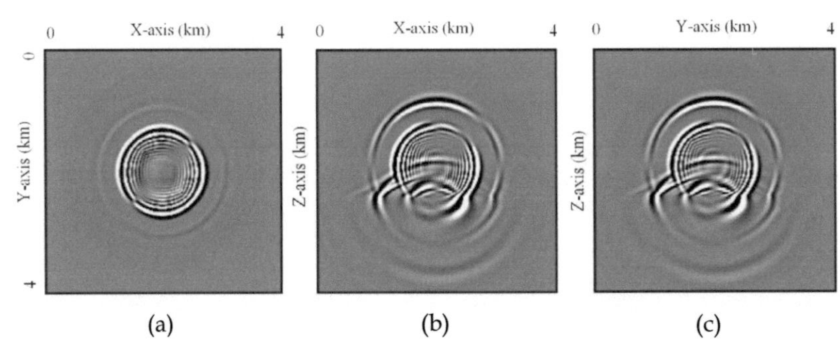

Fig. 19. Snapshots of the seismic wave fields at time 0.6 sec for the u_3 component in the two-layer isotropic medium, generated by the fourth-order SG, for (a) the *xy* plane, (b) the *xz* plane, and (c) the *yz* plane.

5.3 VTI model

In order to investigate the performance of the RK method for the anisotropic case, we simulate the elastic wave propagating in a 3D VTI medium. For this case we consider the following wave equation:

$$\begin{cases} \rho\dfrac{\partial^2 u_1}{\partial t^2} = c_{11}\dfrac{\partial^2 u_1}{\partial x^2} + c_{66}\dfrac{\partial^2 u_1}{\partial y^2} + c_{55}\dfrac{\partial^2 u_1}{\partial z^2} + (c_{12}+c_{66})\dfrac{\partial^2 u_2}{\partial x\partial y} + (c_{13}+c_{55})\dfrac{\partial^2 u_3}{\partial x\partial z} + f_1 \\ \rho\dfrac{\partial^2 u_2}{\partial t^2} = c_{66}\dfrac{\partial^2 u_2}{\partial x^2} + c_{22}\dfrac{\partial^2 u_2}{\partial y^2} + c_{44}\dfrac{\partial^2 u_2}{\partial z^2} + (c_{12}+c_{66})\dfrac{\partial^2 u_1}{\partial x\partial y} + (c_{23}+c_{44})\dfrac{\partial^2 u_3}{\partial y\partial z} + f_2 \\ \rho\dfrac{\partial^2 u_3}{\partial t^2} = c_{55}\dfrac{\partial^2 u_3}{\partial x^2} + c_{44}\dfrac{\partial^2 u_3}{\partial y^2} + c_{33}\dfrac{\partial^2 u_3}{\partial z^2} + (c_{13}+c_{55})\dfrac{\partial^2 u_1}{\partial x\partial z} + (c_{23}+c_{44})\dfrac{\partial^2 u_2}{\partial y\partial z} + f_3 \end{cases} \quad (23)$$

In this experiment, the computational domain is $0 \le x \le 5$ km, $0 \le y \le 5$ km, and $0 \le z \le 5$ km. The elastic constants and the medium density included in equation (23) are c_{11}=26.4 GPa, c_{33}=15.6 GPa, c_{13}=6.11 GPa, c_{44}=4.38 GPa, c_{66}=6.84 GPa, $c_{22} = c_{11}$, $c_{23} = c_{13}$, $c_{55} = c_{44}$, $c_{12} = c_{11} - 2c_{66}$, and ρ=2.17 g/cm³, respectively. The source with the peak frequency f_0=17 Hz is located at the center of the computational domain as defined in equation (21). The spatial and temporal increments are $\Delta x = \Delta y = \Delta z = 25$ m and $\Delta t = 1.0\times 10^{-3}$ sec, respectively, resulting in 3.3 grid points per minimum wavelength because the minimal qS wave velocity is 1.4207 km/sec from the elastic constants and the medium density.

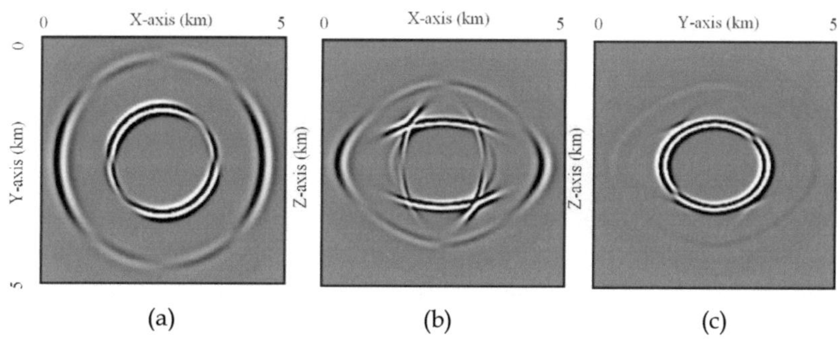

Fig. 20. Snapshots of elastic wave fields at time 0.7 sec for the x direction displacement (u_1) in the VTI medium, generated by the RK for (a) the *xy* plane, (b) the *xz* plane, and (c) the *yz* plane.

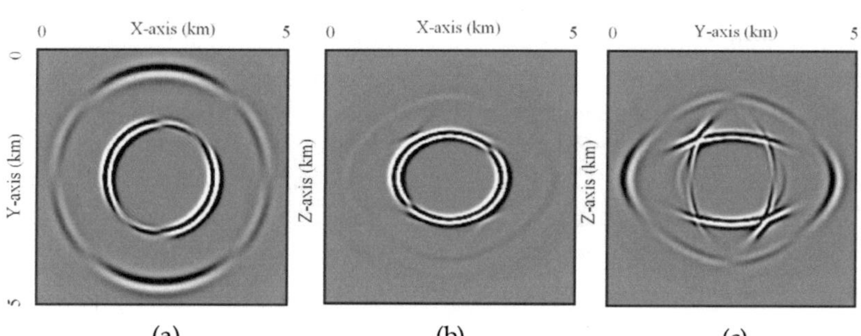

Fig. 21. Snapshots of elastic wave fields at time 0.7 sec for the y direction displacement (u_2) in the VTI medium, generated by the RK for (a) the *xy* plane, (b) the *xz* plane, and (c) the *yz* plane.

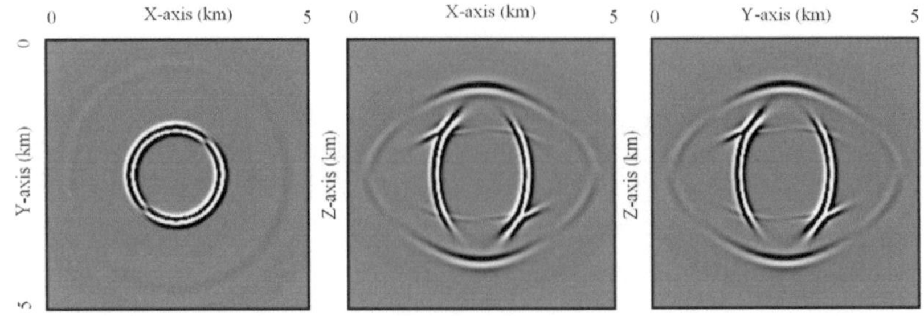

(a) (b) (c)

Fig. 22. Snapshots of elastic wave fields at time 0.7 sec for the z direction displacement (u_3) in the VTI medium, generated by the RK for (a) the *xy* plane, (b) the *xz* plane, and (c) the *yz* plane.

The wave field snapshots for u_1, u_2 and u_3 components at time 0.7 *sec* are shown in Figures 20, 21, and 22. Figure 20 shows the snapshots of the u_1 component in *xy*-, *xz*-, and *yz*-planes, whereas Figures 21 and 22 show the snapshots of u_2 and u_3 components in the three planes. The snapshots of the three displacement components in the *xy* plane (transverse plane), shown in Figures 20(a), 21(a), and 22(a), show that the wavefronts of *P* and *S* waves are a circle in the VTI medium, whereas other snapshots in Figures 20, 21, and 22 show that the wavefronts of *P* and *S* waves are an ellipse and the quasi-*P* (q*P*) and quasi-*SV* (q*SV*) waves show the directional dependence on propagation velocity. The qSV wavefronts have cusps and triplications depending on the value of c_{13} (Faria & Stoffa, 1994). Triplications can be observed in the horizontal component qSV wavefronts in the *xz* plane for the u_1 component (Fig. 20b), in the *yz*-plane for the u_2 component (Fig. 21c), and in the vertical component qSV wavefronts shown in Figures 22(b) and 22(c), respectively. Furthermore, in the VTI medium we can observe that the shear-wave splitting shows in Figures 20(b) and 21(c), and the arrival times of quasi-*SH* and qSV waves are different by comparing Figures 20(c) and 21(b) with Figures 20(b), 21(c), 22(b), and 22(c).

6. Summary

The two-stage RK method for solving 3D acoustic and elastic wave equations in isotropic and anisotropic media is developed via the four-stage fourth-order RK algorithm for solving ordinary differential equations and the high-degree multivariable interpolation approximation. In other words, the time derivatives are approximated via the two-stage fourth-order RK and the high-order space derivatives are calculated using the multivariable interpolation approximation. On the basis of such a structure, we have to first convert these high-order time derivatives to the spatial derivatives, which is similar to the high-order FD or so-called LWC methods (Lax and Wendroff, 1964; Dablain, 1986). However, the fourth-order RK method in approximating the high-order spatial derivatives is different from these high-order FD, LWC, and staggered-grid methods stated previously that only use the wave displacement at some grid points to approximate the high-order spatial derivatives or directly discretizing the original wave equation. This RK method uses simultaneously both the wave displacement and its gradients to approximate the high-order derivatives [see formulae (A3) to (A7)]. In other words, when determining these high-order spatial derivatives included in equation (8) or equation (9), the RK method uses not only the values of the displacement U and the particle velocity W at the mesh point (i, j, k) and its neighboring gridpoints [see equations (A3)–(A7)], but also the values of the gradients of the displacement U and particle velocity W. Based on such a structure, the two-stage RK retains more wavefield information included in the displacement function, the particle velocity, and their gradients. As a result, the new RK can effectively suppress the numerical dispersion and source-generated noises caused by discretizing the wave equations when too-coarse grids are used or models have large velocity contrast between adjacent layers, and has higher spatial accuracy though the RK only uses a local difference operator that three gridpoints are used in a spatial direction.

Numerical dispersion analysis in section 4.1 and wave-field modeling results confirm our conclusion that the RK method has smaller numerical dispersion than the fourth-order LWC and SG methods. At the same time, these numerical results also imply that simultaneously using both the wave displacement, particle velocity, and their gradients to approximate the high-order derivatives is important for decreasing the numerical dispersion caused by the discretization of wave equations because the particle velocity and the gradients of both the wave displacement and the particle velocity include important wave-field information. On the other hand, using these connection relations such as equation (A2) and those omitted in this chapter between the grid point (i, j, k) and its neighboring nodes $(i+p, j+q, k+r)$ $(p, q, r=-1, 0, 1)$ keeps the continuity of gradients. The continuity and high accuracy (fourth-order accuracy in space) of gradients improve automatically the continuity of the stresses that are the linear combinations of gradients or the Hook sum, further resulting in the RK having less numerical dispersion when models have strong interfaces between adjacent layers. It suggests that we should consider the particle velocity and wave-gradient fields and the use of connection relations such as equation (A2), and so on, as we design a new numerical method to solve the 3D acoustic and elastic wave equations.

It appears that the CPU time of the two-stage RK is more than that of the fourth-order LWC and the SG methods, but in fact, because this method yields less numerical dispersion than both the LWC and SG methods, we can afford to increase the temporal increment through using coarser spatial increments to achieve the same accuracy as those of the LWC and the SG methods on a finer spatial grid with smaller time steps. Numerical computational results show that the RK method can also effectively suppress the numerical dispersion and the source-noise as the number of gridpoints in a minimum wavelength is about 3.3. Hence the total CPU time of the RK will not be larger than those of the LWC and the SG methods. As observed in our experiment, the computational speed of the RK is roughly 3.3 times of the fourth-order LWC and the SG on a fine grid to achieve the same accuracy as that of the RK, and the storage space required for the RK is only about 31.3% of the fourth-order LWC and about 27.8% of the fourth-order SG, respectively.

In conclusion, the 3D RK method has the following properties: (1) it can suppress effectively the numerical dispersion and source noise for practically coarse spatial and time steps; (2) it provides extra wave-field information including the particle velocity field and their time derivatives and spatial gradients, so the two-stage RK can be directly extended to solve the two-phase porous wave equations that include the first-order time derivatives such as Biot's porous wave equations (Biot 1956a,b); (3) it can increase greatly the computational efficiency and save storage space if larger spatial and temporal increments are used; (4) it only uses the local difference operator to obtain the high-order spatial accuracy. We initiate possible, more applications of the RK method in large-scale acoustic or seismic modeling, reverse time migration, and inversion based on the acoustic-wave equation, despite the computation time and memory requirements are the bottle-neck for their vast applications.

7. Appendices

7.1 Appendix A: evaluation of high-order derivatives

In order to numerically solve equation (8), we need to compute the high-order spatial derivatives $(\partial^{q+l+m}V / \partial x^q \partial y^l \partial z^m)_{i,j,k}^n$ $(2 \le q+l+m \le 3)$ so that the time advancing of the 3D RK equation (9) is implemented. To do this, following the local interpolation methods (Yang et al., 2007, 2010), we introduce the local interpolation function of spatial increments Δx, Δy, and Δz in the *x*, *y*, and *z* directions as follows:

$$G(\Delta x, \Delta y, \Delta z) = \sum_{r=0}^{5} \frac{1}{r!}(\Delta x \frac{\partial}{\partial x} + \Delta y \frac{\partial}{\partial y} + \Delta z \frac{\partial}{\partial z})^r V\,, \tag{A1}$$

which defines the interpolation relations between the grid point (*i*, *j*, *k*) and its 26 neighboring nodes such as (*i*, *j*, *k*+1), (*i*, *j*, *k*-1), (*i*, *j*+1, *k*+1), (*i*, *j*+1, *k*), (*i*, *j*+1, *k*-1), (*i*, *j*-1, *k*+1), (*i*, *j*-1, *k*), (*i*, *j*-1, *k*-1), (*i*+1, *j*, *k*+1), (*i*+1, *j*, *k*-1), (*i*+1, *j*+1, *k*+1), (*i*+1, *j*+1, *k*), (*i*+1, *j*+1, *k*-1), (*i*+1, *j*-1, *k*+1), (*i*+1, *j*-1, *k*), (*i*+1, *j*-1, *k*-1), (*i*+1, *j*, *k*), (*i*-1, *j*, *k*+1), (*i*-1, *j*, *k*-1), (*i*-1, *j*+1, *k*+1), (*i*-1, *j*+1, *k*), (*i*-1, *j*+1, *k*-1), (*i*-1, *j*-1, *k*+1), (*i*-1, *j*-1, *k*), (*i*-1, *j*-1, *k*-1), and (*i*-1, *j*, *k*). For example, at the grid point (*i*-1, *j*-1, *k*), we have the following interpolation relations:

$$\begin{aligned}
&[G(-\Delta x,-\Delta y,0)]_{i,j,k}^n = V_{i-1,j-1,k}^n\,,\\
&\left[\frac{\partial}{\partial \Delta x}G(-\Delta x,-\Delta y,0)\right]_{i,j,k}^n = \left(\frac{\partial}{\partial x}V\right)_{i-1,j-1,k}^n,\\
&\left[\frac{\partial}{\partial \Delta y}G(-\Delta x,-\Delta y,0)\right]_{i,j,k}^n = \left(\frac{\partial}{\partial y}V\right)_{i-1,j-1,k}^n,\\
&\left[\frac{\partial}{\partial \Delta z}G(-\Delta x,-\Delta y,0)\right]_{i,j,k}^n = \left(\frac{\partial}{\partial z}V\right)_{i-1,j-1,k}^n.
\end{aligned} \tag{A2}$$

Similarly, the rest 100 connection relations at other 25 neighboring nodes can be easily written.

From the 104 relations, we have similar approximation formulae as in the cited reference (Yang et al., 2010) to approximate the high-order spatial derivatives included in equation (8) or equation (9). For convenience, we list these approximation formulae used in the 3D RK method as follows

$$\left(\frac{\partial^2 V}{\partial g^2}\right)_{i,j,k}^n = \frac{2}{(\Delta g)^2}\delta_g^2 V_{i,j,k}^n - \frac{1}{2\Delta g}\left(E_g^1 - E_g^{-1}\right)\left(\frac{\partial V}{\partial g}\right)_{i,j,k}^n, \tag{A3}$$

$$\begin{aligned}
\left(\frac{\partial^2 V}{\partial g \partial e}\right)_{i,j,k}^n = &\frac{1}{2\Delta g}\left(E_g^1 - E_g^{-1}\right)\left(\frac{\partial V}{\partial e}\right)_{i,j,k}^n + \frac{1}{2\Delta e}\left(E_e^1 - E_e^{-1}\right)\left(\frac{\partial V}{\partial g}\right)_{i,j,k}^n\\
&- \frac{1}{4\Delta g \Delta e}\left(E_g^1 E_e^1 + E_g^{-1}E_e^{-1} - E_g^1 E_e^{-1} - E_g^{-1}E_e^1\right)V_{i,j,k}^n,
\end{aligned} \tag{A4}$$

$$\left(\frac{\partial^3 V}{\partial g^3}\right)_{i,j,k}^n = \frac{15}{2(\Delta g)^3}\left(E_g^1 - E_g^{-1}\right)V_{i,j,k}^n - \frac{3}{2(\Delta g)^2}\left(E_g^1 + 8I + E_g^{-1}\right)\left(\frac{\partial V}{\partial g}\right)_{i,j,k}^n, \tag{A5}$$

$$\begin{aligned}\left(\frac{\partial^3 V}{\partial g^2 \partial e}\right)_{i,j,k}^n &= \frac{1}{2\Delta g \Delta e}\left(-E_g^1 E_e^1 - E_g^{-1}E_e^{-1} + E_g^1 + E_g^{-1} - 2\delta_e^2\right)\left(\frac{\partial V}{\partial g}\right)_{i,j,k}^n \\ &+ \frac{1}{(\Delta g)^2}\delta_e^2\left(\frac{\partial V}{\partial e}\right)_{i,j,k}^n \\ &+ \frac{1}{4(\Delta g)^2 \Delta e}\left(5E_g^1 E_e^1 - 5E_g^{-1}E_e^{-1} + E_g^1 E_e^{-1} - E_g^{-1}E_e^1\right)V_{i,j,k}^n \\ &+ \frac{1}{4(\Delta g)^2 \Delta e}\left(-6E_g^1 + 6E_g^{-1} - 4E_e^1 + 4E_e^{-1}\right)V_{i,j,k}^n,\end{aligned} \tag{A6}$$

$$\begin{aligned}\left(\frac{\partial^3 V}{\partial x \partial y \partial z}\right)_{i,j,k}^n &= \frac{1}{4\Delta y \Delta z}\left(E_y^1 E_z^1 + E_y^{-1}E_z^{-1} - E_y^1 E_z^{-1} - E_y^{-1}E_z^1\right)\left(\frac{\partial V}{\partial x}\right)_{i,j,k}^n \\ &+ \frac{1}{4\Delta x \Delta z}\left(E_x^1 E_z^1 + E_x^{-1}E_z^{-1} - E_x^1 E_z^{-1} - E_x^{-1}E_z^1\right)\left(\frac{\partial V}{\partial y}\right)_{i,j,k}^n \\ &+ \frac{1}{4\Delta x \Delta y}\left(E_x^1 E_y^1 + E_x^{-1}E_y^{-1} - E_x^1 E_y^{-1} - E_x^{-1}E_y^1\right)\left(\frac{\partial V}{\partial z}\right)_{i,j,k}^n \\ &- \frac{1}{4\Delta x \Delta y \Delta z}\left(E_x^1 E_y^1 E_z^1 + E_x^{-1}E_y^{-1}E_z^{-1} + E_x^{-1}E_y^{-1}E_z^1 - E_x^1 E_y^1 E_z^{-1}\right)V_{i,j,k}^n \\ &- \frac{1}{4\Delta x \Delta y \Delta z}\left(E_x^{-1}E_y^1 E_z^{-1} - E_x^1 E_y^{-1}E_z^1 + E_x^1 E_y^{-1}E_z^{-1} - E_x^{-1}E_y^1 E_z^1\right)V_{i,j,k}^n.\end{aligned} \tag{A7}$$

where $g = x, y, z$ in formulae (A3) and (A5), and $g, e = x, y, z$ and $g \neq e$ in equations (A4) and (A6). $V_{i,j,k}^n$, $\left(\frac{\partial V}{\partial x}\right)_{i,j,k}^n$, $\left(\frac{\partial V}{\partial y}\right)_{i,j,k}^n$, $\left(\frac{\partial V}{\partial z}\right)_{i,j,k}^n$, and $\left(\frac{\partial^{q+l+m} V}{\partial x^q \partial y^l \partial z^m}\right)_{i,j,k}^n$ denote $V(i\Delta x, j\Delta y, k\Delta z, n\Delta t)$, $\frac{\partial}{\partial x}V(i\Delta x, j\Delta y, k\Delta z, n\Delta t)$, $\frac{\partial}{\partial y}V(i\Delta x, j\Delta y, k\Delta z, n\Delta t)$, $\frac{\partial}{\partial z}V(i\Delta x, j\Delta y, k\Delta z, n\Delta t)$, and $\frac{\partial^{q+l+m}}{\partial x^q \partial y^l \partial z^m}V(i\Delta x, j\Delta y, k\Delta z, n\Delta t)$, respectively. These notations δ_z^2 and E_z^1 in equations (A3) to (A7) are the second-order central difference operators and displacement operators in the z-direction, which are defined by

$$\delta_z^2 V_{i,j,k}^n = V_{i,j,k+1}^n - 2V_{i,j,k}^n + V_{i,j,k-1}^n, \quad E_z^1 V_{i,j,k}^n = V_{i,j,k+1}^n, \text{ and } E_z^{-1}V_{i,j,k}^n = V_{i,j,k-1}^n.$$

Other operators such as δ_x^2, E_x^1, E_x^{-1} in the x-direction and δ_y^2, E_y^1, E_y^{-1} in the y-direction are defined similarly.

7.2 Appendix B: derivation of stability criteria

For the 3D homogeneous case, to obtain the stability condition of the two-stage RK method under the condition of $\Delta x = \Delta y = \Delta z = h$, we consider the 3D acoustic wave equation. Substituting the following solution

$$\overline{V}_{j,l,q}^{n} = \begin{pmatrix} V \\ \partial_x V \\ \partial_y V \\ \partial_z V \end{pmatrix}^{n} \exp\left[i(k_1 jh + k_2 lh + k_3 qh)\right] \tag{B1}$$

into the 3D RK method (9) together with relations (A3)-(A7), we can obtain the following equation

$$\overline{V}^{n+1} = G\overline{V}^{n} . \tag{B2}$$

In equation (B1), k_1, k_2 and k_3 are the components of the wave-number $k=(k_1, k_2, k_3)^T$ and G is the growth matrix, whose detail expression is omitted because of its complex elements.

We assume that λ_1, λ_2, …, and λ_p are the eigenvalues of G. We know that the scheme with the growth matrix G is stable only if $|\lambda_j| \le 1, j = 1,2,\cdots,p$ are satisfied. From which, we can obtain the stability criterion of the RK method for the 3D homogeneous case as follows

$$\alpha \le \alpha_{\max} \approx 0.577, \tag{B3}$$

where $\alpha_{\max}$ denotes the maximum value of the Courant number defined by $\alpha = c_0 \Delta t / \Delta x$, with the acoustic velocity being c_0.

Similarly, we can easily obtain the stability criteria (11) and (12) for the 1D and 2D cases.

7.3 Appendix C: derivation of the dispersion relation

To investigate and optimize the dispersion error, we derive the dispersion relation of the 3D RK method. For this, following the analysis methods presented in Dablain (1986) and Yang et al. (2006), we substitute the harmonic solution

$$\overline{V}_{j,l,q}^{n} = \overline{V}^{0} \cdot \exp(i(\omega n\tau) \cdot \exp\left[i(k_1 jh + k_2 lh + k_3 qh)\right] \tag{C1}$$

into the 3D RK equation (9), we can obtain the following linear equations about $\overline{V}^0$

$$\exp(i(\omega n\tau) \cdot \overline{V}^{0} = G \cdot \overline{V}^{0} , \tag{C2}$$

where $\overline{V}^0 = (V^0, \partial_x V^0, \partial_y V^0, \partial_z V^0)^T$, ω is the angular frequency, and G is also the growth matrix. From (C2), we can obtain the following dispersion equation:

$$Det[\exp(i(\omega n\tau) \cdot I - G] = 0. \tag{C3}$$

Using the dispersion relation (C3), we obtain the ratio of the numerical velocity c_{num} to the phase velocity c_0 as follows

$$R = \frac{c_{num}}{c_0} = \frac{\omega\tau}{2\pi\alpha S_p} = \frac{\gamma}{2\pi\alpha S_p}, \tag{C4}$$

where α is the Courant number, $S_p = h/\lambda_{num} = h\omega/2\pi c_{num}$ is the spatial sampling ratio, and $\gamma = \omega\tau$ satisfies the dispersion equation (C3).

For comparison, here we also present the dispersion relation of the fourth-order staggered-grid (SG) scheme (Moczo et al., 2000). Using the definition of the spatial sampling ratio S_p and the Courant number α, we can obtain the following dispersion relation of the SG method through a series of derivation:

$$R = \frac{c_{num}}{c_0} = \frac{\omega\tau}{2\pi\alpha S_p} = \frac{\arcsin\left(\alpha\sqrt{x_1^2 + x_2^2 + x_3^2}\right)}{\pi\alpha S_p}, \tag{C5}$$

where

$$x_i = \frac{9}{8}\sin\theta_i - \frac{1}{24}\sin 3\theta_i, \ i = 1,2,3,$$

$$\theta_1 = \pi S_p \cos\varphi \sin\delta,$$

$$\theta_2 = \pi S_p \sin\varphi \sin\delta,$$

$$\theta_3 = \pi S_p \cos\delta,$$

in which $0 \le \delta < \pi$, and $0 \le \varphi < 2\pi$.

8. Acknowledgments

The authors acknowledge support provided by the National Science Fund for Distinguished Young Scholars of China (Grant No. 40725012). They also express their gratitude to other members of the Computational Geophysics Laboratory for their support.

9. References

Aki, K. & Richards, P. G. (1980). *Quantitative Seismology: Theory and Methods*, W.H. Freeman & Co., ISBN 0-935702-96-2, San Francisco, CA.

Biot, M. A. (1956a). Theory of propagation of elastic waves in a fluid-saturated porous solid, I. Low-frequency range, *J. Acoust. Soc. Amer.*, Vol. 28, No. 2, pp. 168-178. ISSN 0001-4966

Biot, M. A. (1956b). Theory of propagation of elastic waves in a fluid-saturated porous solid, II. Higher-frequency range. *J. Acoust. Soc. Amer.*, Vol. 28, No. 2, pp. 179-191. ISSN 0001-4966

Blanch, J. O. & Robertsson, J. O. A. (1997). A modified Lax-Wendroff correction for wave propagation in media described by Zener elements, *Geophys. J. Int.*, Vol. 131, No. 2, (November 1997), pp. 381-386, ISSN 0956-540X

Booth, D. C. & Crampin, S. (1983). The anisotropic reflectivity technique: anomalous arrives from an anisotropic upper mantle, *Geophys. J. R. astr. Soc.*, Vol. 72, No. 3, (March 1983), pp. 767-782, ISSN 0016-8009

Carcione, J. M. (1996). Wave propagation in anisotropic, saturated porous media: plane wave theory and numerical simulation, *J. Acoust. Soc. Amer.*, Vol. 99, No. 5, (May 1996), pp. 2655-2666, ISSN 0001-4966

Carcione, J. M. & Helle, H. B. (1999). Numerical solution of the poroviscoelastic wave equation on a staggered mesh, *J. Comput. Phys.*, Vol. 154, No. 2, (September 1999), pp. 520-527. ISSN 0021-9991

Carcione, J. M., S. Valle, S. & Lenzi, G. (1999). GPR modelling by the Fourier method: improvement of the algorithm, *Geophys. Prospect.*, Vol. 47, No. 6, pp. 1015-1029, ISSN 0016-8025

Chen, X. F. (1993). A systematic and efficient method of computing normal modes for multi-layered half-space, *Geophys. J. Int.*, Vol. 115, No. 2, (November 1993), pp. 391-409, ISSN 0956-540X

Dablain, M.A. (1986), The application of high-order differencing to the scalar wave equation: Geophysics, Vol. 51, No. 1, (January 1986), pp. 54-66, ISSN 0016-8033

de Hoop, A.T. (1960). A modification of Cagniard's method for solving seismic pulse problems, *Appl. Sci. Res.*, Vol. 8(B), No. 1, pp. 349–356, ISSN 0365-7140

Dimitri, K. & Jeroen, T. (2003). A perfectly matched layer absorbing boundary condition for the second-order seismic wave equation, *Geophys. J. Int.*, Vol. 154, No. 1, (June 2003), pp. 146–153, ISSN 0956-540X

Dumbser, M., Käser, M. & Toro, E. F. (2007). An arbitrary high-order Discontinuous Galerkin method for elastic waves on unstructured meshes - V. Local time stepping and p-adaptivity, *Geophys. J. Int.*, Vol. 171, No. 2, (November 2007), pp. 695-717, ISSN 0956-540X

Faria, E. L. & Stoffa P. L. (1994). Traveltime computation in transversely isotropic media, Geophysics, Vol. 59, No. 2, (February 1994), pp. 272-281, ISSN 0016-8033

Fei, T. & Larner, K. (1995). Elimination of numerical dispersion in finite difference modeling and migration by flux-corrected transport, *Geophysics*, Vol. 60, No. 6, (November-December 1995), pp. 1830–1842, ISSN 0016-8033

Fornberg, B. (1990). High-order finite differences and pseudo-spectral method on staggered grids, *SIAM J. Numer. Anal.*, Vol. 27, No. 4, (August 1990) pp. 904-918, ISSN 0036-1429

Holberg, O. (1987). Computational aspects of the choice of operator and sampling interval for numerical differentiation in large-scale simulation of wave phenomena, *Geophysical Prospecting*, Vol. 35, No. 6, (July 1987), pp. 629-655, ISSN 0016-8025

Igel, H., Mora, P. & Riollet, B. (1995). Anisotropic wave propagation through finite-difference grids, *Geophysics*, Vol. 60, No. 4, (July-August 1995), pp. 1203-1216, ISSN 0016-8033

Kelly, K., Ward, R., Treitel, S. & Alford, R. (1976). Synthetic seismograms: A finite-difference approach, *Geophysics*, Vol. 41, No. 1, (February 1976), pp. 2–27, ISSN 0016-8033

Kondoh, Y., Hosaka, Y. & Ishii, K. (1994). Kernel optimum nearly-analytical discretization algorithm applied to parabolic and hyperbolic equations, *Computers Math. Appl.*, Vol. 27(3), No. 3, (Feburay 1994), pp. 59–90, ISSN 0898-1221

Komatitsch, D. & Vilotte, J. P. (1998). The spectral element method: an efficient tool to simulate the seismic response of 2D and 3D geological structures, *Bull. Seism. Soc. Am.*, Vol. 88, No. 2, (April 1998), pp. 368-392, ISSN 0037-1106

Kosloff, D. & Baysal, E. (1982). Forward modeling by a Fourier method, *Geophysics*, Vol. 47, No. 10, (October 1982), pp. 1402-1412, ISSN 0016-8033

Lax, P. D. & Wendroff, B. (1964). Difference schemes for hyperbolic equations with high order of accuracy, *Commun. Pure Appl. Math. Vol.* 17, No. 3, (August 1964), pp. 381–398, ISSN

Mizutani, H., Geller, R. J. & Takeuchi, N. (2000). Comparison of accuracy and efficiency of time-domain schemes for calculating synthetic seismograms, *Phys. Earth Planet. Interiors*, Vol. 119, No. 1-2, (April 2000), pp. 75–97, ISSN 0031-9201

Moczo, P., Kristek, J. & Halada, L. (2000). 3D fourth-order staggered-grid finite-difference schemes: stability and grid dispersion, *Bull. Seism. Soc. Am.*, Vol. 90, No. 3, (June 2000), pp. 587-603, ISSN 0037-1106

Richtmyer, R. D. & Morton, K. W. (1967). Difference Methods for Initial Value Problems, *Interscience*, New York.

Sei, A. & Symes, W. (1995). Dispersion analysis of numerical wave propagation and its computational consequences, *J. Sci. Comput.*, Vol. 10, No. 1, (March 1995), pp. 1-27, ISSN 0885-7474

Vichnevetsky, R. (1979). Stability charts in the numerical approximation of partial differential equations: A review, *Mathematics and Computers in Simulation*, Vol. 21, , pp. 170–177, ISSN 0378-4754

Virieux, J. (1986). P-SV wave propagation in heterogeneous media: Velocity-stress finite-difference method, *Geophysics*, Vol. 51, No. 4, (April 1986), pp. 889-901, ISSN 0016-8033

Wang, S. Q., Yang, D. H. & Yang, K. D. (2002). Compact finite difference scheme for elastic equations, *J. Tsinghua Univ. (Sci. & Tech.)*, Vol. 42, No. 8, pp. 1128-1131. (in Chinese), ISSN 1000-0054

Yang, D. H., Liu, E., Zhang, Z. J. & Teng, J. W. 2002. Finite-difference modeling in two-dimensional anisotropic media using a flux-corrected transport technique, *Geophys. J. Int.*, Vol. 148, No. 2, (Feburary 2002), pp. 320-328, ISSN 0956-540X

Yang, D. H.,. Peng, J. M. Lu, M. & Terlaky, T. (2006). Optimal nearly analytic discrete approximation to the scalar wave equation, *Bull. Seism. Soc. Am.*, Vol. 96, No. 3 , (June 2006), pp. 1114-1130, ISSN 0037-1106

Yang, D. H., Song, G. J., Chen, S. & Hou, B. Y. (2007a). An improved nearly analytical discrete method: An efficient tool to simulate the seismic response of 2-D porous structures, *J. Geophys. Eng.*, Vol. 4, No. 1, (March 2007), pp. 40-52, ISSN 1742-2132

Yang D. H., Song, G. J. & Lu, M. (2007b). Optimally accurate nearly-analytic discrete scheme for wave-field simulation in 3D anisotropic media, *Bull. Seism. Soc. Am.*, Vol. 97, No. 5, (October 2007), pp. 1557-1569, ISSN 0037-1106

Yang, D. H., Teng, J. W., Zhang, Z. J. & Liu, E. (2003). A nearly-analytic discrete method for acoustic and elastic wave equations in anisotropic media, *Bull. Seism. Soc. Am.*, Vol. 93, No. 2, (April 2003), pp. 882–890, ISSN 0037-1106

Yang, D. H. & Wang, L. (2010). A split-step algorithm with effectively suppressing the numerical dispersion for 3D seismic propagation modeling, *Bull. Seism. Soc. Am.*, Vol. 100, No. 4, (August 2010), pp. 1470-1484, ISSN 0037-1106

Zhang, Z. J., Wang, G. J. & Harris, J. M. (1999). Multi-component wavefield simulation in viscous extensively dilatancy anisotropic media, *Phys. Earth planet Inter.*, Vol. 114, No. 1, (July 1999), pp. 25–38, ISSN 0031-9201

Zheng, H.S., Zhang, Z. J. & Liu, E. (2006). Non-linear seismic wave propagation in anisotropic media using the flux-corrected transport technique, *Geophys. J. Int.*, Vol. 165, No. 3, (June 2006), pp. 943-956, ISSN 0956-540X

9

Studies on the Interaction Between an Acoustic Wave and Levitated Microparticles

Ovidiu S. Stoican
National Institute for Laser, Plasma and Radiation Physics - Magurele
Romania

1. Introduction

The electrode systems generating a quadrupole electric field, having both static and variable components, under certain conditions, allow to maintain the charged particles in a well defined region of space without physical solid contact with the wall of a container. This process is sometime called *levitation*. Usually, these kinds of devices are known as *quadrupole traps*. The operation of a quadrupole trap is based on the *strong focusing* principle (Wuerker et al., 1959) used most of all in optics and accelerator physics. Due to the impressive results as high-resolution spectroscopy, frequency standards or quantum computing, the research has been directed mainly to the ion trapping. Wolfgang Paul, who is credited with the invention of the quadrupole ion trap, shared the Nobel Prize in Physics in 1989 for this work. Although less known, an important amount of scientific work has been deployed to develop similar devices able to store micrometer sized particles, so-called *microparticles*. Depending on the size and nature of the charged microparticles to be stored, various types of quadrupole traps have been successfully used as a part of the experimental setups aimed to study different physical characteristics of the dust particles (Schlemmer et al., 2001), aerosols (Carleton et al., 1997; Davis, 1997), liquid droplets (Jakubczyk et al., 2001; Shaw et al., 2000) or microorganisms (Peng et al., 2004). In this paper the main outlines of a study regarding the effect of a low frequency acoustic wave on a microparticles cloud which levitates at normal temperature and atmospheric pressure within a quadrupole trap are presented. The acoustic wave generates a supplementary oscillating force field superimposed to the electric field produced by the quadrupole trap electrodes. The aim of this experimental approach is evaluating the possibility to manipulate the stored microparticles by using an acoustic wave. That means both controlling their position in space and performing a further selection of the stored microparticles. It is known that, as a function of the trap working parameters, only microparticles whose charge-to-mass ratio Q/M lies in a certain range can be stored. Such a selection is not always enough for some applications. In the case of a conventional quadrupole trap where electrical forces act, particle dynamic depends on its charge-to-mass ratio Q/M. Because the action of the acoustic wave is purely mechanical, it is possible decoupling the mass M and the electric charge Q, respectively, from equation of motion. An acoustic wave can be considered as a force field which acts remotely on the stored microparticles. There are two important parameters which characterize an acoustic wave, namely wave intensity and frequency. Both of them can be varied over a wide range so that the acoustic wave mechanical effect can be settled very precisely. The experiments have been focused on the acoustic frequency range around the frequency of the ac voltage applied to the trap electrodes, where resonance effects are expected. Comparisons between experimental results and numerical simulations are included.

2. Linear electrodynamic trap

To store the micrometer sized particles (microparticules), in air, at normal temperature and pressure, the electrodes system shown in Fig.1 has been used. The six electrodes consist of four identical rods (E1, E2, E3, E4), equidistantly spaced, and two end-cap disks (E5, E6). The rod electrodes E2 and E3 are connected to a high ac sinusoidal voltage $V_{ac} = V_0 cos 2\pi f_0 t$. The electrode E4 is connected to a dc voltage U_x while the electrode E1 is connected to the ground. The end-caps electrodes E5, E6 are connected to a dc voltage U_z. Such an electrodes arrangement is known as a *linear electrodynamic trap*. A linear electrodynamic trap is characterized by a simple mechanical layout, confines a large number of microparticles and offers good optical access. For an ideal linear electrodynamic trap, near the longitudinal axis $x, y \ll R$, assuming $L_z \gg R$ and neglecting geometric losses, the electric potential may be expressed approximately as a quadrupolar form (Major et al., 2005; Pedregosa et al., 2010):

$$\phi(x, y, t) = \frac{(x^2 - y^2)}{2R^2}(U_0 + V_0 cos \Omega t) \tag{1}$$

where R is the inner radius of the trap and $\Omega = 2\pi f_0$. The voltage Uz assures the axial stability of the stored particles while the electric field due to the voltage U_x balances the gravity force. In the particular case of the linear electrodynamic trap used in this work, the diameter of the rods and distance between two opposite rods are both 10 mm, therefore R=5 mm. The distance L_z between the end-cap disk electrodes can be varied between 30 mm and 70 mm. During the experiments, the distance L_z has been kept at 35 mm. The dc voltages U_z and U_x can be varied in the range 0-1000V, while the ac voltage V_{ac} is on the order of 1-4kV$_{rms}$ at a frequency from 40 to 100 Hz. The usual values for the voltages applied to trap electrodes are summarized in Table 1. Microparticles cloud is confined in a narrow region along the longitudinal trap

Voltage	Electrode	Range	Frequency
$V_{ac} = V_0 cos 2\pi f_0 t$	E2 and E3	V_0=1-4kV	40-100Hz
U_x	E4	0-1000V	dc
U_z	E5 and E6	0-1000V	dc

Table 1. Summary of the electric voltages applied to the linear trap electrodes.The electrode E1 is connected to ground.

axis (see Fig.2). To avoid the perturbations produced by the air streams the whole trap is placed inside a transparent plastic box. The charged microparticles are stored inside the trap for hours in a quasi interaction free environment. More details on the linear electrodynamic traps mechanical layout can be found in (Gheorghe et al., 1998; Stoican et al., 2001). The motion of a charged particle in a quadrupole electric field is very well known e.g. (Major et al., 2005; March, 1997) and an extensive review are beyond the scope of this paper. Here will be summarized only the basic equations necessary to perform an appropriate numerical analysis of the effect of an acoustic field on the stored microparticles. Taking into account the expression (1) of the electric potential and the presence of a supplementary force due to the

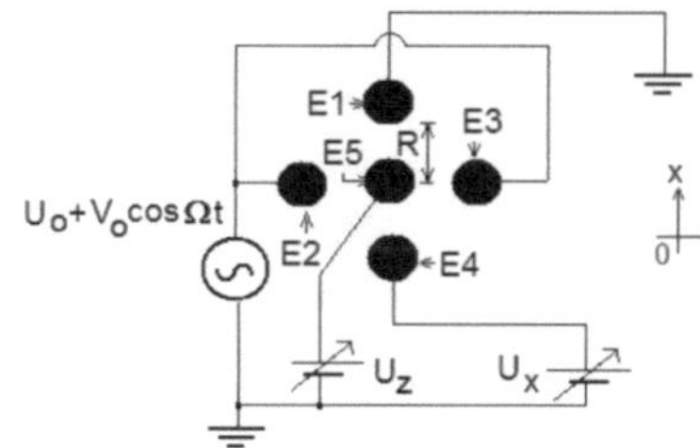

(a) View of the section xy

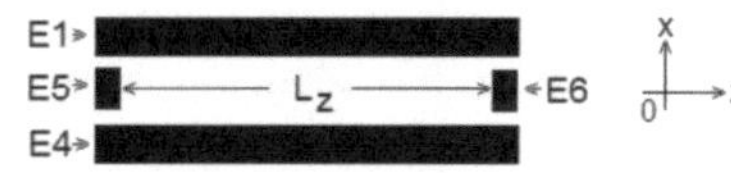

(b) View of the section xz

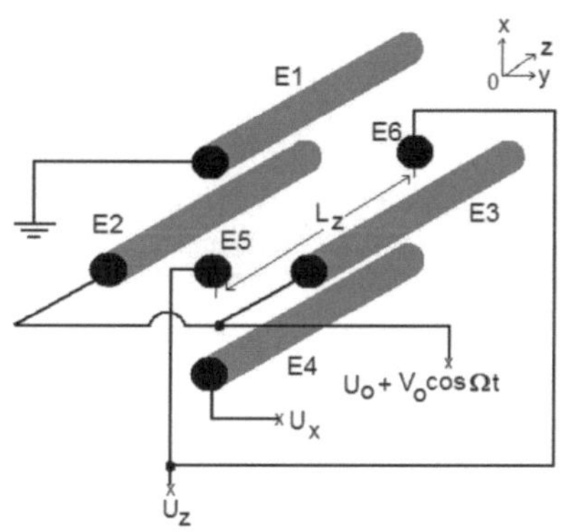

(c) 3D view

Fig. 1. Schematic drawing and electrodes wiring of a linear electrodynamic trap. The drawings are not to scale.

acoustic field, equations of motion in the (x,y) plane for a charged particle of mass M and charge Q, located near the linear trap axis, are:

$$M\frac{d^2x}{dt^2} = -\frac{Qx}{R^2}(U_0 + V_0 cos\Omega t) - k\frac{dx}{dt} + F_{Ax}(t) \tag{2}$$

and

$$M\frac{d^2y}{dt^2} = \frac{Qy}{R^2}(U_0 + V_0 cos\Omega t) - k\frac{dy}{dt} + F_{Ay}(t) \tag{3}$$

The terms $-k\frac{dx}{dt}$ and $-k\frac{dy}{dt}$, describe the drag force exerted on an object moving in a fluid. Assuming that the particles are spherical, according to *Stokes's law*:

$$k = 6\pi\eta(d/2) \tag{4}$$

where d is the diameter of the charged particle while $\eta \approx 1.8\text{x}10^{-5}\text{kgm}^{-1}\text{s}^{-1}$ is the air viscosity at normal pressure and temperature. The time-depended terms $F_{Ax}(t)$ and $F_{Ay}(t)$, stand for the force exerted by the acoustic wave on the stored particle. The Ox is the vertical axis. The microparticles weight has been neglected. Making change of variable $\xi = \Omega t/2$ the equations (2) and (3) can be rewritten as:

$$\frac{d^2x}{d\xi^2} + \delta\frac{dx}{d\xi} + (a_x + 2q_x cos2\xi)x - s_{Ax} = 0 \tag{5}$$

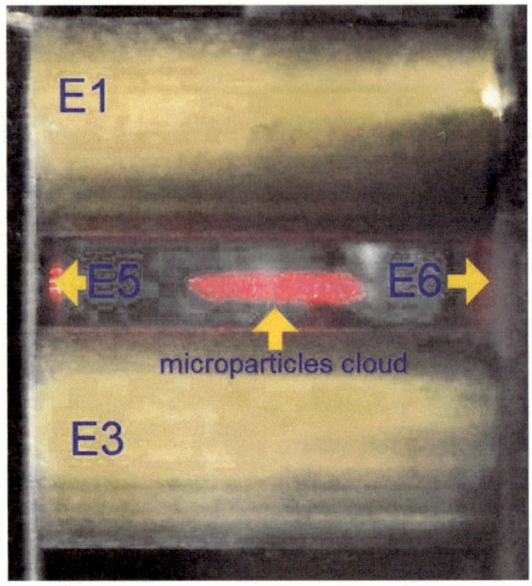

Fig. 2. Microparticles cloud stored along the longitudinal axis of the linear trap

and

$$\frac{d^2y}{d\xi^2} + \delta\frac{dy}{d\xi} + (a_y + 2q_y cos2\xi)y - s_{Ay} = 0 \tag{6}$$

The dimensionless parameters $a_{x,y}$, $q_{x,y}$ and δ are given by:

$$a_x = -a_y = \frac{4QU_0}{M\Omega^2R^2} \tag{7}$$

$$q_x = -q_y = \frac{2QV_0}{M\Omega^2R^2} \tag{8}$$

$$\delta = \frac{6\pi\eta d}{M\Omega} \tag{9}$$

The time dependent functions $s_{Ax}(t)$ and $s_{Ay}(t)$ are given by:

$$s_{Ax}(t) = \frac{4F_{Ax}}{M\Omega^2} \tag{10}$$

$$s_{Ay}(t) = \frac{4F_{Ay}}{M\Omega^2} \tag{11}$$

The pair (a, q) defines the operating point of the electrodynamic trap and determines entirely the characteristics of the particle motion. In the absence of the terms due to the drag force ($-k\frac{dx}{dt}$ and $-k\frac{dy}{dt}$) and the acoustic wave $F_{Ax}(t)$ and $F_{Ay}(t)$, a differential equation of type (5) or (6) is called the *Mathieu equation* (McLachlan, 1947). It can be shown that solutions of a Mathieu equation describe a spatial bounded motion (stable solutions) only for certain regions of the *(a,q)* plane called *stability domains*. This means that, a charged particle can remain indefinitely in the space between the trap electrodes. Additionally, the charged particle trajectory must not cross the electrodes surface implying the supplementary restrictions in its initial position and velocity. One could say that, within the stability domains, a potential barrier arises preventing the stored charged particles to escape out of the trap. As an example, for the first stability domain, if $a_x = 0$, the stable solutions are obtained if $0 < q_x < 0.908$. The first domain stability corresponds to the lowest voltages applied to the trap electrodes. Due to the air drag area of the first stability domain is enlarged so that, depending on the value of δ, the particle can remain inside the trap even if $q_x > 0.908$. Operation within the higher order stability domains is not practical because of very high voltage to be applied across the trap electrodes. As can be seen in (7) and (8) the operating point depends on the electrodynamic trap geometry, electrodes supply voltages characteristics and charge-to-mass ratio of the stored particle. Knowing the operating point of the trap, its dimensions and applied voltages, then charge-to-mass ratio of the stored particle can be estimated. If $\delta = 0, F_{Ax,y} = 0, |a_x|, |a_y|, |q_x|, |q_y| \ll 1$ (*adiabatic approximation*), the differential equations (5) and (6) have the solutions (Major et al., 2005):

$$x(t) = x_0 cos(\omega_x t + \varphi_x)(1 + \frac{q_x}{2} cos\Omega t) \tag{12}$$

$$y(t) = y_0 cos(\omega_y t + \varphi_y)(1 + \frac{q_y}{2} cos\Omega t) \tag{13}$$

where

$$\omega_x = \frac{\Omega}{2}\sqrt{\frac{q_x^2}{2} + a_x} \tag{14}$$

and

$$\omega_y = \frac{\Omega}{2}\sqrt{\frac{q_y^2}{2} + a_y} \tag{15}$$

Under these conditions the motion of a charged particle confined in a quadrupole trap can be decomposed in a harmonic oscillation at frequencies $\omega_i/2\pi$ called "*secular motion*" and a harmonic oscillation at the frequency $f_0 = \Omega/2\pi$ of ac voltage called "*micromotion*" . As a consequence the motional spectrum of the stored particle contains components $\omega_i/2\pi$ and $f_0 \pm \omega_i/2\pi (i = x, y)$. For arbitrary values of the parameters $a_{x,y}$, $q_{x,y}$ and δ, equations (5) or (6) can be numerically solved.

3. Experimental setup

The experimental setup is based on the method described in (Schlemmer et al., 2001) used for a linear trap. The scheme of the experiment is shown in Fig.3. The output beam of a low power laser module (650 nm, 5 mW) is directed along the longitudinal axis (Oz axis) of the linear trap. A hole drilled through one of the end-cap electrode (E6) allows the laser beam illuminating the axial region of the trap where the stored particles density is maximal and the electric potential is well approximated by the relation (1). A photodetector PD placed outside of the trap and oriented normal to the laser beam receives a fraction of the radiation scattered by the stored particles and converts it into an electrical voltage U_{ph} proportional to the incident radiation intensity. To prevent electrical perturbations due to the existing ac high voltage applied to the electrodes trap, the photodetector is encapsulated in a cylindrical shielding box. The effect of the background light is removed by means of an appropriate electronic circuit. The acoustic excitation of the stored microparticles is achieved by a loudspeaker placed next to the trap. The loudspeaker generates a monochromatic acoustic wave with frequency f_A. In this way both electrical field created by the trap electrodes and the force due to the acoustic wave act simultaneously on the stored microparticles. The motion of the stored particles

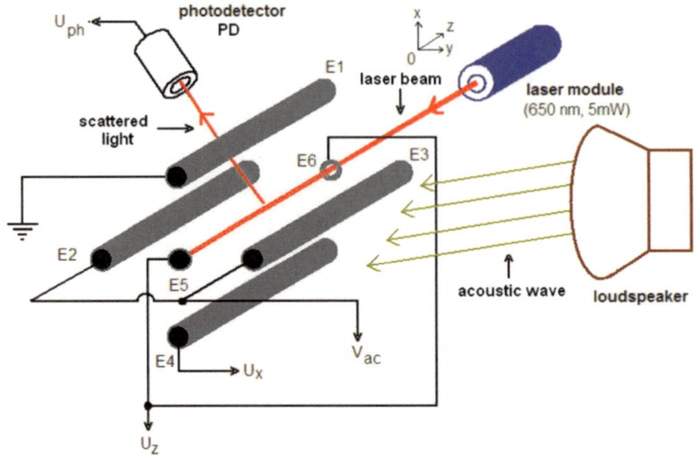

Fig. 3. Schematic of the experimental setup

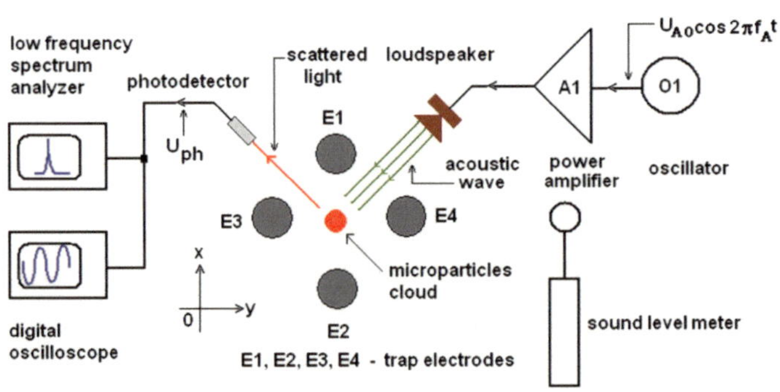

Fig. 4. Block diagram of the measurement chain. The trap electrodes wiring is not shown.

modulates the intensity of the scattered radiation. Therefore the photodetector output voltage U_{ph} contains the same harmonic components. By analysing changes in the structure of the frequency domain spectrum of the voltage U_{ph}, the effect of the acoustic wave on the stored particles can be evaluated. For this purpose a measurement chain whose block diagram is shown in Fig. 4 has been implemented. A digital low frequency spectrum analyser is used to determine the harmonic components of the voltage U_{ph}. The loudspeaker is supplied by the low frequency power amplifier A1 which is driven by the low frequency oscillator O1. The intensity of the acoustic wave is monitored by means of a sound level meter. Both frequency and intensity of the acoustic wave can be varied. A similar version of the experimental setup has been previously described in (Stoican et al., 2008) where preliminary investigations regarding the effect of the acoustic waves on the properties of the microparticles stored in a linear electrodynamic trap particle has been reported.

4. Experimental measurements

The microparticles consist of Al_2O_3 powder, 60-200 μm in diameter, stored at normal pressure and temperature. The working parameters (U_x, U_z and V_0) of the linear trap were chosen so that the magnitude of the harmonic component of the photodetector output voltage U_{ph}

corresponding to f_0 to reach a maximum (Fig. 5a). This operating point of the trap is known as *"spring point"* (Davis et al., 1990). Two typical spectra of the voltage U_{ph} recorded in these experimental conditions are shown in Fig. 6. Only frequencies less than, or equal to $3f_0/2$,

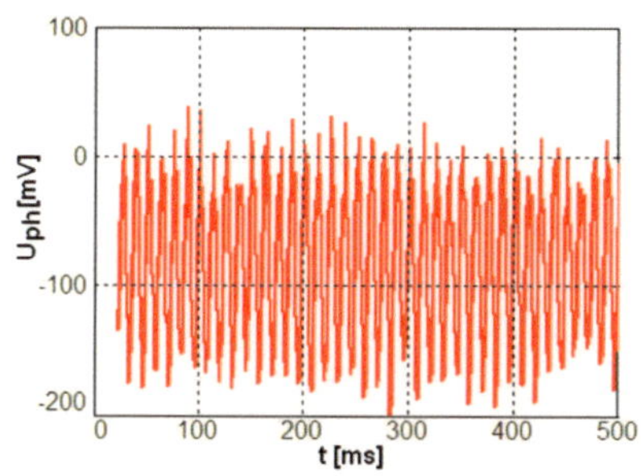

(a) No acoustic excitation

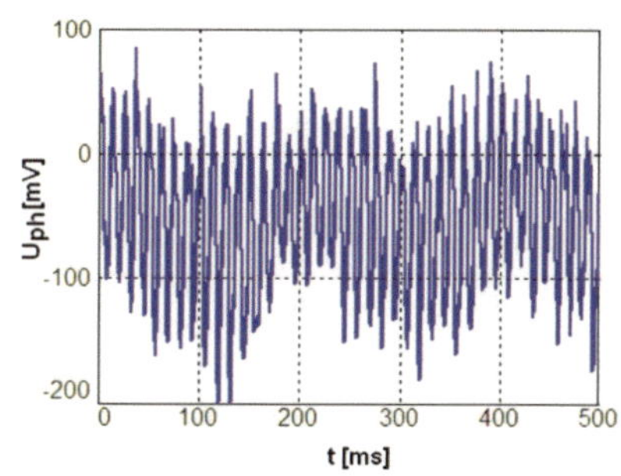

(b) Acoustic excitation, f_A=75Hz. The voltage U_{ph} appears to be amplitude modulated.

Fig. 5. Oscilloscope image representing the time variation of the photodetector voltage output U_{ph} in the absence (a) and presence (b) of the acoustic excitation. Experimental conditions: V_0=3.3kV, U_x=0, U_z=920V, f_0=80Hz. Experimental results.

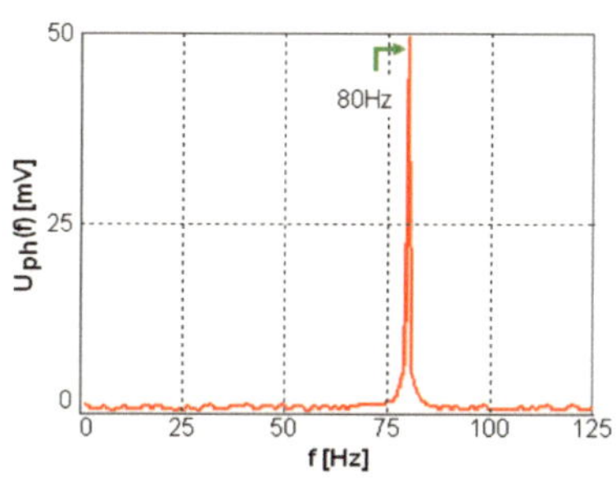

(a) U_z=920V

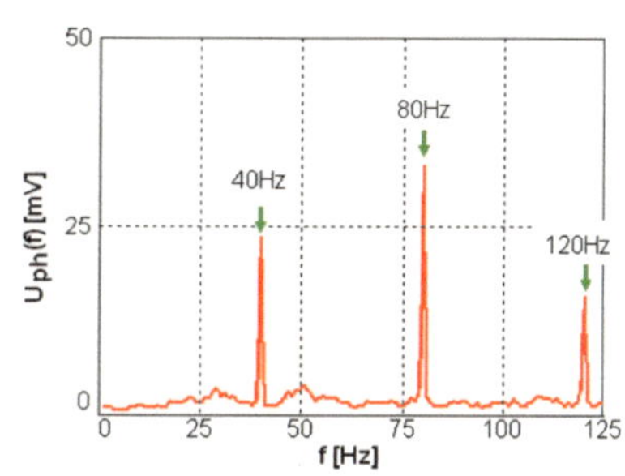

(b) U_z=100V

Fig. 6. Typical spectra of the photodetector output voltage U_{ph} without the acoustic excitation. Experimental conditions: f_0=80Hz, V_0=3.3kV, U_x=0V. Experimental results.

have been considered because upper lines could be caused by the ac voltage V_{ac} waveform imperfections or digital data processing. Also it was necessary to limit the frequency band to keep a satisfactory resolution of the recordings. As it can be seen from Fig. 6, under these conditions, without the acoustic excitation, the spectra of the photodetector output voltage contain only three significant lines, namely $f_0/2$, f_0 and $3f_0/2$ (here 40Hz, 80Hz and 120Hz). Depending on the applied dc voltages and photodetector position some lines could missing. Several spectra of the photodetector output voltage U_{ph} recorded during acoustic excitation of the microparticles at different frequencies f_A are shown in Fig. 7. The measured sound level was about 85dB. By examining the experimental records, it can be seen that the supplementary lines occur in the motional spectrum of the stored particles. As an empirical rule, the frequency peaks due to the acoustic excitation belong to the combinations of the form $f_0 \pm |f_0 - f_A|$ and $n|f_0 - f_A|$ where n=0, 1, 2..., f_A is the frequency of the acoustic field and f_0 is the frequency of the applied ac voltage V_{ac}. The rule is valid both for $f_A < f_0$ and $f_A > f_0$. As seen in Fig. 7a and Fig. 7d, the two spectra are almost identical because $|f_0 - f_A|$=20Hz in both cases. If f_A is close to the f_0 the lowest frequency component $|f_0 - f_A|$ yields an amplitude modulation of the voltage U_{ph} very clearly defined (Fig. 5b). The effect is similar to the beat signal due to the interference of two harmonic signals of slightly different frequencies.

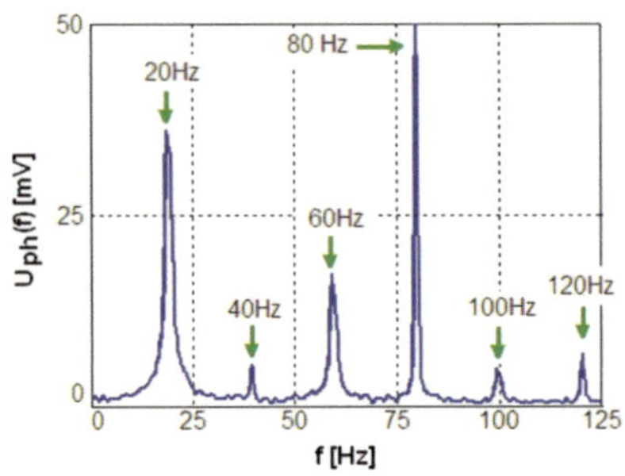

(a) f_A=60Hz

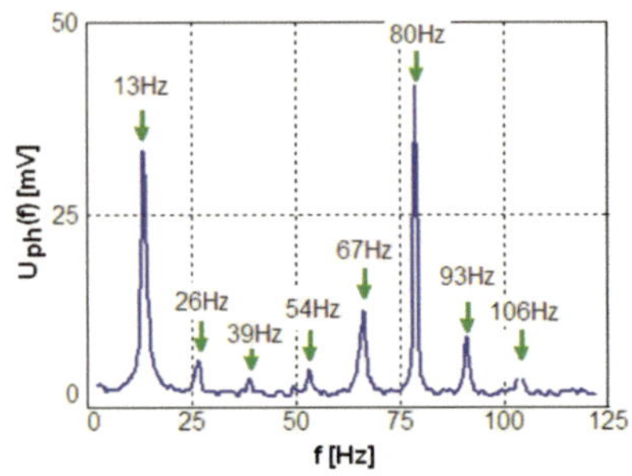

(b) f_A=67Hz

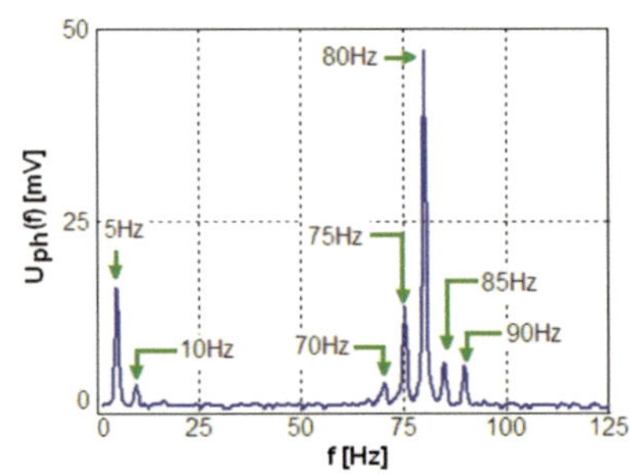

(c) f_A=75Hz

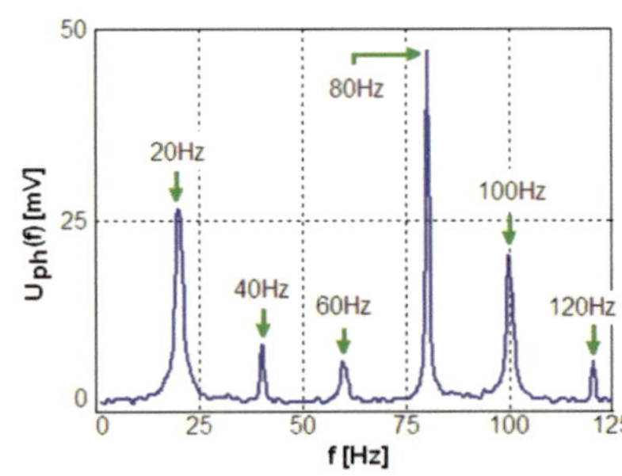

(d) f_A=100Hz

Fig. 7. Spectrum of the photodetector output voltage U_{ph} when the stored microparticles are excited by an acoustic wave at different frequencies f_A. Experimental conditions: f_0=80Hz, V_0=3.3kV, U_x=0V, U_z=920V, sound level $\approx$ 85dB. Experimental results.

5. Numerical analysis of the stored particle motion in an acoustic field

A qualitative interpretation of the experimental results requires a numerical analysis on the motion of the stored particles. For this purpose the differential equations (5) and (6) must be numerically solved. Consequently, it is necessary to express functions $s_{Ax}(t)$ and $s_{Ay}(t)$ from (10) and (11) in terms of quantities which are known or can be experimentally measured. A body subjected to an acoustic wave field, experiences a steady force called *acoustic radiation pressure* and a time varying force caused by the periodic variation of the pressure in the surrounding fluid. The radiation pressure is always repulsive meaning that it is directed as the wave vector. The time varying force oscillates at the frequency of the acoustic wave and its time average is equal to zero. The radiation pressure $\overline{F_R}$ exerted on a rigid spherical particle by a plane progressive wave is derived in (King, 1934) as:

$$\overline{F_R} = \frac{\pi^5 d^6}{\lambda^4} \overline{w} F(\rho_0/\rho_1) \tag{16}$$

where d is the particle diameter, λ is the wavelength of the acoustic wave and $\overline{w}$ represents the acoustic energy density. The parameter $F(\rho_0/\rho_1)$ is called *relative density factor* being given by:

$$F(\rho_0/\rho_1) = \frac{1 + \frac{2}{9}(1 - \rho_0/\rho_1)^2}{(2 + \rho_0/\rho_1)^2} \tag{17}$$

where ρ_0 and ρ_1 are air and particle density, respectively. If $\rho_0/\rho_1 \ll 1$, as in the present case (see bellow), practically $F(\rho_0/\rho_1)$=0.305. For a plane progressive acoustic wave, energy density is (Beranek, 1993; Kinsler et al., 2000):

$$\overline{w} = p_s^2/\rho_0 c^2 \tag{18}$$

where p_s is *sound pressure* and c is the speed of propagation of the acoustic wave. Sound pressure can be estimated according to the formula which defines the *sound pressure level* SPL:

$$SPL = 20log_{10}\frac{p_s}{p_{ref}} \tag{19}$$

where p_{ref}=2 x 10^{-5}N/m^2 is the standard reference pressure. Sound pressure level (SPL) is expressed in dB and can be experimentally measured by using sound level meters. In (King, 1934), as an intermediate result, the velocity of a spherical particle placed in an acoustic wave field is given as:

$$\frac{dr}{dt} = -\frac{A_1}{\alpha^3} k_0 \frac{\rho_0}{\rho_1} \frac{1}{F_1 - iG_1} \tag{20}$$

where $k_0 = \omega/c$, $\alpha = k_0 d/2$, while $\omega = 2\pi f_A$ is the angular frequency of the acoustic wave. If $\rho_0/\rho_1 \ll 1$ and $\alpha \ll 1$ then $F_1(\alpha) \simeq 2/\alpha^3$ and $G_1(\alpha) \simeq -1/3$. The quantity A_1 is a measure of wave intensity. Assuming a monochromatic plane progressive acoustic wave, the amplitude of the acoustic oscillating force exerted on the particle may be written:

$$F_{A0r} = |M\frac{d^2r}{dt^2}| = \frac{\omega M|A_1|}{\alpha^3} k_0 \frac{\rho_0}{\rho_1} \frac{1}{|F_1 - iG_1|} \tag{21}$$

where $|A_1| = 3cv/\omega$. Taking into account expressions for F_1, G_1, $|A_1|$, α and k_0, written above, the relation (21) becomes:

$$F_{A0r} = 2\omega M v \frac{\rho_0}{\rho_1} \tag{22}$$

The quantity v represents the amplitude of the velocity of the surrounding fluid particles (air in this case), which are oscillating due to the acoustic wave, and is related to the sound pressure by the relation:

$$v = \frac{\sqrt{2} p_s}{\rho_0 c} \tag{23}$$

Finally, the amplitude of the oscillating force, considering $\rho_0/\rho_1 \ll 1$ and $\alpha \ll 1$, is:

$$F_{A0r} = \frac{2\sqrt{2}\omega M p_s}{\rho_1 c} \tag{24}$$

Consequently:

$$F_{Ar} = \overline{F_R} + F_{A0r} cos(2\pi f_A t + \beta) \tag{25}$$

As further numerical evaluations will demonstrate the acoustic radiation pressure $\overline{F_R}$ is several order of magnitude less than that of oscillating force amplitude F_{A0r} and has been neglected. In order to simplify theoretical analysis, the weight of the microparticles (i. e. Mg where g=9.8m/s^2) has been also neglected. Experimentally, the microparticles weight are usually compensated by the electric field due to the dc voltage U_x. As a result, considering the experiment geometry (Fig. 4):

$$F_{Ax}(t) = F_{Ay}(t) \approx \frac{\sqrt{2}}{2} F_{A0r} cos(2\pi f_A t + \beta) \tag{26}$$

Therefore the two function $s_{Ax}(t)$ and $s_{Ay}(t)$ takes form:

$$s_{Ax}(t) = s_{Ay}(t) = s_{A0} cos(2\pi f_A t + \beta) = s_{A0} cos(2 \frac{f_A}{f_0} \xi + \beta) \tag{27}$$

where

$$s_{A0} = \frac{8\omega p_s}{\rho_1 c \Omega^2} \tag{28}$$

All quantities existing in formula (28) are known or can be experimentally measured. The operating conditions considered for numerical analysis are summarized in Table 2 . The diameter of the Al_2O_3 microparticles can vary in the range from 60μm to 200μm, as

Parameter	Notation	Value
trap inner radius	R	5 x 10^{-3} m
particle density (Al_2O_3)	ρ_1	3700 kg/m^3
air density	ρ_0	1.2 kg/m^3
speed of sound	c	343 m/s
air viscosity	η	1.8 x 10^{-5} kgm^{-1}s^{-1}
sound pressure level	SPL	85 dB
sound pressure	p_s	0.35 N/m^2
sound frequency	f_A	100 Hz
ac voltage frequency	f_0	80 Hz

Table 2. Operating conditions considered for numerical analysis

before mentioned. The two corresponding limit values of the parameters depending on the microparticles size, are shown in Table 3. The operating conditions taken into account are the same as in Table 2. According to numerical values listed in Table 2 and Table 3, $\rho_0/\rho 1 \ll 1$ and

particle diameter d	6 x 10^{-5} m	2 x 10^{-4} m
particle mass M	4.18 x 10^{-10} kg	1.55 x 10^{-8} kg
particle weight Mg	4.10 x 10^{-9} N	1.51 x 10^{-7} N
$\alpha = k_0 d/2$	5.49 x 10^{-5}	1.83 x 10^{-4}
δ	0.09	0.008
acoustic radiation pressure $\overline{F_R}$	2.82 x 10^{-32} N	3.87 x 10^{-29} N
oscillating force amplitude F_{A0r}	2.08 x 10^{-13} N	7.71 x 10^{-12} N
s_{A0}/R	1.11 x 10^{-6}	1.11 x 10^{-6}

Table 3. The limit values corresponding to the particle possible diameter. Operating conditions are listed in Table 2

$\alpha \ll 1$, which are in good agreement with our previous assumptions. Fourier transforms $X(f)$ and $R(f)$, of the rectangular $x(t)$ and radial $r(t) = \sqrt{x^2(t) + y^2(t)}$ coordinates, respectively, in the range from 0 to 125Hz, without the acoustic excitation, obtained by solving numerically differential equations (5) or (6), for several values of the parameters q_x, are shown in Fig.8. Fourier transforms of $x(t)$ and $y(t)$ are similar. As theory of Mathieu equations ascertains, numerical calculations shows that for $|q_{x,y}| \ll 1$, the motional spectrum contains mainly harmonic components due to the secular motion, namely $\omega_x/2\pi$ and $f_0 \pm \omega_x/2\pi$. In practice, the microparticles trapping in such conditions is difficult to be achieved because the potential barrier is too low so that weak external perturbations can eject stored particles out of the trap. The numerical results show that, close to the limit of the first stability domain, the micromotion becomes important and the motional spectrum contains only a few significant lines, namely at f_0 for $r(t)$ and at $f_0/2$ and $3f_0/2$ for $x(t)$, respectively. Fourier transforms $X(f)$ and $R(f)$, of the rectangular $x(t)$ and radial $r(t)$ coordinates, respectively, in the range from 0 to 125Hz, in the presence of the acoustic excitation, obtained by solving numerically differential equations (5) or (6), for several values of the acoustic wave frequency f_A, are shown in Fig. 9. Simulation conditions are listed in Table 4. Expression of the parameter s_{A0}/R corresponds to SLP=85dB and R=5mm. All the remaining parameters, except acoustic wave frequency f_A, are listed in Table 2 and Table 3 The variation of the normalized time

$f_0[Hz]$	q_x	a_x	δ	s_{A0}/R	$r(0)$	$\dot{r}(0)$	β	time interval[s]
80	0.908	0	0.01	0.11 x $10^{-7} f_A[Hz]$	0	0	0	0-40

Table 4. Conditions taken into account to simulate the motion of the a microparticle in the presence of an acoustic field (Figs.9 and 10)

average, $\bar{r}(t)/R$, of the coordinate r as a function of acoustic wave frequency f_A is shown in Fig 10. As on can see from this figure there are two critical frequencies of the acoustic wave, namely $f_0/2$ and $3f_0/2$, where the amplitude of the particle motion rises sharply. It is interesting to note that this very high growth of the motion amplitude does not occur for f_A=f_0. The microparticles cloud appears to form a *non-linear oscillating system* for which resonance frequencies occur at pf_0/q with p and q integer (Landau & Lifshitz, 1976). The variation of the normalized time average, $\bar{r}(t)/R$, of the coordinate r as a function of $log_{10}(s_{A0}/R)$ at acoustic wave frequency f_A=120Hz is shown in Fig 11. According to (28) and (19) the parameter s_{A0} may be considered a measure of the acoustic excitation because $log_{10}(s_{A0}) = log_{10}p_s + constant$ and sound pressure level SPL depends on $log_{10}p_s$. As on can see from this figure, beyond a threshold value, the average distance from the trap axis grows monotonically with the level of the sound pressure level. Around $s_{A0}/R \simeq 10^{-8}$ there is a threshold value where amplitude of the microparticle motion increases suddenly (the region within the green frame from Fig.11). The transition is not uniform, there are maximum and minimum values of $\bar{r}(t)$. This phenomenon could be interpreted as so called *collapse of resonance* which means that the amplitude of resonance goes to zero for some value of the perturbation amplitude (Olvera, 2001). However numerical result must be regarded with caution. Some hypothesis aimed to simplify mathematical treatment has been introduced. For example it was assumed that the parameter β is equal to zero and it remains constant. In fact β, which signifies the phase shift between ac voltage V_{ac} and acoustic wave, varies randomly because the two oscillations are produced by different generators. Further experiments are necessary to confirm theoretical assertions. All of the numerical results have been obtained by using SCILAB 5.3.1.

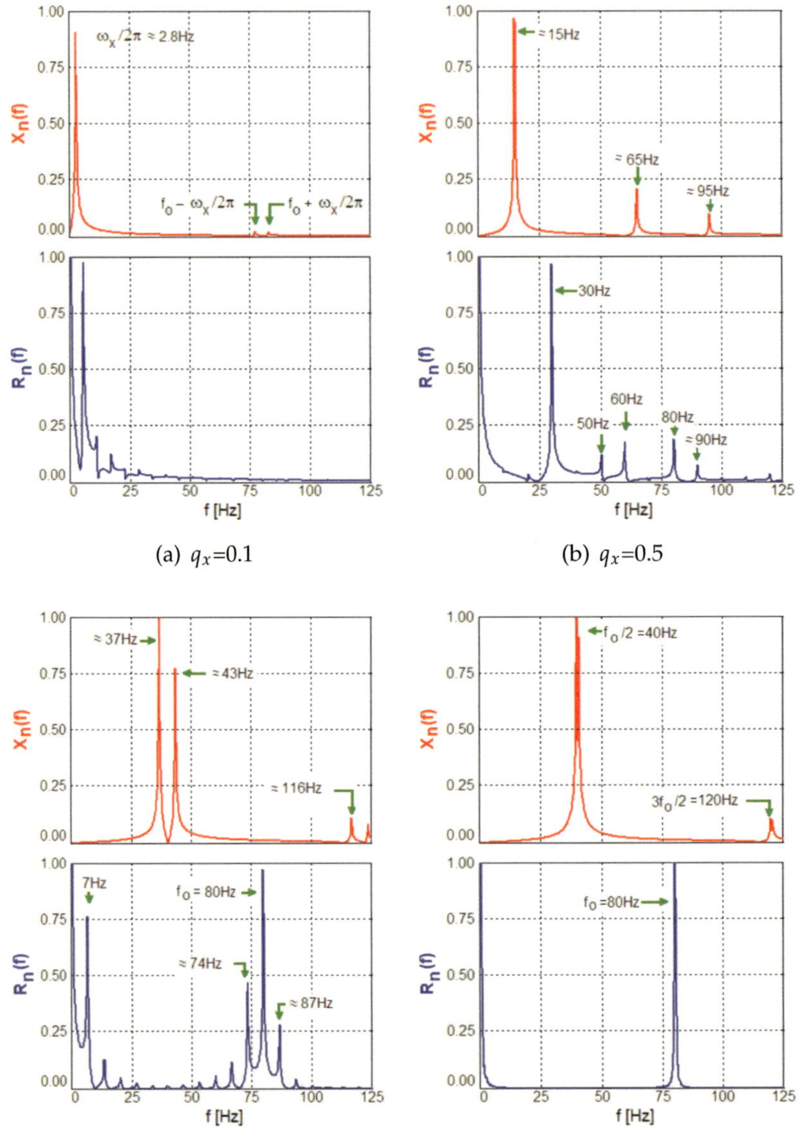

(a) q_x=0.1

(b) q_x=0.5

(c) q_x=0.9

(d) q_x=0.908

Fig. 8. $X_n(f) = X(f)/X_{max}(f)$ (red) and $R_n(f) = R(f)/R_{max}(f)$ (blue) where $X(f)$ and $R(f)$ are Fourier transform of the rectangular $x(t)$ and radial $r(t)$ coordinates, respectively, for $0<f<125$Hz, a_x=0, $f_0 = \Omega/2\pi$=80Hz, δ=0.01, without the acoustic excitation. Numerical result.

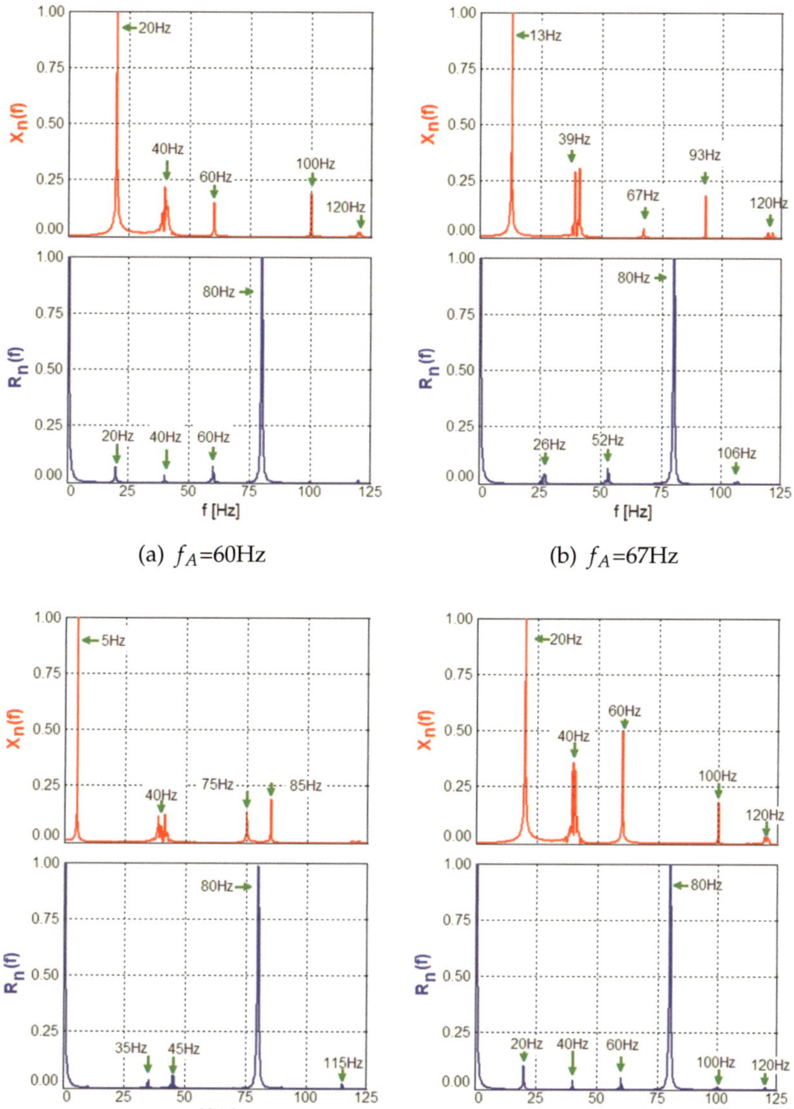

(a) f_A=60Hz (b) f_A=67Hz

(c) f_A=75Hz (d) f_A=100Hz

Fig. 9. $X_n(f) = X(f)/X_{max}(f)$ (red) and $R_n(f) = R(f)/R_{max}(f)$ (blue) where $X(f)$ and $R(f)$ are Fourier transform of the rectangular $x(t)$ and radial $r(t)$ coordinates, respectively, for $0<f<125$Hz, in the presence of the acoustic excitation, for several values of the acoustic wave frequency f_A. The plots correspond to the frequency f_0=80Hz of the ac supply voltage V_{ac} and to an incident sound level of 85dB. A complete list of the simulation conditions is displayed in Table 4. Numerical result.

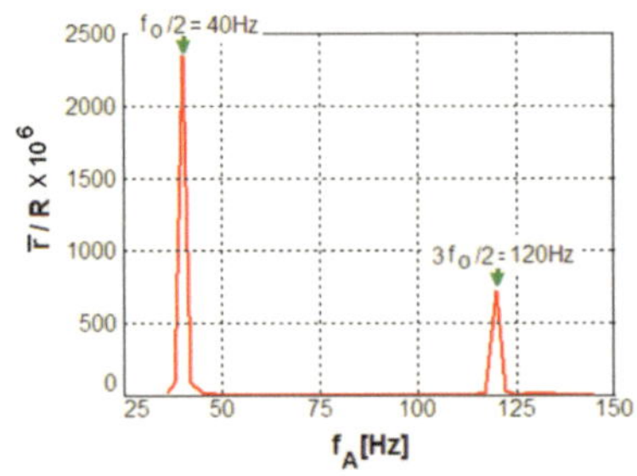

Fig. 10. The normalized time average of the coordinate $r(t)$ as a function of acoustic wave frequency f_A. R represents the inner radius of the linear trap. Simulation conditions are displayed in Table 4. Numerical result.

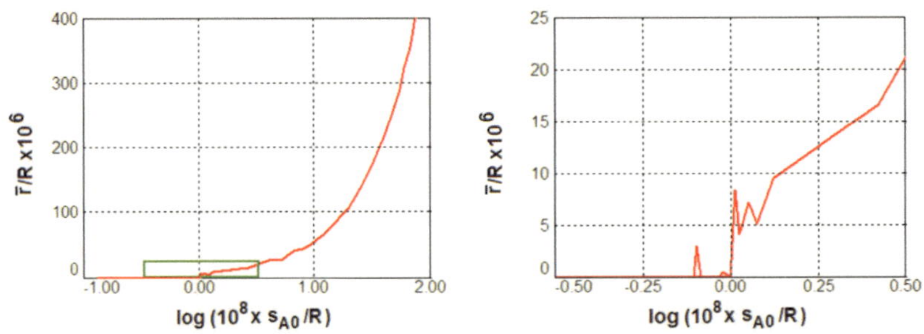

(a) Complete view

(b) Expanded view of the region within the green frame

Fig. 11. The normalized time average of the coordinate $r(t)$ as a function of parameter s_{A0}/R at acoustic wave frequency f_A=120Hz. Parameter s_{A0} is a measure of acoustic excitation($log_{10}(s_{A0}) \sim SPL$). Simulation conditions, except parameter s_{A0}/R, are displayed in Table 4. Numerical result.

6. Conclusion

Although measurements result has been interpreted using an approximative and simplified theory of the linear trap, by comparing the experimental spectra with numerical simulations a good qualitative agreement can be noticed. The spectrum of the laser radiation intensity scattered by the stored microparticles seems to be a superposition of the $r(t)$ and $x(t)$ (or $y(t)$) harmonic components, respectively. Because the experimental spectrum obtained without the acoustic excitation, shown in Fig.6 is similar to that displayed in Fig.8d we may suppose that operating point of the linear trap during the experiment corresponds to $q_x \approx 0.908$ (*i.e.* $Q/M \approx 5.4 \times 10^{-4}$ C/kg cf. (8)). The simulation of the stored particle motion in an acoustic wave has been done taking into account this assumption. By knowing the ratio Q/M and the mass of a microparticle (from Table 3) electric charge Q can be evaluated. Because the microparticles are stored in air at normal pressure, a supplementary restriction on the physical characteristics of the microparticles arises. Assuming that microparticles are spherical and the electric charge is uniform distributed over its surface, the electric field E_d at the surface is:

$$E_d = \frac{Q}{4\pi\epsilon_0(d/2)^2} \tag{29}$$

where ϵ_0=8.854 x 10^{-12} F/m. Electric charge Q results from:

$$Q = M(\frac{Q}{M}) = \frac{4\pi}{3}(\frac{d}{2})^3\rho_1(\frac{Q}{M}) \tag{30}$$

where Q/M is known. The electric field E_d can not exceed a certain threshold value E_{max}, called *dielectric strength*, at which surrounding gas molecules become ionized, so that the condition $E_d < E_{max}$ is necessary. From (29) and (30), the above condition can be written:

$$d < \frac{6\epsilon_0 E_{max}}{(Q/M)\rho_1} \tag{31}$$

The dielectric strength E_{max} varies with the shape and size of the charged body, humidity and pressure of the air. We consider here the dielectric strength of air E_{max} ≈3 x 10^6 V/m. By replacing numerical values in (31), it results that d<69μm. Therefore the diameter of the microparticles which produce the laser beam scattering sensed by the photodetector lies in the range 60-69μm. Distribution of the frequency peaks are almost identical both in experimental recordings and numerical simulation, but their relative heights are different. Besides a simplified theoretical approach, there are a few possible experimental issues. The dc voltages U_x and U_z shift the stored microparticles cloud position relative to the trap axis and, implicitly, to the photodetector aperture. Thus, viewing angle of the photodetector is modified and some harmonic components could not be observed or appear to be very weak. On the other hand, non-uniformity of the frequency characteristic for the common loudspeakers, especially at frequencies below 100Hz, could distort experimental recordings. Numerical estimations demonstrate that, in the described experimental conditions and within the considered range of the acoustic wave frequency, the acoustic radiation pressure $\overline{F}_R$ is very weak compared to the microparticles weight Mg or amplitude of the oscillating force exerted by acoustic wave F_{A0r}(see Table 3). Certainly, the first improvement of theoretical approach would be to add a constant term corresponding to the microparticles weights in differential equations (2) and/or (3), depending on the orientation of the reference system. Like weight, acoustic radiation pressure does not vary in time and can be regarded, mathematically speaking, as a small correction of the microparticles weight. As a result the acoustic radiation pressure is not important here, where only acoustic waves with frequency around frequency f_0 of the ac voltage have been investigated. The acoustic radiation pressure can become significantly at higher frequencies, (see (16)) varying according to f^4. On the other hand, according to the numerical results, if the intensity of the acoustic wave is larger than a threshold value and its frequency has proper values, the amplitude of the stored microparticles motion increases significantly. These frequencies are identical to certain frequency peaks existing in the motional spectrum of the stored particle predicted by the numerical simulations (Fig. 8d) and observed experimentally (Fig.6b). Therefore its action is effective only for the microparticles species whose properties meet certain conditions. In this way, theoretically, the selective manipulation of the stored microparticles by means of the acoustic waves becomes possible.
We gratefully acknowledge material support provided by *Mira Technologies Group SRL - Bucharest, Romania*.

7. References

Beranek, L. L. (1993). *Acoustics*, Acoustical Society of America, New York.

Carleton, K. L., Sonnenfroh, D. M., Rawlins, W. T., Wyslouzil, B. E. & Arnold, S. (1997). Freezing behavior of single sulfuric acid aerosols suspended in a quadrupole trap, *Journal of Geophysical Research* Vol. 102(No. D5): 6025–6033.

Davis, E. J. (1997). A history of single aerosol particle levitation, *Aerosol Science and Technology* Vol. 26(No. 3): 212–254.

Davis, E. J., Buehler, M. F. & Ward, L. T. (1990). The double-ring electrodynamic balance for microparticle characterization,*Review of Scientific Instrument* Vol.61(No. 4): 1281–1288.

Gheorghe, V. N., Giurgiu, L., Stoican, O., Cacicovschi, D., Molnar, R. & Mihalcea, B. (1998). Ordered structures in a variable length a.c. trap., *Acta Physica Polonica A* Vol.93(No. 4): 625– 629.

Jakubczyk, D., Zientara, M., Bazhan, W., Kolwas, M. & Kolwas, K. (2001). A device for light scatterometry on single levitated droplets, *Opto-Electronics Review* Vol. 9(No.4): 423–430.

King, V. L. (1934). On the acoustic radiation pressure on spheres, *Proceedings of the Royal Society (London)* Vol.A147: 212–240.

Kinsler, L. E., Frey, A. R., Coppens, A. B. & Sanders, J. V. (2000). *Fundamentals of Acoustics*, John Wiley and Sons, Inc, New York.

Landau, L. D. & Lifshitz, E. M. (1976). *Mechanics*, Butterworth-Heinemann, Oxford.

Major, F. G., Gheorghe, V. N. & Werth, G. (2005). *Charged Particle Traps, Physics and Techniques of Charged Particle Confinement*, Springer, Berlin Heidelberg New York.

March, R. E. (1997). An introduction to quadrupole ion trap mass spectrometry, *Journal of Mass Spectrometry* Vol. 32: 351–369.

McLachlan, N. W. (1947). *Theory and Application of Mathieu Functions*, Clarendon Press, Oxford.

Olvera, A. (2001). Estimation of the amplitude of resonance in the general standard map, *Experimental Mathematics* Vol.10(No. 3): 401–418.

Pedregosa, J., Champenois, C., Houssin, M. & Knoop, M. (2010). Anharmonic contributions in real rf linear quadrupole traps, *International Journal of Mass Spectrometry* Vol. 290(No. 2-3): 100–105.

Peng, W. P., Yang, Y. C., Kang, M. W., Lee, Y. T. & Chang, H. C. (2004). Measuring masses of single bacterial whole cells with a quadrupole ion trap, *Journal of the American Chemical Society* Vol. 126(No.38): 11766–11767.

Schlemmer, S., Illemann, J., Wellert, S. & Gerlich, D. (2001). Nondestructive high-resolution and absolute mass determination of single charged particles in a three-dimensional quadrupole trap, *Journal of Applied Physics* Vol. 90(No. 10): 5410–5418.

Shaw, R. A., Lamb, D. & Moyle, A. M. (2000). An electrodynamic levitation system for studying individual cloud particles under upper-tropospheric conditions, *Journal of Atmospheric and Oceanic Technology* Vol. 17(No. 7): 940–948.

Stoican, O. S., Dinca, L. C., Visan, G. & Radan, S. (2008). Acoustic detection of the parametrical resonance effect for a one-component microplasma consisting of the charged microparticles stored in the electrodynamic traps, *Journal of Optoelectronics and Advanced Materials* Vol.10(No. 8): 1988–1990.

Stoican, O. S., Mihalcea, B. & Gheorghe, V. N. (2001). Miniaturized microparticle trapping setup with variable frequency, *Roumanian Reports in Physics* Vol. 53(No. 3-8): 275–280.

Wuerker, R. F., Shelton, H. & Langmuir, R. V. (1959). Electrodynamic containement of charged particles, *Journal of Applied Physics* Vol. 30(No. 3): 342–349.

10

Acoustic Waves in Bubbly Soft Media

Bin Liang[1], Ying Yuan[2], Xin-ye Zou[1] and Jian-chun Cheng[1]
[1]*Nanjing University, Nanjing*
[2]*Jiangsu Teachers University of Techonology, Changzhou*
P. R. China

1. Introduction

There exists a special class of solid media called soft media (or weakly compressible media), for which the inequality $\lambda >> \mu$ is satisfied (λ and μ are the Lamé coefficients) [1-6]. Such media with very small shear stiffness are dynamically similar to liquids to a great extent and exhibit strongly a "water-like" characteristic and, therefore, are also associated with the name of "water-like" media. The class of soft media includes many common media in the fields of scientific research and practical applications (e.g. soft rubbers, tissues, or biomimetic materials), and air bubbles are often introduced due to artificial or non-artificial reasons. In the presence of air bubbles in a soft medium, the bubbles will oscillate violently as an acoustic wave propagates in the bubbly medium. It has been proved that only if the ratio $\lambda >> \mu$ is sufficiently large can a bubble in a solid behave like an oscillator with a large quality factor, otherwise oscillation will be damped within a short time comparable to its period [7]. This results in the fact that the acoustical property of a bubbly soft medium is particularly different from that of a usual solid medium containing bubbles. The strong scattering of acoustic waves by bubbles in soft media may lead to some significant physical phenomena stemming from the multiple-scattering effects, such as the localization phenomenon [8-11]. On the other hands, a bubbly soft medium has potential applications to a variety of important situations, such as the fabrication of acoustic absorbent with high efficiency or the clinic application of ultrasonic imaging utilizing the contrast agents [12]. Consequently, it is of academic as well as practical significance to make a comprehensive study on the acoustic waves in a bubbly soft medium.

Attempts to theoretically investigate the bubble dynamics in a soft medium go back many decades [1-4]. Meyer et al have performed the early measurements of the resonance frequency of a bubble in rubbers [1]. The dynamical equation for arbitrarily large radial motion has been derived by Erigen and Suhubi in Rayleigh-Plesset form [2]. Ostrovsky has derived a Rayleigh-Plesset-like equation to describe the nonlinear oscillation of an individual bubble in a soft medium [3,4]. On the basis of the linear solution of the bubble dynamic equation of Ostrovsky, Liang and Cheng have theoretically investigated the acoustic propagation in an elastic soft medium containing a finite number of bubbles [8-11]. By rigorously solving the wave field using a multiple-scattering method, they have revealed the ubiquitous existence of the significant phenomenon of acoustic localization in such a class of media, and identified the "phase transition" phenomenon similar with the order-disorder phase transition in a ferromagnet. For practical samples of bubbly soft media, however, the multiple-scattering method usually fails due to the extremely large number of

bubbles and the strong acoustical nonlinearity in such media. In such cases, a more useful approach is to treat the bubbly soft media as a homogeneous "effective" medium characterized by nonlinear effective parameters, which are remarkably influenced by many factors associated with bubble oscillations such as surface tension, compressibility, viscosity, surrounding pressure, and an encapsulating elastic shell. Emelianov et al have studied the influence of surrounding pressure on the nonlinear dynamics of a bubble in an incompressible medium [5]. Their model has then been extended by Zabolotskaya et al to account for the effects of surface tension, viscosity, weak compressibility, and elastic shell [6]. Based on the bubble dynamics model of Zabolotskaya et al, Liang et al have developed an effective medium method (EMM) to give description of the nonlinear sound propagation in bubbly soft media with these effects accurately included [12]. Compared with the viscosity of soft medium, the resonance of the system introduced by bubbles becomes the most dominant mechanism for acoustic attenuation in a bubbly soft medium which necessarily depends on the parameters of bubbles [13]. With the availability of obtaining the effective acoustical parameters of bubbly soft media, Liang et al have presented an optimization method on the basis of fuzzy logic and genetic algorithm to yield the optimal acoustic attenuation of such media by optimizing the parameters of the size distribution of bubbles [14].
The chapter is structured as follows: we first focus on the phenomenon of acoustic localization in elastic soft media containing finite numbers of bubbles in Section 2. In Section 3, we study the phase transition in acoustic localization which helps to identify the phenomenon of localization in the presence of viscosity. In Section 4, we discuss the EMM to describe nonlinear acoustic property of bubbly soft media. Finally in Section 5, we study how to enhance the acoustic attenuation in soft bubbly media in an optimal manner.

2. Acoustic localization in bubbly soft media

In this section, we theoretically study the sound propagation in an elastic soft medium containing a finite number of bubbles by using a self-consistent method to reveal the existence of acoustic localization under proper conditions. It will be proved that the phenomenon of localization can be identified by properly analyzing the spatial correlation of wave field.

2.1 The model

Consider a longitudinal wave in an elastic soft medium containing air bubbles with small volume fraction β. We shall restrict our attention to the wave propagation at low frequencies, i.e., $kr_0 << 1$. Here $k = \omega / c_l$ and $c_l = \sqrt{(\lambda + 2\mu) / \rho}$ are the wavenumber and the speed of the longitudinal wave, respectively, and ρ is the mass density. The wave is assumed to be of angular frequency ω and emitted from a unit point source located at the origin, surrounded by N spherical air bubbles that are randomly located at $\mathbf{r}_i$ in a not overlapping way with $i = 1, 2, \cdots, N$. For simplicity and without loss of generality, all the bubbles are assumed to be of uniform radius r_0 and randomly distributed within a spatial domain, which is taken as the spherical shape so as to eliminate irrelevant effects due to an irregular edge. Then the radius of the spherical bubble cloud is $R_0 = (N / \beta)^{1/3} r_0$. Here the source is placed inside rather than outside the bubble cloud, which is the only way to isolate the localization effect from boundary effects and unambiguously investigate the problem of whether the transmitted waves can indeed be trapped [15].

2.2 Validation of approximation

As the incident longitudinal wave is scattered by a bubble in soft media, the energy converted into shear wave can be expected negligible, on condition of sufficiently small shear stiffness. Next we shall present a numerical demonstration of the validation of such an approximation by inspecting the scattering cross section of a single bubble in soft media.

The total scattering cross section σ of a single bubble can be expressed as $\sigma = \sigma_L + \sigma_S$, where σ_L and σ_S refer to the contributions of the scattered longitudinal and shear waves, respectively. Therefore it is conceivable that the ratio σ_L / σ_S actually reflects the extent to which the mode conversion will occur as the incident longitudinal wave is scattered by the bubble. The value of σ_L / σ_S can be readily obtained from Eq. (19) in Ref. [16].

A series of numerical experiments has been performed for an individual bubble in a variety of media, for the purpose of investigating how the energy converted into shear wave is affected as the media of matrix varies. Figure 2.1 presents the typical result of the comparison between the ratios σ_L / σ_S versus frequency kr_0 for different media of matrix. Three particular kinds of elastic soft media are considered: Agar-gelatin, plastisol, and gelatin. Aluminum is also considered for comparison. The physical parameters of these materials are listed in Table 2.1. Note that the physical parameters of soft media approximate to those of water except for different values of shear moduli, which seems to be natural for such "waterlike" media.

From Fig. 2.1, we find: (1) the energy of the scattered shear wave is extremely small when compared to that of the scattered longitudinal wave, as long as the ratios λ / μ are sufficiently large. Contrarily the magnitude of the longitudinal components is nearly the same to the shear components in the scattered field for μ comparable with λ. (2) for an individual bubble in different soft media, the energy converted into shear wave is diminished as the shear modulus of the medium of matrix decreases, as expected. (3) moreover, the ratios σ_L / σ_S are greatly enhanced in the proximity of the natural resonance of the individual bubbles, owing to the giant monopole resonance that is the dominant mode of bubble pulsation at low frequencies [17].

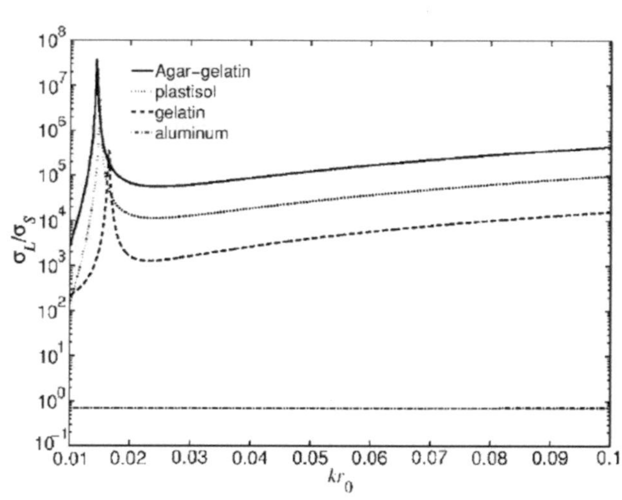

Fig. 2.1 The ratio σ_L / σ_S versus frequency kr_0 for a single bubble in different materials.

Thereby it follows that the energy converted into shear waves is insignificant as the incident longitudinal wave is scattered by a bubble in soft media. The shear component of the scattered field may thus be expected to be negligible as the longitudinal wave propagates in a bubbly soft medium. It should be stressed, however, that such an approximation can only be regard valid under the condition that the propagation of the elastic waves is not completely diffusive. Otherwise there will be an entirely contradicting conclusion that the ratio between the energy of the shear and the longitudinal wave is as large as $2(c_l / c_s)^3$ when the wave field is completely diffusive [18,19]. Here $c_s = \sqrt{2\mu / \rho}$ is the speed of the shear wave.

Such a seeming paradox may be clarified by estimating the ratio between the mean free path (MFP) of the wave and the linear "sample size" R of the random medium. For a classical wave propagating in a random media, the transport MFP is defined as the length over which momentum transfer becomes uncorrelated, while the elastic MFP is the average distance between two successive scattering events [20].

For an elastic wave, the following inequality has to be fulfilled to completely attain the diffusive regime [21]:

$$1/k << l_{T_l} << R, \; 1/\kappa << l_{T_s} << R,$$

where l_{T_l} and l_{T_s} refer to the transport MFPs of longitudinal wave and shear wave respectively, $\kappa = \omega / c_s$ is the wave number of the shear wave. Otherwise the wave propagation will be predominantly ballistic as the ratio between the MFP and the sample size is large [22].

At low frequencies from kr_0 =0 to 0.1, a set of numerical experiments have been carried out to estimate the MFPs of the elastic waves in various bubbly soft media for all the structure parameters employed in the present study. The typical values of the sample size R lie roughly within the range $4\times10^{-4}\text{m} \le R \le 0.8\text{m}$. The results indicate that: (1) for the shear wave in bubbly soft media, the typical values of the transport MFP l_{T_s} and the elastic MFP l_{E_s} of shear wave lie approximately within the range $6\times10^{-3}\text{m} \le l_{T_s}, l_{E_s} \le 4\times10^{5}\text{m}$. For any particular set of structure parameters, both the transport MFP l_{T_s} and the elastic MFP l_{E_s} are much larger than the sample size, i.e., $l_{E_s}, l_{T_s} >> R$. This manifests that the scattering of the low-frequency shear wave by the bubbles in a soft medium is weak and, therefore, the propagation of the shear wave in such a medium is ballistic rather than diffusive. (2) for the longitudinal wave, for most frequencies the range of the typical values of l_{T_l} and l_{E_l} are roughly $4\times10^{-3}\text{m} \le l_{T_l}, l_{E_l} \le 3\times10^{5}\text{m}$ and the relation $l_{E_l}, l_{T_l} >> R$ holds for any particular set of structure parameters, except for a frequency range within which $1\times10^{-6}\text{m} \le l_{T_l} \le 0.8\text{m}$ and the Ioffe-Regel criterion $kl_{T_l} \le 1$ is satisfied for any particular set of structure parameters [23]. As will be explained later, such a frequency range is in fact the region where the phenomenon of acoustic localization occurs. This is consistent with the conclusion regarding a bubbly liquid that the coherent wave dominates the transmission for most frequencies except for the localization region [24].

As a result, the condition to attain a completely diffusive wave field can not be satisfied at low frequencies as the longitudinal wave propagates in a bubbly soft medium. Consequently, the shear component of the scattered field is unimportant when compared to the longitudinal component, and neglect of mode conversion only leads to negligible errors.

The results of the numerical experiments indicate that the numerical error caused by neglecting mode conversion is less than 1% as long as the ratio λ / μ is roughly larger than 10^4. In the following study, we shall assume the scattered field to be totally longitudinal without taking mode conversion into consideration.

Materials	ρ (kg/m³)	λ	μ	λ / μ
Agar-gelatin[a]	1000	2.25GPa	6.35KPa	4×10^5
plastisol[b]	1000	2.55GPa	12.1KPa	2×10^5
gelatin[c]	1000	2.25GPa	39.0KPa	5×10^4
aluminum	2700	111.3GPa	37.1GPa	3

[a]Reference [25].
[b]References [4], [26], and [27].
[c]Reference [6].

Table 2.1. The physical parameters of the materials

2.3 The bubble dynamics

For an individual bubble in soft elastic media, the radial pulsation driven by a plane longitudinal wave is described by a Rayleigh-Plesset-like equation, as follows [3,4]:

$$\frac{d^2U}{dt^2}+\omega_0^2U-\frac{r_0}{c_l}\frac{d^3U}{dt^3}=GU^2+q\left[2U\frac{d^2U}{dt^2}+\left(\frac{dU}{dt}\right)^2\right]-ep_{inc}\,, \tag{2.1}$$

where $U=4\pi(r_{in}^3-r_0^3)/3$ is volume perturbation of an individual bubble with r_{in} being the instantaneous radius, p_{inc} refers to the incident plane wave on the bubble, i.e., the driving force of the bubble pulsation, $\omega_0=\sqrt{\omega_1^2+\omega_2^2}$ is the bubble resonance frequency with $\omega_1=\sqrt{4\mu/\rho r_0^2}$ and $\omega_2=\sqrt{3\rho_a c_a^2/\rho r_0^2}$ corresponding to the Meyer-Brendel-Tamm resonance and the Minnaert bubble resonance, respectively [8], the parameters of G, q, and e are given as $G=(9+2\chi)q\omega_0^2/2$, $q=1/(8\pi r_0^3)$, $e=4\pi r_0/\rho$. Here χ is a coefficient of asymmetry, which must lie in the range $0<\chi\le 1$. For a spherical bubble $\chi=1$. Up to the first order approximation, one obtains the harmonic solution of Eq. (2.1), as follows

$$U=-\frac{4\pi r_0}{\rho(\omega_0^2-\omega^2-i\omega^3 r_0/c_l)}p_{inc}, \tag{2.2}$$

At low frequencies, the acoustic field produced by acoustic radiation from a single bubble is taken to be the diverging spherical wave [6]

$$p_s(\mathbf{r},t)=p_s(\mathbf{r})\exp(-i\omega t)\approx-\frac{1}{4\pi r}\frac{d^2}{dt^2}U\left(t-\frac{r}{c_l}\right), \tag{2.3}$$

We assume that the bubble is the i th bubble located at $\mathbf{r}_i$ and express the incident plane wave as

$$p_{inc}(\mathbf{r},t)=p_{inc}(\mathbf{r})\exp(-i\omega t), \tag{2.4}$$

Substituting Eqs. (2.2) and (2.4) in Eq. (2.3) and discarding the time factor $\exp(-i\omega t)$, one obtains the scattered wave at $\mathbf{r}$ from the i th bubble, denoted by $p_s^i(\mathbf{r})$, as follows:

$$p_s^i(\mathbf{r}) = f p_{inc}(\mathbf{r}_i) G_0(\mathbf{r}-\mathbf{r}_i), \tag{2.5}$$

where $G_0(\mathbf{r}-\mathbf{r}_i) = \exp(ik|\mathbf{r}-\mathbf{r}_i|)/|\mathbf{r}-\mathbf{r}_i|$ is the usual three-dimensional Green's function, f is the scattering function of a single bubble, defined as

$$f = \frac{r_0}{(\omega_0^2/\omega^2 - 1 - ikr_0)}, \tag{2.6}$$

Eqs. (2.5) and (2.6) clearly show that the scattered fields and scattering function of a single bubble in soft media take an identical form as in liquid media, except for different expressions of ω_0 (see Eqs. (2) and (2) in Ref. [24] and also Eq. (2) in Ref. [28]). Compared with the resonance frequency of a bubble in liquid media that includes only the Minnaert contribution ω_2, an additional term ω_1 is present in Eq. (2.6), due to the contribution of shear wave in the solid wall of bubble. In the limiting case $\lambda/\mu \to \infty$, the soft medium becomes a liquid medium, and Eq. (2.6) degenerates to Eq. (4) in Ref. [24] due to the vanishing of ω_1. Despite such an agreement of the results in the form, we have to particularly stress that it is the radial pulsation equation of a single bubble from which the present theory begins, rather than the well-studied scattering function of a bubble employed in Ref. [24].

2.4 The self-consistent formalism

Among many useful formalisms suggested for describing the acoustic scattering by a finite group of random scatterers, the self-consistent method is proved to be particularly effective [29]. This method is based on a genuine self-consistent scheme proposed first by Foldy [30] and reviewed in Ref. [31]. Later Ye et al employed the self-consistent method to investigate the wave propagation in a bubble liquid [24]. In the method, the multiple scattering of waves is represented by a set of coupled equations, and the rigorous results can be obtained by solving the equations. In the present study, the wave propagation in a bubbly soft medium will be solved by using the self-consistent method in an exact manner.

In the presence of bubbles, the radiated wave from the source is subject to multiple scattering by the surrounding bubbles. Hence the total acoustic wave at a spatial point $\mathbf{r}$ is supposed to include the contributions from the wave directed from the source and the scattered waves from all bubbles, i.e.,

$$p(\mathbf{r}) = p_0(\mathbf{r}) + \sum_{i=1}^{N} p_s^i(\mathbf{r}), \tag{2.7}$$

Analogously, the incident wave upon the i th bubble should consist of the direct wave from the source and the scattered waves from all bubbles except for itself, i.e., [31]

$$p_{inc}(\mathbf{r}_i) = p_0(\mathbf{r}_i) + \sum_{j=1, j\neq i}^{N} p_s^j(\mathbf{r}_i), \tag{2.8}$$

At $kr_0 << 1$, the dimension of bubbles is very small when compared to the wavelength, and the plane wave approximation still stands for the incident wave upon an individual bubble. In this situation, the dynamics of an individual bubble can be described by the radial pulsation equation given by Eq. (2.1). Therefore for the i th bubble, according to Eqs. (2.5) and (2.8), the following self-consistent equation should be satisfied:

$$p_s^i(\mathbf{r}) = f\left[p_0(\mathbf{r}_i) + \sum_{j=1,j\neq i}^{N} p_s^j(\mathbf{r}_i)\right] G_0(\mathbf{r}-\mathbf{r}_i), \tag{2.9}$$

Upon incidence, each bubble acts effectively as a secondary pulsating source. The scattered wave from the i th bubble is regarded as the radiated wave and is rewritten as [28]

$$p_s^i(\mathbf{r}) = A_i G_0(\mathbf{r}-\mathbf{r}_i) \tag{2.10}$$

where the complex coefficient A_i refers to the effective strength of the secondary source. Substitution of Eq. (2.10) in Eq. (2.8) yields

$$A_i = f\left[p_0(\mathbf{r}_i) + \sum_{j=1,j\neq i}^{N} A_j G_0(\mathbf{r}_j - \mathbf{r}_i)\right], \tag{2.11}$$

By setting $\mathbf{r}$ in Eq. (2.10) to any bubble other than the i th, this equation becomes a set of closed self-consistent equations which can be expressed as

$$\mathbf{A} = f\mathbf{P} + \mathbf{CA}, \tag{2.12}$$

where $\mathbf{A} = [A_1, A_2, ..., A_N]^T$, $\mathbf{P} = [p_0(\mathbf{r}_1), p_0(\mathbf{r}_2), ..., p_0(\mathbf{r}_N)]^T$, $\mathbf{C}$ is an $N \times N$ matrix, defined as

$$\mathbf{C} = \begin{bmatrix} C_{11} & \cdots & C_{1N} \\ \vdots & \ddots & \vdots \\ C_{N1} & \cdots & C_{NN} \end{bmatrix},$$

where $C_{ij} = (1-\delta_{ij}) f G_0(\mathbf{r}_j - \mathbf{r}_i)$ with δ_{ij} being the Kronecker symbol. It is apparent that $\mathbf{C}$ is symmetric according to the principle of reciprocity. One may readily obtain from Eq. (2.12)

$$\mathbf{A} = f(1-\mathbf{C})^{-1}\mathbf{P}. \tag{2.13}$$

For an arbitrary configuration of the bubble distribution, the acoustic field in any spatial point may thus be solved exactly from Eqs. (2.7), (2.10) and (2.13). It is obvious that the multiple scattering effects have been incorporated during the computation [24].

2.5 The spatial correlation function

With the purpose of identifying efficiently the phenomenon of localization, we consider the spatial correlation that physically describes the interaction between the wave fields at different spatial points. For an arbitrary spatial point $\mathbf{r}$, the wave field is normalized as $T(r) = p(r)/p_0(r)$ to eliminate the uninteresting geometrical spreading factor. Here $r = |\mathbf{r}|$ is the distance from the source point to the spatial point, $p(r)$ refers to the acoustic pressure averaged over the sphere of radius r, and $p_0(r) = \exp(ikr)/r$ refers to the propagating wave radiated from the source in the absence of bubbles.

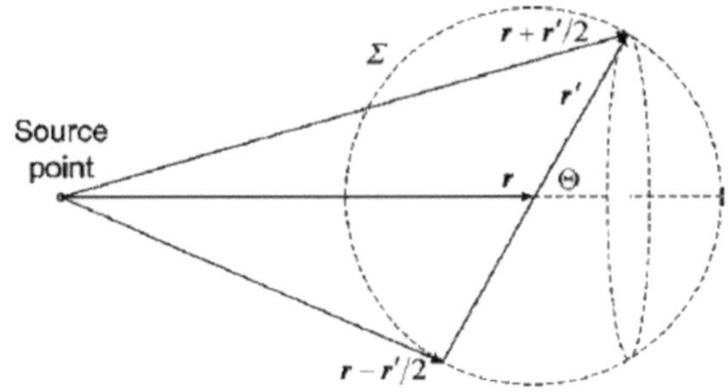

Fig. 2.2. Geometry for spatial correlation function of wave fields at $\mathbf{r}+\mathbf{r}'/2$ and $\mathbf{r}-\mathbf{r}'/2$.

Consider the interaction between the two spatial points $\mathbf{r}+\mathbf{r}'/2$ and $\mathbf{r}-\mathbf{r}'/2$, as shown in Fig. 2.2. The spatial correlation is defined as the average over the sphere $\boldsymbol{\Sigma}$ that is located at $\mathbf{r}$ and of radius $\xi/2$. Here $\xi=|\mathbf{r}'|$ is the distance between $\mathbf{r}+\mathbf{r}'/2$ and $\mathbf{r}-\mathbf{r}'/2$. Note that the normalized wave field $T(r)$ is axially symmetric about $\mathbf{r}$ and depends only on Θ. The average can thus be accomplished by performing the integration with respect to Θ. Then the spatial correlation function is expressed as follows:

$$g(\mathbf{r}+\mathbf{r}'/2,\mathbf{r}-\mathbf{r}'/2)=\frac{\int_0^{\pi}2\pi\langle T(|\mathbf{r}+\mathbf{r}'/2|)T^*(|\mathbf{r}-\mathbf{r}'/2|)\rangle\cdot(\xi/2)^2\sin\Theta d\Theta}{4\pi(\xi/2)^2}$$
$$=\frac{1}{2}\int_0^{\pi}\langle T(|\mathbf{r}+\mathbf{r}'/2|)T^*(|\mathbf{r}-\mathbf{r}'/2|)\rangle\sin\Theta d\Theta \qquad (2.14)$$

where $|\mathbf{r}\pm\mathbf{r}'/2|=\sqrt{r^2+\xi^2/4\pm\xi r\cos\Theta}$, and $\langle\cdot\rangle$ refers to the ensemble average carried over random configuration of bubble clouds.

It is apparent that the preceding definition of the spatial correlation function refers to the average interaction between the wave fields at every pair of spatial points for which the distance is ξ and the center of symmetry locates at $\mathbf{r}$. By using Eq. (2.14) and taking the ensemble average over the whole bubble cloud, then, we define the total correlation function that is a function of the distance ξ so as to describe the overall correlation characteristics of the wave field. In respect that the normalized wave field $T(r)$ is symmetric about the origin, the total correlation function can be obtained by merely performing the integration with respect to r, given as below:

$$C(\xi)=\frac{\int_0^{R_0}4\pi r^2 g(\mathbf{r}+\mathbf{r}'/2,\mathbf{r}-\mathbf{r}'/2)dr}{\int_0^{R_0}4\pi r^2 g(\mathbf{r},\mathbf{r})dr}. \qquad (2.15)$$

2.6 Acoustic localization in bubbly elastic soft media

A set of numerical experiments has been carried out for various bubble radii, numbers and volume fractions. Figure 2.3 presents the typical results of the total transmission and the total backscattering versus frequency kr_0 for bubbly gelatin with the parameters $N=200$, $r_0=1$ mm, and $\beta=10^{-3}$, respectively. The total transmission is defined as $I=\langle|T|^2\rangle$, and

the received point is located at the distance $r = 2R$ from the source. The total backscattering is defined as $\langle |\sum_i^N p_s^i(0)|^2 \rangle$, referring to the signal received at the transmitting source.
It is clearly suggested in Fig. 2.3(a) that there is a region of frequency slightly above the bubble resonance frequency, i.e., approximately between kr_0 =0.017 and 0.077 in this particular case, in which the transmission is virtually forbidden. Within this frequencies range, the Ioffe-Regel criterion is satisfied and a maximal decrease of the diffusion coefficient D roughly by a factor of 10^5 is observed and D can thus be considered having a tendency to vanish, i.e., $D \to 0$. Here the diffusion coefficient is defined as $D = v_t l_{T_l} / 3$ with v_t being the transport velocity that may be estimated by using an effective medium method [32]. Indeed, this is the range that suggests the acoustic localization where the waves are considered trapped [24], confirming the conjectured existence of the phenomenon of localization in such a class of media. Outside this region, wave propagation remains extended. For the backscattering situation, the result shows that the backscattering signal persists for all the frequencies, and an enhancement of backscattering occurs particularly in the localization region. As has been suggested by Ye et al, however, the backscattering enhancement that appears as long as there is multiple scattering can not act as a direct indicator of the phenomenon of localization [28]. In the following we shall thus focus our attention on the transmission that helps us to identify the localization regions, rather than the backscattering of the propagating wave.

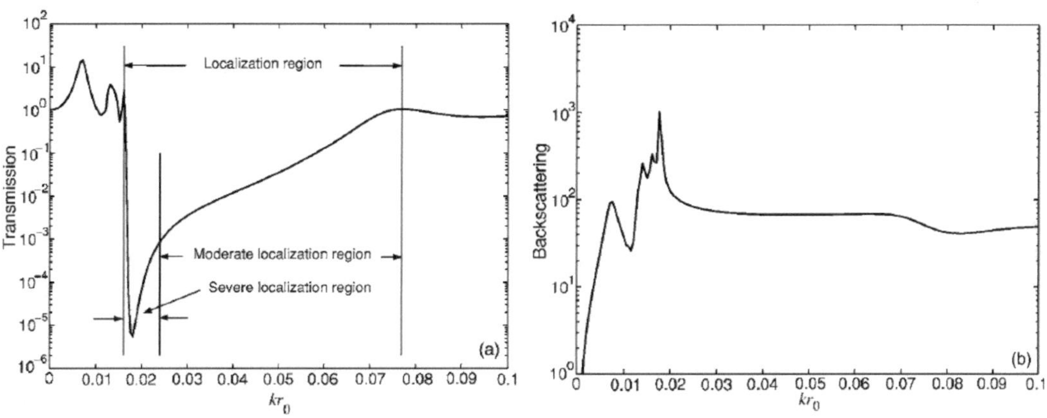

Fig. 2.3. The total transmission (a) and the total backscattering (b) versus frequency kr_0 for bubbly gelatin.

Since the sample size is finite, the transmission is not completely diminished in the localization region, as expected [24]. In this particular case, there exists a narrow dip within the localization region between kr_0 =0.017 and 0.024, hereafter termed severe localization region, in which the most severe localization occurs. The waves are moderately localized between kr_0 =0.024 and 0.077, termed moderate localization region, due to fact that the finite size of sample still enables waves in this region to leak out [15]. We find from Fig. 2.3 that for such systems of internal resonances, the waves are not localized exactly at the internal resonance, rather at parameters slightly different from the resonance. This indicates that mere resonance does not promise localization, supporting the assertion of Rusek et al [33] and Alvarez et al [34].

To identify the phenomenon of localization by inspecting the correlation characteristics of the wave field in bubbly soft media, the total correlation functions are numerically studied for various frequencies and bubble parameters. Figure 2.4 illustrates the typical result of the comparison between the total correlation functions for bubbly gelatin at three particular frequencies chosen as below, within, and above the localization region: kr_0 =0.012, 0.018, and 0.1, referring to Fig. 2.3. Here the parameters of bubbles are identical with those used in Fig. 2.3. Observation of Fig. 2.4 clearly reveals that the total correlation decays rapidly along the distance ξ in the case of kr_0 =0.018, while the decrease of correlation with the increase of ξ is very slow in the cases of kr_0 =0.012 and kr_0 =0.1. Such spatial correlation behaviors may be understood by considering the coherent and the diffusive portions of the transmission. Here the coherent portion is defined as $I_C = |\langle T\rangle|^2$, and the diffusive portion is $I_D = I - I_C$. Figure 2.5 plots the total transmission and the coherent portion versus frequency kr_0 for bubbly gelatin with the parameters used in Fig. 2.4. It is obvious that the coherent portion dominates the transmission for most frequencies, while the diffusive portion dominates within the localization region. This is in good agreement with the conclusion drawn by Ye et al for bubbly liquids (cf. see Fig. 1 in Ref. [24]). As a result, there exist strong correlations between pairs of field points even for a considerable large distance within the non-localized region where the wave propagation is predominantly coherent. Contrarily, within the localization region almost all the waves are trapped inside a spatial domain and the fluctuation of wave field at a spatial point fails in interacting effectively with any other point far from it. These results suggest that proper analysis of the spatial correlation behaviors may serve for a way that helps discern the phenomenon of localization in a unique manner.

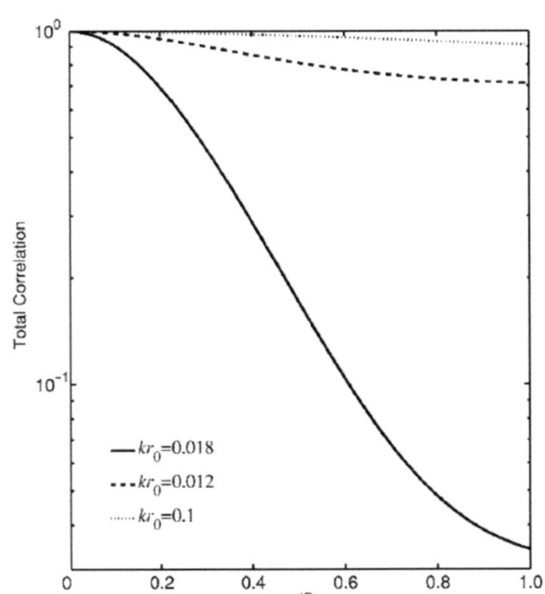

Fig. 2.4. The total correlation versus distance ξ for bubbly gelatin at three particular frequencies chosen as below, within, and above the localization region, respectively.

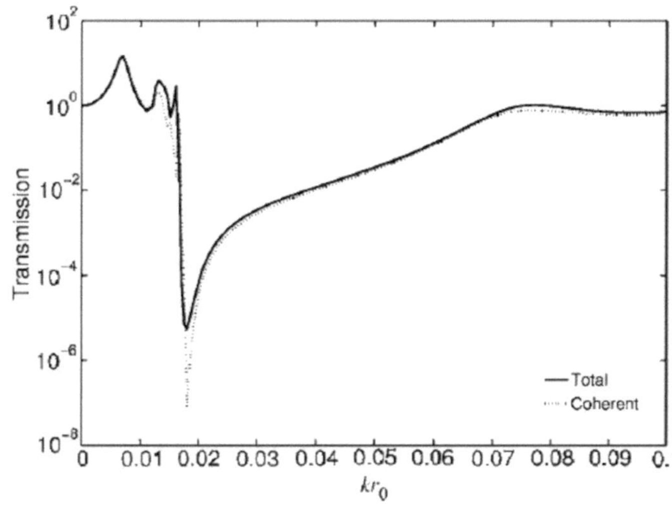

Fig. 2.5. The total transmission and the coherent portion versus frequency kr_0 for bubbly gelatin.

3. Phase transition in acoustic localization in bubbly soft media

In this section, we focus on the localization in bubbly soft medium with the effect of viscosity taken into account, by inspecting the oscillation phases of bubbles rather than the wave fields. It will be proved that the acoustic localization is in fact due to a collective oscillation of the bubbles known as a phenomenon of "phase transition", which helps to identify phenomenon of localization in the presence of viscosity.

3.1 The influence of viscosity on acoustic localization

So far, we have considered the localization property in a bubbly soft medium, which is regarded as totally elastic for excluding the effects of absorption that may lead to ambiguity in data interpretation. In practical situations, however, the existence of viscosity effect may notably affect the propagation of acoustic waves and then the localization characteristics in a bubbly soft medium. Note that the practical sample of a soft medium is in general assumed viscoelastic [6] and the existence of viscosity inevitably causes ambiguity in differentiating the localization effect from the acoustic absorption which might result in the spatial decrease of wave fields as well [36]. In the presence of viscoelasticity, the Lamé coefficients of the soft medium may be rewritten as below:

$$\lambda = \lambda_e + \lambda_v \frac{\partial}{\partial t}, \; \mu = \mu_e + \mu_v \frac{\partial}{\partial t}, \tag{3.1}$$

where λ_e and μ_e are the elastic Lamé coefficients, λ_v and μ_v are viscosity factors given by Kelvin-Voigt viscoelastic model. In the following we shall assume λ_v =0, as is usually done for a soft medium [35]. The viscosity factor μ_v may be manually adjusted in the numerical simulations to inspect the sensibility of the results to the absorption effects.

Note that the acoustic wave is a simple harmonic wave of angular frequency ω. Then the longitudinal wave number in the soft viscoelastic medium becomes a complex number as $\tilde{k} = k' + ik'' = \omega / \tilde{c}_l$. Here the real and the imaginary parts represent the propagation and the attenuation of the longitudinal wave in a soft viscoelastic medium, respectively, and $\tilde{c}_l$ refers to the effective speed of the wave. For the acoustic wave that propagates in a soft viscoelastic medium permeated with bubbles, the influence of the viscosity effect may be ascribed to two aspects: (1) the propagation of the acoustic wave in a soft viscoelastic medium should be described by a series of complex parameters instead of the corresponding real parameters (i.e., $k \to \tilde{k}$, $c_l \to \tilde{c}_l$, etc.) to account for the absorption effects; (2) the dynamical behavior of an individual bubble will be greatly affected by the friction damping of pulsation that results from the viscoelastic solid wall. The incorporation of the effect of acoustic absorption due to viscosity effects amounts to adding a term $\nu \cdot dU / dt$ in the dynamical equation of a single bubble in a soft elastic medium [8]. Here $\nu = 4\mu_v / (\rho r_0^2)$ is a coefficient characterizing the effect of acoustic absorption. By seeking the linear solution of the modified dynamical equation in a same manner as in Section 2.3, one may derive the scattering function f of a single bubble in a soft viscoelastic medium, as follows:

$$f = \frac{r_0}{(\omega_0^2 / \omega^2 - 1 - i\omega r_0 / c_l - i\nu / \omega)}, \tag{3.2}$$

where ω_0 refers to the resonance frequency of an individual bubble in a soft medium. On condition that the soft medium is totally elastic, the expression of the scatter function f degenerates to Eq. (2.6) due to the vanishing of the term $-i\nu / \omega$. In such a case, the acoustic field in any spatial point can thus be solved exactly in a same manner as in Section 2.4.

By rewriting the complex coefficient A_i in Eq. (2.10) as $A_i = |A_i| \exp(i\theta_i)$ with the modulus and the phase physically represent the strength of secondary source and the oscillation phase, respectively. For the ith bubble, it is convenient to assign a two-dimensional unit phase vector, $\boldsymbol{u}_i = \cos\theta_i \hat{x} + \sin\theta_i \hat{y}$ to the oscillation phase of the bubble with $\hat{x}$ and $\hat{y}$ being the unit vectors in the x and y directions, respectively. The phase of emitting source is set to be zero. Thereby the oscillation phase of every bubble is mapped to a two-dimensional plane via the introduction of the phase vectors and may be easily observed in the numerical simulations by plotting the phase vectors in a phase diagram.

In actual experiments, it is the variability of signal that is often easier to analysis [36]. Hence the behavior of the phases of the oscillating bubbles may be readily studied by inspecting the fluctuation of the oscillation phase of bubbles is investigated as well. Here the fluctuation of the phase of bubbles is defined as follows [36]:

$$\delta\theta = \left[\frac{1}{N} \sum_{\mathrm{i}=1}^{N} \theta_{\mathrm{i}}^2 - \bar{\theta}^2 \right]^2,$$

where $\bar{\theta} = \frac{1}{N} \sum_{\mathrm{i}=1}^{N} \theta_{\mathrm{i}}$ is the averaged phase.

3.2 Localization and phase transition in bubbly soft media

Figure 3.1 displays the typical results of the phase diagrams for a bubbly gelatin at different driving frequencies, with the values of viscosity factors manually adjusted to study the influence of the effect of acoustic absorption. Three particular frequencies are employed (See Fig. 2.3): $\omega r_0/c_l$ =0.01 (Fig. 3.1(a), below the localization region), $\omega r_0/c_l$ =0.1 (Fig. 3.1(b), above the localization region), and $\omega r_0/c_l$ =0.02 (Figs. 3.1(c) and (d), within the localization region). In a phase diagram, each circle and the corresponding arrow refer to the three-dimensional position and the phase vector of an individual bubble, respectively. In Figs. 3.1(a-c) we choose the viscosity factor as μ_v =0, i.e., the soft medium that serves as the host medium is assumed totally elastic; while in Fig. 3.1(d) the value of viscosity factor is set to be μ_v = 50P (1P=0.1Pa·s). For a comparison we also examine the spatial distribution of the wave fields and plot the transmissions as a function of the distance from the source in Fig. 3. 2 in cases corresponding to Fig. 3.1. Note that the energy flow of an acoustic wave is conventionally $\boldsymbol{J} \sim |p|^2 \nabla \theta_i$. This mathematical relationship reveals the fact that the gradient of oscillation phases of bubbles is crucial for the occurrence of localization. Apparently, when the oscillation phases of different bubbles exhibit a coherent behavior (i.e. θ_i is a constant) while $|p|$ is nonzero, the acoustic energy flow will stop and the acoustic wave will thereby be localized within a spatial domain [36]. Moreover, such coherence in oscillation phases of bubbles is a unique feature of the phenomenon of localization that results from the multiple scattering of waves, but lacks when other mechanism such as absorption effect dominates, as will be discussed later. Consequently, it should be promising to effectively identify the localization phenomenon by giving analysis to the oscillation phases of bubbles and seeking their ordering behaviors.

It is apparent in Figs. 3.1(a) and (b) that the phase vectors pertinent to different bubbles point to various directions as the driving frequency of the source lies outside the localization region. In other words, the oscillation phases of the bubbles located at different positions in a bubbly soft medium are random in non-localized states. Correspondingly, the curves 1 (thin solid line) and 2 (thin dashed line) in Fig. 3.2 shows that the non-localized waves remain extended and can propagate through the bubble cloud. As observed in Fig. 3.1(c), however, the phase vectors located at different spatial positions point to the same direction when localization occurs, which indicates that the oscillation phases of all bubbles remain constant and the energy flow of the wave stops. The transition from the non-localized state to the localized state of the wave can be interpreted as a kind of "phase transition", which is characterized by the unusual phenomenon that all the bubbles pulsate collectively to efficiently prohibit the acoustic wave from propagating [10]. Such a concept of phase transition is physically consistent with the order-disorder phase transition in a ferromagnet [37]. Note that the phase of emitting source is assumed to be zero in the numerical simulations, i.e., the phase vector at the source points to positive $\hat{x}$ direction, while all the phase vectors in Fig. 3.1(c) point to the negative $\hat{x}$-axis. This means that as the localization occurs, almost all bubbles tend to oscillate completely in phase but exactly out of phase with the source, which leads to the fact that the localized acoustic energies are trapped within a small spatial domain adjacent to the source as shown by the curve 3 (thick solid line) in Fig. 3.2. These numerical results are consistent with the previous conclusions obtained for bubbly water and bubbly soft elastic media [10,36]. Therefore it is reasonable to conclude that such a phenomenon of phase transition is the intrinsic physical mechanism from which the acoustic localization stems.

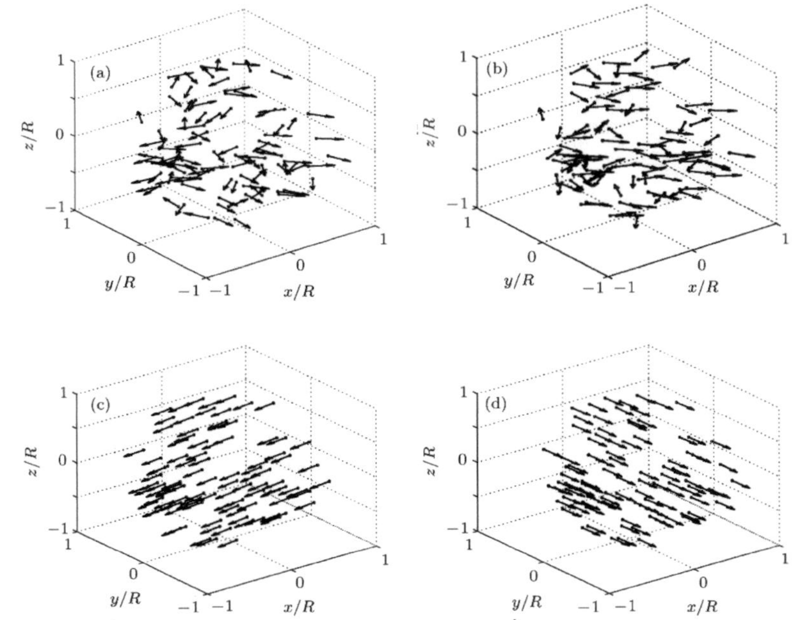

Fig. 3.1. The phase diagrams for the oscillating bubbles in a bubbly gelatin with different structural parameters: (a) $\omega r_0/c_l$ =0.01, μ_v=0; (b) $\omega r_0/c_l$ =0.1, μ_v=0; (c) $\omega r_0/c_l$ =0.02, μ_v=0; (d) $\omega r_0/c_l$ =0.02, μ_v=50P.

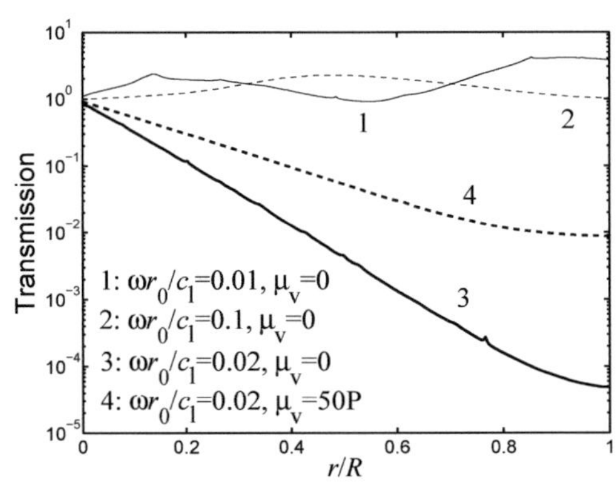

Fig. 3.2. Transmissions versus the distance from the source in a bubbly gelatin with different structural parameters.

Note that the effect of acoustic absorption has been completely excluded in Figs. 3.1(a-c) which may cause ambiguity in identifying the phenomenon of localization. It is thus of much more practical significance to investigate the localization properties in the case where soft medium is assumed viscoelastic, and the corresponding results are shown in the phase diagram given by Fig. 3.1 (d) as well as the comparison between transmissions versus r in Fig. 3.2. As the viscosity factors of the soft medium are manually increased, the phenomena of phase transition can be identified in a bubbly soft viscoelastic medium provided that the driving frequency of acoustic wave falls within the localization region. Meanwhile, exponential decay of the wave fields with respect to the distance from the source is shown by the curve 4 (thick dashed line) in Fig. 3.2. Observation of Fig. 3.1(d) and Fig. 3.2 apparently manifests that, however, the adjustment of the values of the viscosity factors leads to changes of the direction to which all the phase vectors point collectively varies and the decay rates of the transmissions versus r.

For a bubbly soft viscoelastic medium, it is still possible to achieve the acoustic localization since the condition can be satisfied that the oscillation phases of bubbles at any spatial points remain constant, but the extents of localization are necessarily affected by the presence of viscosity effect. It is thus difficult to differentiate the phenomenon of acoustic localization from that of the acoustic absorption without referring to the analysis of the behavior of the phases of bubbles [11]. Notice that in Fig. 3.1, as the viscosity factors are gradually enhanced, the angles between the directions of the phase vectors and the negative x-axis increase. This means that the phase-opposition states between the oscillations of all the bubbles and the source as well as the extents to which the acoustic wave is localized are weaken due to the enhancement of the viscosity. Therefore it may be inferred that the occurrence of phase transition in a bubble soft medium is a criterion for identifying the phenomenon of localization, while the localization extents can be predicted by accurately analyzing the relationship between the oscillation phases of the bubbles and the source.

It is convenient to employ a phase diagram method for observing the collective phase properties of the bubbles and thereby seeking the existence of the phenomenon of phase transition, however the values of the oscillation phase of each bubble could not be directly read via the phase diagrams in a precise manner. We then illustrate the statistical properties of the parameters of $\langle\theta\rangle$ for all the bubbles in Fig. 3.3 for a more explicit observation of the values of oscillation phases of the bubbles. Here $\langle\cdot\rangle$ denotes the ensemble average over random configurations of bubble clouds, $p_\theta(\theta)$ is defined as the probability that the values of $\langle\theta\rangle$ fall between θ and $\theta+\Delta\theta$, i.e., $\theta\le\langle\theta\rangle<\theta+\Delta\theta$, with $\Delta\theta$ referring to the difference between the two neighbor discrete values of θ. And the values of $p_\theta(\theta)$ have been normalized such that the total probability equals 1. In Fig. 3.3 three particular values of viscosity factors are considered: $\mu_v = 0$ (curve 3, thick solid line), 50P (curve 4, thick dashed line), 200P (curve 5, thick dotted line). It is obvious in Fig. 3.3 that: (1) Outside the localization region, as shown by the thin curves 1 (solid line) and 2 (dashed line), the values of oscillation phases $\langle\theta\rangle$ exhibit large extents of randomnesses, which indicates a lack of the above-mentioned collective behavior of the bubble oscillation crucial for the existence of localization, in accordance with the results shown in Figs. 3.1(a) and (b). (2) When the phenomenon of localization occurs, the oscillation phases almost remain constant for bubbles located at different spatial points, which is illustrated by the delta-function shapes of the thick curves 3-5. It is also noteworthy that the oscillation phase of each bubble approximates $-\pi$ in an elastic medium, and that the presence of the viscosity effect does not change such a phenomenon of phase transition but leads to a larger average value of oscillation phases $\langle\theta\rangle$. A monotonic increase of the values of the oscillation phases of bubbles is clearly observed as the viscosity factors are gradually enhanced. In the soft medium with viscosity factor μ_v=50P, the values of $\langle\theta\rangle$ nearly equal -0.45π for all the bubbles, and $\langle\theta\rangle$ approximate -0.15π for the case of μ_v=200P.

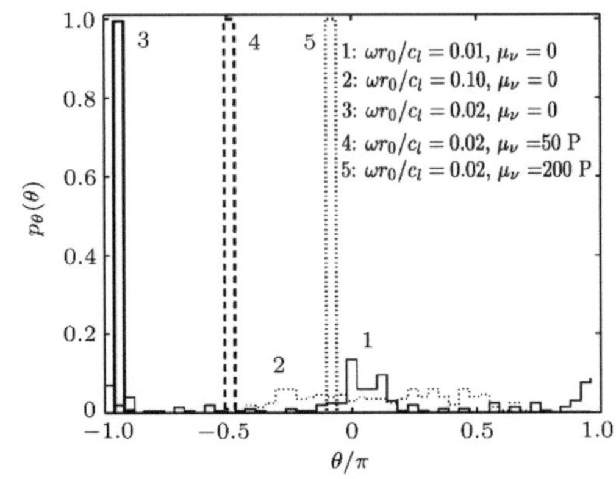

Fig. 3.3. The comparison between the statistical behaviors of the oscillation phases of bubbles in a bubbly gelatin with different structural parameters.

The principal influence of the viscosity effect on the localization property in a bubbly soft medium attributes intrinsically to two aspects of physical mechanism. The localization phenomenon in inhomogeneities had been extensively proved to stem from the important multiple scattering processes between scatterers. In a viscoelastic medium the recursive process of multiple scattering could not be well established due to the effect of acoustic absorption caused by the viscosity, which necessarily impairs the extent to which the acoustic wave can be localized. For an individual bubble pulsating in a viscoelastic medium, on the other hand, the oscillation will be hindered by the friction damping caused by the viscoelastic solid wall. While the bubble in an elastic soft medium can behave like a high quality factor oscillator [2], the increase of viscosity factors will definitely reduce the quality factor that is defined as $Q=\omega/\upsilon$ and then the strength of the resonance response of bubble to the incident wave. This prevents the bubbles from becoming effective acoustic scatterers, which is crucial for the localization to take place [24]. As a result, it is perceivable that the increase of the viscosity effects diminishes the extent to which all bubbles pulsate out of phase with the source, and a complete prohibition of acoustic wave could not be attained.

Figure 3.4 displays the fluctuations of the oscillation phases of bubbles $\delta\theta$ as a function of the normalized frequency $\omega r_0/c_l$ in a bubbly gelatin for four particular values of viscosity factors: μ_v =0, 5P, 50P and 500P. Note also that the fluctuations of the phases approaches zero at the zero frequency limit due to the negligibility of the scattering effect of bubbles. The phenomena of phase transitions can be clearly observed characterized by significant reductions of the fluctuations within particular ranges of frequencies whose locations are in good agreement with the corresponding frequency regions where the localization occurs. This is consistent with the previous results obtained for bubbly water. Moreover, it is apparently seen that the amounts to which the fluctuations $\delta\theta$ decrease can act as reflections of the extents of the acoustic localizations. In a bubbly viscoelastic soft medium, such a phenomenon of phase transition persists within the localization region, while the increase of the value of viscosity factor leads to a weaker reduction of the fluctuation of phases. In the particular case where the viscosity effects are extremely strong, i.e., μ_v =500P, the localization is absent due to the fact that the effects of multiple scattering and the bubble resonance are severely destroyed, and the phenomenon of phase transition could not be identified. The comparison of Figs. 3.1-3.4 proved that the phenomenon of phase transition is a valid criterion of the existence of acoustic localization in such a medium, and the values of the oscillation phases of the bubbles help to determine the extent to which the acoustic waves are localized. Consequently it is fair to conclude that the proper analysis of the oscillation phases of bubbles can indeed act as an efficient approach to identify the phenomenon of acoustic localization in the practical samples of bubbly soft media for which the viscosity effects are generally nontrivial. The important phenomenon of phase transition is an effective criterion to determine the existence of localization, while the extent to which the acoustic wave is localized may be estimated by inspecting the values of the oscillation phases or the reduction amount of the phase fluctuation.

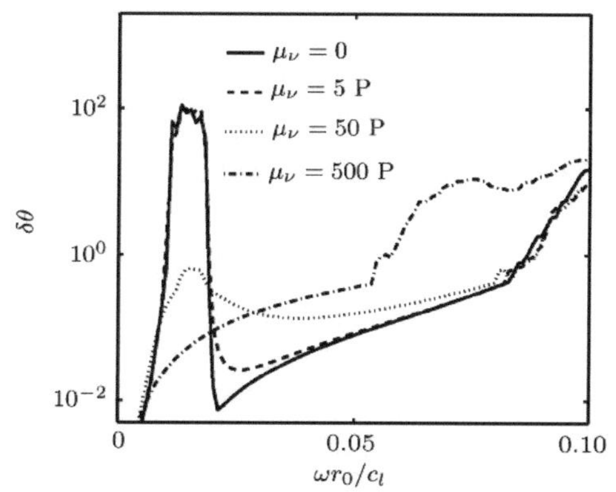

Fig. 3.4. The comparison between the fluctuations of the oscillation phases of bubbles versus frequency in a bubbly gelatin with different values of viscosity factors.

4. Effective medium method for sound propagation in bubbly soft media

In this section, we discuss the nonlinear acoustic property of soft media containing air bubbles and develop an EMM to describe the strong acoustic nonlinearity of such media with the effects of weak compressibility, viscosity, surrounding pressure, surface tension, and encapsulating shells incorporated. The advantages as well as limitations of the EMM are also briefly discussed.

4.1 Bubble dynamics

Consider an encapsulated gas bubble surrounded by a soft viscoelastic medium. When in equilibrium, the gas pressure in the bubble is denoted P_g, and the pressure infinitely far away is P_∞. For the case where the equilibrium pressure equals the surrounding pressure (i.e. $P_g = P_\infty$), the shear stress is uniform throughout the soft medium. Such a case is referred to as an initially unstressed state, for which the equilibrium values of the inner and outer radius of the bubble are designated R_0 and R_{s0}, respectively. In the general case, however, the encapsulated bubble may be pressurized, such that $P_g \neq P_\infty$. Such a case is denoted as a prestressed case due to the fact that a nonuniform shear stress is generated inside the medium to balance the pressure difference. For a prestressed cases we define the equilibrium values of the inner and outer radius as R_1 and R_{s1}, respectively. The geometry is shown in Fig. 4.1. Figure 4.1(a) shows an unstressed case where one has $R_0 = R_1$ and $R_{s0} = R_{s1}$. In the cases where $P_g < P_\infty$, however, it is apparently that the pressure difference between P_g and P_∞ will force the bubble to shrink, and one thus has $R_0 > R_1$ and $R_{s0} > R_{s1}$, as illustrated in Fig. 4.1(b). In contrast, one has $R_0 < R_1$ and $R_{s0} < R_{s1}$ if $P_g > P_\infty$. As the bubble oscillates, the instantaneous values of the inner and outer radius are defined as $R(t)$ and $R_s(t)$, respectively.

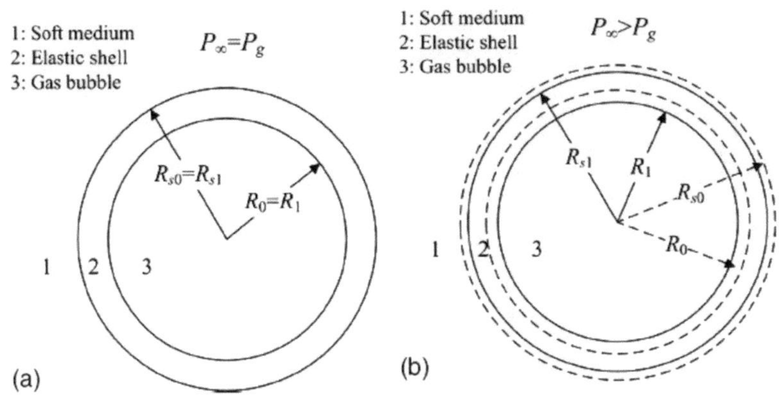

Fig. 4.1. Geometry of an encapsulated gas bubble in a soft medium in (a) an initially unstressed state and (b) a prestressed state.

Zabolotskaya et al [6] has studied the nonlinear dynamics in the form of a Rayleigh-Plesset-like equation for an individual bubble in such a model, and provided the approaches to include the effects of compressibility, surface tension, viscosity, and an encapsulating shell. Note that Eq. (53) in Ref. [6] accounts for the effects of surface tension, viscosity, and shell but applies only to the case of an incompressible medium. Adding the compressibility term $(d^3w/dt^3)/c_m$ that accounts for the radiation damping to the left hand side of this equation, one readily obtain the equation that describes the nonlinear oscillation of a single bubble in a soft medium, as follows:

$$
\begin{aligned}
F(R)R\frac{d^2R}{dt^2}+\frac{3}{2}G(R)\left(\frac{dR}{dt}\right)^2-\frac{1}{c_m}\frac{d^3w}{dt^3}=\frac{1}{\rho_m}\Bigg\{&P_g\left(\frac{R_1}{R}\right)^{3\gamma}-P_\infty \\
&-P_e(R)-\frac{2\sigma_g}{R}-\frac{2\sigma_m}{R_s}-\frac{4}{R}\frac{dR}{dt}\left[\eta_s\left(1-\frac{R^3}{R_s^3}\right)+\eta_m\frac{R^3}{R_s^3}\right]\Bigg\},
\end{aligned}
\tag{4.1}
$$

In the preceding expression, the parameter w is defined as $w = R^3/3$, $P_\sigma = 2\sigma_g/R + 2\sigma_m/R_s$ is the effective pressure due to the surface tension with σ_g and σ_m being the surface tensions at the inner gas-shell interface and the outer shell-medium interface, respectively, and ψ is the dissipation function that is found to be $\psi = 8\pi R(dR/dt)^2\left[\eta_s(1-R^3/R_s^3)+\eta_m R^3/R_s^3\right]$ with η_s and η_m being the shear viscosity coefficients in the shell region and in the medium, respectively, the parameters of $F(R)$ and $G(R)$ are given as

$$F(R)=\frac{\rho_s}{\rho_m}+\left(1-\frac{\rho_s}{\rho_m}\right)\frac{R}{R_s},\quad G(R)=\frac{\rho_s}{\rho_m}+\left(1-\frac{\rho_s}{\rho_m}\right)\left(\frac{4}{3}-\frac{1}{3}\frac{R^3}{R_s^3}\right)\frac{R}{R_s}$$

with ρ_s and ρ_m are the mass densities of the shell and the surrounding medium, respectively, and $P_e(R)$ refers to the effective pressure due to the strain energy stored in shear deformation of both the shell and the medium, defined as [6]

$$P_e(R)=4\int_r^{R_s}\left(\frac{\partial\varepsilon_s}{\partial I_1}+\frac{r^2}{\tilde{r}^2}\frac{\partial\varepsilon_s}{\partial I_2}\right)\left(\frac{r^2}{\tilde{r}^2}-\frac{\tilde{r}^4}{r^4}\right)\frac{dr}{r}+4\int_{R_s}^{\infty}\left(\frac{\partial\varepsilon_m}{\partial I_1}+\frac{r^2}{\tilde{r}^2}\frac{\partial\varepsilon_m}{\partial I_2}\right)\left(\frac{r^2}{\tilde{r}^2}-\frac{\tilde{r}^4}{r^4}\right)\frac{dr}{r},\tag{4.2}$$

where $\varepsilon=\varepsilon(I_1,I_2,I_3)$ refers to the strain energy density with I_1, I_2, I_3 being the principal invariants of Green's deformation tensor, r and $\tilde{r}$ refer to the Eulerian and the Lagrangian coordinates, respectively, the subscripts s and m refer to the shell and for the surrounding medium, respectively.

For the convenience of the following investigation, we will evaluate Eq. (4.1) here in the quadratic approximation by rewriting it into another form for the perturbation in bubble volume defined as $U=4\pi(R^3-R_1^3)/3$. For a soft medium Mooney's constitutive relation [38] is the most widely used model equation and has been adopted by many previous studies regarding the nonlinear dynamics of a bubble in such a medium [3-6]. For facilitating the comparison with the previous studies, therefore, we employ Mooney's relation to evaluate the effective pressure $P_e(R)$, as follows:

$$\varepsilon_p=\mu_p[(1+\chi)(I_1-3)+(1-\chi)(I_2-3)]/4,\qquad p=s,m\tag{4.3}$$

where μ_p is the shear modulus.

Substituting Eq. (4.3) into Eq. (4.2) and expanding $P_e(R)$ to quadratic order, one may derive an analytical approximation of $P_e(R)$, as follows:

$$P_e(R)=P_e(\zeta)=P_e(\zeta_1)+(\zeta-\zeta_1)P_e'(\zeta_1)+(\zeta-\zeta_1)^2P_e''(\zeta_1)/2,$$

where $\zeta=R/R_0$, $\zeta_1=R_1/R_0$, the primes represent derivatives with respect to ζ, and the parameters of $P_e(\zeta)$, $P_e'(\zeta)$, and $P_e''(\zeta)$ are given as below:

$$\begin{aligned}P_e(\zeta)=&\mu_m\int_0^a\left[(1+\chi)(yx^{-1}-y^7x^{-8})+(1-\chi)(y^{-1}-y^5x^{-6})\right]dx\\&+\mu_s\int_a^1\left[(1+\chi)(yx^{-1}-y^7x^{-8})+(1-\chi)(y^{-1}-y^5x^{-6})\right]dx,\end{aligned}\tag{4.4a}$$

$$\begin{aligned}P_e'(\zeta)=&\mu_s\zeta^2\int_0^a\left[(1+\chi)(7y^{10}x^{-8}-y^4x^{-2})+(1-\chi)(y^2+5y^8x^{-6})\right]dx\\&+\mu_m\zeta^2\int_a^1\left[(1+\chi)(7y^{10}x^{-8}-y^4x^{-2})+(1-\chi)(y^2+5y^8x^{-6})\right]dx,\end{aligned}\tag{4.4b}$$

$$\begin{aligned}P_e''(\zeta)=&\mu_s\zeta^4\int_0^a\left[(1+\chi)(4y^7x^{-2}-70y^{13}x^{-8})-(1-\chi)(2y^5+40y^{11}x^{-6})\right]dx\\&+\mu_m\zeta^4\int_a^1\left[(1+\chi)(4y^7x^{-2}-70y^{13}x^{-8})-(1-\chi)(2y^5+40y^{11}x^{-6})\right]dx\\&+2P_e'(\zeta)/\zeta,\end{aligned}\tag{4.4c}$$

where $x = R_0 / \tilde{r}$, $y = R_0 / r$, $a = R_0 / R_{s0}$.
Substituting Eq. (4.4) into Eq. (4.1), one obtains the expansion of Eq. (4.1) to quadratic order in U, as follows:

$$\begin{aligned}\frac{d^2U}{dt^2} + \delta_1 \frac{dU}{dt} + \omega_1^2 U - \frac{R_1}{F_1 c_m}\frac{d^3U}{dt^3} &= G_1 U^2 + \delta_2 U \frac{dU}{dt} \\ &+ H_1\left[\left(\frac{dU}{dt}\right)^2 + 2U\frac{d^2U}{dt^2}\right] - e_1 P_A,\end{aligned} \tag{4.5}$$

where $P_A(t) = P_\infty - P_0$ is the applied acoustic pressure with P_0 being the pressure at infinity in the absence of sound, $\omega_1 = \sqrt{\omega_g^2 + \omega_e^2 - \omega_\sigma^2}$ is the nature frequency of bubble for which the components are given as

$$\omega_g^2 = 3\gamma P_g D_1, \; \omega_e^2 = P_e'(\zeta_1)\zeta_1 D_1, \; \omega_\sigma^2 = 2D_1 R_1^{-1}(\sigma_g + \sigma_m \gamma_R^4),$$

γ is the ratio of specific heats which is chosen as γ=1.4 since the we only consider air bubbles in the present study, δ_1 and δ_2 are the viscous damping coefficients at linear and quadratic order, respectively, defined as

$$\delta_1 = 4D_1\left[\eta_s(1-\gamma_R^3) + \eta_m \gamma_R^3\right], \; \delta_2 = 24 q_1 D_1\left[\eta_s(1-\gamma_R^6) + \eta_m \gamma_R^6\right],$$

G_1 is the nonlinearity coefficient associated with gas compressibility, elasticity, and surface tension that is defined as

$$G_1 = q_1\left[3(\gamma+1)\omega_g^2 + \left(2 - \frac{P_e''(\zeta_1)\zeta_1}{P_e'(\zeta_1)}\right)\omega_e^2 - 4\omega_\sigma^2 + \frac{8\sigma_m R_1}{\rho_m F_1 R_{s1}^4}(1-\gamma_R^3)\right],$$

and the parameters of F_1, H_1, e_1, q_1, γ_ρ, γ_R and D_1 are given as follows
$F_1 = \gamma_\rho + (1-\gamma_\rho)\gamma_R$, $H_1 = q_1 F_1^{-1}\left[\gamma_\rho + (1-\gamma_\rho)\gamma_\rho^4\right]$, $e_1 = 4\pi D_1 R_1^3$, $q_1 = 1/(8\pi R_1^3)$, $\gamma_\rho = \rho_s / \rho_m$, $\gamma_R = R_1 / R_{s1}$, $D_1 = 1/(\rho_m F_1 R_1^2)$.

4.2 Effective medium method
4.2.1 Effective medium

We now study the propagation of a plane acoustic wave in an infinite soft medium containing random encapsulated bubbles, subject to the condition that the volume content of the bubbles is small but the number of bubbles on a scale of wavelength order is large. Then it can be proved that the multiple scattering effects are negligible [39] and the homogeneous approximation well known for liquid containing bubbles can be employed [3]. Consider a small volume element of the medium of length dx_i in the x_i direction in the Cartesian coordinate (i=1, 2, 3) that is sufficiently large to include a number of bubbles. In the present study, we shall focus our attention on the cases where the amplitude of wave is small, for the purpose of investigating the strong physical nonlinearity of such a class of media [3,4]. Then the dynamic nonlinearity is negligible that dominates only on condition that the amplitude of wave is finite. According to the stress-strain relationship and neglecting the contribution of the gas inside the bubbles, the stress tensor may be expressed as (see Ref. [7], pp. 10)

$$\sigma_{ik}^{s} = (1-\beta)\left[K u_{ll}^{s} \delta_{ik} + 2\mu (u_{ik}^{s} - \delta_{ik} u_{ll}^{s} / 3) \right], \tag{4.6}$$

where σ_{ik}^{s} and u_{ik}^{s} are the stress tensor and the strain tensor of the solid phase, β is the total volume fraction of the bubbles, $K = \lambda + 2\mu / 3$ is the bulk modulus, δ_{ik} refers to the Kronecker delta which is defined as

$$\delta_{ik} = \begin{cases} 1, & i = k \\ 0, & i \neq k \end{cases}.$$

On the other hand, the volume element of the bubbly soft medium may be regarded as a volume element of "effective" medium that is homogeneous and is described by effective acoustical parameters. The stress tensor of the effective medium may be given as below:

$$\tilde{\sigma}_{ik} = \tilde{K} \tilde{u}_{ll} \delta_{ik} + 2\tilde{\mu} (\tilde{u}_{ik} - \delta_{ik} \tilde{u}_{ll} / 3), \tag{4.7}$$

where $\tilde{\sigma}_{ik}$ and $\tilde{u}_{ik}$ are the stress tensor and the strain tensor of the effective medium, respectively, $\tilde{K} = \tilde{\lambda} + 2\tilde{\mu} / 3$ is the effective bulk modulus with $\tilde{\lambda}$ and $\tilde{\mu}$ being the effective Lamé coefficients of the effective medium.

4.2.2 Influence of bubble oscillation

As the acoustic wave propagates in the bubbly medium, the volume of the bubbles will change due to the oscillation of the bubbles driven by the acoustic wave. As a result, the variation of the volume element includes the compression of the elastic phase and the variation of the total volume of the bubbles. Then one has

$$\tilde{\theta} = \theta^{s} (1-\beta) + V_t, \tag{4.8}$$

where $\tilde{\theta} = \tilde{u}_{11} + \tilde{u}_{22} + \tilde{u}_{33}$ and $\theta^{s} = u_{11}^{s} + u_{22}^{s} + u_{33}^{s}$ are the volume changes of the effective medium and the elastic phase, respectively, V_t is the variation of the specific volume of bubbles.

As the bubble distorts under the action of the shear deformation, the principal radii of curvature of the surface will change. This will certainly change the effect of surface tension and then the bubble volume. In the present study, however, the bubbles are assumed spherical, and such an effect is then negligible that does not change the nature of the bubble dynamics. Then it is fair to assume approximately that the pure shear deformation of the volume element will not affected by the existence of bubbles. Then one has

$$\sigma_{ik}^{s} = \tilde{\sigma}_{ik}, u_{ik}^{s} = \tilde{u}_{ik}. \quad i \neq k \tag{4.9}$$

Substituting Eq. (4.9) into Eqs. (4.6) and (4.7) yields $\tilde{\mu} = \mu(1-\beta)$.
From Eqs. (4.6) and (4.7) one readily obtains

$$\sigma_{11}^{s} + \sigma_{22}^{s} + \sigma_{33}^{s} = 3(1-\beta)(\lambda + 2\mu / 3)\theta^{s}, \tag{4.10a}$$

$$\tilde{\sigma}_{11} + \tilde{\sigma}_{22} + \tilde{\sigma}_{33} = 3(\tilde{\lambda} + 2\tilde{\mu} / 3)\tilde{\theta} \tag{4.10b}$$

Under the action of an applied force, the element of effective medium is defined to produce the same stress as the element of the bubbly medium. Hence one has

$$\sigma_{11}^{s} + \sigma_{22}^{s} + \sigma_{33}^{s} = \tilde{\sigma}_{11} + \tilde{\sigma}_{22} + \tilde{\sigma}_{33} \tag{4.11}$$

Substituting Eqs. (4.8-4.10) into Eqs. (4.6) and (4.7) yields (for $\lambda >> \mu$)

$$\tilde{C}_{11}\theta^{s} = C_{11}(\theta^{s} - V_{t}) \tag{4.12}$$

where $C_{11} = \lambda + 2\mu$, $\tilde{C}_{11} = \tilde{\lambda} + 2\tilde{\mu}$ are the elastic modulus of the soft medium and the effective medium, respectively.
For the purpose of solving the unknown quantity V_t, it is necessary to obtain the variation of the volume of an individual bubble U which is described by the equation for the oscillation of a bubble given by Eq. (4.5). In Eq. (4.5), the acoustic pressure P_A accounts for the driving force of the oscillation of the bubble. Due to the fact the shear wave does not change the volume of the bubble, the driving force of the bubble oscillation is not affected by the shear wave but determined by the total radial force exerted by the incident wave [3]. According to Eq. (17) in Ref. [3] one has

$$P_{A} = \frac{\lambda + 2\mu / 3}{\lambda + 2\mu}\sigma \simeq \sigma, \tag{4.13}$$

where $\sigma = \tilde{\sigma}_{11}$ is the pressure generated by the incident wave.
Substituting Eq. (4.13) into Eq. (4.5) yields

$$\begin{aligned} \frac{d^2U}{dt^2} + \delta_1\frac{dU}{dt} + \omega_1^2 U - \frac{R_1}{F_1 c_m}\frac{d^3U}{dt^3} &= G_1U^2 + \delta_2 U\frac{dU}{dt} \\ &+ H_1\left[\left(\frac{dU}{dt}\right)^2 + 2U\frac{d^2U}{dt^2}\right] - e_1\sigma, \end{aligned} \tag{4.14}$$

Owing to the fact that Eq. (4.14) is nonlinear only to second order, a potential solution has the form

$$U \approx U_1\exp(i\omega t) + U_2\exp(i2\omega t) + c.c., \tag{4.15a}$$

$$\sigma \approx \sigma_1(\mathbf{r})\exp(i\omega t) + \sigma_2(\mathbf{r})\exp(i2\omega t) + c.c., \tag{4.15b}$$

where σ_1 and σ_2 refer to the linear and the nonlinear waves, respectively, U_1 and U_2 refer to the amplitude of the linear pulsation and the nonlinear response, respectively, $\mathbf{r}$ refers to the three-dimensional space coordinate position of the field point that may be expressed in the Cartesian coordinate as $\mathbf{r} = x_1\hat{i} + x_2\hat{j} + x_3\hat{k}$.
Substituting Eq. (4.15) into Eq. (14) and assuming that $1 > U_1 > U_2$ yield

$$U_1 = g_1\sigma_1,\ U_2 = g_2\sigma_2 + \Gamma\sigma_1^2, \tag{4.16}$$

where

$$g_1 = -\frac{e_1}{(\omega_1^2 - \omega^2) + i(\omega\delta_1 + \omega^3 R_1 / c_m)}, \; g_2 = -\frac{e_1}{(\omega_1^2 - 4\omega^2) + i(2\omega\delta_1 + 8\omega^3 R_1 / c_m)},$$

$$\Gamma = g_1^2 \frac{(G_1 - 3\omega^2 H_1) + i\omega\delta_2}{(\omega_1^2 - 4\omega^2) + i(2\omega\delta_1 + 8\omega^3 R_1 / c_m)},$$

Expanding the displacement vector $\mathbf{u}$, up to second order approximation, as

$$\mathbf{u} = \mathbf{u}_1(\mathbf{r})\exp(i\omega t) + \mathbf{u}_2(\mathbf{r})\exp(i2\omega t) + c.c.,$$

one obtains

$$\sigma_1 = \tilde{C}_{11}^{(1)} \nabla \cdot \mathbf{u}_1, \; \sigma_2 = \tilde{C}_{11}^{(2)} \nabla \cdot \mathbf{u}_2 \tag{4.17}$$

If the size distribution function of the bubbles is specified as $n(R_0)$ (so that $n(R_0)dR_0$ is the number of bubbles with radii from R_0 to $R_0 + dR_0$ in unit volume), the variation of the specific volume of the bubbles V_t is related to the volume variation of an individual bubble U by the relationship

$$V_t = \int U n(R_0) dR_0, \tag{4.18}$$

Expanding V_t to second order approximation as

$$V_t = V_1 \exp(i\omega t) + V_2 \exp(i2\omega t) + c.c., \tag{4.19}$$

one obtains from Eqs. (4.16-4.19)

$$V_1 = V_{g1} \tilde{C}_{11}^{(1)} \nabla \cdot \mathbf{u}_1, \; V_2 = V_{g2} \tilde{C}_{11}^{(2)} \nabla \cdot \mathbf{u}_2 + V_\Gamma \left(\tilde{C}_{11}^{(1)} \nabla \cdot \mathbf{u}_1\right)^2 \tag{4.20}$$

where
$V_{g1} = \int g_1 n(R_0) dR_0$, $V_{g2} = \int g_2 n(R_0) dR_0$, and $V_\Gamma = \int \Gamma n(R_0) dR_0$.
If all the bubbles are of the uniform radius R_0, V_t is related to U by the relationship $V_t = NU$ with $N = 3\beta(4\pi R_0^3)^{-1}$ being the number of bubbles in unit volume. In such cases one has

$$V_{g1} = N g_1, \; V_{g2} = N g_2, \; V_\Gamma = N\Gamma.$$

Expressing the volume change of the solid phase as $\theta^s = \nabla \cdot \mathbf{u}$, one may rewrite Eq. (4.12) as follows:

$$\tilde{C}_{11}^{(1)} \nabla \cdot \mathbf{u}_1 = C_{11}(\nabla \cdot \mathbf{u}_1 - V_1), \; \tilde{C}_{11}^{(2)} \nabla \cdot \mathbf{u}_2 = C_{11}(\nabla \cdot \mathbf{u}_2 - V_2). \tag{4.21}$$

4.2.3 The wave equations

According to Ref. [6], $\rho_m F_1$ is defined as the effective density of the soft medium surrounding the bubbles, the effective density of the effective medium may thus be identified as $\tilde{\rho} = \rho_m F_1(1-\beta) + \rho_g \beta$. Since the dynamic nonlinearity of the medium associated with the finite amplitude of wave has been ignored, the wave equation of the effective medium may be written as follows:

$$\nabla(\tilde{C}_{11}\nabla \cdot \mathbf{u}) - \tilde{\mu}\nabla \times \nabla \times \mathbf{u} = \tilde{\rho}\frac{\partial^2 \mathbf{u}}{\partial t^2}, \tag{4.22}$$

We represent the displacement vector $\mathbf{u}$ in terms of the sum of the potentials, as follows:

$$\mathbf{u} = \nabla\Phi + \nabla \times \mathbf{\Psi}, \tag{4.23}$$

for which the vector potential $\mathbf{\Psi}$ satisfies $\nabla \cdot \mathbf{\Psi} = 0$.
Up to second order approximation, the scalar potential Φ may be written as

$$\Phi \approx \Phi^{(1)} + \Phi^{(2)} + c.c., \tag{4.24}$$

where

$$\Phi^{(1)}(\mathbf{r},t) = \Phi_1(\mathbf{r})\exp(i\omega t), \; \Phi^{(2)}(\mathbf{r},t) = \Phi_1(\mathbf{r})\exp(i2\omega t). \tag{4.25}$$

Substitution of Eqs. (4.23) and (4.24) in Eq. (4.22) yields

$$\tilde{C}_{11}^{(1)}\nabla^2\Phi^{(1)} = \tilde{\rho}\frac{\partial^2\Phi^{(1)}}{\partial t^2}, \; \tilde{C}_{11}^{(2)}\nabla^2\Phi^{(2)} = \tilde{\rho}\frac{\partial^2\Phi^{(2)}}{\partial t^2}, \tag{4.26}$$

$$\tilde{\mu}\nabla^2\mathbf{\Psi} = \tilde{\rho}\frac{\partial^2\mathbf{\Psi}}{\partial t^2}. \tag{4.27}$$

As observed from Eq. (4.27), this equation takes on a non-resonant form and the influence of the existence of the bubbles on the propagation of the shear wave in a bubbly soft medium is insignificant. In the following we shall restrain our attention in the propagation of the compressional wave in such a medium.
Substituting Eqs. (4.20), (4.21), and (4.25) in Eq. (4.26), we arrive at the equations that must be satisfied by the scalar potentials of the first and the second order, as follows:

$$C_{11}\nabla^2\Phi_1 + \tilde{\rho}\omega^2(1 + V_{g1}C_{11})\Phi_1 = 0, \tag{4.28a}$$

$$C_{11}\nabla^2\Phi_2 + 4\tilde{\rho}\omega^2(1 + V_{g2}C_{11})\Phi_2 = C_{11}V_\Gamma\left(\tilde{C}_{11}^{(1)}\right)^2\left(\nabla^2\Phi_1\right)^2, \tag{4.28b}$$

Eqs. (4.28a) and (4.28b) give description of the the propagation of the fundamental and the second harmonics of the compressional wave in a bubbly soft medium, respectively. Note that Eq. (4.28) is derived on the basis of Eq. (4.22) which is expressed as a form of a linear order terms with nonlinear propagation parameters due to the nonlinear oscillation of bubbles. Consequently it is seen that Eq. (4.28b) takes a simple form without any quadratic term involved that represents the dynamic nonlinearity caused by the finite amplitude of wave. In the right hand side of this equation, however, a quadratic term appears that accounts for the transfer of acoustical energy from the fundamental to the second harmonic waves, which results from the strong physical nonlinearity that dominates for a bubbly medium.

4.2.4 One-dimensional case

Now consider a one-dimensional case in which a plane longitudinal wave propagates along the $\hat{x}_1$ direction in a bubbly soft medium. For simplicity while without losing generality, we assume that all the bubbles are of the same equilibrium radius R_0. In such a case Eq. (4.25) becomes

$$\Phi^{(1)}(x_1,t)=\Phi_1(x_1)\exp(i\omega t),\ \Phi^{(2)}(x_1,t)=\Phi_2(x_1)\exp(i2\omega t). \tag{4.29}$$

Using the Kelvin-Voigt viscoelastic model, the Lamé coefficients of the soft viscoelastic medium may be rewritten as

$$\lambda=\lambda_m,\ \mu=\mu_m+\eta_m\partial/\partial t. \tag{4.30}$$

Substitution of Eqs. (4.29) and (4.30) in Eq. (4.28) yields

$$\frac{d^2\Phi_1}{dx_1^2}+(\Lambda_1^2+\tilde{\rho}\omega^2V_{g1})\Phi_1=0, \tag{4.31a}$$

$$\frac{d^2\Phi_2}{dx_1^2}+(\Lambda_2^2+4\tilde{\rho}\omega^2V_{g2})\Phi_2=\tilde{\rho}^2\omega^4V_\Gamma\Phi_1^2, \tag{4.31b}$$

where $\Lambda_1^2=\omega^2\tilde{\rho}/[(\lambda_m+2\mu_m)+i2\omega\eta_m]$, and $\Lambda_2^2=4\omega^2\tilde{\rho}/[(\lambda_m+2\mu_m)+i4\omega\eta_m]$.
We introduce the effective wave numbers defined as complex numbers that can be expressed in terms of real effective wave speeds and effective attenuations, as follows:

$$\tilde{k}_1=\sqrt{\Lambda_1^2+\tilde{\rho}\omega^2V_{g1}}=\omega/c_1-i\alpha_1,\ \tilde{k}_2=\sqrt{\Lambda_2^2+4\tilde{\rho}\omega^2V_{g2}}=2\omega/c_2-i\alpha_2,$$

where $\tilde{k}_i$ refers to the effective wave numbers, c_i and α_i refer to the (real) effective wave speed and the effective attenuation, respectively; and the subscripts i =1, 2 refer to the fundamental wave and the second harmonic wave, respectively.
Assuming $\Phi_1=\Phi_{1A}\exp(-i\tilde{k}_1x_1)$, from Eq. (4.31a) one readily obtains the expressions of c_1 and α_1, as follows:

$$c_1=\left(\frac{-A_1+\sqrt{A_1^2+B_1^2}}{2}\right)^{-1/2},\quad \alpha_1=\omega B_1c_1/2. \tag{4.32}$$

where the parameters of A_2, B_2, and C are given as follows:

$$A_1=\tilde{\rho}\omega^2\left[\frac{\lambda_m+2\mu_m}{(\lambda_m+2\mu_m)^2+4\omega^2\eta_m^2}-\frac{Ne_1(\omega_1^2-\omega^2)}{(\omega_1^2-\omega^2)^2+(\omega\delta_1+\omega^3R_1/c_m)^2}\right],$$

$$B_1=\tilde{\rho}\omega^2\left[\frac{2\omega\eta_m}{(\lambda_m+2\mu_m)^2+4\omega^2\eta_m^2}-\frac{Ne_1(\omega\delta_1+\omega^3R_1/c_m)}{(\omega_1^2-\omega^2)^2+(\omega\delta_1+\omega^3R_1/c_m)^2}\right].$$

It is apparent that the solution of Eq. (4.31b) is supposed to consist of a general solution and a special solution. By invoking the boundary condition that the second harmonic wave should be zero at the beginning, i.e., $\Phi_2\big|_{x_1=0}=0$, one can readily determine the expression of the second harmonic wave as follows:

$$\Phi_{2g} = C\Phi_{1A}^{2}\left[\exp(-i\tilde{k}_2 x) - \exp(-i\tilde{k}_2 x)\right] + c.c., \tag{4.33}$$

where

$$c_2 = \left(\frac{-A_2 + \sqrt{A_2^2 + B_2^2}}{2}\right)^{-1/2}, \quad \alpha_2 = \omega B_2 c_2 . \tag{4.34}$$

where the parameters of A_2, B_2 and C are given as follows:

$$A_2 = \tilde{\rho}\omega^2\left[\frac{\lambda_m + 2\mu_m}{(\lambda_m + 2\mu_m)^2 + 16\omega^2\eta_m^2} - \frac{Ne_1(\omega_1^2 - 4\omega^2)}{(\omega_1^2 - 4\omega^2)^2 + (2\omega\delta_1 + 8\omega^3 R_1 / c_m)^2}\right],$$

$$B_2 = \tilde{\rho}\omega^2\left[\frac{4\omega\eta_m}{(\lambda_m + 2\mu_m)^2 + 16\omega^2\eta_m^2} - \frac{Ne_1(2\omega\delta_1 + 8\omega^3 R_1 / c_m)}{(\omega_1^2 - 4\omega^2)^2 + (2\omega\delta_1 + 8\omega^3 R_1 / c_m)^2}\right].$$

$$C = [(M_1N_1 + M_2N_2) + i(M_2N_1 - M_1N_2)]/(N_1^2 + N_2^2),$$

where

$$M_1 = \tilde{\rho}^2\omega^4 Ne_1^2\left[(G_1 - 3\omega^2 H_1)(A_1^2 - B_1^2) + 2A_1B_1\omega\delta_2\right],$$

$$M_2 = \tilde{\rho}^2\omega^4 Ne_1^2\left[(A_1^2 - B_1^2)\omega\delta_2 - 2A_1B_1(G_1 - 3\omega^2 H_1)\right],$$

$$N_1 = K_1(-4A_1 + A_2) + L_1(B_2 - 4B_1),$$

$$N_2 = L_1(-4A_1 + A_2) - K_1(B_2 - 4B_1),$$

where

$$K_1 = (\omega_1^2 - 4\omega^2)\left[(\omega_1^2 - \omega^2)^2 - (\omega\delta_1 + \omega^3 R_1 / c_m)^2\right] \\ -2(\omega_1^2 - \omega^2)(\omega\delta_1 + \omega^3 R_1 / c_m)(2\omega\delta_1 + 8\omega^3 R_1 / c_m),$$

$$L_1 = \left[(\omega_1^2 - \omega^2)^2 - (\omega\delta_1 + \omega^3 R_1 / c_m)^2\right](2\omega\delta_1 + 8\omega^3 R_1 / c_m) \\ +2(\omega_1^2 - 4\omega^2)(\omega_1^2 - \omega^2)(\omega\delta_1 + \omega^3 R_1 / c_m).$$

In practical, the nonlinearity parameter (B/A) is of particular significance that may be used to define the nonlinearity of the media. In the present study, therefore, we introduce an effective nonlinearity parameter $(B/A)_e$ to describe the extent to which the nonlinearity of a bubbly medium is enhanced by the nonlinear oscillation of bubbles. The value of $(B/A)_e$ may be determined near the natural frequency of bubble, as given below: [39]

$$(B/A)_e = \left| \frac{4C\rho_m c_m^3 (2\alpha_1 - \alpha_2)}{\omega} - 2 \right|$$

It is apparent that the expression of the effective nonlinearity parameter $(B/A)_e$ derived here is identical in form with the one obtained by Ma et al except that their approach only applies to a liquid containing shelled bubbles [39].

Despite the similarity between the EMM and other methodologies which also investigate the wave propagation in inhomogeneities by treating the media as a homogeneous effective medium [40-45], the EMM definitely differs from them in several respects. It is a fundamental distinction that the EMM accounts for the nonlinearity of the bubbly soft medium up to a second-order approximation, whereas most of the previous ones only use linear approximation when homogenizing the medium [40-43], which inevitably loses significant details for bubbly soft media with particularly strong "physical" nonlinearity [3,4]. Second, the EMM permits one to take into consideration the effects of weak compressibility, surface tension, viscosity, surrounding pressure, and an encapsulating elastic shell, which can only be partially accounted for by other methods [44,45]. There are important practical reasons for pursuing more precise results in various engineering situations, for which the incorporation of these effects is apparently necessary. (For a detailed discussion on this topic and a comparison between the application of the EMM and some other methods in different cases serving as simple models of practical situations, see Ref. [12]) Finally, the EMM could apply to three-dimensional cases rather than one-dimensional cases. Most of the relative studies investigate only the wave propagation in an infinite effective medium for which the one-dimensional approximation is sufficient, but it is indispensable to obtain the three-dimensional effective parameters for some practical structures of finite sizes.

It must be stressed, however, that there also exist limitations of the application of the EMM despite its effectiveness. First, the multiple scattering effects have been neglected when we homogenizing the bubbly soft medium, therefore the EMM can not apply to bubbly media with extremely large volume fractions. Second, the EMM is developed under quadratic approximation by employing a simple perturbation approach, and the nonlinearity of medium is studied by inspecting the second harmonic wave with no harmonics of orders higher than 2 involved. Finally, the present model could not enable full incorporation of all the practical effects that affect the acoustical properties of a bubbly medium, such as the buckling of bubbles [46]. These problems will be the focus of a future study.

5. Optimal acoustic attenuation of bubbly soft media

In this section, we present an optimization method on the basis of fuzzy logic (FL) and genetic algorithm (GA) to obtain the optimal acoustic attenuation of a longitudinal wave in a bubbly soft medium by optimizing the parameters of size distribution of bubbles. This optimization method can be used to design acoustic absorbent with uniformly high acoustic attenuation within the frequency band of interest, without the precise mathematical model required.

5.1 Acoustic attenuation in bubbly soft media

The oscillation of an air bubble in a soft medium is special, due to the fact that only if the ratio λ/μ is sufficiently large can this bubble behave effectively as a resonant oscillator [7]. When compared with the viscoelasticity of the medium, the resonance of the system introduced by bubbles becomes the most dominant mechanism for acoustic attenuation [13]. For a bubbly soft medium, it is apparent that the acoustic properties are affected by all the structural parameters, of the bubbles and of the medium. By employing the EMM presented in Section 4, we can accurately predict the acoustic parameters of a bubbly soft medium for arbitrary structural parameters. In this situation, one may expect to enhance the acoustic attenuation of such a medium in an optimal manner with the aid of a fast computer.

Consider the one-dimensional propagation of a longitudinal wave in an infinite bubbly soft medium with small volume fraction Φ_b. On condition that the bubbles are not very densely packed, the multiple scattering effects are negligible, and the acoustic properties of such a bubbly soft medium can be described by using the EMM.

If all the bubbles are of uniform radius r_0, the bubble volume fraction will be $\Phi_b=4\pi N_b(r_0)^3/3$ with N_b being the number of bubbles per unit volume. When the bubble sizes are not uniformly distributed, the volume fraction is related to the distribution function $n(r)$ of bubble sizes, as follows:

$$\Phi_b = 4\pi\int_0^{\infty} n(r)r^3 dr / 3 \tag{5.1}$$

where $n(r)dr$ is the number of bubbles per unit volume having a radius between r and $r+dr$. For simplicity, the bubbles in the soft medium are assumed to be free bubbles (no encapsulating shells), the effects of surface tension and the ambient pressure are neglected, and acoustic nonlinearity of the bubbly soft medium are not taken into account. Then the effective acoustic attenuation of longitudinal wave in the bubbly soft medium can be derived from Eq. (4.32), as follows:

$$\alpha_1' = A_1'\left(-B_1'/2+\sqrt{B_1'^2/4+A_1'}\right)^{-1/2}, \tag{5.2}$$

where the parameters of A and B are given as follows:

$$B_1' = \int_0^{\infty}\frac{n(r)e(r)\rho_0}{\chi_1(r)}\left(\cos\phi(r)-\frac{2\mu_v\omega\sin\phi(r)}{\lambda+2\mu}\right)dr+\frac{4\mu_v\omega A_1'-\omega^2\rho_0}{\lambda+2\mu},$$

$$A_1' = \int_0^{\infty}\frac{n(r)e(r)\rho_0\sin\phi(r)}{2\chi_1(r)}dr+\frac{2\mu_v\omega^3\rho_0}{2(\lambda+2\mu)+8\mu_v^2\omega^2},$$

where

$$\chi_1(r)=\left[\left(1-\frac{\omega_0^2}{\omega^2}\right)^2+\left(\frac{\omega r}{c_l}+\frac{4\mu_v}{\omega\rho_0 r^2}\right)^2\right]^{1/2},\quad \phi(r)=\tan^{-1}\left[\left(\frac{4\mu_v}{\omega\rho_0 r^2}+\frac{\omega r}{c_l}\right)\left(1-\frac{\omega_0^2}{\omega^2}\right)^{-1}\right],$$

where ρ_0 is the mass density, μ_v is the lossy factor given by the Kelvin-Voigt viscoelastic model, c_l is the velocity of the longitudinal elastic wave.

In general, the enhancement of acoustic attenuation is equivalent to regularly providing sufficient acoustic attenuation in the frequency range of interest. Due to the resonance of the system introduced by bubbles, the acoustic attenuations exhibit a remarkable enhancement effect near the bubble resonant frequencies, and there exist resonance peaks in the spectral domain [40]. We consider the acoustic attenuation caused by the oscillation of the bubbles, and neglect the contribution of the viscosity to the acoustic attenuation. The resonance frequency can be decreased by reducing the shear modulus of the medium or enlarging bubbles, and the acoustic attenuation will be collectively enhanced as Φ_b increases [45]. However, it is impractical to unlimitedly increase the volume fraction and dimension of bubbles. The strength of the bubbly medium will be weakened if the bubbles are too densely packed or oversize, and oversize bubbles are not feasible for a practical medium of finite size. It is of interest to provide regularly high acoustic attenuation in targeted frequency range while minimizing the volume fraction and dimension of bubbles.

To decrease the resonance location, it is more effective to reduce the shear modulus of the medium than to merely enlarge bubbles. Besides, the acoustic attenuation will also be enhanced as Φ_b remains constant while the shear modulus reduces [45]. Hence we choose silicone that has low shear modulus as the medium for which the mechanical parameters are: ρ_0=1000kg/m^3, the velocity of the longitudinal and the shear elastic wave are c_l=1700m/s and c_s=20m/s, respectively [47]. The lossy factor is chosen as μ_v=80P. On the other hand, it has been proved that the nonuniform distribution of bubble sizes has an averaging effect tends to increase the acoustic attenuation over a wider frequency range and result in a much broader resonance peak [40]. In what follows, therefore, distribution of bubble sizes is introduced and the probability density function of normal distribution is employed to describe the distribution function. Due to the peak-broadening effects of size distribution, together with the amplitude-enhancing effects of volume fraction, one may hope to obtain an optimal acoustic attenuation for a bubbly soft medium by choosing the structural parameters appropriately. This leads to the necessity of some optimization method.

For such a problem with multiple adjustable parameters, a full-space search method will not be practical, and a global optimization method is expected to be effective [48]. The success of an optimization method depends to a great extent on the definition of a proper objective function. For such a problem, however, it may be difficult to mathematically create an appropriate objective function in traditional ways, since the ability of acoustic attenuation of a medium is usually evaluated qualitatively. With the purpose of avoiding such mathematical efforts, we will define the objective function by using FL that bases on decision rules rather than mathematical equations and describe linguistically the relationship between input and output [49,50], and use a GA that can locate the global optimum despite that the objective function is built without knowing its clear mathematical model [51-53].

5.2 Numerical example

In the following we will exemplify a numerical case for enhancing the acoustic attenuation of the bubbly soft medium in an optimal manner. As an example, we intend to obtain uniformly effective acoustic attenuation for longitudinal wave propagating within the bubbly soft medium, in a broad frequency range at intermediate frequencies. And the following requirement is proposed:

1. The bubbly medium can attenuate longitudinal wave by no less than 10dB/cm, in a frequency range as broad as possible within the intermediate frequency range of [5KHz, 800KHz].
2. The wave at the frequency of 5KHz should be effectively attenuated.
3. The acoustic attenuations in targeted frequency range must be uniform.

This quantitative requirement serves for the goal of the optimization. The effectiveness of optimization method will be eventually evaluated in terms of the extent to which the requirement is fulfilled.

For a particular medium, the large and the small bubbles contribute to the acoustic attenuation at low and high frequencies, respectively. It is thus possible to increase acoustic attenuation at low frequencies as well as extend the width of resonance peak, by introducing the size distribution of large and small bubbles and tuning up their parameters properly. Then the distribution function $n(r)$ is given as below:

$$n(r)=\begin{cases} n_1(r)+n_2(r), & r_L \le r \le r_U \\ 0, & \text{elsewhere} \end{cases}, \tag{5.3}$$

where r_U (r_L) refers to the radius of the largest (smallest) bubble in the medium, $n_1(r)$ and $n_2(r)$ refer to the number densities of large and small bubbles respectively, as follows:

$$n_1(r)=n_0 R_b (r_2/r_1)^3 \exp[-(r/r_1-1)/(2\sigma_{b1})]^2,\ \ n_2(r)=n_0 \exp[-(r/r_2-1)/(2\sigma_{b2})]^2,$$

where the value of $R_b(r_2/r_1)^3$ represents the ratio of number density of large bubbles to small bubbles, r_j and σ_{bj} (j=1,2) refers to the center and the width of size distribution, respectively. The value of n_0 can be easily determined from the relationship given by Eq.(5.2). Now the bubble parameters are the only adjustable parameters affecting the acoustic attenuation of the bubbly medium, including Φ_b and the parameters of distribution function $n(r)$. It is apparent that the objective function to be created is a multiple inputs problem and the input variables consist of all these adjustable parameters, i.e., Φ_b, $r_{L,U}$, $r_{1,2}$, R_b, $\sigma_{b1,2}$.

It is obvious that the ability of acoustic attenuation of the bubbly medium mostly depends on the location and shape of the lowest resonance peak in spectral domain. To describe the location and width of this resonance peak, we introduce two parameters $f_0=f_L$ and $W_b=f_U-f_L$ defined as the lowest effective attenuation frequency and the effective attenuation bandwidth, respectively. Here f_U (f_L) is the upper (lower) limit of a frequency band within which acoustic attenuation of any frequency is more than a threshold value α_t (α_t =10dB/cm), and here does not exist a $f_L' < f_L$, such that f_L' satisfies the condition as well. And a standard deviation function Σ is introduced to scale the degree of regularity of attenuation, as follows:

$$\Sigma=\sqrt{\int_{f_L}^{f_U}[\alpha_1'(f)-\alpha_t]^2 df/(f_U-f_L)}, \tag{5.4}$$

where $\alpha_1'(f)$ refers to acoustic attenuation at the frequency of f. It is apparent that the introduced parameters f_0, W_b and Σ can be easily obtained from the acoustic attenuation predicted by EMM and describe quantitatively the characteristic of the lowest resonance peak in spectral domain.

By using FL, we set up a fuzzy inference system (FIS), for which the parameters f_0, W_b and Σ are chosen as the input parameters and the explicit output is defined as s ($0 \le s \le 100$).

There are three membership functions for f_0: "low", "intermediate" and "high". And there are three membership functions for W_b as well: "narrow", "average" and "broad". Similarly Σ consists of three conditions of degree of deviation denoted by "small", "ordinary" and "large". The membership functions of the inputs are built on a simple Gaussian curve due to its smoothness in varying.

Three inputs are captured consisting of f_0, W_b and σ, and the fuzzy relation between the fuzzy inputs and the required output s are shown by the following inference rules:

Rules 1: If (f_0 is "high") or (W_b is "narrow") and (σ is "large") then (s is "bad")

Rules 2: If (f_0 is "intermediate") and (W_b is "average") and (σ is "ordinary") then (s is "mediocre")

Rules 3: If (f_0 is "low") and (W_b is "average") and (σ is "ordinary") then (s is "good")

Rules 4: If (f_0 is "intermediate") and (W_b is "broad") and (σ is "ordinary") then (s is "good")

Rules 5: If (f_0 is "intermediate") and (W_b is "average") and (σ is "small") then (s is "good")

Rules 6: If (f_0 is "intermediate") and (W_b is "broad") and (σ is "small") then (s is "very good")

Rules 7: If (f_0 is "low") and (W_b is "average") and (σ is "small") then (s is "very good")

Rules 7: If (f_0 is "low") and (W_b is "broad") and (σ is "ordinary") then (s is "very good")

Rules 8: If (f_0 is "low") and (W_b is "broad") and (σ is "small") then (s is "excellent")

The above inference rules relate these inputs to the output s consisting of five membership functions: "bad", "mediocre", "good", "very good", "excellent". The triangular membership function is adopted because this membership representation shows boundary clearly.

It is apparent that the mapping of the multiple input parameters (f_0, W_b and Σ) to the output s can be conveniently constructed, by defining the fuzzy rules as a set of linguistic rules according to the aforementioned requirement, without knowing the clear mathematical model. Then the value of output s gives a quantitative description of the extent to which the qualitative requirement is met. A bubbly soft medium of better acoustic attenuation will correspond to an output of larger value. With the aid of the FIS, we readily define an objective function corresponding to this nine-input, one-output problem. Mathematically speaking, this objective function based on FL may not be completely precise, and the clear mathematical model is not visible. But it is readily guaranteed that the acoustic attenuation ability of a bubbly soft medium is evaluated strictly by the decision rules, which is the unique advantage of FL for such a problem.

By defining an objective function for mapping the multiple inputs properly to a clear output, the optimal enhancement of acoustic attenuation ability amounts to an optimization problem of generating a maximal output by tuning up the inputs. Such an optimization is performed by employing GA optimizer. The objective function and the output s are regarded as the fitness function and the fitness, respectively. The nine input variables are encoded as the chromosome. GA optimizer searches for the optimum of fitness function by adjusting the bubble parameters and seeking the most proper proportion. To guarantee the physical feasibility, a set of constraints of the variables are applied, as follows:

$$0 < \Phi_b \le 5\%,\ 10\mu\text{m} \le r_L \le r_1 \le r_2 \le r_U \le 2\text{mm},\ 0 < R_b \le 10,\ 0 < \sigma_{b1}, \sigma_{b2} \le 1 .$$

In the process of GA optimization, the number of population and maximal number of generation are chosen as 80 and 500, respectively, the crossover and mutation ratio are set to 0.8 and 0.05, respectively [53].

As a result, the bubbly silicone with optimized structural parameters has a value of fitness as high as 99.2. Correspondingly, the lower and upper limits of the effective attenuation band are $f_L \approx 5$KHz and $f_U \approx 800$KHz, respectively. Then one has $f_0 \approx 5$KHz and $W_b \approx 795$KHz. The optimal acoustic attenuation versus frequency is plotted in Fig. 5.1. Figure 5.2 displays the corresponding size distribution function $n(r)$. The numerical result shows that the goal of optimization is attained perfectly, which is indicated numerically by the fitness and illustrated graphically in Fig. 5.1. The bubbly medium with optimized structural parameters can effectively attenuate longitudinal waves in an intermediate frequency range of [5KHz, 800KHz] with an acoustic attenuation approximating a constant value of 10(dB/cm). As shown by the results of the optimization process, such a bubbly medium may be applied to design broadband acoustic absorbent at intermediate frequencies with high efficiency. Compared with acoustic absorbent designed by using traditional method, the acoustic absorbent designed by using the present optimization method has broader attenuation band and higher efficiency. Moreover, the width and the location of its attenuation band may be conveniently controlled due to the adjustability of the objective function and the optimizer.

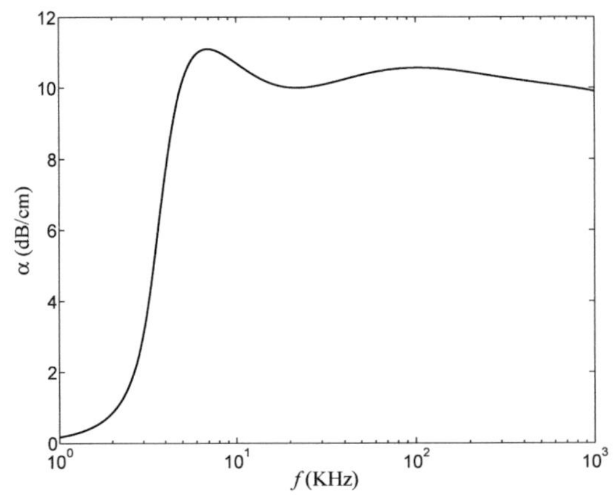

Fig. 5.1. Optimal acoustic attenuation α versus frequency for the bubbly silicone.

To study the necessity and efficiency of the optimization method, we also consider other three cases with no optimization applied. The structural parameters of these four cases are listed in Table I and the comparison of attenuation curves are displayed in Fig. 5.3. Observation of Fig. 5.3 and Table 5.1 shows that: (1) Case 4 has a better acoustic attenuation and a higher value of fitness than any of the other cases, which means that the relative values of fitness describe effectively the acoustic attenuation of the bubbly medium in accord with the requirement. (2) Size distribution effect helps to ameliorate the acoustic attenuation, which is proved by comparing case 3 with cases 1 and 2. But the comparison between cases 3 and 4 shows that optimal acoustic attenuation can not be guaranteed by size distribution with random parameters. (3) The optimization method is efficient and essential in enhancing the acoustic attenuation of the bubbly medium.

Case	Φ_b (%)	r_L(μm)	r_U(mm)	r_1(mm)	r_2(μm)	R_b	σ_{b1}	σ_{b2}	s
1	1.4	1.7	1.7	□	□	□	□	□	39.6
2	1.4	23	23	□	□	□	□	□	41.2
3	0.5	40	1.1	0.5	200	0.08	0.4	0.1	63.6
4	1.4	17	1.8	1.7	23	1.2	0.7	0.5	99.2

Table 5.1 The structural parameters of the four different cases.

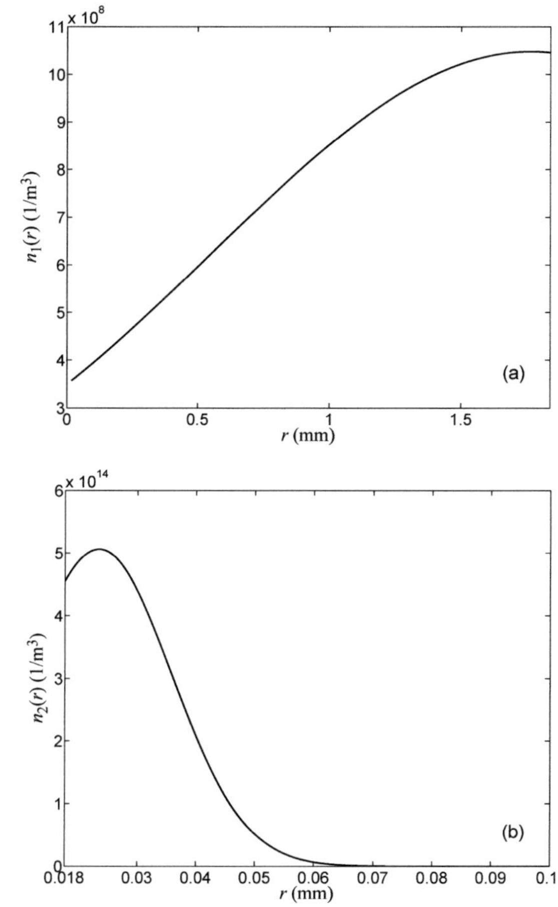

Fig. 5.2. Number densities of large (a) and small (b) bubbles in the bubbly silicone with optimal acoustic attenuation.

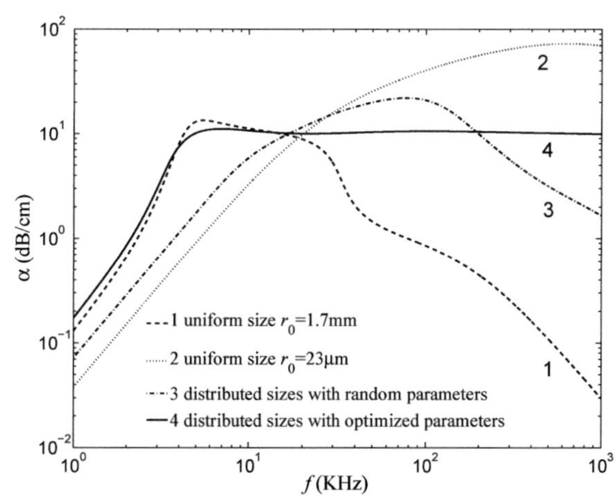

Fig. 5.3. Comparison of acoustic attenuations versus frequency for the four different cases.

6. Conclusions

In this chapter, we first consider the acoustic propagation in a finite sample of bubbly soft elastic medium and solve the wave field rigorously by incorporating all multiple scattering effects. The energy converted into shear wave is numerically proved negligible as the longitudinal wave is scattered by the bubbles. Under proper conditions, the acoustic localization can be achieved in such a class of media in a range of frequency slightly above the resonance frequency. Based on the analysis of the spatial correlation characteristic of the wave field, we present a method that helps to discern the phenomenon of localization in a unique manner. Then we taken into consideration the effect of viscosity of the soft medium and investigate the localization in a bubbly soft medium by inspecting the oscillation phases of the bubble. The proper analysis of the oscillation phases of bubbles is proved to be a valid approach to identify the existence of acoustic localization in such a medium in the presence of viscosity, which reveals the existence of the significant phenomenon of phase transition characterized by an unusual collective behavior of the phases.

For infinite sample of bubbly soft medium, we present an EMM which enables the investigation of the strong nonlinearity of such a medium and accounts for the effects of weak compressibility, viscosity, surrounding pressure, surface tension, and encapsulating shells. Based on the modified equation of bubble oscillation, the linear and the nonlinear wave equations are derived and solved for a simplified 1-D case. Based on the EMM which can be used to conveniently obtain the acoustic parameters of bubbly soft media with arbitrary structural parameters, we present an optimization method for enhancing the acoustic attenuation of such media in an optimal manner, by applying FL and GA together. A numerical simulation is presented to manifest the necessity and efficiency of the optimization method. This optimization method is of potential application to a variety of situations once the objective function and optimizer are adjusted accordingly.

7. References

[1] E. Meyer, K. Brendel, and K. Tamm, J. Acoust. Soc. Am. 30, 1116 (1958).
[2] A. C. Eringen and E. S. Suhubi, *Elastodynamics* (Academic, New York, 1974).
[3] L. A. Ostrovsky, Sov. Phys. Acoust. 34, 523 (1988).
[4] L. A. Ostrovsky, J. Acoust. Soc. Am. 90, 3332 (1991).
[5] S. Y. Emelianov, M. F. Hamilton, Yu. A. Ilinskii, and E. A. Zabolotskaya, J. Acoust. Soc. Am. 115, 581 (2004).
[6] E. A. Zabolotskaya, Yu. A. Ilinskii, G. D. Meegan, and M. F. Hamilton, J. Acoust. Soc. Am. 118, 2173 (2005).
[7] L. D. Landau and E. M. Lifshits, *Theory of Elasticity* (Pergamon, Oxford, 1986).
[8] B. Liang and J. C. Cheng, Phys. Rev. E. 75, 016605 (2007).
[9] B. Liang, Z. M. Zhu, and J. C. Cheng, Chin. Phys. Lett. 23, 871 (2006).
[10] B. Liang, X. Y. Zou, and J. C. Cheng, Chin. Phys. Lett. 26, 024301 (2009).
[11] B. Liang, X. Y. Zou, and J. C. Cheng, Chin. Phys. B 19, 094301 (2010).
[12] B. Liang, X. Y. Zou, and J. C. Cheng, J. Acoust. Soc. Am. 124, 1419 (2008).
[13] G. C. Gaunaurd and J. Barlow, J. Acoust. Soc. Am. 75, 23 (1984).
[14] B. Liang, X. Y. Zou, and J. C. Cheng, Chin. Phys. Lett. 24, 1607 (2007).

[15] K. X. Wang and Z. Ye, Phys. Rev. E 64, 056607 (2001).
[16] C. F. Ying and R. Truell, J. Appl. Phys. 27, 1086 (1956).
[17] G. C. Gaunaurd, K. P. Scharnhorst and H. Überall, J. Acoust. Soc. Am. 65, 573 (1979).
[18] D. M. Egle, J. Acoust. Soc. Am. 70, 476 (1981).
[19] R. L. Weaver, J. Acoust. Soc. Am. 71, 1608 (1982).
[20] K. Busch, C. M. Soukoulis, and E. N. Economou, Phys. Rev. B 50, 93 (1994).
[21] A. A. Asatryan, P. A. Robinson, R. C. McPhedran, L. C. Botten, C. Martijin de Sterke, T. L. Langtry, and N. A. Nicorovici, Phys. Rev. E 67, 036605 (2003).
[22] S. Gerristsen and G. E. W. Bauer, Phys. Rev. E 73, 016618 (2006).
[23] C. A. Condat, J. Acoust. Soc. Am. 83, 441 (1988).
[24] Z. Ye and A. Alvarez, Phys. Rev. Lett. 80, 3503 (1998).
[25] S. Catheline, J. L. Gennisson, and M. Fink, J. Acoust. Soc. Am. 114, 3087 (2003).
[26] Z. Fan, J. Ma, B. Liang, Z. M. Zhu, and J. C. Cheng, Acustica 92, 217 (2006).
[27] I. Y. Belyaeva and E. M. Timanin, Sov. Phys. Acoust. 37, 533 (1991).
[28] Z. Ye, H. Hsu, and E. Hoskinson, Phys. Lett. A 275, 452 (2000).
[29] A. Alvarez, C. C. Wang, and Z. Ye, J. Comp. Phys. 154, 231 (1999).
[30] L. L. Foldy, Phys. Rev. 67, 107 (1945).
[31] A. Ishimaru, *Wave Propagation and Scattering in Random Media* (Academic press, New York, 1978).
[32] B. C. Gupta and Z. Ye, Phys. Rev. E 67, 036606 (2003).
[33] M. Rusek, A. Orlowski, and J. Mostowski, Phys. Rev. E 53, 4122 (1996).
[34] A. Alvarez and Z. Ye, Phys. Lett. A 252, 53 (1999).
[35] C. C. Church, J. Acoust. Soc. Am. 97, 1510 (1995).
[36] C. H. Kuo, K. K. Wang and Z. Ye, Appl. Phys. Lett. 83, 4247 (2003).
[37] H. J. Feng and F. M. Liu, Chin. Phys. B 18, 1574 (2009).
[38] M. Mooney, J. Appl. Phys. 11, 582 (1940).
[39] J. Ma, J. F. Yu, Z. M. Zhu, X. F. Gong, and G. H. Du, J. Acoust. Soc. Am. 116, 186 (2004).
[40] G. C. Gaunaurd, H. Überall, J. Acoust. Soc. Am. 71, 282 (1982).
[41] G. C. Gaunaurd and W. Wertman, J. Acoust. Soc. Am. 85, 541 (1989).
[42] A. M. Baird, F. H. Kerr, and D. J. Townend, J. Acoust. Soc. Am. 105, 1527 (1999).
[43] D. G. Aggelis, S. V. Tsinopoulos, and D. Polyzos, J. Acoust. Soc. Am. 116, 3343 (2004).
[44] B. Qin, J. J. Chen, and J. C. Cheng, Acoust. Phys. 52, 490 (2006).
[45] B. Liang B, Z. M. Zhu, and J. C. Cheng, Chin. Phys. 15, 412 (2006).
[46] D. H. Trivett, H. Pincon, and P. H. Rogers, J. Acoust. Soc. Am. 119, 3610 (2006).
[47] A. C. Hennion, and J. N. Decarpigny, J. Acoust. Soc. Am. 90, 3356 (1991).
[48] L. F. Shen, Z. Ye, and S. He, Phys. Rev. B. 68, 035109 (2003).
[49] H. Li, and M. Gupta, *Fuzzy logic and intelligent systems* (Boston, Kluwer Academic Publishers, 1995).
[50] H. J. Zimmermann, L. A. Zadeh, and B. R. Gaines, *Fuzzy sets and decision analysis* (Amsterdam, North-Holland, 1984).
[51] M. Mitchell, *An Introduction to Genetic Algorithms* (Cambridge, MIT Press, 1996).

[52] Y. Xu et al, Chin. Phys. Lett. 22, 2557 (2005).
[53] Y. C. Chang, L. J. Yeh, and M. C. Chiu, *Int. J. Numer. Meth. Engng*. 62, 317 (2005).

11

Inverse Scattering in the Low-Frequency Region by Using Acoustic Point Sources

Nikolaos L. Tsitsas
Department of Mathematics, School of Applied Mathematical and Physical Sciences, National Technical University of Athens, Athens Greece

1. Introduction

The interaction of a point-source spherical acoustic wave with a bounded obstacle possesses various attractive and useful properties in direct and inverse scattering theory. More precisely, concerning the direct scattering problem, the far-field interaction of a point source with an obstacle is, under certain conditions, stronger compared to that of a plane wave. On the other hand, in inverse scattering problems the distance of the point-source from the obstacle constitutes a crucial parameter, which is encoded in the far-field pattern and is utilized appropriately for the localization and reconstruction of the obstacle's physical and geometrical characteristics.

The research of point-source scattering initiated in (1), dealing with analytical investigations of the scattering problem by a circular disc. The main results for point-source scattering by simple homogeneous canonical shapes are collected in the classic books (2) and (3). The techniques of the low-frequency theory (4) in the point-source acoustic scattering by soft, hard, impedance surface, and penetrable obstacles were introduced in (5), (6), and (7), where also explicit results for the corresponding particular spherical homogeneous scatterers were obtained. Moreover, in (5), (6), and (7) simple far-field inverse scattering algorithms were developed for the determination of the sphere's center as well as of its radius. On the other hand, point-source near-field inverse scattering problems for a small soft or hard sphere were studied in (8). For other implementations of near-field inverse problems see (9), and p. 133 of (10); also we point out the point-source inverse scattering methods analyzed in (11).

In all the above investigations the incident wave is generated by a point-source located in the exterior of the scatterer. However, a variety of applications suggests the investigation of excitation problems, where a layered obstacle is excited by an acoustic spherical wave generated by a point source located in its interior. Representative applications concern scattering problems for the localization of an object, buried in a layered medium (e.g. inside the earth), (12). This is due to the fact that the Green's function of the layered medium (corresponding to an interior point-source) is utilized as kernel of efficient integral equation formulations, where the integration domain is usually the support of an inhomogeneity existing inside the layered medium. Besides, the interior point-source excitation of a layered sphere has significant medical applications, such as implantations inside the human head for hyperthermia or biotelemetry purposes (13), as well as excitation of the human brain by the neurons currents (see for example (14) and (15), as well as the references therein). Several physical applications of layered media point-source excitation in seismic wave propagation, underwater acoustics, and biology are reported in (16) and (17). Further chemical, biological and physical applications motivating the investigations of interior and exterior scattering problems by layered spheres are discussed in (18). Additionally, we note that, concerning the experimental realization and configuration testing for the related applications, a point-source field is more easily realizable inside the limited space of a laboratory compared to a plane wave field.

To the direction of modeling the above mentioned applications, direct and inverse acoustic scattering and radiation problems for point source excitation of a piecewise homogeneous sphere were treated in (19).

This chapter is organized as follows: Section 2 contains the mathematical formulation of the excitation problem of a layered scatterer by an interior point-source; the boundary interfaces of the adjacent layers are considered to be $\mathcal{C}^2$ surfaces. The following Sections focus on the case where the boundary surfaces are spherical and deal with the direct and inverse acoustic point-source scattering by a piecewise homogeneous (layered) sphere. The point-source may be located either in the exterior or in the interior of the sphere. The layered sphere consists of N concentric spherical layers with constant material parameters; $N-1$ layers are penetrable and the N-th layer (core) is soft, hard, resistive or penetrable. More precisely, Section 3.1 addresses the direct scattering problem for which an analytic method is developed for the determination of the exact acoustic Green's function. In particular, the Green's function is determined analytically by solving the corresponding boundary value problem, by applying a combination of Sommerfeld's (20), (21) and T-matrix (22) methods. Also, we give numerical results on comparative far-field investigations of spherical and plane wave scattering, which provide quantitative criteria on how far the point-source should be placed from the sphere in order to obtain the same results with plane wave incidence. Next, in Section 3.2 the low-frequency assumption is introduced and the related far-field patterns and scattering cross-sections are derived. In particular, we compute the low-frequency approximations of the far-field quantities with an accuracy of order $\mathcal{O}((k_0a_1)^4)$ (k_0 the free space wavenumber and a_1 the exterior sphere's radius). The spherical wave low-frequency far-field results reduce to the corresponding ones due to plane wave incidence on a layered sphere and also recover as special cases several classic results of the literature (contained e.g. in (2), and (5)-(7)), concerning the exterior spherical wave excitation of homogeneous small spheres, subject to various boundary conditions. Also, we present numerical simulations concerning the convergence of the low-frequency cross-sections to the exact ones. Moreover, in Section 3.3 certain low-frequency near-field results are briefly reported.

Importantly, in Section 4 various far- and near-field inverse scattering algorithms for a small layered sphere are presented. The main idea in the development of these algorithms is that the distance of the point source from the scatterer is an additional parameter, encoded in the cross-section, which plays a primary role for the localization and reconstruction of the sphere's characteristics. First, in Section 4.1 the following three types of far-field inverse problems are examined: (i) establish an algorithmic criterion for the determination of the point-source's location for given geometrical and physical parameters of the sphere by exploiting the different cross-section characteristics of interior and exterior excitation, (ii) determine the mass densities of the sphere's layers for given geometrical characteristics by combining the cross-section measurements for both interior and exterior point-source excitation, (iii) recover the sphere's location and the layers radii by measuring the total or differential cross-section for various exterior point-source locations as well as for plane wave incidence. Furthermore, in Section 4.2 ideas on the potential use of point-source fields in the development of near-field inverse scattering algorithms for small layered spheres are pointed out.

2. Interior acoustic excitation of a layered scatterer: mathematical formulation

The *layered scatterer* V is a bounded and closed subset of $\mathbb{R}^3$ with $\mathcal{C}^2$ boundary S_1 possessing the following properties (see Fig. 1): (i) the interior of V is divided by $N-1$ surfaces S_j (j=2,...,N) into N annuli-like regions (layers) V_j (j=1,...,N), (ii) S_j are $\mathcal{C}^2$ surfaces with S_j containing S_{j+1} and $\text{dist}(S_j, S_{j+1}) > 0$, (iii) the layers V_j (j=1,...,$N-1$), are homogeneous isotropic media specified by real wavenumbers k_j and mass densities ρ_j, (iv) the scatterer's core V_N (containing the origin of coordinates) is penetrable specified by real wavenumber k_N and mass density ρ_N or impenetrable being soft, hard or resistive. The exterior V_0 of the scatterer V is a homogeneous isotropic medium with real constants k_0 and ρ_0. In any layer V_j the Green's second theorem is valid by considering the surfaces S_j as oriented by the outward normal unit vector $\hat{\mathbf{n}}$.
The layered scatterer V is excited by a time harmonic ($\exp(-i\omega t)$ time dependence) spherical acoustic wave, generated by a point source with position vector $\mathbf{r}_q$ in the layer V_q (q=0,...,N). Applying Sommerfeld's method (see for example (20), (21), (22)), the primary spherical field $u^{pr}_{\mathbf{r}_q}$, radiated by this point-source, is expressed by

$$u^{pr}_{\mathbf{r}_q}(\mathbf{r}) = r_q \exp(-ik_q r_q)\frac{\exp(ik_q|\mathbf{r}-\mathbf{r}_q|)}{|\mathbf{r}-\mathbf{r}_q|}, \qquad \mathbf{r} \in \mathbb{R}^3\backslash\{\mathbf{r}_q\}, \tag{1}$$

where r_q=$|\mathbf{r}_q|$. We have followed the normalization introduced in (5), namely considered that the primary field reduces to a plane wave with direction of propagation that of the unit vector $-\hat{\mathbf{r}}_q$, when the point source recedes to infinity, i.e.

$$\lim_{r_q\to\infty} u^{pr}_{\mathbf{r}_q}(\mathbf{r}) = \exp(-ik_q\hat{\mathbf{r}}_q\cdot\mathbf{r}). \tag{2}$$

The scatterer V perturbs the primary field $u^{pr}_{\mathbf{r}_q}$, generating *secondary fields* in every layer V_j. The respective secondary fields in V_j ($j \neq q$) and V_q are denoted by $u^{j}_{\mathbf{r}_q}$ and $u^{sec}_{\mathbf{r}_q}$. By Sommerfeld's method, the total field $u^{q}_{\mathbf{r}_q}$ in V_q is defined as the superposition of the primary and the secondary field

$$u^{q}_{\mathbf{r}_q}(\mathbf{r}) = u^{pr}_{\mathbf{r}_q}(\mathbf{r}) + u^{sec}_{\mathbf{r}_q}(\mathbf{r}), \quad \mathbf{r} \in V_q \setminus \{\mathbf{r}_q\}. \tag{3}$$

Moreover, the total field in V_j ($j \neq q$) coincides with the secondary field $u^{j}_{\mathbf{r}_q}$.
The total field $u^{j}_{\mathbf{r}_q}$ in layer V_j satisfies the Helmholtz equation

$$\Delta u^{j}_{\mathbf{r}_q}(\mathbf{r}) + k_j^2 u^{j}_{\mathbf{r}_q}(\mathbf{r}) = 0, \tag{4}$$

for $\mathbf{r} \in V_j$ if $j \neq q$ and $\mathbf{r} \in V_q\backslash\{\mathbf{r}_q\}$ if $j = q$.
On the surfaces S_q and S_{q+1} the following transmission boundary conditions are required

$$\begin{aligned} & u^{q-1}_{\mathbf{r}_q}(\mathbf{r}) - u^{sec}_{\mathbf{r}_q}(\mathbf{r}) = u^{pr}_{\mathbf{r}_q}(\mathbf{r}), && \mathbf{r} \in S_q \\ & \frac{1}{\rho_{q-1}}\frac{\partial u^{q-1}_{\mathbf{r}_q}(\mathbf{r})}{\partial n} - \frac{1}{\rho_q}\frac{\partial u^{sec}_{\mathbf{r}_q}(\mathbf{r})}{\partial n} = \frac{1}{\rho_q}\frac{\partial u^{pr}_{\mathbf{r}_q}(\mathbf{r})}{\partial n}, && \mathbf{r} \in S_q \end{aligned} \tag{5}$$

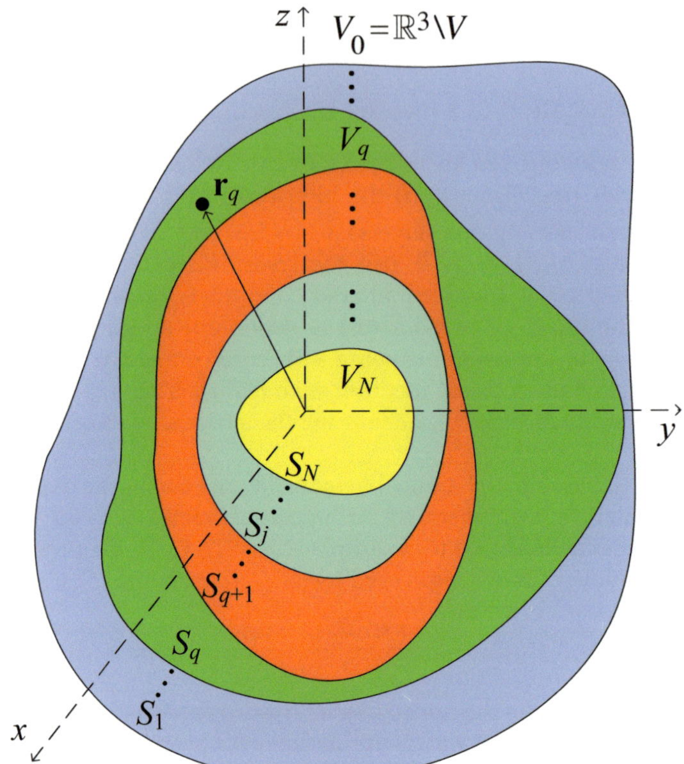

Fig. 1. Typical cross-section of the layered scatterer V.

$$u_{\mathbf{r}_q}^{q+1}(\mathbf{r}) - u_{\mathbf{r}_q}^{sec}(\mathbf{r}) = u_{\mathbf{r}_q}^{pr}(\mathbf{r}), \qquad \mathbf{r} \in S_{q+1}$$

$$\frac{1}{\rho_{q+1}} \frac{\partial u_{\mathbf{r}_q}^{q+1}(\mathbf{r})}{\partial n} - \frac{1}{\rho_q} \frac{\partial u_{\mathbf{r}_q}^{sec}(\mathbf{r})}{\partial n} = \frac{1}{\rho_q} \frac{\partial u_{\mathbf{r}_q}^{pr}(\mathbf{r})}{\partial n}, \quad \mathbf{r} \in S_{q+1} \tag{6}$$

Furthermore, on the surfaces S_j $(j \neq q, q+1, N)$ the total fields must satisfy the transmission conditions

$$u_{\mathbf{r}_q}^{j-1}(\mathbf{r}) - u_{\mathbf{r}_q}^{j}(\mathbf{r}) = 0, \quad \mathbf{r} \in S_j$$

$$\frac{1}{\rho_{j-1}} \frac{\partial u_{\mathbf{r}_q}^{j-1}(\mathbf{r})}{\partial n} - \frac{1}{\rho_j} \frac{\partial u_{\mathbf{r}_q}^{j}(\mathbf{r})}{\partial n} = 0, \quad \mathbf{r} \in S_j \tag{7}$$

For a penetrable core (7) hold also for j=N. On the other hand, for a soft, hard and resistive core the total field on S_N must satisfy respectively
the Dirichlet

$$u_{\mathbf{r}_q}^{N-1}(\mathbf{r}) = 0, \quad \mathbf{r} \in S_N \tag{8}$$

the Neumann

$$\frac{\partial u_{\mathbf{r}_q}^{N-1}(\mathbf{r})}{\partial n} = 0, \quad \mathbf{r} \in S_N \tag{9}$$

and the Robin boundary condition

$$\frac{\partial u_{\mathbf{r}_q}^{N-1}(\mathbf{r})}{\partial n} + ik_{N-1}\lambda u_{\mathbf{r}_q}^{N-1}(\mathbf{r}) = 0, \quad \mathbf{r} \in S_N \quad (\lambda \in \mathbb{R}). \tag{10}$$

The first of Eqs. (5), (6), (7) and Eq. (8) represent the continuity of the fluid's pressure, while the second of Eqs. (5), (6), (7) and Eq. (9) represent the continuity of the normal components of the wave's speed. Detailed discussion on the physical parameters of acoustic wave scattering problems is contained in (4).

Since scattering problems always involve an unbounded domain, a radiation condition for the total field in V_0 must be imposed. Thus, $u_{\mathbf{r}_q}^0$ must satisfy the Sommerfeld radiation condition (10)

$$\frac{\partial u_{\mathbf{r}_q}^0(\mathbf{r})}{\partial n} - ik_0 u_{\mathbf{r}_q}^0(\mathbf{r}) = o(r^{-1}), \quad r \to \infty \tag{11}$$

uniformly for all directions $\hat{\mathbf{r}}$ of $\mathbb{R}^3$, i.e. $\hat{\mathbf{r}} \in S^2 = \{\mathbf{x} \in \mathbb{R}^3, |\mathbf{x}| = 1\}$. Note that a primary spherical acoustic wave defined by (1) satisfies the Sommerfeld radiation condition (11), which clearly is not satisfied by an incident plane acoustic wave.

Besides, the secondary $u_{\mathbf{r}_0}^{sec}$ and the total field $u_{\mathbf{r}_q}^0$ in V_0 have the asymptotic expressions

$$u_{\mathbf{r}_0}^{sec}(\mathbf{r}) = g_{\mathbf{r}_0}(\hat{\mathbf{r}})h_0(k_0 r) + \mathcal{O}(r^{-2}), \quad r \to \infty \tag{12}$$

$$u_{\mathbf{r}_q}^0(\mathbf{r}) = g_{\mathbf{r}_q}(\hat{\mathbf{r}})h_0(k_0 r) + \mathcal{O}(r^{-2}), \quad r \to \infty \quad (q > 0) \tag{13}$$

where $h_0(x)=\exp(ix)/(ix)$ is the zero-th order spherical Hankel function of the first kind. The function $g_{\mathbf{r}_q}$ is the *q-excitation far-field pattern* and describes the response of the scatterer in the direction of observation $\hat{\mathbf{r}}$ of the far-field, due to the excitation by the particular primary field $u_{\mathbf{r}_q}^{pr}$ in layer V_q.

Moreover, we define the *q-excitation differential* (or *bistatic radar*) *cross-section*

$$\sigma_{\mathbf{r}_q}(\hat{\mathbf{r}}) = \frac{4\pi}{k_0^2}|g_{\mathbf{r}_q}(\hat{\mathbf{r}})|^2, \tag{14}$$

which specifies the amount of the field's power radiated in the direction $\hat{\mathbf{r}}$ of the far field. Also, we define the *q-excitation total cross-section*

$$\sigma_{\mathbf{r}_q} = \frac{1}{k_0^2}\int_{S^2} |g_{\mathbf{r}_q}(\hat{\mathbf{r}})|^2 ds(\hat{\mathbf{r}}), \tag{15}$$

representing the average of the amount of the field's power radiated in the far-field over all directions, due to the excitation of the layered scatterer V by a point-source located in layer V_q. Thus, $\sigma_{\mathbf{r}_q}$ is the average of $\sigma_{\mathbf{r}_q}(\hat{\mathbf{r}})$ over all directions. We note that the definition (15) of $\sigma_{\mathbf{r}_q}$ extends that of the scattering cross-section (see (5) of (8) or (17) of (5)) due to a point-source at $\mathbf{r}_0 \in \mathbb{R}^3 \setminus V$.

Finally, we define the *absorption* and the *extinction cross-section*

$$\sigma_{\mathbf{r}_q}^{\mathrm{a}} = \frac{\rho_0}{\rho_{N-1}k_0}\mathrm{Im}\left[\int_{S_N} u_{\mathbf{r}_q}^{N-1}(\mathbf{r})\frac{\overline{\partial u_{\mathbf{r}_q}^{N-1}(\mathbf{r})}}{\partial n}ds(\mathbf{r})\right], \tag{16}$$

$$\sigma^{\mathrm{e}}_{\mathbf{r}_q} = \sigma^{\mathrm{a}}_{\mathbf{r}_q} + \sigma_{\mathbf{r}_q}. \tag{17}$$

The former determines the amount of primary field power, absorbed by the core V_N (since all the other layers have been assumed lossless) and the latter the total power that the scatterer extracts from the primary field either by radiation in V_0 or by absorption. Clearly, $\sigma^{\mathrm{a}}_{\mathbf{r}_q} = 0$ for a soft, hard, or penetrable lossless core, and $\sigma^{\mathrm{a}}_{\mathbf{r}_q} \geq 0$ for a resistive core.
We note that scattering theorems for the interior acoustic excitation of a layered obstacle, subject to various boundary conditions, have been treated in (23) and (24).

3. Layered sphere: direct scattering problems

The solution of the direct scattering problem for the layered scatterer of Fig. 1 cannot be obtained analytically and thus generally requires the use of numerical methods; for an overview of such methods treating inhomogeneous and partially homogeneous scatterers see (25). However, for spherical surfaces S_j, the boundary value problem can be solved analytically and the exact Green's function can be obtained in the form of special functions series. To this end, we focus hereafter to the case of the scatterer V being a *layered sphere*. By adjusting the general description of Section 2, the spherical scatterer V has radius a_1 and surface S_1, while the interior of V is divided by $N-1$ concentric spherical surfaces S_j, defined by $r = a_j$ (j=2,...,N) into N layers V_j (j=1,...,N) (see Fig. 2). The layers V_j, defined by $a_{j+1} \leq r \leq a_j$ (j=1,...,$N-1$), are filled with homogeneous materials specified by real wavenumbers k_j and mass densities ρ_j.

3.1 Exact acoustic Green's function

A classic scattering problem deals with the effects that a discontinuity of the medium of propagation has upon a known incident wave and that takes care of the case where the excitation is located outside the scatterer. When the source of illumination is located inside the scatterer and we are looking at the field outside it, then we have a radiation and not a scattering problem. The investigation of spherical wave scattering problems by layered spherical scatterers is usually based on the implementation of T-matrix (22) combined with Sommerfeld's methods (20), (21). The T-matrix method handles the effect of the sphere's layers and the Sommerfeld's method handles the singularity of the point-source and unifies the cases of interior and exterior excitation. The combination of these two methods leads to certain algorithms for the development of exact expressions for the fields in every layer. Here, we impose an appropriate combined Sommerfeld T-matrix method for the computation of the exact acoustic Green's function of a layered sphere. More precisely, the primary and secondary acoustic fields in every layer are expressed with respect to the basis of the spherical wave functions. The unknown coefficients in the secondary fields expansions are determined analytically by applying a T-matrix method.
We select the spherical coordinate system (r,θ,ϕ) with the origin O at the centre of V, so that the point-source is at r=r_q, θ=0. The primary spherical field (1) is then expressed as (19)

$$u^{pr}_{\mathbf{r}_q}(r,\theta) = \frac{1}{h_0(k_q r_q)} \begin{cases} \sum_{n=0}^{\infty}(2n+1)j_n(k_q r_q)h_n(k_q r)P_n(\cos\theta),\ r > r_q \\ \sum_{n=0}^{\infty}(2n+1)h_n(k_q r_q)j_n(k_q r)P_n(\cos\theta),\ r < r_q \end{cases}$$

where j_n and h_n are the n-th order spherical Bessel and Hankel function of the first kind and P_n is a Legendre polynomial.

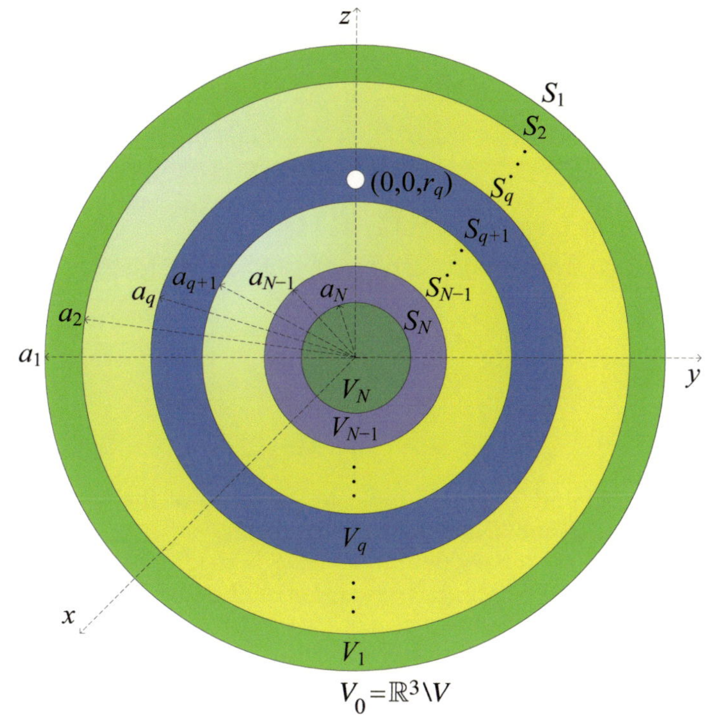

Fig. 2. Geometry of the layered spherical scatterer.

The secondary field $u^j_{\mathbf{r}_q}$ in V_j (j=1,...,$N-1$) is expanded as

$$u^j_{\mathbf{r}_q}(r,\theta) = \sum_{n=0}^{\infty}(2n+1)\frac{h_n(k_q r_q)}{h_0(k_q r_q)}\left(\alpha^j_{q,n} j_n(k_j r) + \beta^j_{q,n} h_n(k_j r)\right)P_n(\cos\theta), \tag{18}$$

where $\alpha^j_{q,n}$ and $\beta^j_{q,n}$ under determination coefficients. The secondary field in V_0 has the expansion (18) with j=0 and $\alpha^0_{q,n}=0$, valid for $r \geq a_1$, in order that the radiation condition (11) is satisfied. On the other hand, since zero belongs to V_N, the secondary field in a penetrable core V_N has the expansion (18) with j=N and $\beta^N_{q,n}$=0, valid for $0 \leq r \leq a_N$.
By imposing the boundary conditions (7) on the spherical surfaces S_j, we obtain the transformations

$$\begin{bmatrix} \alpha^j_{q,n} \\ \beta^j_{q,n} \end{bmatrix} = \mathbf{T}^j_n \begin{bmatrix} \alpha^{j-1}_{q,n} \\ \beta^{j-1}_{q,n} \end{bmatrix} \tag{19}$$

The 2×2 transition matrix $\mathbf{T}^j_n$ from V_{j-1} to V_j, which is independent of the point-source's location, is given by

$$\mathbf{T}^j_n = -ix_j^2 \begin{bmatrix} h'_n(x_j)j_n(y_j) - w_j h_n(x_j) j'_n(y_j) & h'_n(x_j)h_n(y_j) - w_j h_n(x_j)h'_n(y_j) \\ w_j j_n(x_j) j'_n(y_j) - j'_n(x_j) j_n(y_j) & w_j j_n(x_j) h'_n(y_j) - j'_n(x_j) h_n(y_j) \end{bmatrix},$$

where $x_j = k_j a_j$, $y_j = k_{j-1}a_j$, $w_j = (k_{j-1}\rho_j)/(k_j \rho_{j-1})$.

Since $\alpha_{q,n}^{0}$=0, successive application of (19) for j=1,...,q leads to

$$\begin{bmatrix} \alpha_{q,n}^{q} \\ \beta_{q,n}^{q} + \frac{j_n(k_q r_q)}{h_n(k_q r_q)} \end{bmatrix} = \mathbf{T}_{q,n}^{+} \begin{bmatrix} 0 \\ \beta_{q,n}^{0} \end{bmatrix} \tag{20}$$

where

$$\mathbf{T}_{q,n}^{+} = \mathbf{T}_n^{q} ... \mathbf{T}_n^{2} \mathbf{T}_n^{1} \quad . \tag{21}$$

In a similar way successive application of (19) for j=q+1,...,$N-1$ gives

$$\begin{bmatrix} \alpha_{q,n}^{N-1} \\ \beta_{q,n}^{N-1} \end{bmatrix} = \mathbf{T}_{q,n}^{-} \begin{bmatrix} \alpha_{q,n}^{q} + 1 \\ \beta_{q,n}^{q} \end{bmatrix} \tag{22}$$

where

$$\mathbf{T}_{q,n}^{-} = \mathbf{T}_n^{N-1} ... \mathbf{T}_n^{q+2} \mathbf{T}_n^{q+1} \quad . \tag{23}$$

The superscripts + in (20) and $-$ in (22) indicate approach of the layer V_q, containing the point-source, from the layers above and below respectively.
Then, the coefficient of the secondary field in layer V_0 is determined by combining (20) and (22) and imposing the respective boundary condition on the surface of the core V_N, yielding

$$\begin{aligned} \beta_{q,n}^{0} = & \Big\{ j_n(k_q r_q) \Big[f_n(k_{N-1} a_N) (\mathbf{T}_{q,n}^{-})_{12} + g_n(k_{N-1} a_N) (\mathbf{T}_{q,n}^{-})_{22} \Big] \\ & - h_n(k_q r_q) \Big[f_n(k_{N-1} a_N) (\mathbf{T}_{q,n}^{-})_{11} + g_n(k_{N-1} a_N) (\mathbf{T}_{q,n}^{-})_{21} \Big] \Big\} \cdot \\ & \Big\{ h_n(k_q r_q) \Big[f_n(k_{N-1} a_N) (\mathbf{T}_{q,n}^{-} \mathbf{T}_{q,n}^{+})_{12} + g_n(k_{N-1} a_N) (\mathbf{T}_{q,n}^{-} \mathbf{T}_{q,n}^{+})_{22} \Big] \Big\}^{-1} \end{aligned} \tag{24}$$

where

$$f_n = j_n, \qquad j_n', \quad \text{and} \quad j_n' + i\lambda j_n$$
$$g_n = h_n, \qquad h_n', \quad \text{and} \quad h_n' + i\lambda h_n$$

for a soft, hard and resistive core respectively, while for a penetrable core we obtain

$$\beta_{q,n}^{0} = \frac{j_n(k_q r_q)(\mathbf{T}_n^{N} \mathbf{T}_{q,n}^{-})_{22} - h_n(k_q r_q)(\mathbf{T}_n^{N} \mathbf{T}_{q,n}^{-})_{21}}{h_n(k_q r_q)(\mathbf{T}_n^{N} \mathbf{T}_{q,n}^{-} \mathbf{T}_{q,n}^{+})_{22}} . \tag{25}$$

Moreover, by using the above explicit expression for $\beta_{q,n}^{0}$ we see that the coefficients $\alpha_{q,n}^{j}$ and $\beta_{q,n}^{j}$, describing the field in layer V_j (j=1,2,...,N), are determined by successive application of the transformations (19).
By using the above method we recover (for q=0 and N=1) classic results of the literature, concerning the scattered field for the exterior point-source excitation of a homogeneous sphere (see for example (10.5) and (10.70) of (2) for a soft and a hard sphere).
A basic advantage of the proposed method is that the coefficients $\beta_{q,n}^{0,N+1}$ of the secondary field in the exterior V_0 of an N+1-layered spherical scatterer with penetrable core may be obtained directly by means of the coefficients $\beta_{q,n}^{0,N}$ of the corresponding N-layered scatterer by means of an efficient recursive algorithm (19).

Furthermore, for any type of core the q-excitation far field pattern is given by

$$g_{\mathbf{r}_q}(\theta) = \frac{1}{h_0(k_q r_q)} \sum_{n=0}^{\infty} (2n+1)(-i)^n \beta_{q,n}^0 h_n(k_q r_q) P_n(\cos\theta). \tag{26}$$

This expression follows by (12), (13) and (18) for j=0 and by taking into account the asymptotic expression $h_n(z) \sim (-i)^n h_0(z)$, $z \to \infty$. By (14) and (26) we get the q-excitation bistatic radar cross-section

$$\sigma_{\mathbf{r}_q}(\theta) = 4\pi r_q^2 \frac{k_q^2}{k_0^2} \left| \sum_{n=0}^{\infty} (2n+1)(-i)^n \beta_{q,n}^0 h_n(k_q r_q) P_n(\cos\theta) \right|^2. \tag{27}$$

Now, by combining (15) with (26) and using the Legendre functions orthogonality properties ((27), (7.122) and (7.123)), we get the expression of the q-excitation total cross-section

$$\sigma_{\mathbf{r}_q} = 4\pi r_q^2 \frac{k_q^2}{k_0^2} \sum_{n=0}^{\infty} (2n+1) |\beta_{q,n}^0 h_n(k_q r_q)|^2. \tag{28}$$

Next, we will give some numerical results concerning the far-field interactions between the point-source and the layered sphere. In particular, we will make a comparative far-field investigation of spherical and plane wave scattering, which provides certain numerical criteria on how far the point-source should be placed from the sphere in order to obtain the same results with plane wave incidence. This knowledge is important for the implementation of the far-field inverse scattering algorithms described in Section 4.1 below.

Figs. 3a, 3b, and 3c depict the normalized 0-excitation total cross-section $\sigma_{\mathbf{r}_0}/2\pi a_1^2$ as a function of $k_0 a_1$ for a soft, hard, and penetrable sphere for three different point-source locations, as well as for plane wave incidence. The cross-sections for spherical wave scattering are computed by means of (28). On the other hand, by using (28) and taking into account that $h_n(k_0 r_0) \sim (-i)^n h_0(k_0 r_0)$, $r_0 \to \infty$, we obtain

$$\sigma = \frac{4\pi}{k_0^2} \sum_{n=0}^{\infty} (2n+1) |\beta_{0,n}^0|^2,$$

which is utilized for the cross-sections computations due to plane wave scattering.

The cross-section curves reduce to those due to plane wave incidence for large enough distances between the point-source and the scatterer. The critical location of the point-source where the results are almost the same with those of plane wave incidence depends on the type of boundary condition on the sphere's surface. In particular, for point-source locations with distances of more than 8, 5, and 8 radii a_1 from the center of a soft, hard, and penetrable sphere, the results are the same with the ones corresponding to plane wave incidence. We note that those cross-section curves of Figs. 3a and 3b referring to plane wave incidence on a soft and hard sphere coincide with those of Figs. 10.5 and 10.12 of (2).

The 0-excitation total cross-section $\sigma_{\mathbf{r}_0}$ increases as the point-source approaches the sphere ($r_0 \to a_1$) and hence the effect of the spherical wave on the sphere's far-field characteristics increases compared with that of the plane wave. Moreover, $\sigma_{\mathbf{r}_0}$ of a penetrable sphere as a function of $k_0 a_1$ is very oscillatory (see Fig. 3c). These oscillations are due to the penetrable material of the sphere and hence do not appear in the cases of soft or hard sphere. Besides, Fig. 3c indicates that $\sigma_{\mathbf{r}_0}$ of a penetrable sphere is oscillatory also as $k_0 a_1 \to \infty$, while for the other cases converges rapidly.

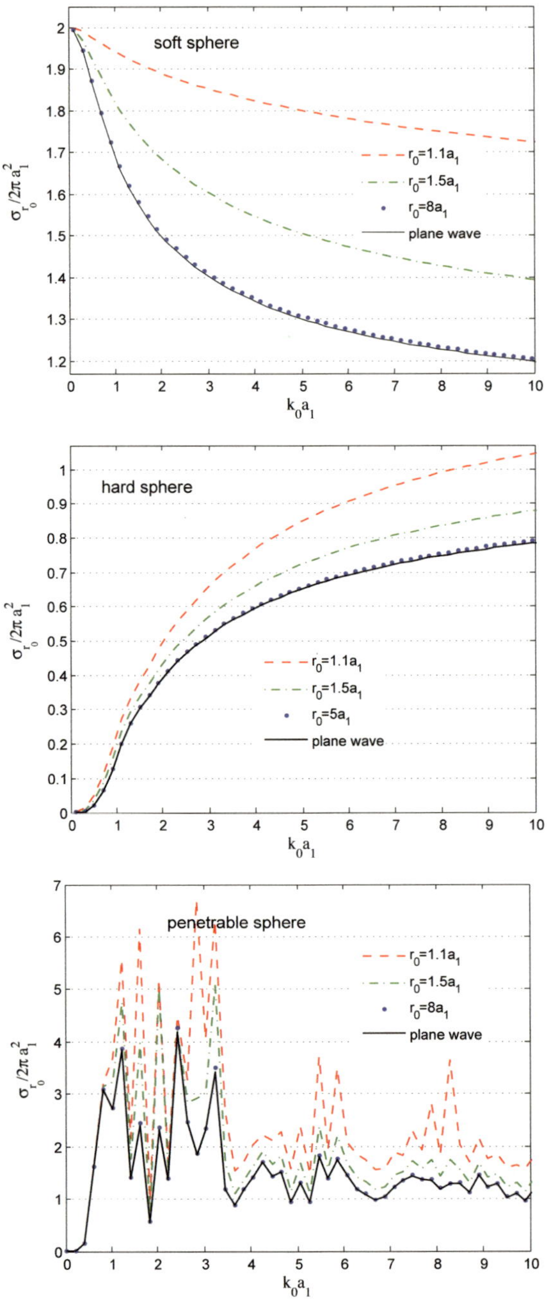

Fig. 3. Normalized 0-excitation total cross-section $\sigma_{\mathbf{r}_0}/2\pi a_1^2$ as a function of $k_0 a_1$ for (a) a soft, (b) a hard, and (c) a penetrable (η_1=3, ϱ_1=2) sphere for various point-source locations and plane wave incidence

3.2 Far-field results for a small layered sphere

All formulae derived up to now are exact. Now, we make the so-called low-frequency assumption $k_0 a_1 \ll 1$ for the case of 3-layered sphere with any type of core, that is assume that the radius a_1 of the sphere is much smaller than the wavelength of the primary field. In order to establish the low-frequency results, we use the following dimensionless parameters

$$\xi_1 = a_1/a_2, \quad \xi_2 = a_2/a_3, \quad \varrho_i = \rho_i/\rho_0, \quad \eta_i = k_i/k_0, \quad \kappa = ik_0 a_1,$$

where ξ_1 and ξ_2 represent the thicknesses of layers V_1 and V_2 and ϱ_i and η_i the relative with respect to free space mass densities and refractive indices of layers V_i (i=1,2,3).
The following two cases are analyzed: (i) exterior excitation by a point-source located at (0,0,r_0) with $r_0 > a_1$, lying in the exterior V_0 of the sphere, and (ii) interior excitation by a point-source located at (0,0,r_1) with $a_2 < r_1 < a_1$, lying in layer V_1 of the 3-layered sphere. We define also

$$\tau_i = a_1/r_i, \qquad d = r_1/a_2 \qquad (i = 0, 1),$$

where τ_0 represents the distance of the exterior point-source from the sphere's surface and τ_1 and d the distances of the interior point-source from the boundaries of layer V_1.
Now, we distinguish the following cases according to the types of the core V_3.
(a) V_3 *is soft*
By using the asymptotic expressions of the spherical Bessel and Hankel functions for small arguments ((26), (10.1.4), (10.1.5)), from (24) we obtain

$$\beta^0_{q,0} = \mathcal{S}^1_{q,0}\kappa + \mathcal{S}^2_{q,0}\kappa^2 + \mathcal{S}^3_{q,0}\kappa^3 + \mathcal{O}(\kappa^4), \qquad \kappa \to 0 \qquad (q = 0, 1) \tag{29}$$

$$\beta^0_{0,n} \sim \frac{(k_0 a_1)^{2n+1}}{i(2n+1)c_n^2}\mathcal{S}_{0,n}, \quad \beta^0_{1,n} \sim \frac{\eta_1^n (k_0 a_1)^{2n+1}}{ic_n^2}\mathcal{S}_{1,n}, \qquad k_0 a_1 \to 0 \qquad (n \geq 1) \tag{30}$$

where

$$c_n = 1 \cdot 3 \cdot 5 \cdots (2n-1), \quad c_0 = 1,$$

and the quantities $\mathcal{S}^j_{q,0}$ (j=1,2,3) and $\mathcal{S}_{q,n}$ depend on the parameters ξ_1, ξ_2, ϱ_1, ϱ_2, d and are given in the Appendix of (19).
In order to calculate the q-excitation far-field patterns with an error of order κ^4, the coefficients $\beta^0_{q,0}$, $\beta^0_{q,1}$ and $\beta^0_{q,2}$ are sufficient. For $\kappa \to 0$, (26) gives the approximation of the q-excitation far-field patterns

$$g_{\mathbf{r}_0}(\theta) = \mathcal{S}^1_{0,0}\kappa + \left[\mathcal{S}^2_{0,0} + \tau_0 \mathcal{S}_{0,1} P_1(\cos\theta)\right]\kappa^2 + \left[\mathcal{S}^3_{0,0} - \mathcal{S}_{0,1}P_1(\cos\theta) - \frac{\tau_0^2}{3}\mathcal{S}_{0,2}P_2(\cos\theta)\right]\kappa^3 + \mathcal{O}(\kappa^4) \tag{31}$$

$$g_{\mathbf{r}_1}(\theta) = \mathcal{S}^1_{1,0}\kappa + \left[\mathcal{S}^2_{1,0} + 3\tau_1 \mathcal{S}_{1,1} P_1(\cos\theta)\right]\kappa^2 + \left[\mathcal{S}^3_{1,0} - 3\eta_1\mathcal{S}_{1,1}P_1(\cos\theta) - \frac{5}{3}\tau_1^2\mathcal{S}_{1,2}P_2(\cos\theta)\right]\kappa^3 + \mathcal{O}(\kappa^4) \tag{32}$$

By (28) the calculation of the q-excitation total cross-sections to the same accuracy requires only $\beta^0_{q,0}$ and $\beta^0_{q,1}$, giving for $\kappa \to 0$ the approximations

$$\sigma_{\mathbf{r}_0} = 4\pi a_1^2 \left\{ (\mathcal{S}^1_{0,0})^2 + k_0^2 a_1^2 \left[(\mathcal{S}^2_{0,0})^2 - 2\mathcal{S}^1_{0,0}\mathcal{S}^3_{0,0} + \frac{\tau_0^2}{3}(\mathcal{S}_{0,1})^2 \right] \right\} + \mathcal{O}(\kappa^4) \tag{33}$$

$$\sigma_{\mathbf{r}_1} = 4\pi a_1^2 \left\{ (\mathcal{S}^1_{1,0})^2 + k_0^2 a_1^2 \left[(\mathcal{S}^2_{1,0})^2 - 2\mathcal{S}^1_{1,0}\mathcal{S}^3_{1,0} + 3\tau_1^2(\mathcal{S}_{1,1})^2 \right] \right\} + \mathcal{O}(\kappa^4). \tag{34}$$

(b) *V_3 is hard*
The respective results are as follows

$$\beta^0_{q,0} = \mathcal{H}^1_{q,0}\kappa + \mathcal{H}^2_{q,0}\kappa^2 + \mathcal{H}^3_{q,0}\kappa^3 + \mathcal{O}(\kappa^5), \quad \kappa \to 0 \qquad (q = 0,1) \tag{35}$$

$$\beta^0_{0,n} \sim \frac{n}{n+1}\frac{(k_0 a_1)^{2n+1}}{i(2n+1)c_n^2}\mathcal{H}_{0,n}, \quad \beta^0_{1,n} \sim \frac{\eta_1^n (k_0 a_1)^{2n+1}}{i(n+1)c_n^2}\mathcal{H}_{1,n}, \quad k_0 a_1 \to 0 \quad (n \geq 1), \tag{36}$$

where the quantities $\mathcal{H}^j_{q,0}$ (j=1,2,3) and $\mathcal{H}_{q,n}$, depending on ξ_1, ξ_2, ϱ_1, ϱ_2, d, are given in the Appendix of (19).
Now, by using (26) and (28) we obtain the low-frequency expansions of the q-excitation far-field patterns and total cross-sections as $\kappa \to 0$

$$g_{\mathbf{r}_0}(\theta) = \frac{\tau_0}{2}\mathcal{H}_{0,1}P_1(\cos\theta)\kappa^2 + \left[\mathcal{H}^3_{0,0} - \frac{\mathcal{H}_{0,1}}{2}P_1(\cos\theta) - \frac{2\tau_0^2}{9}\mathcal{H}_{0,2}P_2(\cos\theta) \right]\kappa^3 + \mathcal{O}(\kappa^4), \tag{37}$$

$$g_{\mathbf{r}_1}(\theta) = \mathcal{H}^1_{1,0}\kappa + \left[\mathcal{H}^2_{1,0} + \frac{3}{2}\tau_1\mathcal{H}_{1,1}P_1(\cos\theta) \right]\kappa^2 + \left[\mathcal{H}^3_{1,0} - \frac{3}{2}\eta_1\mathcal{H}_{1,1}P_1(\cos\theta) + \frac{5}{9}\tau_1^2\mathcal{H}_{1,2}P_2(\cos\theta) \right]\kappa^3 + \mathcal{O}(\kappa^4), \tag{38}$$

$$\sigma_{\mathbf{r}_0} = \pi a_1^2 \frac{\tau_0^2}{3}(\mathcal{H}_{0,1})^2(k_0 a_1)^2 + \mathcal{O}(\kappa^4) \tag{39}$$

$$\sigma_{\mathbf{r}_1} = 4\pi a_1^2 \left\{ (\mathcal{H}^1_{1,0})^2 + k_0^2 a_1^2 \left[(\mathcal{H}^2_{1,0})^2 - 2\mathcal{H}^1_{1,0}\mathcal{H}^3_{1,0} + \frac{3}{4}\tau_1^2(\mathcal{H}_{1,1})^2 \right] \right\} + \mathcal{O}(\kappa^4). \tag{40}$$

(c) *V_3 is penetrable*
From (25) we have

$$\beta^0_{1,0} = \mathcal{P}^1_{1,0}\kappa + \mathcal{P}^2_{1,0}\kappa^2 + \mathcal{P}^3_{1,0}\kappa^3 + \mathcal{O}(\kappa^4), \quad \kappa \to 0 \tag{41}$$

$$\beta^0_{0,n} \sim \frac{ink_0^{2n+1}a_1^{2n+1}}{c_n^2(2n+1)}\mathcal{P}_{0,n}, \quad \beta^0_{1,n} \sim \frac{i\eta_1^n (k_0 a_1)^{2n+1}}{c_n^2}\mathcal{P}_{1,n} \quad k_0 a_1 \to 0 \quad (n \geq 1), \tag{42}$$

where the quantities $\mathcal{P}^j_{1,0}$ (j=1,2,3) and $\mathcal{P}_{q,n}$ depend on the parameters ξ_1, ξ_2, ϱ_1, ϱ_2, ϱ_3, d and are given in the Appendix of (19).

From (26) we obtain for $\kappa \to 0$ the approximation of the q-excitation far-field patterns and total cross-sections

$$g_{\mathbf{r}_0}(\theta) = \kappa^2(\kappa - \tau_0)\mathcal{P}_{0,1}P_1(\cos\theta) + \frac{2}{3}\kappa^3\tau_0^2\mathcal{P}_{0,2}P_2(\cos\theta) + \mathcal{O}(\kappa^4), \tag{43}$$

$$g_{\mathbf{r}_1}(\theta) = \mathcal{P}_{1,0}^1\kappa + \left[\mathcal{P}_{1,0}^2 - 3\tau_1\mathcal{P}_{1,1}P_1(\cos\theta)\right]\kappa^2$$
$$+ \left[\mathcal{P}_{1,0}^3 + 3\eta_1\mathcal{P}_{1,1}P_1(\cos\theta) + \frac{5}{3}\tau_1^2\mathcal{P}_{1,2}P_2(\cos\theta)\right]\kappa^3 + \mathcal{O}(\kappa^4), \tag{44}$$

$$\sigma_{\mathbf{r}_0} = -4\pi a_1^2\kappa^2\frac{\tau_0^2}{3}(\mathcal{P}_{0,1})^2 + \mathcal{O}(\kappa^4) \tag{45}$$

$$\sigma_{\mathbf{r}_1} = 4\pi a_1^2\left[(\mathcal{P}_{1,0}^1)^2 + k_0^2a_1^2\left((\mathcal{P}_{1,0}^2)^2 - 2\mathcal{P}_{1,0}^1\mathcal{P}_{1,0}^3 + 3\tau_1^2(\mathcal{P}_{1,1})^2\right)\right] + \mathcal{O}(\kappa^4). \tag{46}$$

For a small layered sphere with a resistive core the corresponding far-field results are given in (19).

We note that: (i) for a soft core the leading order terms of the far-field patterns $g_{\mathbf{r}_q}$ and total cross-sections $\sigma_{\mathbf{r}_q}$ are of order κ^1 and κ^0 independently of q, (ii) for a hard, resistive, and penetrable core the leading order terms of $g_{\mathbf{r}_q}$ and $\sigma_{\mathbf{r}_q}$ are of order κ^2 for q=0 and κ^1 and κ^0 for q=1. Thus, the interaction characteristics of spheres with hard, resistive, and penetrable cores change drastically for point-sources embedded inside the sphere. This last fact is exploited appropriately in certain inverse scattering algorithms, described below.

Now, concerning certain reductions of the above derived low-frequency far-field results, we note that the point-source incident field (1) for $r_0 \to \infty$ ($\tau_0 \to 0$) reduces to the plane wave (2). Thus, the preceding far-field results for $\tau_0 \to 0$ reduce to the respective ones, corresponding to the plane wave incidence on a 3-layered sphere with soft, hard, and penetrable core V_3 respectively. We remark that the leading term approximations of the 0-excitation cross-sections (39) and (45) for a hard and penetrable core are zero for $\tau_0 = 0$. Hence the interaction of a layered sphere with hard and penetrable core with an incident low-frequency field is weaker when the incident field is a plane wave than when it is a spherical wave.

Moreover, concerning the spherical wave excitation of a 2-layered sphere with various types of cores, the far-field results are derived by the respective above ones for $\xi_2 = 1$, $\varrho_2 = \varrho_1$, $\varrho_3 = \varrho_2$, and $\eta_2 = \eta_1$. On the other hand, the derived low-frequency far-field results recover (for $\xi_1 = \xi_2 = 1$, $\varrho_1 = \varrho_2 = 1$, $\eta_1 = \eta_2 = 1$) several results of the literature, concerning the exterior spherical wave excitation of 1-layered (homogeneous) small spheres, subject to various boundary conditions; for more details see (19). Besides, by letting also $\tau_0 \to 0$ they recover classic low-frequency far-field results for plane wave incidence; see (10.28) of (2) and (7.33), (7.35), (7.42), (7.53), (7.55) and (7.66) of (4).

Importantly, the convergence of the low-frequency total cross-section to the exact one plays a significant role in the inverse scattering algorithms described in Section 4 below. To this end, here we will investigate numerically the convergence of the low-frequency q-excitation total cross-section to the exact one for a 3-layered sphere with various types of cores. Figs. 4 and 5 depict the exact and low-frequency normalized q-excitation total cross-section $\sigma_{\mathbf{r}_q}/2\pi a_1^2$ as a function of k_0a_1 for a 3-layered sphere with soft and hard core. The exact total cross-sections are computed by means of (28), while the low-frequency ones by (33), (34), (39), and (40).

For small k_0a_1 the convergence of the low-frequency to the exact total cross-section is excellent for all types of cores with weak and strong coatings subject to both interior and exterior

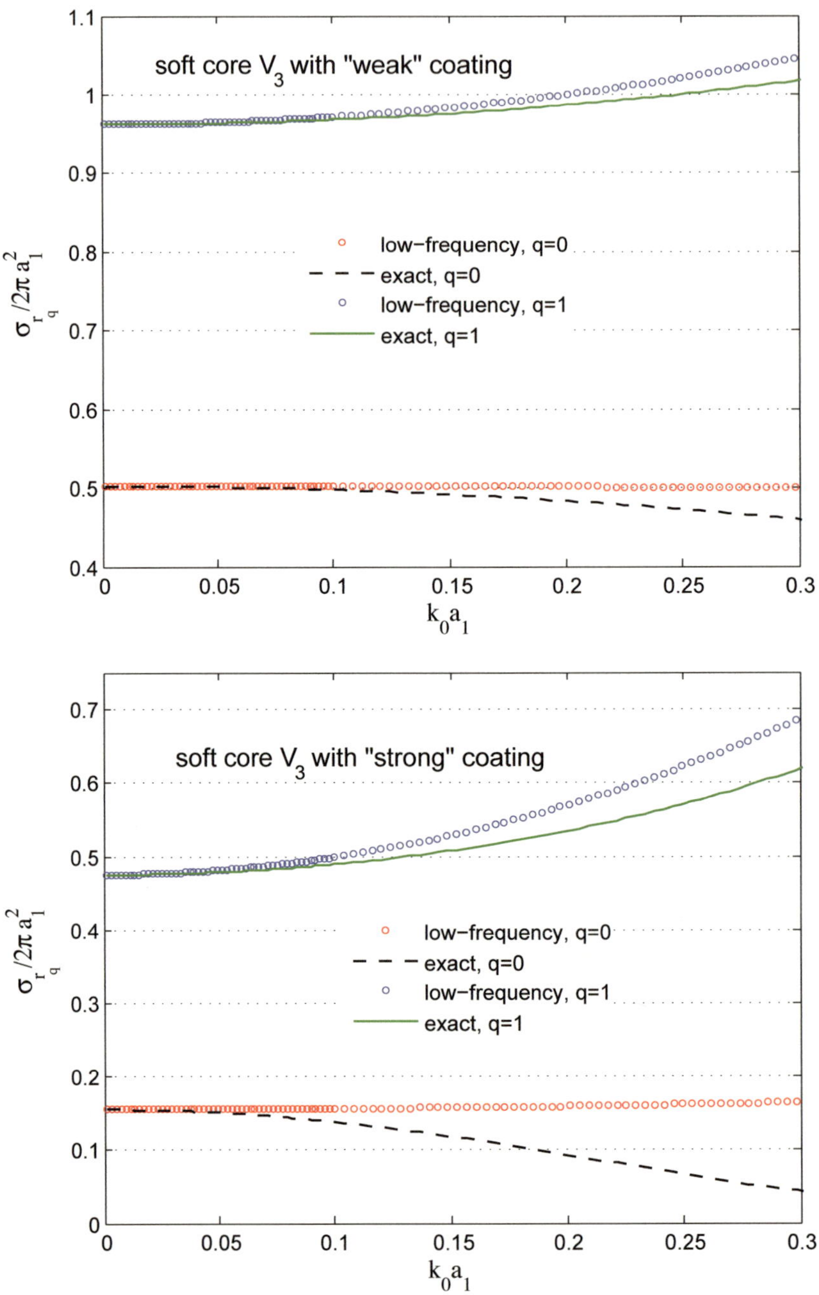

Fig. 4. Exact and low-frequency q-excitation total cross-section $\sigma_{\mathbf{r}_q}/2\pi a_1^2$ as a function of $k_0 a_1$ for a 3-layered sphere (a_2=0.5a_1, a_3=0.25a_1) with soft core with (a) weak (η_1=1.4, η_2=1.6, ϱ_1=1.2, ϱ_2=1.4), and (b) strong (η_1=2, η_2=3, ϱ_1=3, ϱ_2=2) coating. The two point-source locations are $r_0 = 1.1a_1$ and $r_1 = 1.1a_2$.

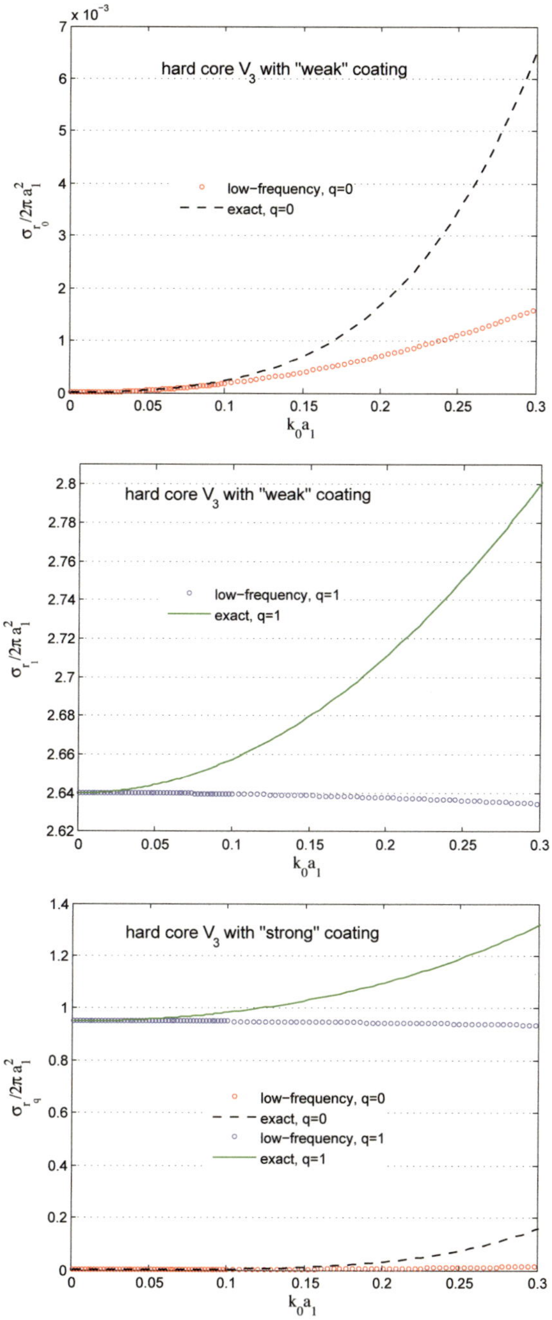

Fig. 5. Exact and low-frequency q-excitation total cross-section $\sigma_{\mathbf{r}_q}/2\pi a_1^2$ as a function of $k_0 a_1$ for a 3-layered sphere with hard core with (a) and (b) weak, and (c) strong coating.

excitation. However, even for $k_0 a_1$ lying outside the low-frequency region, good convergence is also achieved.

In the low-frequency region, the 1-excitation total cross-sections are much larger than the 0-excitation ones. This is the reverse situation to that occurring outside the low-frequency region, as discussed above, where the 0-excitation cross-section is larger that the 1-excitation. So, interior and exterior excitations constitute control mechanisms of the far-field intensity, depending on the desired application.

3.3 Near-field results for a small layered sphere

We consider a 2-layered sphere with soft core excited by an exterior source and still continue to adopt the low-frequency assumption $k_0 a_1 \ll 1$, that is we assume that the outer radius of the layered sphere, a_1, is much smaller than the wavelength of the primary field. More precisely, we suppose that $k_0 a_1$, $k_0 r_0$, $k_1 a_2$ and $k_1 r_q$ are all small (the waves are long compared to all geometrical lengths) whereas k_1/k_0 and ρ_1/ρ_0 are not assumed to be small. We define and use the following dimensionless parameters

$$\xi = a_1/a_2, \quad \varrho = \rho_1/\rho_0, \quad \eta = k_1/k_0, \quad \kappa = k_0 a_1, \quad \tau_0 = a_1/r_0, \quad d_1 = r_1/a_2 = \xi/\tau_1. \tag{47}$$

We use the asymptotic expressions of the spherical Bessel and Hankel functions for small arguments ((26), (10.1.4), (10.1.5)) to obtain the approximations of the elements of the transition matrix (19), as $\kappa \to 0$, by means of which we derive approximations for the field coefficients $\alpha^j_{0,n}$ and $\beta^j_{0,n}$. Then we estimate the secondary field at the source point, using

$$u^{sec}_{\mathbf{r}_0}(r_0, 0) = h_0(k_0 r_0) \sum_{n=0}^{\infty} (2n+1) \beta^0_{0,n} [\mathscr{H}_n(k_0 r_0)]^2, \tag{48}$$

where

$$\mathscr{H}_n(w) = h_n(w)/h_0(w).$$

Moreover, from (30), we get

$$\beta^0_{0,n} \sim \frac{\kappa^{2n+1}}{i(2n+1)c_n^2} \mathcal{S}_{0,n}(\xi, \varrho), \qquad \kappa \to 0, \tag{49}$$

where

$$\mathcal{S}_{0,n}(\xi, \varrho) = \frac{\mathcal{B}_n(\xi) + \varrho\, n \mathcal{A}_n(\xi)}{\Delta_n(\xi, \varrho)}, \quad \Delta_n(\xi, \varrho) = \mathcal{B}_n(\xi) - \varrho\,(n+1)\mathcal{A}_n(\xi), \tag{50}$$

$$\mathcal{A}_n(x) = 1 - x^{2n+1}, \quad \mathcal{B}_n(x) = n(x^{2n+1} + 1) + 1.$$

Now, since

$$\mathscr{H}_n(w) \sim c_n w^{-n} \qquad \text{as} \qquad w \to 0,$$

the small-κ approximation of (48) combined with (49) gives the following approximation of the secondary field at the external point-source's location

$$u^{sec}_{\mathbf{r}_0}(r_0, 0) = -\tau_0 \exp(ik_0 r_0) \sum_{n=0}^{\infty} \mathcal{S}_n(\xi, \varrho)\, \tau_0^{2n} + \mathcal{O}(\kappa), \tag{51}$$

as $k_0a \to 0$. Note that, contrary to the case of low-frequency far-field results (see for example (19)), in the near-field every term of the infinite series contributes to the leading order $\mathcal{O}(1)$ behaviour. Now, by means of (51) we get

$$|u_{\mathbf{r}_0}^{sec}(r_0, 0)| = \left| \tau_0 \sum_{n=0}^{\infty} \mathcal{S}_n(\xi, \varrho)\, \tau_0^{2n} \right| + \mathcal{O}(\kappa^2), \tag{52}$$

as $k_0a \to 0$.

4. Layered sphere: inverse scattering problems

The low-frequency far- and near-field results, presented in Sections 3.2 and 3.3, are now utilized for the development of far- and near-field inverse scattering algorithms for the localization and reconstruction of the sphere's characteristics.

4.1 Far-field inverse scattering problems

The low-frequency realm offers a better environment for inverse scattering, compared to the exact yet complicated far-field series solutions (26) and (28), since the corresponding low-frequency far-field patterns are much more tractable. In particular, we develop inverse scattering algorithms for the determination of the geometrical and physical characteristics of the 3-layered sphere and the point-source, based on low-frequency measurements of the 0- and 1- excitation total cross-section. Additionally, we emphasize that the embedding of the point-source inside the scatterer offers additional essential information for the problem's characteristics, which cannot be given by a point-source outside the sphere.

4.1.1 Determination of the point-source's location

We determine the location of the point-source for given geometrical and physical characteristics of the sphere. This type of problem is expected to find applications in the determination of the layer that the neuron currents radiate in investigations of the human brain's activity (15). An inverse scattering algorithm is established for the hard core case.
We use the 0- and 1- excitation total cross-sections (39) and (40) of a 3-layered sphere with hard core. The decision on whether the point-source lies in the interior or the exterior of the sphere is based on the following *algorithmic criterion*:
Measure the leading term (of order $(k_0a_1)^0$) in the low-frequency expansion of the total cross-section. If this measurement is zero, then from (39), the point-source is outside the sphere. On the other hand, if this measurement is not zero then from (40) the point-source is inside the sphere.
After determining the layer that the point-source is lying, we can also obtain its position r_q (q=0,1). For exterior excitation, we compute the position r_0 from the measurement of the leading order term of the 0-excitation total cross-section (39)

$$m_0 = \pi k_0^2 \frac{a_1^6}{3r_0^2} (\mathcal{H}_{0,1})^2,$$

while for interior excitation, the measurement of the second order term of the 1-excitation total cross-section (40)

$$m_1 = 4\pi k_0^2 a_1^4 \left[(\mathcal{H}_{1,0}^2)^2 - 2\mathcal{H}_{1,0}^1 \mathcal{H}_{1,0}^3 + \frac{3a_1^2}{4r_1^2} (\mathcal{H}_{1,1})^2 \right]$$

gives the position r_1 of the point-source.

4.1.2 Determination of the layers material parameters

We determine the mass densities of the sphere's layers for given center's coordinates and layers radii of the sphere. Inverse scattering algorithms are established for a sphere with penetrable core.

Consider two point-source locations (0,0,b_0) and (0,0,b_1) with known distances $b_0 > a_1$ and $a_2 < b_1 < a_1$ from the layered sphere's center (that is the first point-source lies in the exterior and the second one in the interior of the scatterer). First, measure the leading order low-frequency term m_0 of the 0-excitation total cross-section (45) for a point-source at (0,0,b_0)

$$m_0 = \frac{4\pi k_0^2 a_1^6}{3b_0^2} (\mathcal{P}_{0,1})^2, \tag{53}$$

and then the leading m_1 and the second order term m_2 of the 1-excitation total cross-section (46) for a point-source at (0,0,b_1)

$$m_1 = 4\pi \frac{b_1^2}{\varrho_1^2}, \tag{54}$$

$$m_2 = 4\pi k_0^2 a_1^4 \left((\mathcal{P}_{1,0}^2)^2 - 2\mathcal{P}_{1,0}^1 \mathcal{P}_{1,0}^3 + \frac{3a_1^2}{b_1^2} (\mathcal{P}_{1,1})^2 \right). \tag{55}$$

Eq. (54) provides the mass density ϱ_1 of layer V_1. Eqs. (53) and (55) constitute a 2×2 non-linear system, the solution of which provides the densities ϱ_2 and ϱ_3 of layers V_2 and V_3.

4.1.3 Determination of the sphere's center and the layers radii

We determine the center's coordinates and the layers radii of the 3-layered sphere for given mass densities and refractive indices of the sphere's layers. We establish the inverse scattering algorithms for a sphere with soft core.

Choose a Cartesian coordinate system $Oxyz$ and five point-source locations (0,0,0), (ℓ,0,0), (0,ℓ,0), (0,0,ℓ), and (0,0,2ℓ) with unknown distances b_1, b_2, b_3, b_4 and b_5 from the layered sphere's center. The parameter ℓ represents a chosen fixed length. First, measure the leading order low-frequency term m_0 of the 0-excitation total cross-section (33) for a point-source located at the origin

$$m_0 = 4\pi a_1^2 (\mathcal{S}_{0,0}^1)^2 \tag{56}$$

Then, measure the second order low-frequency term m_j of the 0-excitation total cross-section (33) for each point-source location

$$m_j = 4\pi k_0^2 a_1^4 \left[(\mathcal{S}_{0,0}^2)^2 - 2\mathcal{S}_{0,0}^1 \mathcal{S}_{0,0}^3 + \frac{a_1^2}{3b_j^2} (\mathcal{S}_{0,1})^2 \right] \quad (j = 1, \ldots, 5)$$

Measurability techniques permitting the isolation of the individual measurements m_0 and m_j from the total cross-section measurements as well as measurement sensitivity aspects are discussed in (5).
Next, measure the second order cross-section term m_6 for a point-source far away from the sphere ($\tau \to 0$) (namely for plane wave incidence)

$$m_6 = 4\pi k_0^2 a_1^4 \left[(\mathcal{S}_{0,0}^2)^2 - 2\mathcal{S}_{0,0}^1 \mathcal{S}_{0,0}^3 \right] \tag{57}$$

We define the dimensionless quantities

$$\gamma_j = \frac{\ell}{\sqrt{m_j - m_6}} = \sqrt{\frac{3}{4\pi}} \frac{\ell}{k_0 a_1^2 \mathcal{S}_{0,1}} \frac{b_j}{a_1}. \tag{58}$$

Now, we have seven equations with the eight unknowns a_1, a_2, a_3, and b_j. The 8-th equation is derived by the law of cosines

$$b_5^2 = 2\ell^2 + 2b_4^2 - b_1^2, \tag{59}$$

and thus from (58) and (59) we get

$$\left(\frac{b_j}{\ell}\right)^2 = \frac{2\gamma_j^2}{\gamma_5^2 - 2\gamma_4^2 + \gamma_1^2},$$

which determine the distances b_j. The intersection point of the four spheres centered at (0,0,0), (ℓ,0,0), (0,ℓ,0), (0,0,ℓ) with determined radii b_1, b_2, b_3, b_4 coincides with the center of the layered spherical scatterer.
Finally, (56), (57), and (58) for j=1 constitute the 3×3 non-linear system

$$a_1 \mathcal{S}_{0,0}^1 = \sqrt{m_0/4\pi}$$

$$a_1^4 \left[(\mathcal{S}_{0,0}^2)^2 - 2\mathcal{S}_{0,0}^1 \mathcal{S}_{0,0}^3 \right] = m_6/(4\pi k_0^2)$$

$$a_1^3 \mathcal{S}_{0,1} = (\sqrt{3} b_1 \ell)/(\sqrt{4\pi} k_0 \gamma_1),$$

which gives as solutions the layers radii a_1, a_2, and a_3.

4.2 Near-field inverse scattering problems

The derived low-frequency near-field expansions in Section 3.3 are now utilized to establish inverse scattering algorithms for the determination of the geometrical characteristics of the piecewise homogeneous sphere. More precisely, the near-field inverse problem setting is as follows: we know the sphere's radius, a_1, as well as the location of its center and we are interested in estimating the core's radius, a_2. Note that for problems, where we need to determine also the sphere's radius and its center, appropriate modifications of the far-field algorithms (to the present context of near-field measurements) presented in Section 6.1 of (19) may be applied.

Consider a 2-layered sphere with soft core. We suppose that the radius a_1 and the density ϱ are known and we will estimate the core's radius a_2 by using a single measurement of the near-field due to an exterior point-source. The corresponding near-field at the exterior point-source's location is given by (52). The series appearing in (52), namely

$$\mathscr{S}(\xi,\tau_0,\varrho)=\sum_{n=0}^{\infty}\frac{\mathcal{B}_n(\xi)+\varrho\, n\mathcal{A}_n(\xi)}{\mathcal{B}_n(\xi)-\varrho\,(n+1)\mathcal{A}_n(\xi)}\tau_0^{2n}, \tag{60}$$

may be expressed with the aid of Gauss hypergeometric functions, if we consider that the relative mass density ϱ of the coating's layer V_1 is close to 1. For this case, the series (60) has the following properties

$$\mathscr{S}(\xi,\tau_0,1)=\frac{1}{\xi}\sum_{n=0}^{\infty}\frac{1}{2n+1}\left(\frac{\tau_0^2}{\xi^2}\right)^n=\frac{1}{2\tau_0}\ln\left(\frac{\xi+\tau_0}{\xi-\tau_0}\right), \tag{61}$$

$$\begin{aligned}\frac{\partial\mathscr{S}}{\partial\varrho}(\xi,\tau_0,1)&=\frac{1}{\xi^2}\sum_{n=0}^{\infty}\frac{n\xi^{2n+1}+n+1}{2n+1}(1-\xi^{2n+1})\left(\frac{\tau_0^2}{\xi^4}\right)^n\\&=\frac{1}{\xi^2}F\left(2;\frac{1}{2};\frac{3}{2};\frac{\tau_0^2}{\xi^4}\right)-\frac{\tau_0^2}{3}F\left(2;\frac{3}{2};\frac{5}{2};\tau_0^2\right)-\frac{1}{2\tau_0}\ln\left(\frac{\xi+\tau_0}{\xi-\tau_0}\right),\end{aligned} \tag{62}$$

where $F\equiv {}_2F_1$ is the Gauss hypergeometric function (26). Now, by considering the best linear approximation around the fixed point $\varrho=1$, from (60)-(62), we get

$$\begin{aligned}\mathscr{S}(\xi,\tau_0,\varrho)&\simeq\mathscr{S}(\xi,\tau_0,1)+\frac{\partial\mathscr{S}(\xi,\tau_0,1)}{\partial\varrho}(\varrho-1)\\&=\frac{1}{2\tau_0}\ln\left(\frac{\xi+\tau_0}{\xi-\tau_0}\right)(2-\varrho)-\left[\frac{1}{\xi^2}F\left(2;\frac{1}{2};\frac{3}{2};\frac{\tau_0^2}{\xi^4}\right)-\frac{\tau_0^2}{3}F\left(2;\frac{3}{2};\frac{5}{2};\tau_0^2\right)\right](1-\varrho),\end{aligned}$$

which in combination with (52) and (60) gives

$$\begin{aligned}|u_{\mathbf{r}_0}^{sec}(r_0,0)|=\Bigg|&\frac{2-\varrho}{2}\ln\left(\frac{\xi+\tau_0}{\xi-\tau_0}\right)\\&-\left[\frac{\tau_0}{\xi^2}F\left(2;\frac{1}{2};\frac{3}{2};\frac{\tau_0^2}{\xi^4}\right)-\frac{\tau_0^3}{3}F\left(2;\frac{3}{2};\frac{5}{2};\tau_0^2\right)\right](1-\varrho)\Bigg|+\mathcal{O}(\kappa^2).\end{aligned} \tag{63}$$

The radius a_2 of the core of the 2-layered sphere may now be determined by means of (63) as follows. We measure the leading order term in the above low-frequency expansion of $|u_{\mathbf{r}_0}^{sec}(r_0,0)|$, and we obtain a non-linear equation with respect $\xi=a_1/a_2$. Since, the radius a_1 is known, the solution of this equation provides the core's radius a_2.
Similar near-field algorithms may be developed for a 2-layered sphere with hard, and penetrable core.

5. Acknowledgement

The author's work was supported by the State Scholarships Foundation, while he was a post-doctoral research scholar.

6. References

[1] D. S. Jones, A new method for calculating scattering with particular reference to the circular disc, *Commun. Pure Appl. Math.* 9 (1956) 713–746.

[2] J. J. Bowman, T. B. Senior, P. L. Uslenghi, *Electromagnetic and Acoustic Scattering by Simple Shapes*, North Holland Publ. Co., 1969.

[3] D. S. Jones, *Acoustic and Electromagnetic Waves*, Oxford University Press, 1986.

[4] G. Dassios and R. Kleinman, *Low Frequency Scattering*, Clarendon Press, 2000.

[5] G. Dassios and G. Kamvyssas, Point source excitation in direct and inverse scattering: the soft and the hard small sphere, *IMA J. Appl. Math.* 55 (1995) 67–84.

[6] G. Dassios and G. Kamvyssas, The impedance scattering problem for a point-source field. The small resistive sphere, *Quart. J. Mech. Appl. Math.* 50 (1997) 321–332.

[7] G. Dassios, M. Hadjinicolaou and G. Kamvyssas, Direct and Inverse Scattering for Point Source Fields. The Penetrable Small Sphere, *Z. Angew. Math. Mech.* 79 (1999) 303–316.

[8] C. Athanasiadis, P.A. Martin and I.G. Stratis, On spherical-wave scattering by a spherical scatterer and related near-field inverse problems, *IMA J. Appl. Math.* 66 (2001) 539–549.

[9] C. J. S. Alves, and P. M. C. Ribeiro, Crack detection using spherical incident waves and near-field measurements. *Boundary Elements XXI*, edited by C. A. Brebbia and H. Power (WIT Press, Southampton, 1999) 355–364.

[10] D. Colton and R. Kress, *Inverse acoustic and electromagnetic scattering theory*, Springer-Verlag, 1992.

[11] R. Potthast, *Point-Sources and Multipoles in Inverse-Scattering Theory*, Chapman and Hall/CRC, 2001.

[12] J. Coyle, Locating the support of objects contained in a two-layered background medium in two dimensions, *Inverse Problems* 16 (2000) 275–292.

[13] J. Kim and Y. Rahmat-Samii, Implanted Antennas Inside a Human Body: Simulations, Designs, and Characterizations, *IEEE Trans. Microwave Theory Tech.* 52 (2004) 1934–1943.

[14] G. Dassios, A. S. Fokas and F. Kariotou, On the non-uniqueness of the inverse magnetoencephalography problem, *Inverse Problems* 21 (2005) L1–L5.

[15] G. Dassios, On the hidden electromagnetic activity of the brain, in: Mathematical Methods in Scattering Theory and Biomedical Engineering, Proceedings of the Seventh International Workshop, World Scientific Publishing Co., 2006, pp. 297–303.

[16] K. Aki and P. G. Richards, *Quantitative Seismology. Theory and Methods*, Freeman, 1980.

[17] A. E. Yagle and B. C. Levy, Application of the Schur algorithm to the inverse problem for a layered acoustic medium, *J. Acoust. Soc. Am.* 76 (1984) 301–308.

[18] A. Moroz, A recursive transfer-matrix solution for a dipole radiating inside and outside a stratified sphere, *Annals of Physics* 315, (2005) 352–418.

[19] N. L. Tsitsas, and C. Athanasiadis, Point source excitation of a layered sphere: direct and far-field inverse scattering problems, *Quart. J. Mech. Appl. Math.* 61, (2008) 549–580.

[20] A. Sommerfeld, *Partial Differential Equations in Physics*, Academic Press, 1949.

[21] J. A. Stratton, *Electromagnetic Theory*, McGraw-Hill, 1941.

[22] N. L. Tsitsas and C. Athanasiadis, On the scattering of spherical electromagnetic waves by a layered sphere, *Quart. J. Mech. Appl. Math.* 59 (2006) 55–74.

[23] C. Athanasiadis, N. L. Tsitsas, Scattering theorems for acoustic excitation of a layered obstacle by an interior point source, *Stud. Appl. Math.* 118 (2007), 397–418.

[24] N. L. Tsitsas, C. Athanasiadis, On the interior acoustic and electromagnetic excitation of a layered scatterer with a resistive or conductive core, *Bull. Greek Math. Soc.* 54 (2007) 127–141.
[25] W. C. Chew, *Waves and Fields in Inhomogeneous Media*, IEEE Press, New York, 1995.
[26] M. Abramowitz and I. A. Stegun. *Handbook of Mathematical Functions*, Dover, 1965.
[27] I. S. Gradshteyn and I. M. Ryzhik. *Table of Integrals, Series and Products*, Academic Press, 2000.